AUDIO PRODUCTION TIPS

Audio Production Tips: Getting the Sound Right at the Source provides practical and accessible information detailing the production processes for recording today's bands. By demonstrating how to 'get the sound right at the source', author Peter Dowsett lays the appropriate framework to discuss the technical requirements of optimizing the sound of a source. Through its coverage of critical listening, pre-production, arrangement, drum tuning, gain staging and many other areas of music production, *Audio Production Tips* allows you to build the wide array of skills that apply to the creative process of music production. Broken into two parts, the book first presents foundational concepts followed by more specific production advice on a range of instruments.

Key features:

- Important in-depth coverage of music theory, arrangement and its applications.
- Real-life examples with key references to the author's music production background.
- Presents concepts alongside the production of a track captured specifically for the book.
- A detailed companion website that includes audio, Pro Tools session files of the track recording process, and videos with accompanying audio that can be examined in the reader's DAW.

Please visit the accompanying companion website, available at www.audio productiontips.com, for resources that further support the book's practical approach.

Peter Dowsett is a British audio engineer known for his experience with many facets of the music industry including studio engineering, mixing, mastering and live sound. At Metropolis Studios, he has worked with clients including Pharrell Williams, Snoop Dogg, Black Eyed Peas, Will.I.Am, Rick Ross, and Nick Jonas. His work has been synced on ABC and Channel 5 and featured on cover CDs for *Metal Hammer* magazine. He has received extensive international radio and television airplay.

    In a live sound environment Peter has toured with Ugly Kid Joe, Fozzy, Twisted Wheel and Beholder and has engineered major festivals including Download, Sonisphere, Wacken Polish Woodstock, and Lockerse. He has also worked FOH or monitors at Coventry Kasbah, Birmingham Institute, and Nuneaton Queens Hall for The Darkness, Buzzcocks, Foals, Miles Kane, Dodgy, Ash, Cast, Chase and Status, Pulled Apart by Horses, and We are the Ocean.

AUDIO PRODUCTION TIPS

Getting the Sound Right at the Source

PETER DOWSETT

First published 2016
by Focal Press
711 Third Avenue, New York, NY 10017

and by Focal Press
2 Park Square, Milton Park, Abingdon, Oxon OX14 4RN

Focal Press is an imprint of the Taylor & Francis Group, an informa business

© 2016 Taylor & Francis

The right of Peter Dowsett to be identified as author of this work has been asserted by him in accordance with sections 77 and 78 of the Copyright, Designs and Patents Act 1988.

Audio track 'Take Her Away' copyright © 2012 Ghoulish Records and Luna Kiss

All images, videos, audio examples and Pro Tools session data except where stated are created and owned by Focal Press.

All rights reserved. No part of this book may be reprinted or reproduced or utilized in any form or by any electronic, mechanical, or other means, now known or hereafter invented, including photocopying and recording, or in any information storage or retrieval system, without permission in writing from the publishers.

Notices
Knowledge and best practice in this field are constantly changing. As new research and experience broaden our understanding, changes in research methods, professional practices, or medical treatment may become necessary.

Practitioners and researchers must always rely on their own experience and knowledge in evaluating and using any information, methods, compounds, or experiments described herein. In using such information or methods they should be mindful of their own safety and the safety of others, including parties for whom they have a professional responsibility.

Product or corporate names may be trademarks or registered trademarks, and are used only for identification and explanation without intent to infringe.

Library of Congress Cataloging in Publication Data
Dowsett, Peter.
　Audio production tips: getting the sound right at the source/by Peter Dowsett.
　　pages cm
　1. Popular music—Production and direction. 2. Sound recordings—Production and direction. 3. Sound—Recording and reproducing. I. Title.
　ML3470.D68 2016
　781.49—dc23
　2015012786

ISBN: 978-1-138-80737-2 (pbk)
ISBN: 978-1-138-80736-5 (hbk)
ISBN: 978-1-315-75112-2 (ebk)

Typeset in Utopia and Univers
by Florence Production Ltd, Stoodleigh, Devon, UK

Printed and bound in the United States of America by Sheridan Books, Inc. (a Sheridan Group Company).

CONTENTS

	Acknowledgements	*vi*
	Introduction	*vii*
	Contributors	*x*
	Preface	*xi*
1	Production Philosophies, Your Ears and Critical Listening	1
2	Project Management and Pre-Production	23
3	Basic Music Theory	35
4	Song Structure, Lyrics and Melody	111
5	Advanced Music Theory	141
6	Arrangement and Orchestration	173
7	Demystifying Recording Levels	219
8	Microphone, Pre-Amp and Live Room Choice	237
9	The Critical Listening Environment	253
10	Decision-Making and General Recording Techniques	283
11	Tracking Drums	309
12	Tracking Electric and Bass Guitar	355
13	Tracking Other Instrumentation	415
14	Tracking Vocals	479
	Epilogue	*511*
	Appendix	*513*
	Index	*517*

ACKNOWLEDGEMENTS

First of all, I would like to extend a big 'thank you' to my amazing girlfriend Mădălina Mocanu for her invaluable help with the graphic design, line drawings, camera work and video editing in this project, and for her unwavering support throughout. Dale Driver and Chris Porter for additional video editing, grading, intro sequence design and camera work. Edward Cottrill for his advice as a writer and proofreading. Matthew Cotterill and Leon Cooke for helping to arrange this course. Paul Kent and Nathan Morris at Flipside Studios, Coventry, and Dutch Van Spall and Sean McCabe at Strawhouse, Rugby, for the use of their studios and additional support. Dave Robilliard for line drawings. Megan Ball, Mary LaMacchia and the rest of the team at Focal Press. My parents, sister and extended family for giving me the time, financial backing and unequivocal patience to be able to develop my skills to the level they are today. Last but not least, I would like to thank Wil, Chris, James and Ross of Luna Kiss for not only allowing me to use their track 'Take Her Away' in this programme, but also for their trust and patience as the whole tracking, recording and mastering process took several times longer than usual.

INTRODUCTION

The modern music listener demands perfection; the modern musician demands perfection. To add to this, most unsigned bands are less competent musicians and demand more than can be achieved in the time available. So what does this mean? In short, it means you have to be in this business for the love of it. You have to demand more from yourself than others. It means that you have to be willing to go the extra mile to get results. It means that you have to become a leader, and a confident 'people person' who can get the best from all sorts of personalities. If you are just after fame, fortune and accolades, then you are better off becoming a musician. With long and underpaid studio days in small claustrophobic spaces with little to no ventilation, if you don't have complete passion and dedication for the art of music production, you'll find this career frustrating and demeaning, and you will most likely give up.

To add to this, there is always going to be someone who doesn't like what you do. You can't please everyone all the time, and trying to do so is likely to result in something bland. As long as the band, label and management are happy, no one else matters. The fact is, when you first start out in this business, you will probably have as many failures as successes (if not more). Do not let this rock you; we all had to start somewhere. This advice is particularly important in this day and age, where social media makes it very easy for people to comment publicly. Many people will think that they can do a better job than you, and some of them can (these are the people you want close). However, I've also witnessed a ton of people who talk nonsense. You need to be able to stay positive and remain driven to succeed in this industry, and this is one of the toughest hurdles to overcome.

I've already mentioned that the industry has changed. Opportunities to work hard within an organization and make your way up from being the 'tea-boy' are dwindling. What hasn't changed, however, is that you need to push yourself – get yourself into the right places and make opportunities happen. I don't care what you have to do – make people aware of what you do and what you can offer them! My route into the industry was to use almost my entire maintenance loan while at university to buy a modest recording set-up. I then established myself with bands by getting a job in a small local venue, which I did by telling the venue's promoter I'd do the work for cheaper than the guy currently doing it. From there, I could get to know all of the local bands and offer them cheap or even free recording sessions. This soon got the attention of a local recording studio – and I was in. There was a large amount of bravado involved. You have to believe in yourself and what you are doing. Put yourself out there regardless of the knocks your pride and ego will take if you fall flat on your face.

I am not trying to put anyone off here, but I've seen many friends and associates give up, many of whom were very talented. This job is not for the faint of heart. That said, music production has provided some of the most invigorating and enthralling moments of my life. This is what I dedicate myself to wholeheartedly, and after years of training, practical experience and trial and error, it is paying dividends.

Knowledge is often required from people working at the same level as you. The natural tendency for some people is be scared of competition or scared to share knowledge. However, with persistence, you will find people that also believe that 'we are stronger together'. One of the best things I ever did to further my skills as a mix engineer was to start a local mix-off club. The premise of a mix-off is that a person shares the individual WAV files of a track that they have recorded, and other people have the opportunity to mix it and then compare and contrast their results. We used to set a time limit of five hours per mix and then get together one night to do the comparisons (followed by some beers and usually a celebratory curry). To get this started, I emailed approximately 20 producers in my local area about the idea. Most of them seemed interested. Some clearly appeared threatened and scared, while others lacked confidence in their own ability. Over the space of a few weeks, a core of five to six people formed, and we began holding monthly mix-off events. After the first meeting, it became very clear that within the group, we had widely varying degrees of skill in terms of mix engineering. However, we all learnt roughly the same amount from each other. It was slightly competitive, but it was infinitely more friendly, informative and groundbreaking. As a result of starting the mix-off group, I ended up based in the same complex as all of these other engineers, running a communal studio with our own post-production rooms within the complex.

Long before you think about recording anything, there are many things to consider that the disorganized, inexperienced and underprepared can easily overlook or omit.

Ever heard the phrase 'preparation is the key to success'? Well, this is true in music production too; this preparation is called pre-production.

The first half of this course provides a solid foundation, including information on: music production philosophies, audio perception, critical listening skills, music theory and also the work you do with a band before you even get them into a recording studio. This content is extremely important because it will help you build the skills required to cultivate a vision for material that you are working with. Without this, you won't know what you want to hear and therefore you will not know how to get the sound right at source. The second half of this course gives much more specific production advice on a range of instruments that will allow you to achieve the best possible capture.

FURTHER RESOURCES

To keep this course concise, it cannot possibly cover topics such as room acoustics in the level of detail discussed in educational material designed solely for that subject matter. The initiative is with you to delve as far down the rabbit hole as you can. Exploring the literature found in the bibliography is a great way to enhance your knowledge on each subject matter!

There are also a number of accompanying resources included with this course. Where this is the case, you will see the following box containing further resources, with the icons below indicating the type of resource:

> **Further Resources**   www.audioproductiontips.com/source/resources
>
> 📖 Further reading (book, e-book, etc.) (1)
> 💻 Downloadable resource (website, text, software program, etc.) (2)
> 🔊 Audio (song, sound clip, voice instruction, etc.) (3)
> 🎬 Video (4)

The number in brackets after each entry indicates where the full details of that resource can be found in the index on the following web page:

www.audioproductiontips.com/books/getting-the-sound-right-at-source/resources

I have provided the resources in an online index in this way in order to ensure links are updated regularly as they change with the natural ebb and flow of the Web.

CONTRIBUTORS

Along with the many people I've already outlined, I've also had the support of some of the industry's leading plug-in manufacturers:

Celemony: www.celemony.com

Digidesign: www.digidesign.com

IK Multimedia: www.ikmultimedia.com

Lexicon: www.lexiconpro.com

Massey: www.masseyplugins.com

Pro Audio DSP: www.proaudiodsp.com

Slate Digital: www.slatedigital.com

Sonalksis: www.sonalksis.com

SoundToys: www.soundtoys.com

Waves: www.waves.com

PREFACE

MISSION STATEMENT

From this point on, I will be referring to this book as a course; this is to make it clear that the focus is on your personal journey and that you *need* to put the concepts into practice to gain the most from it. This course is the product of my journey. My experience, however, is not only drawn from studio experience: many of my biggest epiphanies have occurred while touring and doing live sound for several recognized bands. My intention in writing this course is to help bridge the gap between higher education and real-world experience. It is aimed at those of you who have a basic knowledge of studio engineering but need to learn the practicalities of the job. This course is designed to make you aware of the skills needed to produce a recording session, but the specifics of a digital audio workstation (DAW) will not be covered. This course is also written to give insight into areas of the job that are often glossed over. Think of this as your very own studio internship.

This is not an introduction to sound or the devices found in a recording studio – those aspects are covered very well in a plethora of existing texts. You should be familiar with these issues, as well as some of the scientific fundamentals of sound such as amplitude, frequency, wavelength and phase, in order to get the most from this course. If you need to know such information, you can find it in *Modern Recording Techniques* by David Miles Huber. To keep electronics and maths to a minimum, I've also placed much of the more peripheral theory in the appendices, which will be referred to where necessary – please at least attempt to read them.

This course is structured in a palatable way rather than a purely chronological one. Some parts of this journey might feel like they miss 'pieces of the jigsaw'. However, this is intentional so that more crucial elements are memorized before attempting to build on them. However, if you have any questions after finishing the course, feel free to email me.

You will learn in whichever way suits you. There is in-depth written theory, practical examples, tasks and video content. Much of the video content features myself in the studio working with a signed band on a track that you will become familiar with from the practical examples.

In some of the video examples, the shots to camera are unscripted and were made during the session. This is a real recording session and the resulting recording has been released commercially.

The track was recorded on a shoestring budget and I also self-imposed some rules regarding the equipment and environment it was recorded in to more closely match the conditions that you're likely to be working in. Recording the track at Abbey Road might have been more impressive but it wouldn't be representative of the struggles that you will face on a day-to-day basis. What this course should prove is that you don't need to have the best equipment to produce commercial-grade results.

You even get all of the completed Pro Tools session files to peruse at your own leisure, and remix yourself. If you use an alternative DAW, there will be consolidated audio files for you to import into the software of your choice.

Remember, though, this is your internship, and not applying yourself to the set tasks or examples will massively inhibit your progress. Don't just read or watch – engage, experience and immerse.

RETHINKING MUSIC PRODUCTION

In today's music industry climate, it is more difficult than ever to become a top-level engineer or producer. Even the biggest studios are starting to struggle to survive. In the last decade, there has been a huge boom in the project studio market, and there are also an ever-growing number of establishments offering music technology courses. With such a skew between supply and demand, it is currently harder than ever to even get an internship at a respectable studio, let alone make a breakthrough and be involved with the making of a hit record. However, some people still manage it, so why can't one of those be you?

Having left university, I was lucky enough through determination (and the exaggeration of my own skill level) to get an engineering job in a local studio. No, this was not an internship or an assisting job; I was given a brief tour of the routing of the studio and left to my own devices to run the sessions. While this naïve and perhaps ignorant approach opened some doors for me where those less cocksure may have failed, it still gave me a rude awakening. My misguided belief was that, having completed my higher education course in Music Production, I had developed all of the skills required to succeed in the music industry straight away. However, learning the basic technical elements is only part of the battle. The social skills required for record production and the speed at which you need to work, as well as the development of your own critical listening skills, are difficult to teach in a formal environment. These skills are best learnt through practice and experience – the sort of practice and experience that is learnt with an internship. As I've previously stated, these internships are getting more and more sought after, and with such competition you really need to give yourself the best possible chance to impress prospective employers or clients.

I want to take what you think you know about music production and turn it on its head. Other engineering courses and textbooks imply that the technical considerations of microphone choice, placement and signal processing are the most important factors while recording. I am sorry to say that this isn't true. I don't want to imply that these things aren't very important – they are, and they need to be learnt. The problem with the majority of these textbooks is that they would have you assume that the players themselves are ideal, as well as their choice of accessories, tone of the source and the musicians' own tonal decisions.

In reality, this is rarely the case. Even if the musician has a good idea of what he or she wants and has the right equipment to achieve it, there is usually some tweaking to be done from the engineer/producer to optimize his or her signal chain for recording purposes.

If the sound from the source is not optimized before it hits the microphone, then you will always be playing catch-up and having to process the signal more than is necessary. Sometimes budgets will mean that some 'corner-cutting' is unavoidable; at other times the creative direction will shift during the recording session. Finally, dare I say it, there will be times where you make mistakes or don't pick up on potential issues during the recording phase. Relax, you are only human, plus we have more tools than ever to fix elements after the fact. However, for the most natural results, the aim should be to get the sound right at source.

These factors are seldom written about, and engineers/producers generally have to learn this on the job. Luckily for you, this course makes these sonic considerations and covers topics such as drum tuning and guitar selection extensively.

To make matters even worse, manufacturers' marketing, Internet forums and YouTube tutorials lead you to believe that the particular brand of equipment you use is the most important factor (usually the stuff for which you would have to remortgage your house to purchase). This plays into our desire for a magic-bullet solution – a reason why you are not to blame for the ineffectiveness of your work.

Before you blame your gear, take a look at what you might be doing wrong. To summarize how important your technical equipment is, I'd say that the right choice of great equipment will undoubtedly improve your productions, but they will never make or break it.

I am sorry to shatter this illusion, but you will thank me for it later. I'd estimate that the recording gear you use accounts for 10 per cent or less of a recording's overall quality.

Some of us like fast cars, big houses and lavish holidays; I knew I was destined to be working in music production because the latest model of compressor 'gave me kicks'. This type of gear lust is natural, but it is important not to overstate the equipment's importance. Despite my more realistic views on the importance of the equipment, I am still guilty of overspending from time to time.

You should also experiment with all the different job roles in music production: engineering, editing, producing, mixing and mastering. Each role has traits that allow certain personality types to thrive. For instance, engineering, editing and mastering are much more methodical processes. Having an acute attention to detail is very, very important to the mastery of these skills. For instance, if the tracking engineer starts worrying too much about performance, vibe or arrangement, then he or she will most likely step on the toes of the producer and also fail to choose the optimal microphone and the positioning required to properly capture the performance.

Production and mixing, however, have a much more creative nature. This is not to say the previously mentioned jobs are not creative, but, in the grand scheme of things, the producer and mix engineer are the visionaries of the recording project. Just to further reiterate: I do not perceive the producer and mix engineer exclusively as the most important pieces of the jigsaw – it is the tracking engineer's job to help make the producer's sonic goal a reality.

These days, most people starting out in the recording industry will start out recording small, unsigned bands or their own bands. In these circumstances, you will expose yourself to all of the roles in the record production process.

This is a great way to 'cut your teeth' and allow you to work out which elements of the recording process suit your personality.

For me, the production and mixing phase is what interests me most. I work fast and do things aggressively. I can also be a little clumsy and lack the perseverance to find that elusive sweet spot for a particular source. Conversely, my studio assistant often talks to me about spending hours finding the perfect compressor settings or tweaking mic placements. His skills complement mine, and vice versa, which is why we work so well together. This latter personality type will find it tough to churn out a mix with 100 audio tracks in a day, and it will be difficult for him or her to juggle the overseeing of many arrangement/composition or effects changes during tracking.

PRODUCTION PHILOSOPHIES, YOUR EARS AND CRITICAL LISTENING

1.1 PRODUCTION PHILOSOPHIES

As an engineer/producer, your number-one priority is improving your own skills, which is done by listening critically to sources and memorizing their sonic attributes. This first chapter is about how to develop a mindset that accelerates your learning in these areas, and also gives an introduction on how to listen critically. Remember that critical listening for a producer never stops; you never cease expanding your repertoire of internalized sounds.

I am not going to descend into spiritualistic teaching, but there are definitely a few different mindsets that I find incredibly useful to adopt during different parts of the music production process. I feel it is crucial to express these before beginning this course in earnest because these mindsets affect every choice you make during the recording process and also the direction you take in your own career.

1.1.1 There Are No 'Magic-Bullet' Solutions

A lot of people look for a magic-bullet solution to a problem. Questions such as, 'What is the best EQ preset for a kick drum?', 'Which is the best vocal mic for metal?', or 'How do I record a drum kit?' all miss the most fundamental mindset of the music production jigsaw. Having some rules of thumb is not a problem, but you should remember that *each problem is different.* The player's style, instrument, effects, the microphone and its placement, the room, and the preamp all have an effect on each other. Plus, the tonality of the other instruments, the genre and the way that the arrangement progresses also help to sculpt your decision-making process. There are no straightforward answers to any of those questions. The best way to achieve results quickly is to memorize how different rooms, equipment and placements sound, and apply that knowledge to what you want to achieve (aka *critical listening*).

Even though there are no magic-bullet solutions, there are aspects that can be improved fairly quickly and which provide dramatic improvements to the quality of your work. This involves improving your monitoring environment; improving your monitoring should be your second-highest priority (after improving your own skill set).

Whether you are tracking, mixing or mastering, every decision you make comes from information given to you by your monitoring environment. For instance, if you think that you need to give an electric guitar part more treble so that it pierces through the mix better, you have arrived at that decision because of the way that it sounds through your speakers in that particular room.

Put simply, your perception of the mix is the direct result of how your speakers and room combine to give you the sound you hear.

Problems occur when your critical listening environment is inaccurate; this is the number-one reason why results don't translate well into other environments. Improving your monitoring environment will improve the quality of your output more than any microphone or plug-in purchase possibly could.

You can improve your monitoring environment in many ways:

- tweak speaker positioning;
- upgrade your speakers;
- acoustically treat your room; and
- move into a bigger room with fewer inherent acoustic issues.

If your budget is small, you will want to start by prioritizing improvements to your listening environment. Information on how to do this is included in Chapter 9: The Critical Listening Environment. I've refrained from presenting this information too early in the book because it is a little heavy on physics and mathematical principles – I want to have you hooked first! However, if you don't already have a basic recording set-up, you may want to read this chapter first.

Before moving on to some of my music production philosophies, I want to further desensitize you to the importance of gear by stressing the importance of another acoustics-related factor: your recording environment.

Just like your critical listening room, the place where you record the music you create also imparts a tonal quality to your productions. I would much rather record through mid-level equipment in a superb live room than through top-end equipment in a poorly dimensioned or treated live room. In Chapter 10: Decision-Making and General Recording Techniques, we will discuss some ideas regarding general mic placement and how the quality of the live room might affect this.

The general rule is: the more imperfect the live room, the harder you will work placing the microphones, and the more you will have to process the signal to make it sound good. This point is particularly pertinent to instruments that rely heavily on capturing ambience. The tonal characteristics and its associated reverb time is called *room tone*. Drums, acoustic guitars and classical string instruments are all known to benefit dramatically from a controlled and acoustically optimized room.

Because of this, and also the start-up costs required to properly mic a drum kit, I'd recommend building a good relationship with a couple of local well-equipped recording studios that you can dry-hire (i.e. without an engineer) to record such instruments and build that cost into your budget.

1.1.2 Everything for a Reason

One aspect of music production (and life in general, really) that I'd lost track of a little was to actually think about what you need to do before you do it. I realized that I was stuck in a workflow, adding plug-ins, setting up the same mics, editing drums in the same way, all out of routine. I first noticed this behavioural pattern in my use of dynamic processing. I'd stopped listening to what was required and had started adding compressors and limiters without considering the effect of such processing. Now while I believe that 90 per cent of the time, dynamic control is needed, I should never have taken this as a given.

- Listen
- Analyse
- Process

In that order!

1.1.3 Vision Means Progress

Whatever your role is in a production, what really counts is your vision. This relates heavily to the principal above (listen, analyse, process) and also the development of your critical listening skills. You are no good to anyone if you don't have a vision. For the producer and mixer, this vision comes with the territory, but even the tracking engineer should have vision. For instance, the producer might say, 'I want this electric guitar to sound like The Smiths,' and the engineer must then work out the guitar, amp, effects and microphone combination to make it work. This takes vision. As long as you are constantly developing your analytical skills and doing everything for a reason, your ability to anticipate what a track needs will improve. With commitment, this improvement will grow at an astonishing rate.

1.1.4 Some Golden Oldies

If you have read a few things about music production, the chances are you will have heard these pearls of wisdom.

Get the Sound Right at Source

You wondered when I'd get to this, didn't you? From the outset, aim to press record only when everything sounds as you intend it to. Then every subsequent decision will be made wisely and precisely. Remember, before you start to think about microphone selection and position, think about the player, instrument, amplifier and any in-line effects. Before you start using hardware EQs and compressors, experiment with microphone selection and placement. The general rule of thumb is that the earlier you correct an issue, the more natural the sound will be. Do not underestimate the power of correct mic placement. It is often overlooked, but in my experience it is far more important to find the 'sweet spot' than it is to have the best microphone, pre-amp and character compressor.

It's Not All about the Equipment

One of the big questions I get asked a lot is about specific gear or software. A lot of the time, I rely on a trusty £80 Shure SM57, not some £3,000 boutique microphone. I am also a Pro Tools user – why? Because I learnt it, inside out, first. That is the only reason I choose to use it; all the major DAWs are capable of getting commercial-grade results. So stop worrying about your equipment and concentrate on developing your skills to maximize the quality of your results. Besides, until you have sorted out your room and monitoring, any gear purchase is not the right way to spend your money. The chances are, if you can't afford to buy a piece of equipment, you aren't at the right stage of growth to warrant its purchase. So don't make excuses for bad results!

If It Sounds Good, It Is Good

This was a phrase coined by Duke Ellington (along with a similar incarnation by Joe Meek), and whichever way you look at it, this phrase is true. In one sense, it could mean that if it sounds right, who cares what it looks like? In the DAW age, we are used to looking at graphical representations of the processing we are doing to the signal. It is human nature to want things to be neat and tidy. Fight this urge and learn to use your ears. That is not saying that you should never take note of the graphical representations. Use them to reinforce what your ears are already telling you, rather than overruling them. In another sense, this phrase could mean that you should know when to break the rules. For instance, on a recent mix that I completed, I was tweaking the Little Labs IBP phase alignment tool on a ribbon mic that I'd been using in conjunction with an SM57 on a distorted electric guitar. I rotated the phase until it sounded right for the mix. It turned out the signals were slightly out of phase, resulting in a bottom end dropout that 'hollowed' out the mix and gave the bass guitar more clarity and space. This could be viewed as bad practice, and that I should have preserved phase and used an EQ to cut those unwanted frequencies, but it sounded right, and I followed this instinct. If this example sounds like gobbledygook, then you are reading the right book!

Don't Leave It to Fix It in the Mix

A by-product of the Internet generation is the prevalence of online mix tutorials. Sure, learning to mix properly is fun and pivotally important to the overall result. However, these recording sites often give a lot of attention to mixing, thus inadvertently shifting importance from creating better sounds at source to problem-solving at mixdown. The reality is that mixing is no more important than any of the other stages in music production; it is just the time where the recording process is usually judged to be a success or failure.

If you are recording and mixing your own projects, you need to be aware that a lot of your problems at mixdown will actually be caused by inherent errors in the tracking and arrangement stages.

A fantastic mix starts with fantastic tracking. Your favourite records started out with some of the finest tracking engineers and producers at the helm. Without them, the mix engineers wouldn't look anywhere near as good as they do.

Half the Job Is Managing Social Situations

Finally, it shouldn't be underestimated how much of your job is reliant on your social skills. This is not only to entice people to record with you, but also to keep morale up and get the best out of the performers. Recording can bring the best and worst out in a band, and you need to know the right thing to say at the right time. This might be a white lie to keep morale high; it might be a diplomatic way of telling a band member to change a part; it might be an encouraging or entertaining word that re-energizes the mood and gets the band 'vibing' again. In whatever way possible, you need to get the best out of these people, and if you can predict and ease potential conflicts before they happen, so much the better.

A perfect example of this was on a recent record I produced. The lead guitarist was a fantastic talent and a very inventive player, but he loved virtuosity and was somewhat attached to his selection of gear. How do you go about explaining to a guitarist that instead of playing scintillating solos on an Ibanez through a high-gain amp, he should strip back to a Les Paul on a crunch setting and play with more feel, because it would be better for the song? First, I had to work out what the rest of the band thought about my ideas and also his playing style.

Once I realized that the other band members were in agreement with my analysis of the situation, I had to analyse whether this person was going to be hesitant or reluctant to change. Hesitance and/or reluctance to trust the producer's instincts can be born of an elitist attitude, insecurity or the player simply wanting his or her personality to be heard. In this case, it was the last option (though, most often, it is one of the other two). To get the most out of him, I instructed the rest of the band to leave the studio while we worked on his lead parts. I sat him down and told him that I believed the songs are the most important part of any record, and that, in my opinion, the majority of the songs didn't suit that style of playing. Note: in my opinion, it's always better for the producer to be the bad guy than to risk dividing band members against each other. As expected, he was averse to change because he had spent so much time perfecting those riffs and so much of his own hard-earned cash buying that equipment. I responded by saying that I was not trying to dumb him down, and that there would be areas where he could unleash some virtuosity, but that he should at least try to embrace my ideas and vision before rejecting them. I also mentioned that I would not try to veto his ideas and wanted the record's direction to be equally down to the band as myself (remember, I already knew that the rest of the band shared my vision). Once we got two songs in, he sat down next to me between takes and told me that I was right. He became the standout performer on the record, and played with tons of feel. In one particular song, he was able to keep his lead solo 100 per cent intact, and because a lot of his virtuosity had been stripped back, it made the solo a highlight on the album.

The lesson here is that there are a million variables when handling a situation, and I think that you have to trust your gut instinct when making decisions. If you truly want what is best for the song, then it's hard to fail, providing you leave your ego at the door. If your idea is not successful, be open, honest and unafraid to admit that it was a misjudgement. This is not a sign of weakness; it is a sign of strength.

1.2 THE PRODUCTION TEAM'S MISSION

The true skill of an audio engineer/producer is *not* that he or she knows how to use a mixing console or can operate a DAW. You can learn the basics of many of those things in a few days (although proficiency and speed is another issue altogether). The more important skills to admire and appreciate are your production idols' deep understanding of:

- Creating a strong vision.
- Knowledge of the genre(s) of music he or she works in.
- A seemingly innate knack for knowing why a song might be weak (in reality, this is built from a large amount of experience and practice). This could include problems with composition, technical skill of members, arrangement, lack of rhythmic development or layering, tone selection or many, many other things.
- A good grasp of whether a frequency or dynamic misbalance would be best resolved by engineering alterations (mic placement, amp choice, EQ, etc.) or arrangement alteration (adding/removing instrumentation).
- The technical knowledge of how to get the envisioned tones from the player, instrument and/or amplifier. This could be an alteration of a player's technique, amplifier settings, instrument choice, etc.
- Knowledge of the 'sound' of technical equipment to further hone tonality.

All of these skills mean that the production team can oversee the recording process so that the song's construction meets a few conditions:

- The music, lyrics and form are all congruent with the intended message.
- The layering is such that it progresses with the message and the listener's interest is maintained throughout the piece.
- The arrangement allows the track to 'breathe' (i.e. it is transparent and isn't cluttered or muddy) (the mix engineer also plays a part in this).
- The technical decisions regarding the capturing of tone make the track easier to mix. For instance, using lots of vintage analogue equipment for a 'sixties sound' or using a stone room, reverse gate or synth kick for an 'eighties drum sound'.
- Technical decisions are made to align with the sentiments of the genre. For instance, the rawness of an old-school punk track could mean the production team would likely prefer to capture much of the instrumentation live, use bright overhead mics on the drum kit and keep the guitar tones rather distorted, jagged, edgy and unpolished.
- On a more clinical business level, the production team needs to make sure that the outcome of the production aligns with the intended audience that the record label wants to sell it to.

1.2.1 The Production Team

Before we get into the course in earnest, I want to define the stages involved with the creation of a song. The music production process is usually made up of three phases: tracking, mixing and mastering. In total, these phases typically involve at least four personnel: a tracking engineer, a producer, a mix engineer and a mastering engineer. In practice, these phases may start to overlap each other a little, and on smaller budgets these stages may be carried out by the same person. However, it is still important to understand the unique purpose of each process. Effectively, each represents a specific type of quality control:

1. *Tracking.* This is the process of getting the song recorded to a recallable medium. Tracking got its nickname from the days of analogue tape where each instrument is captured individually and given its own track on tape; these days, the channels in a DAW can also be called 'tracks'. The balance and sound of each channel can be recalled for adjustment later.
2. *Mixing.* This is the part of the process where each of the individual tracks in a recording is blended to create a version of the song where all of the individual sounds gel as well as possible to create the best possible 'soundscape' and 'journey'. The end result of this phase is the 'mix'.
3. *Mastering.* This is the process where each of the mixes is processed and organized so that a complete product (usually an album) can be manufactured. Another important aspect of the mastering process is as a last level of quality control over the sonic quality of each mix so that the product sounds as good as it can on a variety of playback mediums (such as large club loudspeakers, hi-fi systems, laptop speakers, headphones and earbuds, etc.).

Many sonic problems occur when key decisions are deferred until later in the music production process. If the tracking engineer/producer is wary of making a wrong decision or simply tries to keep everything neutral for the mix engineer, it forces the mix engineer to resort to more drastic use of processing such as EQ and compression to create balance. The problem with this is that it is a rather less organic way to manipulate sound compared to techniques within your control while tracking, such as adapting playing technique, instrument selection, room choice, and mic choice and placement. Despite the time and best efforts of the mix engineer, if he or she is put in the position of having to compensate for a lack of direction and focus in the earlier stages of tracking, it will usually result in a less natural-sounding end product. For instance, if you want a lo-fi garage rock sound, why not record it in a small room with thrashy cymbals and gritty guitars? And why not leave it slightly looser and edgier in the process? The same principle also applies between the mixing and the mastering engineers – you are usually better off to fix an issue as early as possible.

To get the best possible recording, it means that you need to *get the sound right at source.* At this point, I think it is important to consider what I mean by this. I am *not* saying that the tracking engineer should EQ and compress the track on input to make it sound as good as it possibly can; this is the mix engineer's job. I simply mean that the tracking engineer should do everything in his or her power to get as close to the vision in his or her head as early in the production chain as possible. If the same effect can be achieved with guitar, amp selection or mic/pre-amp choice and placement rather than adding 6 dB of EQ to the top end, it is more likely to be the better option. This means that the playing style is more important than the mic choice/placement, and in turn the A/D conversion is less important than the mic choice/placement and so forth. A rough rule of thumb is:

player → instrument → mic choice/placement → pre-amp → A/D converters

Of course, like so many creative endeavours, there are caveats and these rules are actually more like guidelines. These guidelines are often bent or even broken by experienced engineers. This is because experienced engineers have the basic tonal attributes so well internalized that they can experiment with ways to manipulate the chain to produce a sound 'even larger than life'. As you develop and gain more experience, you will also start to know when to abide by the guidelines and when not to.

Back in the 1950s and early 1960s, a track would be performed live because engineers did not have the technology to capture and play back individual tracks in the way we do today. This forced engineers to make such decisions at source. These days, due to the many stages of production, it is too easy to leave the decision to someone else. Please don't do this! If you are the tracking engineer or producer, any track you give to your mix engineer should be in such a state so as to guide the mix engineer towards realizing the initial production concept without the need for him or her to guess what your intentions were for the track.

1.3 YOUR EARS AND THE DECIBEL

Whatever we choose to do in life, our decisions are always based around the clues that our senses give us. Understanding the way that our auditory system works makes music production easier. Included in this section is a brief explanation of how our ears work, the measurements we use to express audio volume and how it relates to our ears' perception of loudness, distance and direction.

1.3.1 Your Ears

The anatomy of the ear is very sophisticated but it can be broken down into three distinct parts:

- the outer ear;
- the middle ear; and
- the inner ear.

Another central function of the ears is to help balance. For the sake of this simple description, aspects of the ear that deal with pressurization and balance are omitted.

The outer ear consists of the fleshy visible part of the ear, called the pinna, the ear canal and the eardrum (tympanic membrane). The pinna collects and funnels sound waves into the ear canal to be processed by the middle ear. When sound waves hit the eardrum, it causes it to vibrate sympathetically to the sound wave.

The middle ear is an air-filled cavity behind the eardrum that includes three bones (ossicles): the hammer (or malleus), anvil (or incus) and stirrup (or stapes). The hammer is connected to the eardrum, and it transmits the vibrations produced by the eardrum to the anvil and stirrup. Through this section of the ear, the vibrations from the eardrum are amplified nearly 30 times. This can be dangerous with loud sounds, particularly those with extreme transients such as gunshots.

Finally, there's the inner ear. After the ossicles, the vibrations are passed to the oval window, exciting fluid contained in the cochlea, which is a pea-sized, snail-shaped organ containing two fluid-filled chambers and microscopic 'hairs' that act as receptors. Once the oval window is excited, the fluid flows around the cochlea, in turn exciting the hairs and resulting in electrochemical impulses being sent to the brain that are interpreted as sound. Our ears have a form of self-protection that helps protect damage inside the cochlea (called oval widow contact breakers) but it usually takes 20–75 ms for it to kick in, making extreme transients still extremely dangerous for our hearing. Despite this self-protecting mechanism, loud volume levels can easily lead to serious damage (we will shortly discuss how to look after our hearing).

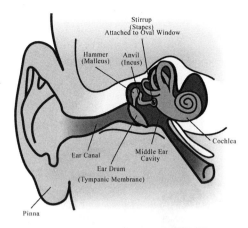

Figure 1.1 Anatomy of the ear[1]

1.3.2 The Decibel

As we have just discovered, the human ear is an extremely sensitive mechanism, capable of detecting and interpreting a huge range of frequencies and volumes. The approximate energy range of human hearing is 10,000,000,000,000:1.[2] If I were to express differences in acoustical pressure between a quiet sound and a comparatively loud sound using a linear representation, I would have to use a widely varying and cumbersome range of values, which would not be easy for the human brain to comprehend let alone calculate. Therefore, we need to find an alternative way of representing these changes in intensity, one that allows us to easily comprehend and communicate them between engineers and equipment.

First, we need a mathematical function that reduces large numeric values into smaller, more user-friendly ones. To do this, we use a logarithm that takes a linear system and makes its values increase exponentially. This is also convenient as our hearing perceives loudness in a vaguely exponential way (more on this later).

The bel is a logarithmic expression of power ratio, which had been developed to represent telephone transmission loss. These values were still a little too cumbersome, so decibels (one tenth of a bel) were used to express changes in intensity in audio. (More information on the mathematical background of decibels can be found in Appendix A.)

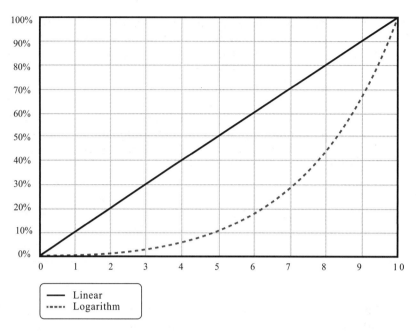

Figure 1.2 Linear and logarithmic curves

The use of decibels is still found in electronics and optics, so they are in no way exclusive to sound engineers. However, from this point forward, I will be talking about its application in audio engineering.

A decibel is a unitless expression of a power or signal-level ratio. This would suffice if we only had to express the change between two levels – for instance, telling your assistant to turn the bass guitar up 3 dB.

In practical terms, though, we sometimes need to base our decisions around absolute values. For example, we generally want to record at a level that will not introduce distortion, while also avoiding undesirable noise. To express this safe operating level as a decibel, we would need to find an absolute value that we deem safe and base our decibel scale around that point. This is called a reference value.

Many different decibel scales exist for many different applications, and all of these are based around a chosen reference value. Several of these decibel scales are just for audio purposes, and in Chapter 7: Demystifying Recording Levels we will discuss the many different analogue and digital scales that are commonly used in recording equipment. For now, though, let's just discuss broader types of decibel scales used in audio.

Types of Decibel Scale for Sound

Now that we know that a decibel is the expression of a change in power ratio, let's examine the dB scales for the common audio applications:

Electrical signal. The meters in newer analogue consoles are expressions of level change in relation to a 'safe' operating level. Common scales for signal level or voltage in the analogue domain are dBu, dBm or dBV.

Digital audio levels. Once an analogue signal is converted to be stored or processed with a computer, we need a different scale type to keep a 'safe' level in the digital domain. Common scales in digital audio are dBFS and dBFS RMS.

Sound pressure level (SPL). This expresses level change between the pressure of air (20 micropascals, or 20 {mu}Pa) and the pressure created by a sound wave at a particular fixed point. So 0 dB SPL is the pressure level of air and also roughly equates to the limit of human sensitivity, without factoring age or exposure to loud music. Usually when you see decibels expressed in terms of physical volume, it is in relation to dB SPL as it is easier to measure than sound intensity.

Sound intensity. This is an expressed change in pressure based on not only a fixed point, but also how sound spreads over an area around the sound source. So sound intensity is defined as power per unit area. You should not become too concerned with sound intensity because it is more difficult to calculate. It is used most often in theoretical debates and psychoacoustics rather than as a measured value in real-world situations.

To muddy the waters even more, decibels in terms of electrical power, digital audio levels, SPL and acoustic intensity are often misused or used interchangeably.

Just remember that each of the three are different scales because they use a different type and level of variable as their reference. Therefore, they should be treated as different entities.

Below are some good rules of thumb to remember when dealing with decibels:

- A doubling of power in an electrical circuit is described as a 3 dB increase.
- A doubling of voltage results in a 6 dB increase in power.
- A doubling of acoustic intensity (power) is described as a 3 dB increase.
- A doubling of SPL results in a 6 dB increase in acoustic intensity (power).
- A doubling in 'perceived loudness' is described as a 6–10 dB increase.

In practice, though, once the initial level has been set, an engineer needs to be able to express the change between two states and not absolute values. When dealing with the change in level of an instrument in a mix, you would reference that change in the electrical signal in your mixing desk or the digital scale in your DAW.

In terms of monitoring levels in your studio, you would reference it against SPL but express the change in terms of electrical change. An example of this would be: 'Can you turn the bass up by 3 dB?' This would relate to either a 3 dBu increase on an analogue mixing console or 3 dBFS in the metering of your DAW. Again, there's more about these types of scales in Chapter 7: Demystifying Recording Levels.

1.3.3 Loudness Perception

Now that the types of dB used in terms of sound are cleared up, it should be easy to relate SPL and electrical power to loudness, right? Unfortunately not.

The problem is the human perception of loudness does not match these scales exactly.

Even though our hearing is broadly logarithmic much like the decibel scale, what arrives at our ears and what we hear aren't always the same. So let's discuss how volume affects our hearing's frequency response and how we localize sound sources.

When dealing with perceived loudness, each successive perceived increase in volume requires a larger increase in power. In addition to this, there are other factors that affect our perception of loudness:

1. The human ear's sensitivity to different frequencies is not equal. Human hearing is tuned by nature to be very sensitive to fundamental frequencies in the human voice.
2. Additionally, as volume increases, our sensitivity between different frequencies starts to even out and naturally compress.
3. The decay of sound in the time domain plays a big part in how we perceive loudness. This is why measuring average levels are equally important as peak ones.
4. As you age or are subjected to loud environments, your hearing will become less sensitive. This is usually most apparent in the extreme high frequencies, but musicians and others who are exposed to loud environments on a regular basis can incur a variety of hearing problems.

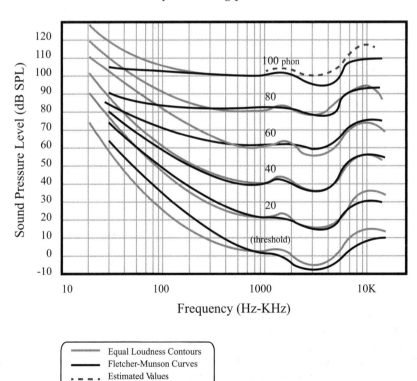

Figure 1.3 Fletcher-Munson curves, or equal loudness contours[3]

What this means is that perceived volume is rather subjective and measuring sound intensity as experienced by an individual is rather difficult. The relationship between our ears' sensitivity to different frequencies at differing volumes was explored by Harvey Fletcher and Wilden A. Munson, and resulted in the famous Fletcher-Munson curves.[4] The Fletcher-Munson research has been improved on over the years and today is often cited as the equal loudness contours. These curves have a profound effect on how we produce, mix and master music, which relates to how loud we should monitor while making music. The reasons why the equal loudness contours are important will become clear throughout this course.

Because 'loudness' is a subjective feeling that is not easily quantifiable, people often relate to it in terms of change in power as expressed in decibels. A good rule of thumb when using decibels in relation to perceived loudness is that roughly every 6 dB increase in electrical power has the effect of doubling the perceived volume.

1.3.4 Sound Localization

Sound localization is the process of determining the location of a sound's source. To enable us to perceive the direction and depth of a source, the brain analyses subtle differences between the sounds that arrive at each ear. These differences comprise acoustic intensity, high-frequency attenuation and time delay.

Localization can be represented by a three-dimensional position: the horizontal angle, the vertical angle, and the distance (for static sounds) or velocity (when its position is moving).

Localization of Sounds on the Horizontal Plane

The horizontal position (left–right) is identified by:

- The amplitude of high-frequency sounds between the ears (called the shadow effect). This is called the *inter-aural level difference (ILD)*.
- The difference between the sound's arrival between the ears. This is called the *inter-aural time difference (ITD)*.

Depending on where the source is located, our head impedes the level of higher frequencies that can reach the ear furthest from the source; this changes the acoustic intensity and spectral qualities of the sound, and further aids the brain in distinguishing the spatial origin of the sound.

The sound's frequency content plays a huge part in the mechanism used to discern the direction of sound. With sounds above 1.6 kHz, inter-aural level difference is the primary tool for horizontal plane localization. Due to their smaller wavelengths and higher number of cycles between reaching each ear, calculating phase differences in high-frequency sounds is inaccurate and fallible. Calculating direction based on level differences, however, is more accurate as higher frequencies are more easily absorbed by the environment and the head itself, leading to a more accurate indication of direction.

The inverse is true for low frequencies, so the brain primarily uses inter-aural time difference to determine direction for sounds below 800 Hz. Because low frequencies are diffracted around the head rather than absorbed, level differences between low frequencies are not significant enough to accurately locate sounds. However, the distance between the ears of an average human is less than half a wavelength below 800 Hz, meaning that phase differences can accurately identify the direction of the sound. For sounds between 800 Hz and 1.6 khz, the brain uses a combination of these mechanisms to help localize sounds. It is also worth noting that sounds below 80 Hz are usually defined as 'non-directional'. This is because the phase difference between the ears becomes too small to discern direction; both ITL and ITD fail to accurately evaluate direction, leaving a sense of omnidirectionality.

Localization of Distance

Localization on the distance plane is identified by:

- Loss of high frequencies. The level attenuation tends to be more drastic in the high frequencies, as higher frequencies are more easily absorbed by the environment.
- Ratio of the direct sound to the reverberated (reflected) sound.
- The time difference between the arrival, at the listening position, of the direct wave and its first strong reflection.
- Loss of amplitude.

Localization of Sounds on the Vertical Plane

The localization of sounds on the vertical plane (up–down) is less accurate than processing sounds on the horizontal plane or distances. In fact, your ability to discern the angle of sounds above and below your head comes from a complex set of spectral notches created in the signal by the ridges on the pinna (the fleshy visible part of the ear). The brain then pattern-matches the signal to look for these specific frequency-based notches to identify the angle at which the sound is coming from. Although the cues from the pinna are the most important, some other spectral notches created by the head and the rest of the body also play a role in vertical localization. The way in which our auditory system responds to these spatial cues are called head-related transfer functions (HRTFs).

1.3.5 The Haas Effect

As well as loudness and depth perception, the human auditory system is also accurate at distinguishing between distinct sounds and those caused by 'quick' reflections in the listening environment. These reflections are called *early reflections*.

Helmut Haas discovered that we can discern the directionality of a direct sound, even in the presence of reflections up to 10 decibels louder than the source. This principle is known as *the Haas effect*. The human auditory system combines all reflections within 20–30 milliseconds that have a similar intensity into a single perceptual whole. This allows the brain to process multiple different sounds at once. Any sounds outside of this time period will be perceived as distinct sounds.

1.4 EAR FATIGUE

As well as the immediate effect that volume has on frequency response (Fletcher-Munson curves), there are also some other negative side effects from continually listening to music that become apparent in the medium term. These negative effects are called ear fatigue. Ear fatigue most often occurs during mixing. However, it can also be an undiagnosed evil in the tracking stages.

Telltale signs of ear fatigue include:

- You feel you can no longer make reasonable judgements about what you are working on.
- You keep going round in circles, adjusting parameters only to then return them to their original state.
- You start to feel that everything you have done is wrong, and everything you try makes the song worse.
- Stress, frustration and irritability.

Ear fatigue is often associated with overexposure to loud volumes. While this certainly does accelerate ear fatigue, it is not the only cause. Our brain is tuned to perceive contrast in our environment, not elements that remain the same. When you listen to the same track for a long time, your brain starts to get used to what it is hearing and starts to like the sound, even if you are far from finished. It is good, therefore, to get into a routine whereby you regularly change the listening environment. You can do this by:

- Taking regular breaks to let your ears 'reset'. A 15-minute break every two or three hours is recommended.
- Comparing your work to commercial releases to gain perspective (this is called referencing).
- Switching between different sets of monitors.
- Briefly turning the music up or down for a couple of minutes before returning it to the original volume.
- Listening from outside the room, in the car, or in any other environment.
- If you have been working on the track for many hours, just go home and come back the next day.

To improve your working conditions, morale and also your decision-making, I recommend that you only work up to eight hours per day while mixing, and only do eight to 10 hours per day while tracking. As general fatigue sets in, you will start to allow performance and technical issues to be committed to the recording that you wouldn't have settled on while fresh. Not only that, but the musicians will also be tired and see their performances falter, and then there is the knock-on effect that long working days have on subsequent days. Working for longer than these periods is often a false economy and usually results in having to redo the offending parts.

1.4.1 Protecting Your Hearing

As we already know, our ears are very sensitive to a wide range of sounds, but they are also fairly fragile. Any prospective audio engineer needs to take serious care of his or her ears. After all, your hearing is the only piece of equipment that can't be replaced if it breaks!

Audio engineers should be vigilant of any levels over 90dB SPL, and limit exposure to such environments. The higher above 90dB the SPL is, the less time you should spend in those environments. Levels over 118dB cause most people discomfort, and levels of 140dB are often referred to as the threshold of pain. If you work often with heavy machinery, live music venues or other noise-polluted environments, hearing protection is a must. I do a lot of live sound engineering, so I take custom moulded ear plugs with me wherever I go. If you are serious about protecting your ears, you should even wear them in nightclubs.

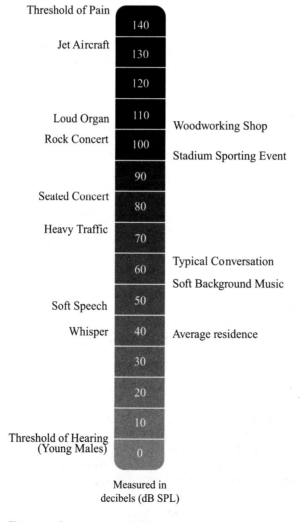

Figure 1.4 Breakdown of dB SPL levels of common environments[5]

Most hearing loss is caused by damaging the microscopic hairs in the inner ear. Common signs that you are exposing yourself to excessive SPLs are:

- Tinnitus. This is ringing in your ears. While tinnitus starts off as a temporary condition, it can quickly become permanent if action isn't taken.
- Having to shout over music to be heard.
- Blocked ear. A constant feeling like your ears are blocked can also signify the early signs of hearing loss.

Permanent hearing damage can occur from acoustic trauma. This is when the ear is exposed to sudden transient heavy noise over the threshold of pain (140dB). Things such as gunshots and explosions are common causes of acoustic trauma. When dealing with music, however, our ears have safety mechanisms to reduce the risk of permanent hearing loss from loud music. This safety mechanism, called *temporary threshold shift*, desensitizes your ear and causes temporary hearing loss. This means that short doses of extremely loud music are unlikely to cause permanent damage. However, repeated exposure will mean that it takes longer for the negative effects to diminish. If exposure is continued without sufficient recovery time hearing loss will become permanent; this is called *permanent threshold shift*.

1.5 MONITORING LEVELS

To avoid the pitfalls of ear fatigue, you should make sure that you monitor at levels where your ears' frequency response is flatter, while also ensuring your monitoring level won't cause damage to your ears. It is therefore important to have a set reference level that you can easily return to time and time again. This also has other benefits while producing music.

When you monitor too low, you end up raising individual channels louder and louder until you run out of room to turn up any further. This causes several potential issues:

- You record at too high levels, so that you can make the instrument you are tracking more audible.
- You run a serious risk of clipping your entire mix.
- You find yourself constantly bringing up and then lowering levels of everything, and potentially ruining some of the finely tuned effects balances.
- It is easy to start over-compressing signals.

If you monitor too high, you are still likely to hit individual channels at a normal level (as, visually, people are scared of hitting meters too low). When this happens, you are likely to run into the following problems:

- Internal balances in your mix may be skewed at normal listening levels.
- Your ears will be naturally compressing the sound. Therefore, you will be underusing compressors and setting percussive instruments such as drums too high.
- Ear fatigue will cause you to make wrong decisions. You could blow up monitor speakers and also potentially damage your hearing.

The level at which you want to monitor depends upon the size of the room you are in. You can measure the monitoring level with an SPL meter. In smaller listening environments, 75–80 dB SPL would be optimal, which is a reasonable level to work at for bass response and also to avoid ear fatigue or damage for sessions under eight hours per day. In a larger environment such as those you would find in a renowned commercial studio, you can afford to monitor at 85 dB SPL at the listening position. This is not to say that you can't alter your listening volume – short blasts at higher and lower volumes are great to hear potential mistakes. Just remember to return to the reasonable level again quickly. After we have discussed metering scales in Chapter 7: Demystifying Recording Levels, we will examine a fail-safe way of calibrating your monitoring levels.[6]

1.6 CRITICAL LISTENING

It is all well and good knowing how to use the technical equipment, but it is absolutely no use if you don't know what you should be listening for. The following tips are designed to help you get started with using and trusting your ears. Becoming better at listening is a skill that you will spend your entire lifetime improving; in fact, once you start to analyse tones, it will quickly become very fulfilling.

Don't expect to be a hotshot at critical listening after reading this section. In fact, it should do the opposite: It should show you how far you still have to go. This whole course is focused around critical listening and many of the forthcoming chapters are focused around the tonality of the instruments you will encounter while recording bands, as well as the key variables that alter their sound. This includes exercises and examples targeted at these specific instruments. Rather than just taking my word as gospel, you should experiment with tweaking as many of the attributes that alter tone as possible, and then memorize the changes they make. This might sound difficult or daunting at first, but it will quickly become automatic.

I can't possibly teach you how to listen critically to a particular instrument without first teaching the specifics of that instrument. For this reason, I want to stress that this section is *only* about getting you to ask the right questions. I am effectively giving you the keys to the car, but you still have to learn to drive it. The best part is, even if you are a seasoned driver, you can get into a new model and have to learn the nuances of that particular vehicle. As well as regarding working with new instruments, this analogy also holds well for working with new acts, particularly in genres you aren't as familiar with. The only answer for this is to practise listening critically and building your personal experience. I suggest you just embrace and enjoy the challenge rather than try to fight the fact that you won't become a master overnight.

Over time, you will start to internalize instruments' natural timbres without consciously thinking about it. You will also realize how the different engineering tools can enhance (or weaken) their tonality. Once you know how the instrument should sound acoustically, as well as how you want it to sound for that particular track, your decision-making processes will become more about fulfilling your vision rather than trial and error. The ability to create and fulfil a vision is the most important skill an audio engineer or producer possesses.

It takes a certain type of concentration to really start paying attention to the fine details of a track. A great way to start developing the critical listening skills required to produce or engineer music is by trying to reverse-engineer a song. If I played you a track right now, could you make a good educated guess about the type of guitar, effects and amp used on a guitar line?

If the answer is yes, can you speculate about the type of microphone(s) and its position and distance from the amp, and the mic pre-amp used?

The rabbit hole goes even deeper – we haven't even talked about EQ, compression, panning, effects or any of the other key skills that an audio engineer must master.

Once you can listen to a track and outline the different parts and how they were probably recorded, you will start to make informed production decisions on your own projects and give focus to your creativity.

I, for one, began producing bands already knowing plenty about EQ, compression, reverb and delay, but had little insight into how it 'should' sound. I hadn't built up that crucial skill of critical listening.

In my opinion, it was the lack of emphasis and time given to critical listening on my university and college courses that left a fundamental hole in my early development. You can learn what EQ is in a day; you can learn the basic operation of a DAW in a week. But learning to use your ears takes persistent repetition, years of practice and a cycle of learning that never ends.

Once you have started to listen critically, you will know soon enough if you are truly the right kind of person to become an audio engineer. If you find yourself listening to a song for the first time and are naturally analysing and dissecting it, then this is the job for you!

Also, are you good enough to detect the nuances of pitch, tone and timing, and tell the musician when it isn't quite right? If not, how can you even begin to produce professional results?

Before you even think about recording any band, you should evaluate whether you confidently know what is required in the genre they are working in. If you don't, then you have a lot of critical listening and research to do!

So how can you actually practise listening critically?

If you are completely new to critical listening, some of the following exercises may go over your head at first. I advise that you keep coming back to these exercises regularly to give yourself perspective on what areas you need to hone, as well as help you to realize how much you have improved. You know it's not a bad thing to pat yourself on the back from time to time!

EXERCISES

1. Listen to songs and work out every layer of instrumentation and how it affects the emotive nature of the track. Also observe how layering is used to progressively build up the track.

2. Once you feel you are able to pick out all of the parts to a song, start trying to pick out the type of equipment used (e.g. 'Les Paul guitar through a Marshall amp with a close-mic'd Shure SM57 and Neve pre-amp').

3. If you are comfortable with exercises 1 and 2 above, then start thinking about EQ, compression, panning and effects, and their

impact on the depth, space and width of the song.

4. Take any instrument you can get your hands on and learn to tune it quickly and accurately, purely by ear (this will take longer than you think).

5. Watch 'average' local bands and listen for timing mistakes. Try to analyse why that mistake has happened (e.g. poor technique with the drummer's right foot; adrenaline forcing the band to play too fast; the bass player's inconsistency in his finger style). Once you are proficient at noticing the types of timing errors, think about the ways you would combat them in the studio. This could be as simple as making a drummer practise a certain technique before you record, or by using tools such as Pro Tools' 'Elastic Audio' to edit the bass in time. (There will be more on editing later!)

6. Sing along to tracks. One of the handiest skills to gain is the ability to help a singer to pitch correctly. The easiest way of doing this is by actually being able to sing the right note back to him or her. Okay, learning to sing is a time-consuming task but I am not asking you to become Freddie Mercury – just being able (and confident enough) to hold a note is adequate.

7. Hang out with people better than you. By whatever means necessary, access a reputable recording studio and listen to the artist(s) and the critical feedback the engineer or producer gives in a session. Can you hear these same issues?

8. Use specifically developed audio training programs that focus on the more technical aspects of critical listening such as: frequencies, distortion, reverb and more. See further resources below.

Further Resources www.audioproductiontips.com/source/resources

📖 Dave Moulton, *Golden Ears*. (1)

📖 Jason Corey, *Audio Production and Critical Listening: Technical Ear Training*. (2)

💻 Harman's *How to Listen* program. (3)

1.7 MY VISION FOR 'TAKE HER AWAY'

During the rest of this course, we are going to be dealing with the results of a real recording session. We will also be dissecting the decisions I made during the recording process. A lot of these decisions were right; some were questionable. Some I'd do differently now; others became a model I've used in future recording projects. As I mentioned above, I am not afraid to acknowledge when I think my decisions might have been better made. Hopefully this is a radical honesty that helps show that music producers (even the top ones) still make a lot of mistakes. Just as you are learning from this course material, so was I when recording this track. There is a therapeutic nature to getting your thoughts across on paper. There is also no greater way to solidify your opinions. So while you are learning from the course content, just take some time to write down any key notes that are important to you.

The track I produced for this course is called 'Take Her Away' by the band Luna Kiss, who happen to be on my own record label Ghoulish Records. It is a seven-minute-long progressive rock masterpiece.

Further Resources www.audioproductiontips.com/source/resources

 Luna Kiss – 'Take Her Away'. Here is the fully finished track that the course in this book is based upon. (4)

I could have chosen an easier project to use for the course, but I believe that the best way to learn is to throw yourself in at the deep end. This track has plenty of variety, whether it be the tempo changes, key changes or production/mixing effects. It will present opportunities to discuss more creative production techniques that are much easier to show than to write about.

As I mentioned earlier, I believe that a lot of music production is about having a vision. Before we move on, it is a good time to mention my vision for this track.

Having heard Luna Kiss rehearse this track, I felt it would benefit from a very regimented sense of timing. The song had many tempo and key changes – if left too loose, the song would sound disjointed. So I decided to record the band separately and tighten the drums and bass quite rigidly using Elastic Audio. I also decided that to leave a little more of a natural groove to the track (and to stop it sounding too clinical), I would use very little editing on the electric guitar and minimize vocal editing and tuning. This is a technique I often also use with bands with a less technically gifted rhythm section, but in the case of Luna Kiss it was purely down to technical concerns regarding the tempo/metre variations. The band have some basic experience with DAWs, so a few weeks before recording I set them to work creating a tempo/metre map for the song and I also got James the drummer practising to a click track.

In terms of vibe and overall tone, I decided that I was going to strive for a production style somewhere between the warmth, dynamics and analogue sound of Pink Floyd and the hard-limited aggressive and modern mix style of The Mars Volta. The former would be a little more difficult to achieve considering the smaller recording environment and the self-restricted lack of high-end analogue hardware (including no mixing console). Remember, I decided to do this so that you can all see that you can create great results from more modest gear, so no more excuses, guys!

The band and I then briefly spoke about their backline and we decided it was all of sufficient quality to record apart from the bass amp, to which we concluded a plain DI and a virtual Sans-Amp would be sufficient. I have on many occasions hired backline equipment and instruments for recording sessions – remember, get it right at source. I also realized from speaking to the band that the song was going to use many voice samples and some trippy effects in the middle section of the song, so we trawled the Internet ahead of recording for some ideas.

Now that I've discussed the vision for the track, Chapter 2: Project Management and Pre-Production will delve into how a vision is created during pre-production and also gives some basic advice on how to manage projects.

NOTES

1. Analog ear, accessed March 2014, http://en.wikipedia.org/wiki/Analog_ear.
2. David Miles Huber and Robert E. Runstein, *Modern Recording Techniques* (Focal Press, 2013), 59.
3. ISO 226:2003, Acoustics – Normal equal-loudness-level contours.
4. H. Fletcher and W.A. Munson, 'Loudness, its definition, measurement and calculation', *Journal of the Acoustic Society of America* 5, 82–108 (1933).
5. Sound transmission, accessed March 2014, www.neurophys.wisc.edu/~ychen/textbook/sound_transmission.html.
6. Bob Katz, *Mastering Audio: The Art and the Science* (Focal Press, 2007).

PROJECT MANAGEMENT AND PRE-PRODUCTION

In my first few years of music production, I worked as an engineer in a small commercial studio where I didn't know what I'd be doing from one day to the next and had to just deal with projects as they arrived. Major-label bands often have the budget for them to write/experiment in the studio but sadly most bands do not have this luxury. As well as this, big commercial studios have studio managers whose job it is to make sure that expectations are handled, needs are met and the engineers/producers have everything they need. The reality on the lower end of the spectrum is that you don't have these luxuries unless you make it a priority. While sometimes you might hit a home run because you got on well with the band and they had all the right ideas and equipment, the reality for me was that, more often than not, the end result was subpar and avoidably so. Once I started working freelance, I prioritized organization and planning and found that, with just a couple of hours of pre-production and planning before a session, the end results were considerably better.

We will discuss pre-production soon, but first let's talk about managing a project.

2.1 PROJECT MANAGEMENT

As well as good communication skills and a solid musical/engineering background, as a producer you will have to manage a project well. This includes many tasks that you might not naturally think of:

- arranging pre-production;
- specifying a realistic amount of studio time for the band/label's needs;
- educating the band on recording processes;
- organizing schedules;
- keeping the band on schedule;
- hiring any equipment needed;
- booking a studio for the budget/sound required;
- sourcing interns/assistants; and
- meeting label deadlines.

If that isn't enough, in some cases extra jobs will be inherited too:

- maintaining/fixing equipment;
- acting like a babysitter for the band;
- counselling; and
- instrumental tuition.

A producer/engineer on smaller budgets really will have to be a master of all of these elements – plus the fact you are most likely having to engineer/produce/mix and master the record.

These days, I choose to start educating potential clients when first meeting them. This is not a teacher/student type of relationship, more an informal chat that has pearls of wisdom interspersed between chit-chat. I often invite prospective clients up to my post-production facility to 'have a coffee and a chat'. This chat usually involves finding out how much experience the band has in studio environments, their expectations, influences and also potential budget. The more groundwork you do before hitting record, the less likely you are to be in for any shocks.

While some bands are highly organized and realistic with their aspirations and expectations, others will sometimes have rather unrealistic expectations or misconceptions, such as:

- 'I want to record 10 songs in a day (and want them to be commercial-release quality).'
- They don't have the budget to record properly (but somehow hope and/or expect engineers to work for less than they deserve).
- They believe they are better musicians than they are and therefore theorize that it will miraculously take them less time.
- 'Our music is already perfect, we need no production work.'
- 'Our <insert badly maintained/budget equipment here> is suitable to get the best results.'
- The band has no real idea on how long is spent in the mixing/mastering stages (or in the worst cases, they don't even know what those stages are).
- 'I want to record live' (but the band are rather uninspiring/sloppy musicians).

Of course, with a select few bands, some of the above may actually be true for them, so don't just stifle their ideas – but more often than not, they are red flags. So you are best erring on the side of caution when negotiating. For instance, when I encounter these issues, I often say things such as:

> Well, I usually recommend a rough guideline of one song per day to track and one song per day to mix. That is quite different to recording 10 songs in a day, so if you were to do that, you would have to really have your material together. With no time for any overdubs, what you will most likely get out of this is a rough demo. Let me come to a rehearsal and I will advise you what I think is best for you, with no obligation. My advice is it is usually good to do fewer songs well than to rush 10 songs.

After I have basically fed them a few questions to see how realistic their expectations are and given them an honest first impression of their expectations, I usually spend a bit of time crunching some numbers with them so that they can see how the process breaks down. Telling them the amount of time for each part of the process helps inexperienced bands see all the work that is done behind the scenes to make a successful recording.

I also work out a rough timetable of how I envisage the session breaking down per instrument. This has several advantages:

- It gives the engineering and production team clear targets to work towards.
- It gives the band an idea of scheduling and also allows them to realize when they need to work a little harder.
- It gives the production team scope to let the band know in ample time if they think the project will take longer than originally planned and give the band the choice of compromising quality or paying for the extra time (provided it is the band at fault for the extra time needed).
- As vocals are often recorded after the instrumentation, when the session wasn't mapped ahead of time the least amount of time was often given to the vocals – which ironically are normally the most important part of most songs.

If the band are in my proximity, I will offer to go to a rehearsal and just give them some honest, no-strings-attached feedback. If they are outside of my locality, then I offer to listen to some rehearsal room demos or try to make it to a gig near me (if they have one).

A further organizational tip is to consider keeping notes on the progress of your session. During the tracking of a larger project, it is often easy to lose track of your progress. Having a daily itinerary as well as a spreadsheet or whiteboard around to mark when milestones have been reached is advisable. Most of the time when recording albums, I will create a row for each song and a column for each instrument and then mark it once I have recorded each part.

In the days of tape, you would also have to keep a track sheet, which is a document that keeps track of which instrument is on each channel of a multitrack. In the digital age, this is not as much of an issue. Digital audio workstations (DAWs) such as Pro Tools, Logic, etc. have their own issues regarding workflow, though, and it is still of paramount importance to be able to navigate through your sessions easily. Because of this, I recommend getting used to a particular colouring, ordering and naming system in your DAW. Despite the fact that tape isn't used as often today, track sheets can be still very useful for the producer, engineers and assistants to note any specific techniques, mic selection, positioning or even just to map out the inputs so that the assistant can patch the microphones correctly into the studio's wall box.

Next, we will look at pre-production in depth.

Further Resources

www.audioproductiontips.com/source/resources

 Audio production tips track sheet word file. (5)

Figure 2.1 Album tracking spreadsheet

Long Road Ghosts - Return Of The Underdogs

Track	Drums	Bass	Rhythm Guitar	Lead Guitar	Synths	Samples	Vox	BV
E.G.G.B.I.T.E								
Matches								
W4N								
2 Steps								
Get Closer								
R.O.T.U								
Stones								
Million Times								
Collab								
Plastic Gun								
Couldn't Sleep								

- Not Complete
- Complete
- N/A

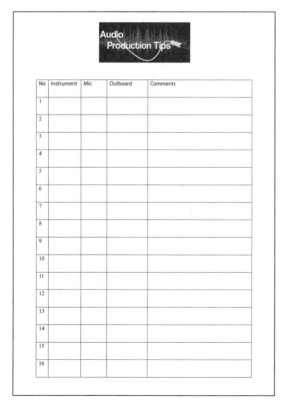
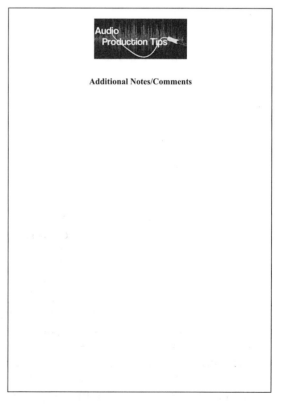

Figure 2.2 Audio production tips track sheet

2.2 PRE-PRODUCTION

Pre-production encompasses everything you do with a band before you enter the recording session. It is a huge part of the recording process and something that is often overlooked by smaller studios, inexperienced freelancers and also the project studio owner. If, as producer, you can spend some time familiarizing yourself with the band and the material they wish to record, you'll have the opportunity to fine-tune their parts and core arrangements before entering the studio. Good pre-production can save a band hours of studio time and alleviates many potential recording problems before even beginning the session. As recording budgets are dwindling, very few artists can afford to write or fully experiment in a studio environment; therefore, the importance of pre-production cannot be understated. Effective pre-production can result in a cleaner, fresher recording and inspire a much more relaxed, stress-free recording environment. It also has a positive effect on the morale of the engineer and/or producer during the session as there is a more defined goal from the offset.

However, this assumes that you, as the producer, have a solid enough vision to work out what could be better and also the experience to suggest some viable alternatives. One of the biggest weaknesses in my early productions was the fact that I hadn't critically listened to enough music to suggest viable alternatives and didn't have the music theory knowledge to verbalize that efficiently to the musician. Although I often got a 'strange feeling' that an element in a track didn't work, though. When I got this feeling, I didn't know what to do with it, and this is where some basic music theory knowledge *and* experience become important.

2.2.1 What Are You Listening for in Pre-Production?

When you hear a band play live during pre-production, you should be inspired and have ideas for how the track could be improved. If you don't, you might need to look at your analytical and critical listening skills again. Here are some common pitfalls that can lead to a lack of creativity for a producer:

1. Do you actually know what you want? If you are unsure what it is, you may lose confidence and fail to make relevant suggestions and establish a clear vision. If/when this happens, listen to more commercial releases in the genre you are working in or research the production techniques of the genre.
2. Inexperienced musicians will often stick to very simple structures and chord progressions. To improve this (if required), you should develop your own compositional skills (learn to write songs and song parts) to help them make their songs keep the listener's interest.
3. Working on an arrangement is not always about adding more; it can often be about decluttering and changing parts that are ineffective.
4. Also, a song should not show up a band member's lack of ability. If necessary, you should adapt the part to suit a player's style and skill level, preferably in pre-production, but if necessary during the recording stage. Sometimes you can add interest in another instrument or voice if you have simplified another layer.

5. Sometimes there will be member(s) of the band that are very good at arrangement and production themselves. A lot of the time this will be from a creative perspective rather than a technical one, and your production role might be to facilitate/translate between the tracking engineer and the particular band member(s).

2.2.2 Pre-Production Processes

Outlined below are the things that I do as a matter of course before any recording session. Usually, these should be done two weeks prior to the recording sessions to allow time to undertake the following steps.

Listen to the Band's Other Recordings

The very first thing I do when taking a booking is to listen to the band's music and ask them what they want from their next set of recordings, whether it is one song, an EP or an album. It is always important to create and agree on a shared vision for the direction of the tracks. This usually involves discussing their influences, their inspirations sonically and lyrically outside of music, and also the production and mix elements they want their own songs to emulate. This may seem obvious but you'd be surprised at how much this stage gets overlooked when having to deal with many different bands in a short period of time.

Get to Know the Band

Once a shared direction is established, I usually arrange to meet the band in a social environment. It is often best to go to one of their gigs, not only because you will get to grips with the band's hierarchy, but it also gives you a great opportunity to hear the songs that they are looking to record. My aims from this meeting are as follows:

- Gauge the personalities of the band, particularly members that have a leadership role. You will usually find one or two main forces in the band that control most of the writing, promotion and organizational tasks. These are the people that you need to develop a strong bond with initially. These are the people that you discuss your production ideas with – the people you keep close. You will then usually find a couple of members that are much more passive about the whole process. This doesn't mean that they don't care about the outcome of the product – it just means that they usually are not so driven (sometimes lazy) and that they also accept their backseat role in the band. These are the people that you have a laugh with, relax with and, when they need it, make sure that they understand that their part in the band is noticed.
- Write notes on the songs they wish to record and think about how to improve them. Note down any weaknesses in the technical ability of the band themselves and weaknesses in arrangement, instrumentation, tone selection and choice of gear. I like to find out the key signatures of the songs, then also work out the tempo and time signatures (I do this with my iPhone at the gig itself, using a metronome app).

Go to a Rehearsal with the Band

Assuming you have already made notes on the songs themselves, you can now go straight into helping the band iron out the kinks in their arrangements. If not, you might want to go to two rehearsals – one to make the notes outlined above, and one to take action on them. The beauty of taking action on these parts before hitting the studio is that you give the musicians time to figure out the best ways to take into consideration the changes you are putting forward to them. This literally saves hours of studio time and also helps ease the stress levels between band and producer. Although I would like to write about the art of arranging a song, for now let's just concentrate on the technicalities of capturing the sound. There will be more in-depth information on arrangement later in Chapter 6: Arrangement and Orchestration.

Get the Drummer to Play to a Click

You'd be surprised quite how many bands (and particularly drummers) have problems playing to a click track. In my early days of production, I had to abandon using a metronome in a short recording session because the drummer could not handle playing to a click. In reality, it doesn't take a drummer much practice to get their head around playing to a metronome. It will actually improve their playing all round, but realistically it still takes a couple of solid days' rehearsal to become completely at ease with it. For the best results, prepare click tracks in WAV or MP3 format in the tempos of the tracks that they are recording and give them to the band to use in rehearsals. This forces the whole band to take note of the metronome.

Make Tonal Suggestions

I've found that bands have more difficulty getting good instrument tone than they do with their musical arrangements. Before the recording session, point out weaknesses in the band's sound and suggest changes that will benefit their music and your shared vision. This could include many things, such as: instrument/amp selection, instrument tunings, vocal coaching, drum treatment and playing style. A producer needs to be a master of vision and be able to offer knowledge and guidance in all of these areas, and, where needed, to call in professionals to make further improvements to the sound. Don't worry if you feel ill-equipped to make tonal suggestions just yet; we will be discussing the different tonal options available in the proceeding chapters, so be patient.

Suggest Compositional Changes

Upon hearing the track for the first time, you should weigh up how well the music serves the song and, if necessary, begin to develop ideas that will improve it. This could be a whole range of alterations, such as:

- alter the key;
- change the tempo;
- change a specific chord to heighten an emotional state;
- simplify a part to make it easier to play;
- transpose an instrument to avoid masking (frequencies clashing);

- change the instrument a part is played on; and
- modify the structure to give the song a more suitable journey.

The thought of suggesting these kinds of changes might make you feel out of your depth right now. However, the rest of this course is designed to help you to start making these sort of decisions and feel comfortable in doing so.

Any composition, arrangement and orchestration changes should not necessarily be set in stone. You may find that some of the ideas you noted in pre-production do not work as well as expected. Also, in a multitrack recording where the song is built from the ground up and additional parts are overdubbed, your vision may start to evolve past what you initially envisioned. This will have an impact on the arrangement. For instance, an inspired synthesizer recorded towards the end of the production process may change the way that you want to layer the track, or affect which parts you want to draw attention to in the mix.

From a mixing engineer's perspective, the producer may have overdone the amount of instrumental layers, and it is down to you to pick and choose the best combination of parts for the track. Overdoing the orchestration or arrangement can be a waste of valuable time, but it is definitely better to give the mixing engineer options, rather than leave blatant sonic holes in the production.

Source Replacement Equipment Where Necessary

An audio engineer's job is primarily 'getting the sound right at source'. If this is done, the production, mixing and mastering will be much easier and also take half the time. If a band has broken, cheap or unsuitable equipment, you will be fighting a losing battle from the start.

Consumables and the Studio Environment

As well as using the best instruments, amplifiers, microphones and pre-amps available to you, ensuring all of your consumables are fresh will help you achieve the best sound possible. I recommend reskinning the drum kit the morning of the session and restringing and setting up guitars the night before. New drumsticks are preferable, and also get your vocalist to bring some throat sweets and some honey and lemon. Other guidelines could include:

- vocalists avoiding cold water and dairy products (as it affects the vocal cords);
- no alcohol or drugs immediately before/during the session; and
- suitable room temperature.

Set up a Session File Ahead of the Session

As you already know the song(s) the band are choosing to record, and you've also estimated tempos and worked out time signature and key signature details, there is nothing stopping you preparing the DAW session before the band arrive (including audio tracks, busses and headphone mixes). This is particularly useful if you don't have an assistant and it takes a bit of pressure off you on the first day of recording.

In a rock production such as 'Take Her Away', the arrangement of instruments is a huge part of the jigsaw that makes a powerful and entertaining track. Much of the track's excitement comes from the interaction and layering of the key instruments – drums, bass, guitars and vocals – and the use of effects. In other genres of music, instruments such as synthesizers, strings, programmed drums and brass may take a more central role. However, this course is on rock production, so we will focus on the key instruments of that genre. Fortunately, the principles of building a successful arrangement remain very similar in other genres too.

2.3 SONG ANALYSIS

So we have discussed that new ideas or alterations should be brought up early to give the musicians time to think about their options. But what makes an effective song?

Good tracks are the result of good mixes, good mixes are the result of good productions, and good productions are the result of well composed, arranged and orchestrated works.

Quite often, two or three very simple composition, arrangement or orchestration changes can make a song drastically better. There is a range of reasons why a song might be ineffective, but it usually boils down to boredom. While there are many things you can do in production to hold interest, in my experience the best tracks are the ones that sound great even while stripped down to their bare bones without any fancy production values or tons of overdubs.

The general gist here is that extra production and overdubs should be adding width, depth and thickness to aid the progression of a song, but they should not be of primary interest.

I always view a track like I do a good roller coaster:

'Just as they become accustomed, give them something unexpected.'

Now this doesn't necessarily mean that there should be tons of tempo, key or instrument changes; in fact, drastic changes can make a listener uneasy, but it means providing variety and progression throughout the track so that the listener feels like he or she is on a journey.

A track should flow like a story, and a typical trick is to end a story climactically. In terms of arrangement, this could be achieved by using the most layers and memorable hooks in the final sections. The story may, however, not be one of resolution; it may not be a triumphant conclusion. There are some great compositional tools out there to help portray certain emotions; spending some time learning these will only serve to improve the end product. While there are some good rules of thumb regarding arrangement, you should always follow the story, feel the emotion and work around these. Just because 99 per cent of songs have a similar structure doesn't mean that yours has to. The song, rather than the current commercial climate or expectation, should always dictate the arrangement.

2.4 DOES A PRODUCER NEED MUSIC THEORY?

Music theory may not be the most fun topic (as it is often presented in a dull way), but it is a seriously undervalued tool in the music producer's arsenal.

I have had many people ask me whether they need to know music theory to be a producer. While you can get by without it, I wouldn't recommend that. Music theory allows you to communicate with musicians in their own language – which will save time in the session, allow you to gain their trust more quickly and make them more open to your ideas. You don't necessarily have to be fluent in all aspects of the language, but a good working knowledge of the key concepts will save time and effort, and musicians will trust your judgement more quickly. Learning music theory will also speed up your own development by allowing you to:

- Memorize concepts more easily.
- Learn from others' mistakes. Music theory wasn't just plucked from thin air; it is based on guidelines for the creation of Western music that has been evolving over hundreds of years.
- The melodic and harmonic knowledge associated with music theory will help you to program any extra instrumentation that you may wish to add.
- The rhythmic knowledge associated with music theory will help you use the grid in your DAW to manipulate and edit audio more accurately and efficiently.

The old cliché is:

'You need to know the rules to be able to break them.'

While this is sensible, I would argue that the following is more apt:

'By knowing the theoretical concepts, we can find the vision in our head more quickly and accurately.'

Many of you reading this may have an established knowledge of music theory. If you do, still read these chapters because it will not only serve as a reminder, but it also approaches these theoretical ideas in a way that is more beneficial for music producers. This includes information on the harmonic series, which is seldom taught to music theory students, but is pivotally important to audio engineers.

To make the forthcoming music theory and practical examples convenient to the DAW generation, it focuses around MIDI piano roll and the tempo grid in your DAW rather than in traditional music notation.

2.4.1 Composition, Arrangement and Orchestration

When we are dealing with the act of creating music, there are in fact three common terms that are used to help break down the process. These are often confused and used interchangeably; in reality, they are three distinct phases that often happen concurrently or overlap. In the case of an orchestral score, these might be three different processes, but in the case of most bands all three processes are happening at once.

- *Composition* is the act of creating the new piece of music. This is the lyrical, chord structure and melodic parts of the piece.
- *Arranging* is the act of adapting which instruments and voices should be used within the piece. Arranging can also involve some alterations to the melodies, chord progressions, harmonies and overall structure to best fit the new scope.
- *Orchestration* is the process of deciding which instruments or voices to use for a particular work. In the case of bands, much of the key orchestration work is done at the same time as the composition and is restricted to the instruments that the band members play live.

When producing, I only use the term 'arrangement' with clients, for two reasons:

- *Simplicity.* Many bands would get confused with the terms anyway and use them interchangeably.
- *Psychological preference.* Sometimes bands are far more open to the term arrangement. Saying that the composition of the song might benefit from a tweak is often perceived negatively, and bands take it to heart and become uptight and protective. Orchestration, on the other hand, often triggers thoughts of overproduction, and bands again become uptight and protective.

The next three chapters will discuss the music theory involved with music production, followed by a chapter on arrangement and orchestration. They will teach you the key music theory that can aid the production of a song. It is beneficial for a producer to have a good knowledge of composition – better still by writing his or her own music – so that the decisions are made from practical experience. These chapters are not structured to teach you how to write great songs, as this depends on the many precedents set in different genres. However, it is designed to give you knowledge to implement techniques where necessary. Listening critically to commercial tracks in the genre you are producing will help you make decisions on the path you take.

It is time for a little disclaimer because I will mention 'working within the precedents of the genre' many times during this course and it might be easy to misconstrue this advice. Specific music traits, patterns and techniques exist in all genres. These nuances will start to seep into your unconscious and by breaking certain conventions the music will somehow sound wrong despite being theoretically solid. Of course, pushing the boundaries and experimenting is a crucial part of the art of music production and should be encouraged because without it, music would just stand still. At the same time, throwing a sub-bass synthesizer into a jazz track would seem crazy to most and is likely to create an end result that most people would find uncomfortable. It's another situation where the old cliché of 'having to know the rules to be able break them' applies. I certainly recommend and encourage you to start to experiment with genres you aren't comfortable in. This is great for personal development and job satisfaction. As well as this, you can take little bits of inspiration from new places and apply them to genres you are familiar with. This will allow you to be more creative, try new concepts and help do something that does feel fresh and push boundaries. Just remember, there are limits to what will be accepted by the artists and the general public. When I mention conforming to genre precedents, I am simply stating that you should be vigilant with regard to these unwritten rules of the style of music.

BASIC MUSIC THEORY

3.1 INTRODUCTION TO COMPOSITIONAL ELEMENTS

Like most elements of life, sound and music can be broken down into mathematics. Just like with decibels in Chapter 1: Production Philosophies, Your Ears and Critical Listening, humans have, over time, developed systems to help express music in ways that are more easily comprehensible to the task we are trying to achieve than numbers alone would be. Over the course of the next pages, we will come to understand why we find the interaction between specific tones pleasing or displeasing. For now, though, let's concentrate on the key elements that help us to express music.

To properly dissect music, we need to be aware of three elements: rhythm, melody and harmony.

- *Rhythm* is the name given to the pattern of sounds over time. Rhythm is the element that gives the song its drive and moves it forward. Even a simple tune has a flow to the notes over time, meaning that all music must have a sense of rhythm.
- *Melody.* The music in Western cultures has evolved to focus around a lead melody. A melody is a succession of single musical notes that the listener considers as a single 'tune'.
- *Harmony* is the use of two or more simultaneous notes to create texture. The simplest harmony would be a main melody plus another part. However, modern compositions feature a range of other notes often grouped together in a chord structure.

To be able to properly understand and manipulate audio in a DAW, we need to be aware of these three core ingredients of music, how they relate to our Western notation system and how they relate to an audio engineer's analysis of frequencies.

We audio engineers actually speak two distinct languages when discussing music. In essence, both of these languages express the same attribute (pitch) but are trying to achieve a different goal. The first is the classic audio engineering terminology, which is a more mathematical language relating to frequency. We use this to help express and optimize the complex tonal attributes of a sound (e.g. 'There is a build-up of 100 Hz in that bass guitar tone'). The second is by musical notation, where we express pitch in relation to a musical note value. We use this language to help display and express pitch and time in a way that can be easily reproduced by a musician or programmed in a DAW.

In simple terms, a musical note is just a reference to a certain fundamental frequency, but framing a frequency as a letter helps to simplify the task. Instead of saying, 'Can you play 440 Hz?' a musician would say, 'Can you play an A?' – or, more specifically, 'Can you play A above middle C?' (based around the middle position of a piano). Writing a whole score as a series of frequencies is rather daunting and inefficient. We will discuss the way that we hear pitch and how we assign it a letter shortly, but first let's discuss rhythm.

3.2 RHYTHM

Rhythm is the one truly indispensable element of music (music can exist without melody, which is the case with primitive percussive music). In our DAWs, we have a time-based grid so that we can see how music breaks down over time. As the music plays, we can see the notes scroll from left to right over time. As previously stated, it is rhythm that gives music its sense of motion – its drive. In our DAW, we use a metronome or click track to generate a pulse that keeps a musician in time.

Rhythm is difficult to break down in an absolute way because it is something a person feels as much as learns. The human mind naturally wants to find patterns, and with rhythm the mind will tend to find patterns even when there isn't one – for instance, how a person can perceive a set of identical clock-ticks as 'tick-tock, tick-tock'. The fact that the human mind is always looking for patterns means that any normal individual will have some sort of natural sense of rhythm; you could ask a random person to tap his or her feet along to a piece of simple pop music and he or she will probably be able to do it. We will call the timing of the foot taps the *pulse* of the music.

The problem with this as a concept is that not everyone would perceive the pulse in exactly the same way. The practical result of this would mean that not

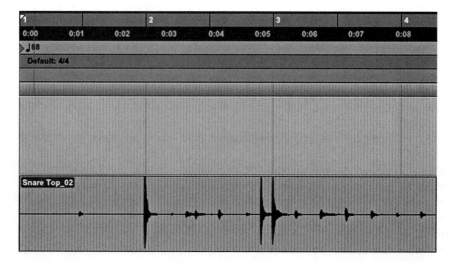

Figure 3.1
Pro Tools grid

everyone would be tapping their feet identically. Some songs will have a much more concrete sense of groove, whereas others can be interpreted differently. This happens because different individuals will find either a different subdivision of the pulse or feel that the pulse is accented in a different way. Let's take the chorus of 'Rock with You' by Michael Jackson, for instance. This is an example of a simple pop drum rhythm. When listening to the chorus, the vast majority of people would feel the pulse based on the groove provided by the kick drum and snare drum. However, you could also feel the pulse as a smaller subdivision as heard by the faster-paced hi-hat.

Ultimately, whether you locked in with the groove of the kick and snare or the hi-hat doesn't matter as long as you could keep in time with it. However, if you were to write the piece out in musical notation, you would want to notate it using a subdivision that would make the score the easiest to read. In most situations, this would be to use the pace that is implied by the majority of the instruments.

In the case of 'Rock with You', it is much more sensible to think of the pulse as found in the kick and snare groove. This is because it more closely resembles the overall pace, which is implied by the bass guitar and vocals; it also has the added benefit of being much easier to count than the groove on the hi-hat. The verse of the track does not use the smaller subdivision of notes on the hi-hat, so this further establishes the implied pace as the one provided by the kick and snare.

When setting up a session in your DAW, you need to make sure the grid is suitable for the song you are recording – including making sure that the grid follows any significant rhythmic changes in the music. Doing so will make it easier to program and edit audio. A lot of the options that can be selected are based on the rhythmic concepts found in musical notation. So to understand how to set up your session correctly, we need to look into how rhythm is written on a score. Before we go into detail, I need to define a few variables:

Beat. First, we need to assign a unit for time in music; this is called a beat. A beat can be considered a single pulse at the subdivision that best dictates the overall pace of the music.

Tempo. Tempo is the number of beats that occur each minute, which is the variable that defines the overall pace of the music. Tempo is abbreviated as BPM (beats per minute).

Pulse. The words beat and pulse are often used interchangeably. I would like to assert that the pulse is the pattern of beats over time. I would also like to state that in some occurrences, a pulse might be created at a smaller subdivision than the basic beat.

Note length. As we saw with 'Rock with You' earlier, different elements of a track can play different subdivisions. In fact, a rhythm can even contain multiple subdivisions within the same part. This means just a single unit for time is not enough – we need to be able to work with multiples and divisions of the main beat to create the different lengths of time we need for each situation. For this reason, we have a note length convention to help define these multiplications and divisions, which we will discuss shortly.

Bar (measure). A bar or measure groups beats in a convenient way to help simplify how we express phrases of music. Bars give musicians a more usable

Figure 3.2 Example of a 4/4 time signature as found in sheet music

reference than having to count every single beat. For instance, a musician might say, 'That phrase is four bars long' or, 'The note changes to C on the third beat of the fourth bar'. When you are dealing with phrases that repeat a number of times, or long musical pieces, you can see why grouping music into bars can come in handy.

Accents. Not all beats are equal and it is human nature to emphasize certain beats more than others. This accenting is inherent in music itself – without it, music wouldn't have such a great sense of motion. Remember, though, people's perceptions of the accent may be different to your own.

Metre. The metre of a piece of music expresses how a pulse is grouped and accented, which can give a piece very different results in groove or feel. A metre can be duple, triple or quadruple.

Time signature. When a piece of music is written in sheet music, its metre will be specified in a time signature. A time signature helps show how a piece of notation should be interpreted. It comprises two whole numbers (no decimals or fractions), one above the other. The top number specifies the metre; the bottom number denotes which note division is being represented. There are two main types of time signature: simple and compound.

3.2.1 Rhythmic Constructs

Defining rhythm is definitely a 'chicken and egg' style conundrum, so some of these definitions might still seem a little unclear. If that is the case, don't worry – there are plenty of practical examples and exercises coming that will tie it all together for you. However, to be able to give some practical examples, we need to go into more depth on each of these variables.

Before I do this, I need throw in one more caveat: the vast majority of popular music is in one time signature called *common time* (or 4/4). Common time is a metre that contains four beats in a bar and can be counted:

	Bar 1				Bar 2				Bar 3		
|1	2	3	4	|1	2	3	4	|1	2	. . . . . .

As common time is used so frequently, the naming conventions behind note length is often described and taught in relation to this time signature, [4/4]. You need to remember that when the music is not in common time, it could have implications for how many beats each note type lasts for.

Note Length

As we mentioned earlier, to be able to notate rhythms, we will need to be able to subdivide notes into smaller values than the core beats. In popular music, you will see a variety of symbols that signify different note divisions. It is important to remember that note length dictates the size relationship between each note type, which is absolute regardless of time signature. The number of beats each note value lasts varies depending on the time signature.

Table 3.1

Notation Symbol	British Name	American Name	Equivalent of	Number of Beats in Common Time
o	Semibreve	Whole note	(not important right now)	4 beats/pulses
𝅗𝅥	Minim	Half note (1/2)	Half of a semibreve	2 beats/pulses
♩	Crotchet	Quarter note (1/4)	A quarter of a semibreve / Half of a minim	1 beat/pulse
♪	Quaver	Eighth note (1/8)	An eighth of a semibreve / A quarter of a minim / Half of a crotchet	Half a beat/pulse
𝅘𝅥𝅯	Semiquaver	Sixteenth note (1/16)	A sixteenth of a semibreve	A quarter of a beat/pulse

Table 3.1 shows the note length symbols, the British and American naming conventions, the size of the notes in relation to each other and the number of beats each one lasts for in common time.

As well as these notes, there are technically further divisions to thirty-second notes and sixty-fourth notes; these are seldom seen in popular music. To work out the lengths of these, carry on dividing the fractions.

Finally, you will also see note lengths with a dot next to them; these are called *dotted notes*. A dotted note adds another half of its total value – for example, a dotted minim (half note) would be the equivalent of three crotchets (quarter notes) and last for three beats in common time.

The British convention is perhaps more difficult to remember, but it has the advantage that it is not so easy to misinterpret the two distinct concepts we are working with:

1. How many beats are in a bar.
2. How notes can be divided.

Let's look at the American convention in common time so that I can show you what I mean. In common time, a whole note lasts a whole bar, a half note lasts half a bar, a quarter note lasts one quarter of a bar and so on. 1 bar = 1 whole note = 2 half notes = 4 quarter notes, etc. Therefore, it would be easy to presume that the naming convention tells you how the notes fit into bars. The problem is this changes when the time signature changes. In a time signature of 2/4, a full bar would be represented by a half note or two quarter notes, and in 6/8 a full bar would be represented by a dotted half note or two dotted quarter notes. The reasons for this will be explained soon, but for now it suffices to say that a whole note does not always mean it is the length of a full bar. Remember, the note length convention is actually explaining the sizes of each note type in relation to each other.

In your DAW, you can control the note length that is viewable in the grid. This is great to easily edit audio or insert MIDI notes of the correct value.

In Pro Tools, you can find the grid settings to the right of the timer.

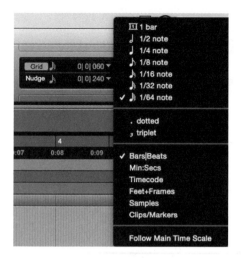

Figure 3.3
Pro Tools grid settings

Metre

Let's consider the application of metres. For now, we will ignore the implication of the lower number found in the time signature.

A metre shows how notes are grouped into bars:

1. If you see a two on the top, there are two beats in a bar, which is called a *duple metre*.
2. If you see three on the top, there are three beats in a bar, which is called a *triple metre*.
3. If you see a four on top, there are four beats in a bar, which is called a *quadruple metre*.

You will never see a one on the top of a time signature. This is because creating a bar of one is rather redundant, as it doesn't simplify the reading of music from a score. Metres that contain other numbers like sixes, sevens, nines, etc. and beyond are usually either a duple, triple, or quadruple metre, although a quintuple metre containing groups of five can sometimes occur. We will discuss how to work out the type of metre for these larger numbers in the time signature section later, as well as how the accents are grouped in each metre.

Time Signature

This will be the section where it should all start making a bit more sense. We have outlined how the top number shows the number of beats in a bar in the simplest cases; in more complex cases, we need to be aware of what the bottom number is telling us to get the full picture of how the bar is structured and how notes are accented. The bottom number tells you the subdivision of the notes in their default state. Therefore, when you see a time signature written, you can express it in a sentence as follows:

'There are <insert top number> <insert note type as described by bottom number> found in a bar.'

The sentence for common time (4/4) would be:

'There are four crotchets (quarter notes) found in a bar.'

Counting Rhythm

To work out a rhythm, it is often counted aloud. From this point on, you will see many examples where I express how you would count various rhythms. I will introduce the convention I use in this course now, using common time (4/4) as an example.

In common time, each beat found in the bar is literally counted out loud (***One***, two, *three*, four) and the subdivisions of that beat are represented by basic phonetic sounds. Each beat in common time is a crotchet (quarter note) in length. Subdivisions of a quaver (eighth note) are counted by saying the word

Table 3.2

Crotchets (quarter notes)	One				two				three				four			
Quavers (eighth notes)	One		and		two		and		three		and		four		and	
Semi-quavers (sixteenth notes)	One	e	and	a	two	e	and	a	three	e	and	a	four	e	and	a

'and'. Subdivisions of a semiquaver (sixteenth note) are counted by saying three sounds: 'e', 'and', 'a'. For example:

Set a metronome to 80 bpm and start counting each number on each click of the metronome and practise the other subdivisions by saying the other phonetics spaced evenly between the clicks.

Any time signature where notes or pulses in a bar can be divided in two like this is called a *simple time signature*.

The problem is we don't always want a rhythm where the notes regularly divide into twos. When we have a song that prominently uses triplet rhythms, there is a more efficient way of organizing the information. This is where we get into compound time. Before we move on, I quickly want to define what a triplet is. A *triplet* is a rhythm that is divisible by three rather than two, which gives the music a distinctive rhythm, which you can count as:

One and a two and a three and a four and a

A time signature where the default state is divisible by three is called a *compound time signature*. Rather than standard note lengths, compound time signatures use dotted note lengths to allow them to be divisible by three. Remember, a dotted crotchet (quarter note) is equal to three quavers (eighth notes). However, using dotted notes creates three problems:

1. It would make music notation less legible and therefore make sight-reading more difficult.
2. We can't notate dotted notes in the time signature (as we can't use decimals or fractions).
3. Having a dotted-note pulse is ambiguous and doesn't show the distinctive groove of triplets.

To illustrate the third point, let's show the dotted crotchets (quarter notes), and below that is its subdivision into three quavers (eighth notes).

	Bar 1			Bar 2		
Crotchets (quarter notes)	One		two	One		two
Quavers (eighth notes)	One and a	two and a	One and a	two and a

If you count these, you can tell that the top pulse gives no indication that there is a triplet feel. Therefore, an inexperienced musician would usually find it a lot easier to lock in and keep in time with the triplet feel using a metronome with the quaver (eighth note) pulse rather than a dotted crotchet (quarter note) pulse, which doesn't emphasize the triplet feel.

For all these reasons, simple time signatures use a crotchet (quarter note) as the bottom number in the time signature, and compound time signatures will be indicated by using a quaver (eighth note) as the bottom number.

Just to hammer home the point once more, when you listen to a metronome with the session set to a compound time signature, you would normally be better off setting the pulse to quavers (eighth notes) so that the instrumentalists keep time better. In some DAWs, you may need to set the metronome to do this; in others, its default state will automatically change the pulse to a quaver (eighth note) when a compound time signature is used. Despite the number of pulses you have set the metronome to, the true number of beats in the bar is still dictated by the dotted crotchet (quarter note) value. This means that:

- 6/8: is a duple metre so has two dotted crotchets beats per bar. The default pulse would usually be set to six quavers (eighths) per bar.
- 9/8: is a triple metre so has three dotted crotchets beats per bar. The default pulse would usually be set to produce nine quavers (eighths).
- 12/18: is a quadruple metre so has four dotted crotchets beats per bar. The default pulse would usually be set to 12 quavers (eighths).

There is one more piece to the rhythmic puzzle to explain: *How do we accent each pulse?* From this point on, we will notate accents using the following convention:

- Strong accents will use **bold and italic** text.
- Medium-stressed beats will use just *italic* text.
- Weak beats (unstressed) will use normal (unformatted) text.

The first beat of a bar is called a *downbeat* and is strongly accented. As well as strong accents, there can also be some less strong accents based on how the beats are grouped. As we will shortly discover, the grouping of accents is dictated by the type of time signature. Whenever a beat is not accented, it is called an *upbeat*. If there are subdivisions of notes that are not occurring on a beat at all (like some of the hi-hat notes in 'Rock with You'), these would be called *off-beats* and are unaccented. Here is the configuration of accents in each of the metre types:

Simple Time

Duple metre: ***Strong***, weak
Triple metre: ***Strong***, weak, weak
Quadruple metre: ***Strong***, weak, *medium*, weak

Compound Time

Duple metre: (***Strong***, weak, weak) (*medium*, weak, weak)
Triple metre: (***Strong***, weak, weak) (*medium*, weak, weak) (*medium*, weak, weak)
Quadruple metre: (***Strong***, weak, weak) (*medium*, weak, weak) (*medium*, weak, weak) (*medium*, weak, weak)

The brackets in the compound examples indicate each beat.

3.2.2 Popular Time Signatures

Now we can use the theoretical knowledge of simple/compound time, duple/triple/quadruple metres, and accents to fully describe each time signature. Let's now break down some time signatures, their accent patterns, and give popular music examples.

2/4: March Time

2/4 is a *simple-duple* time signature, containing two crotchet (quarter) notes in one pulse group (*One*, two). A march will have a metre of two; it will sound similar to a metre of four but the accent will feel different. Here is how a metre of two would be expressed in written form:

```
|Bar 1         | Bar 2         | Bar 3         |

|One    two    |One    two     |One    two     |......
```

> **Further Resources** www.audioproductiontips.com/source/resources
>
> 🔊 Accented metronome of 2/4 at 160 bpm. (6)

Even though they are similar to metres of four, metres of two are seldom seen in popular music and are more commonly used by military bands. 2/4 time is called a 'march' because it represents the 'right foot, left foot' pattern of walking, and military bands have been used since antiquity to keep troops marching in time with one another in order to maintain formation and organization.

An example of a metre of two in popular music is 'Russians' by Sting, used presumably because the lyrical content of the song is heavily political.

EXERCISES

1. Try counting this rhythm out loud over the metronome.

2. Next, emphasize the *one* by saying it louder.

3. Now try to count this out over 'Russians'.

3/4: Waltz Time

3/4 is a *simple-triple* time signature, containing three crotchet (quarter) notes in one pulse group (*One*, two, three). You would count this as:

Bar 1			Bar 2			Bar 3		
One	two	three	**One**	two	three	**One**	two	three ......

Further Resources
www.audioproductiontips.com/source/resources

◀ Accented metronome of 3/4 at 90 bpm. (7)

Metres of three are not as common in popular music as metres of four. A couple of popular musical examples of a metre of three are 'Kiss from a Rose' by Seal and 'Piano Man' by Billy Joel. A 3/4 time signature can easily be confused with a 6/8 time signature and is sometimes open to personal interpretation. For instance, there could be an argument for both the popular music examples above to be considered as 6/8.

However, it is important to remember that if you are hearing 6/8, it would assert a slower tempo than 3/4, due to the subdivision and mismatching metres (3/4 is triple and 6/8 is duple). Below is how the beats would line up. Again, the numbers show the beats and the phonetics show the subdivision of the beat:

	Bar 1			Bar 2			Bar 3			Bar 4		
3/4:	**One**	two	three	**One**	two	three	**One**	two	three	**One**	two	three
	Bar 1						Bar 2					
6/8:	**One**	and	a	*two*	and	a	**One**	and	a	*two*	and	a

Due to the examples feeling a little more upbeat in tempo, I prefer to analyse both 'Kiss from a Rose' and 'Piano Man' as 3/4. The kick drum in the verses of 'Kiss from a Rose' also lends itself to a 3/4 interpretation.

EXERCISES

1. Try counting this rhythm out loud over the metronome.

2. Next, emphasize the *one* by saying it louder.

3. Now try to count this out over 'Kiss from a Rose'.

4/4: Common Time

4/4 is a *simple-quadruple* time signature, containing four crotchet (quarter) notes in two pulse groups (***One***, two, *three*, four). I imagine each bar as a story with two halves: the first half I think of as the strong half as it contains the heavily pronounced beat one (*1*, 2) and the second half I think of as the weak half because it contains the less pronounced downbeat on three (*3*, 4).

|Bar 1 | Bar 2 | Bar 3
|***One*** two *three* four |***One*** two *three* four |***One*** two *three* ......

> **Further Resources** www.audioproductiontips.com/source/resources
>
> Accented metronome of 4/4 at 90 bpm. (8)

There are countless examples of common time, including 'Rock with You' by Michael Jackson and 'Hey Jude' by The Beatles.

EXERCISES

1. Try counting this rhythm out loud over the metronome.

2. Next, emphasize the *one* by saying it louder, and also the *three* but to a lesser extent.

3. Now try to count this out over 'Rock with You'. If you are doing this correctly, you should find the kick aligning with beats *one* and *three*, then the snare aligning with two and four.

6/8 Time

6/8 is a *compound-duple* time signature, which contains two dotted crotchet (quarter) notes. The pulse is often expressed in quaver (eighth) notes in a grouping of two sets of three pulses per bar (**One** and a) (*two* and a).

|Bar 1 |Bar 2
|**One** and a   *two* and a |**One** and a   *two* and a ...

> **Further Resources** www.audioproductiontips.com/source/resources
>
> 🔊 Accented metronome of 6/8 at 80 bpm with dotted crotchet pulse. (9)
>
> 🔊 Accented metronome of 6/8 at 80 bpm with quaver pulse. (10)

Popular examples of 6/8 are 'Nothing Else Matters' by Metallica and 'We Are the Champions' by Queen.

EXERCISES

- Try counting this rhythm out loud over the metronome.
- Next, emphasize the *one* by saying it louder, and also the *two* but to a lesser extent.
- Now try to count this out over 'Nothing Else Matters'. If you are doing this correctly, you should find yourself aligning to the guitar arpeggio.

9/8 Time

9/8 is a *compound-triple* time signature, which contains three dotted crotchet (quarter) notes. The pulse is often expressed in quaver (eighth) notes in a grouping of three sets of three pulses per bar (**One** and a) (*two* and a) (*three* and a).

> **Further Resources** www.audioproductiontips.com/source/resources
>
> 🔊 Accented metronome of 9/8 at 80 bpm with dotted crotchet pulse. (11)
>
> 🔊 Accented metronome of 9/8 at 80 bpm with quaver pulse. (12)

9/8 examples are hard to come by in rock and pop music. However, 9/8 is fairly popular in Motown and soul music. An example of 9/8 is 'Fool for You' by The Impressions (Curtis Mayfield).

EXERCISES

1. Try counting this rhythm out loud over the metronome.

2. Next, emphasize the *one* by saying it louder, and also the *two* and *three* but to a lesser extent.

3. Now try to count this out over 'Fool for You'. If you are doing this correctly, you should find yourself with the *one* falling over the heavily accented first beat and counting the same rhythm over the hi-hat.

12/8 Time

12/8 is a *compound-quadruple* time signature that contains four dotted crotchet (quarter) notes. The pulse is often expressed in quaver (eighth) notes in a grouping of four sets of three pulses per bar (**One** and a) (*two* and a) (*three* and a) (*four* and a).

> **Further Resources**   www.audioproductiontips.com/source/resources
>
> 🔊 Accented metronome of 12/8 at 80 bpm with dotted crotchet pulse. (13)
>
> 🔊 Accented metronome of 12/8 at 80 bpm with quaver pulse. (14)

An example of 12/8 would be 'Everybody Hurts' by REM. You could count this in 6/8 but the chord changes are happening every 12 pulses, which makes me feel it more as 12/8 than a 6/8 rhythm where chord changes happen every two bars. Another example of 12/8 time is 'Why Does It Always Rain on Me?' by Travis.

EXERCISES

1. Try counting this rhythm out loud over the metronome.

2. Next, emphasize the *one* by saying it louder, and also the *two, three and four* but to a lesser extent.

3. Now try to count this out over 'Everybody Hurts'. If you are doing this correctly, you should find yourself aligning to the guitar arpeggio with the chord changing after each bar.

Irregular Time Signatures

Although the above time signatures are the most common, popular tracks do sometimes sway from this, so let's quickly discuss those. All the examples so far can be labelled as regular time signatures as they are duple, triple or quadruple metres (i.e. they can be grouped in even blocks of two, three or four).

Irregular time signatures will either have a block of five, making it a quintuple metre, or have uneven blocks. With these irregular time signatures with uneven blocks, you can divide the pulse groups a number of ways. In the examples below, you should listen to the tracks to decide how you think the pulses are grouped and accented:

5/4
One block of five (*One*, two, three, four, five). An example is 'Take Five' by The Dave Brubeck Quartet.

7/8 or 7/4
Two blocks of two and one of three in whichever order you see fit, for example:

- (*One*, two) (*three*, four) (*five*, six, seven)
- (*One*, two, three) (*four*, five) (*six*, seven)
- (*One*, two) (*three*, four, five) (*six*, seven)

This rhythm is found in 'Money' by Pink Floyd. You are more likely to use 7/4 as the track's time signature when the tempo is slower and 7/8 when the tempo is faster.

3.2.3 Internalizing Rhythm

Now that I have discussed how time signatures are defined, I want you to experiment with creating different rhythms patterns over a metronome to get used to hearing the subdivisions. The following exercises will further hone your skills at recognizing time signatures and their subdivisions.

Before we get to these exercises, I want you to get used to drawing notes in the piano roll of your DAW and routing them through a virtual instrument to make audio come out of your monitors. If you are unfamiliar with the basic workings of MIDI in your DAW, you have some homework to do!

Further Resources　　　　　　　　　　　www.audioproductiontips.com/source/resources

 MIDI playback with a virtual instrument in Pro Tools. (15)

Throughout this portion of the chapter, we are going to be doing a lot of exercises where you are manipulating MIDI notes so that you can listen critically to rhythmic, melodic and harmonic elements of music theory. Once you start listening critically to these elements, you will start to hear them appearing in every song that you listen to and be able to use that to your advantage when tweaking the arrangements of songs that you are producing.

EXERCISES

1. Listen to some of your favourite songs and work out the potential time signature and tempo used.

2. Using your chosen DAW, set up a metronome and experiment with changing the time signature and tempo.

3. Experiment with changing the subdivision of the pulse.

4. Use a 4/4 time signature at 80 bpm and set the grid in your DAW to quarter notes. Place a note on each segment of the grid at a velocity of 96 and listen to the rhythm over the top of the metronome.

5. Take your rhythm from the previous example and increase the velocity of the first beat to 127 and listen again.

6. Now increase the velocity of the third beat to 110 and listen again.

7. Repeat the same grid experiments with different time signatures and tempos. Remember to make sure that you follow the correct strong, medium and weak beats for the time signature you are using.

8. Experiment with double-time rhythms by placing eighth notes over a quarter-note metronome pulse at different tempos.

3.3 THE HARMONIC SERIES

After breaking down rhythm, we have explored the variable of time and specified how notes can be given a length. However, to have melody or harmony, we need to apply a specific pitch or set of pitches. On the surface, the concept of a pitched note seems very simple, but in reality what we hear is much more complex. Musicians aren't quite as concerned about the complex timbral nuances and the scientific make-up of the sound we hear; they are more concerned with hitting the right notes at the right time, with the desired emotion. As engineers, though, we really need to get to grips with finer details of the sound we hear. Therefore, an engineer must be able to relate to pitches by both frequencies and letter. As I have assumed some prior music technology experience, I am not going to discuss how the frequency-based system works, or concepts such as sine wave, wavelength, amplitude, etc. If you don't know this, please research it.

> **Further Resources** www.audioproductiontips.com/source/resources
>
> 📖 David Miles Huber and Robert E. Runstein, *Modern Recording Techniques*. (16)

However, I will discuss the way we express music by letters in Western music. First, I want to introduce you to a concept called the *harmonic series*.

When we play a note on a piano, we don't hear a 'pure' tone (e.g. in the case of A above middle C on a piano, we do not hear a pure 440 Hz tone). A pure tone would be a sine wave, and this is pretty much impossible to create in an acoustic instrument. Instead, we hear a complex tone of 440 Hz in combination with a series of additional distinct frequencies above 440 Hz, which are called *overtones* or *partials*.

The lowest frequency in a waveform (in this case, 440 Hz) is called the *fundamental frequency*. As well as being the lowest frequency, the fundamental usually has the strongest amplitude. Because of this, rather than perceiving the fundamental and the individual overtones as separate, we perceive the sound as the pitch of the fundamental frequency, and hear the overtones as complex tonal colour or *timbre*. The amplitudes and frequencies of these overtones are what give instruments their distinct sounds, or timbre.

The series of overtones found above the fundamental are not random; many of the overtones are mathematically related and are found throughout the natural world. The relationship between these frequencies is the cornerstone of music. Pitched instruments, such as guitar, piano and flute, produce overtones that are closely related to the fundamental. Some instruments produce overtones that have little relation to the fundamental; these are the instruments from which it is difficult to hear a definite pitch, for instance the cymbals on a drum kit. However, let's focus on the related overtones. The fundamental and its related overtones are called *harmonics*.

Before moving on, I want to clear up any confusion between the use of the terms *harmonics* and *overtones*. The term harmonics refers to all of the overtones including the fundamental, whereas an overtone is any of the partials but not the fundamental itself. This means that the first harmonic is the fundamental frequency and the first overtone is actually the second harmonic.

When we delve deeper into harmonics, we find that all of these pitches are in fact multiples of the fundamental frequency.

So at the concert pitch of 440 Hz, we would hear harmonics at:

First harmonic: 440 × 1 = 440 Hz

Second harmonic: 440 × 2 = 880 Hz

Third harmonic: 440 × 3 = 1,320 Hz

Fourth harmonic: 440 × 4 = 1,760 Hz

Fifth harmonic: 440 × 5 = 2,200 Hz

Sixth harmonic: 440 × 6 = 2,640 Hz

Seventh harmonic: 440 × 7 = 3,080 Hz

Eighth harmonic: 440 × 8 = 3,520 Hz

What this means is whenever you pluck a guitar string, it vibrates not only at its full length, but at all of its integer divisions.

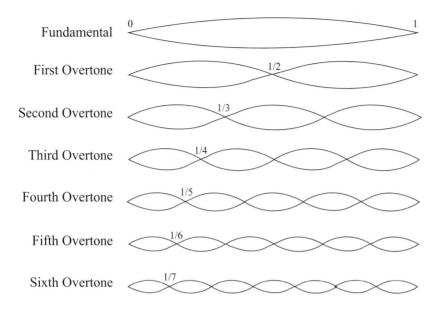

Figure 3.4 The harmonic series

This concept can be visualized most clearly with stringed instruments (see Figure 3.4).

Due to the construction material and the mechanisms used, not all musical instruments create the same strengths of overtones. In fact, some instruments don't even produce all of the harmonics. The reasons why this is the case is not necessarily important, but understanding the way that the harmonic complexity changes our perception of timbre is pivotally important. For instance, a clarinet is open at one end and closed at the other. This means it produces only the odd-numbered overtones, which, along with its reed mouthpiece, creates its signature 'dark' sound. On the other hand, brass instruments have a distinctly rich and bright timbre but can become a little harsh and 'unpleasant'. This is because they have strong overtones that continue high up the harmonic series. Other instruments have added complexity in their timbre because their overtones don't exactly match the theoretical harmonic values. In the case of a piano, its mechanism means that its overtones get progressively sharper than perfect harmonics. When it comes to learning to engineer, we just memorize the timbre of each instrument that we deal with, rather than learning why its harmonic pattern is the way it is. It is easy to work out why these tonal attributes majorly affect our engineering choices such as microphone placement; most of this course is designed around working out the best choices for each application.

What may not be so obvious is that the harmonic series also affects our production and compositional choices. A good understanding of the harmonic series will help you to join all the dots between the engineering and compositional worlds. The harmonic series is the root of what we find pleasant or unpleasant in music.

The harmonics closer to the fundamental sound more 'pleasant' to our ears, and the harmonics further from it sound more 'unpleasant' and 'jarring'. Our whole musical system is based around this. Later in this chapter, we will express the harmonic series as ratios and their equivalent note values. This will tie together how our musical notation system, frequencies and the harmonic series are inseparably linked.

Just a word of warning, though: It would be easy to assume from the preceding sentences that you should avoid 'unpleasant' sounds. However, they are not necessarily bad – they are just one of our many tools we can use to create emotion. Just like in life, we have a mixture of good and bad experiences, and we need the bad experiences to grow and give perspective to the good. If we go through a series of strongly pleasant tones and then move to one that is less pleasant, we get contrast. Your ear naturally wants it to become pleasant again; therefore, a strong pull towards another note is created. I like to think of music as a condensed version of a life experience playing out. Some songs remind you of your darkest hours and others remind you of your biggest triumphs, but they will all have some twists and turns – a journey from A to B. Without the unpleasant, we have no contrast. Humans enjoy songs about loss as well as about triumph. The songs focusing on loss will use more unpleasant tones and stay on them longer than those about triumph. Effective composition is about selecting these tones wisely, so that the music portrays emotion and has movement without becoming too unpleasant.

What we are trying to do as producers and engineers is create the most emotive and efficient version of a piece that we can. When engineering/producing, we are working on two layers simultaneously:

1. The sonic aspect of recording instruments – the way that the player, instrument, recording gear and its use (placement, levels, EQ, etc.) interact. This is the frequency-based language.
2. The way that the arrangement itself interacts – the structure, the notes found within it, its layering, the congruence of the lyrical content and the emotion set with the instrumentation. This is the notation-based language.

Both of these relate to the harmonic series. What we are trying to do with our selection of recording gear and placement is to emphasize the natural transients and harmonics in a way that creates the correct illusion of space and tone. In popular music, this is usually a perfect and larger-than-life representation of what is going on when a band plays. In jazz, this would be a more accurate representation of a band playing live. With arrangement, we are using the harmonic series to create and reinforce an emotional message.

These two aspects have to work together; if the spatial illusion we have created doesn't reinforce the emotional message, the production will be deemed a failure, despite the fact that it might have been cleanly captured. For instance, a raw blues track would most likely sound very wrong if it used sample-enhanced drums and heavily autotuned vocals.

Our journey as a music producer/engineer involves memorizing the theoretical aspects of both of these processes. If we know what a 'dynamic microphone positioned off-axis from a guitar speaker' sounds like, we can begin to make decisions based on it. The same rings true for a producer looking to optimize a set of chords: If we know what a 'dominant 7th resolving to a 1st chord' sounds like, then we can begin to use them wisely. I know this sounds complicated, but don't worry – the rest of this course will help you answer these scenarios.

We often view these two processes separately, but there are definitely crossover points. For instance, if I have an electric guitar tone that is too bright, we could change the spatial illusion we create by:

1. changing guitar/amp;
2. playing with fingers rather than plectrum;
3. correcting the player's technique;
4. using a different string type;
5. adjusting tone control on the guitar amp;
6. moving the microphone position; and
7. changing the microphone or pre-amp.

But these are not the only solutions. The problem may actually be that our attention is drawn to the treble-heaviness of the guitar by a deficiency in the arrangement. How do you know that:

- The pitch the guitar player has chosen isn't just too high?
- The other instruments are not filling out the lower frequencies in the way they should?

Here lies the constant yin-yang relationship between the technical considerations of a recording engineer and the arrangement concerns of the producer.

A lot of this course will be focused on the tonal aspects of the recording process. However, this section is about understanding the compositional elements of a successful recording. Remember, getting the sound right at source includes creating the right emotions in the first place.

3.4 THE WESTERN NOTATION SYSTEM

Now that we have discussed the harmonic series, let's examine how and why we express music the way we do today.

If we break melody down to its core, we could say that music gets its emotion from the combination of:

(a) the timing in which notes or groups of notes are played; and
(b) the distances in pitch between notes or groups of notes.

We call the distance between two notes an interval. The ancient Greek philosopher Pythagoras is credited with discovering that there is a highly mathematical relationship between the intervals we find pleasing and the purity of the ratio between pitches. The ancient Greeks didn't think of music in terms of frequency and this was long before the discovery of the harmonic series. These were simply observations made by subdividing vibrating strings to create 'pleasing' sounds. However, the ratios Pythagoras noted as being pleasing were in fact the same ratio relationships between the first few overtones in the harmonic series and its fundamental frequency:

- First overtone/second harmonic = 2:1 (an octave)
- Second overtone/third harmonic = 3:2 (which we will come to call a perfect fifth)
- Third overtone/fourth harmonic = 4:3 (which we will come to call a perfect fourth)

To clarify the relationship of the above ratios with frequency, let's use 440Hz as our fundamental frequency. An octave (2:1) is therefore 880Hz, a perfect fifth (3:2) is 660Hz and a perfect fourth (4:3) is 586.67Hz.

At this time, these were the only intervals found pleasant enough to be used in music. Over thousands of years, Western cultures have become accustomed to and accepting of more intervals. This means that both the harmonic series and our cultural upbringing play a part in our perception of what intervals are pleasant.

3.4.1 12-Tone Equal Temperament and the Chromatic Scale

As our culture has evolved, so have the classification and tuning systems for music. Over time, there have been many classification and tuning systems used. All of these systems are a compromise between:

(a) The overall number of notes, so that the majority of intervals remain pleasant and instruments remain easy to play.
(b) The practicality of being able to tune instruments so that they could create a set of pleasing intervals from any starting position.
(c) The purity of tone (i.e. the pleasantness created by perfect intervals).

Today, we use a 12-note-per-octave classification system called the *chromatic scale* and a tuning system called the *12-tone equal temperament tuning* (12tET). This tuning system is where the 12 notes are all an equal pitch distance apart. This is extremely efficient because the interval relationships in any key (starting pitch) are identical, meaning that pieces can be moved in pitch freely. However, the downside of 12tET is that these interval values vary slightly from the perfect ratios found in the harmonic series. This makes the overall sound of 12tET a little less satisfying than tunings containing more perfect ratios.

Throughout this musical evolution we have been building on previous precedents, utilizing more notes and creating more efficient tuning systems. Before the emergence of 12tET, several other tuning systems were utilized. The 12 named notes we use today came from previous systems where intervals were created using only one perfect frequency ratio. It was this use of only one perfect ratio that began the quest of greater uniformity of intervals between keys to aid free movement. It was the ratio of 3:2 (a fifth) that was deemed optimal. With this ratio, you can take 12 steps before returning to a note comparable with the first (i.e. an octave relationship). The historically significant *Pythagorean tuning* system utilized intervals of 3:2. This meant that there were far more perfect ratios found in Pythagorean tuning, and therefore it can often sound sweeter than 12-tone equal temperament. However, a major complication arose with this system because the series of 12 intervals of 3:2 do not round off to precisely an octave; it actually exceeds it by a small ratio known as a *Pythagorean comma*. To fix the issue, the interval between the last two notes of an octave had to be shortened, creating a rather unpleasant-sounding interval called the 'wolf interval'.

As time went on and we became more accepting of the use of many different intervals and the movement between keys during a piece, this wolf interval became restricting. Hence, 12-tone equal temperament eventually arose, built on several earlier attempts to address the wolf interval such as *well temperament*. The 12-tone equal temperament system means that the notes (other than octaves) are technically slightly out of tune with true harmonics, but all are close enough for the human ear to barely perceive it (unless played back-to-back with pure intervals).

So why would you intentionally use a system that makes notes out of tune? Well, the 12-tone equal temperament system has an ace up its sleeve: It is extremely convenient, as it allows instruments to play in all keys without tuning problems.

Each of the 12 notes describes the pitch we perceive (i.e. its fundamental frequency). This might seem like a very small number of notes compared to the number of different pitches we can hear. However, these 12 notes are cyclical, so once you have ascended or descended through them, the circle restarts. The gap between two notes that are a full circle apart is called an octave. The A above middle C is 440Hz. The note an octave above this A would also be called A and would have a frequency of 880Hz. The note an octave below 440Hz would be 220Hz. Over time, our culture has developed to consider the relationship between notes that have a relationship of an octave or multiple octaves (a ratio of 2:1) as equivalent. To put it more crudely, these interval of octaves are so harmonically similar that they are considered 'the same'. This is despite the fact that the human ear can determine which is the higher and lower in pitch when comparing two notes an octave apart.

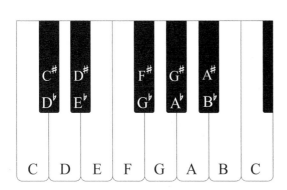

Figure 3.5 Musical notes shown on a piano

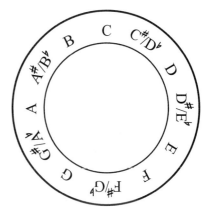

Figure 3.6 Chromatic scale circle

The 12 notes in our circle are called the chromatic scale and are represented by the first seven letters of the alphabet, A, B, C, D, E, F and G (these are the white keys on a piano). The other five notes are found between these (the black notes on the piano) and are sharpened and flattened versions of their neighbours. Sharp and flat notes allow us to raise or lower the pitch of a note so that we can create different scales on the same root, or move a certain scale to a different root and still maintain its interval pattern. This will be explained further when we talk about scales, modes and keys. Black notes are represented by the nearest alphabetical note as well as a sharp (♯) or flat (♭) symbol. When a note is sharpened, it is raised in pitch; when a note is flattened, it is lowered in pitch.

As you can see from Figure 3.5, the location of black notes is between each white note, except between E–F and B–C. You can represent a black note as a sharp or a flat. For example, F♯ and G♭ are the same note on the keyboard. To decide whether you verbalize it as F♯ or G♭ is dependent on its context (primarily the role of the note within a particular key). Again, we will discuss this in more depth when we talk about scales.

Because of its cyclical nature, the chromatic scale is often pictured as a circle:

To be able to implement 12-tone equal temperament in a way that is accurate and reproducible, we need a reference pitch to tune our other notes around. As it happens, A above middle C is the note chosen, and all instruments in an ensemble tune this note to a frequency of 440Hz. This is called *concert pitch*.

3.4.2 Expression of Musical Values

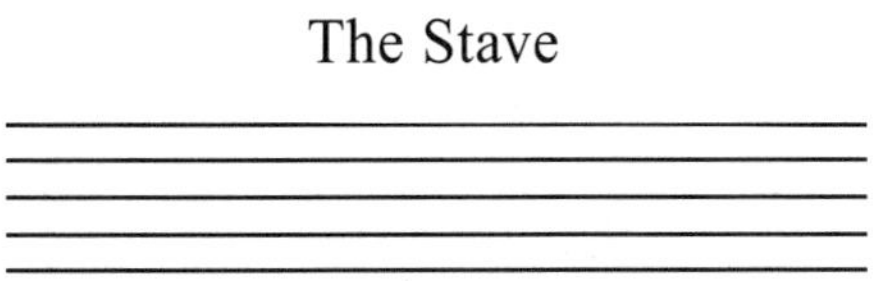

The Stave

Figure 3.8 The stave

When we aren't dealing in classical notation and there is a need to specify a particular pitch, note names are often cited in relation to a full-sized piano's MIDI note (e.g. 'A0' is the lowest A (the first note on a piano) and 'C4' is middle C).

Representing note information on paper is done through *sheet music*. The system uses a five-line *stave* (British) or *staff* (American).

A piece of sheet music will usually begin with a *clef*, which is a symbol that indicates the positioning of pitched notes on the stave. Pitch is shown by placement of notes in relation to the lines of the stave and clef used. The duration of notes is shown with the note values we discussed in the rhythm section. The two most common types of clef are called *treble clef* and *bass clef*.

Figure 3.9 Treble clef and G4

The treble clef is used for instruments with a higher register. The treble clef is based around the note found on the second line from the bottom of the stave (which, in MIDI terms, is G4). Figure 3.9 shows the treble clef symbol and the position of G4 on the stave.

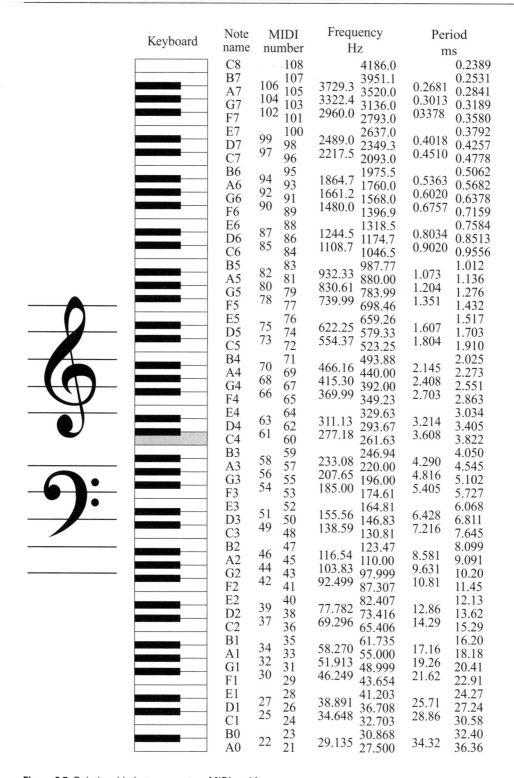

Figure 3.7 Relationship between notes, MIDI and frequency

Figure 3.10 Treble clef notes

From here, we can figure out the other note names simply by going forward or backward through the musical alphabet: A, B, C, D, E, F, G, which gives us:

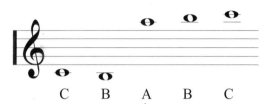

Figure 3.11 Treble clef ledger lines

If we need more notes above or below the stave, we add *ledger lines*, which extend the range of the stave. Middle C (C4) is the first ledger line below the stave in treble clef.

Figure 3.12 Bass clef and F3

The bass clef is used for instruments with a lower register. Bass clef works on the same principles, but is based around the note F3, which is found on the second line from the top.

Figure 3.13 Bass clef notes

Working from here gives the notes found in Figure 3.13.

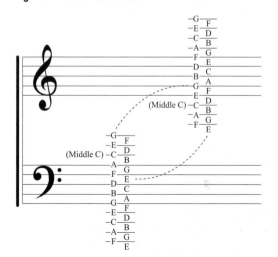

Figure 3.14 Notes in bass and treble clefs

In the bass clef, middle C is again found on a ledger line, this time above the stave. Figure 3.14 illustrates all the notes found in the bass and treble clefs.

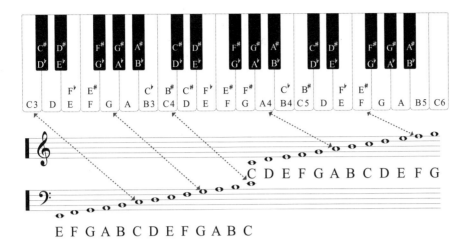

Figure 3.15 White notes represented on stave notation

Figure 3.15 shows the placement of the white notes on a piano between C3 (an octave below middle C) and G5.

You can also represent any of the black keys on a piano by using the ♯ or ♭ modifier. The following figure shows an F♯ note on the treble clef.

As well as a clef, there is usually a key signature listed on the stave, which specifies certain notes are flat or sharp throughout the piece, unless otherwise indicated. The reason why it is advantageous to outline this at the start of a piece is because of scales, which we will be discussing in the next section.

As previously mentioned, we will be dealing with pitch in relation to MIDI notes on piano roll in your DAW, so I won't be concentrating on musical notation. However, knowing the basics of stave notation can come in very handy. If you want to work with classically trained musicians on a daily basis, it is worth delving into this subject more deeply and spending a bit of time working on becoming proficient at sight-reading.

Figure 3.16 F♯5 on treble clef

Figure 3.17 Key signature of G major showing one sharp (F♯)

3.5 INTERVALS, MELODY AND SCALES

One musical note is not a song; it is simply a pitch. Once you get several notes played with an interesting timing, you start to interpret those notes not just as a set of pitches, but as a tune. The way different intervals lead to the building and resolution of tension, combined with timing and the timbre of the instrumentation, creates emotion. This concept of tension-resolution is absolutely key to music, both melodically and harmonically; we will discuss this in depth further into this chapter. For now, it is sufficient to know that much of the allure of music is the complex and imaginative ways that melody, harmony and rhythm combine to give a sense of pay-off when the music returns to a familiar place; often this is referred to as 'going home'.

Even though modern songs are immensely complex structures, even a simple tune will convey some sort of emotion. To start to understand what makes a tune interesting and how to make it express certain emotions, we need to explore intervals, melodies and scales.

3.5.1 Intervals

Earlier on, we mentioned that an interval was the gap in pitch between notes. To express intervals efficiently, we need a way to explain the distance between pitches. We could express them as ratios but again this is very cumbersome. Therefore, they are typically expressed in one of two ways:

- by the amount of steps in the chromatic scale between notes; and
- by their positions in relation to a scale being used.

To express intervals in steps of the chromatic scale, we need to know that the gap between a note and the very next note is called a semitone, and the gap between two consecutive notes is called a tone.

To express an interval in terms of its position within a scale, we need to represent each jump in pitch by a number or name. In most Western scales, there are eight notes. We will discuss the specific values of the eight notes later. But for now, all we need to be aware of is that the jumps within the scale being used are called scale degrees, and can be numbered or named as follows:

Degree	Name
1st	Tonic
2nd	Supertonic
3rd	Mediant
4th	Subdominant
5th	Dominant
6th	Submediant
7th	Leading Tone/Subtonic
1st (8th)	Tonic

3.5.2 Selecting Sympathetic Notes

Now that we know how notes and intervals are named, we need to realize that not all notes in the chromatic scale sound good in combination with each other. The job of the chromatic scale is to give a composer enough variation in notes to create an articulate and effective piece in any key. Doing so means having more notes than will be sympathetic in a single key.

First, let's look at how the notes found in the harmonic series can be expressed in relation to the chromatic scale and, more specifically, notes on a piano. This will start to show us how certain notes are more pleasing than others.

Let's take a G2 note near the bottom of the piano, which is 97.9989Hz. For ease of mathematics, I will round it up to 100Hz. This is rather crude and inaccurate compared to their true values, but exploring the harmonic series more accurately is more of a concern to a mathematician or physicist than it is a music producer. In practice, rounding up will leave a note slightly sharp of G at concert pitch, but it is easily close enough to demonstrate my point.

Now, let's look at the first 10 harmonics of this particular G:

Harmonic number	Frequency (Hz)	Note in 12tET
1	100	G2
2	200	G3
3	300	D4
4	400	G4
5	500	B4
6	600	D5
7	700	F5
8	800	G5
9	900	A5
10	1,000	B5

The most important principle to remember about the harmonic series is that notes that appear closer to the fundamental are more closely related to the fundamental and sound more sympathetic in combination with it (i.e. they feel more 'pleasing' to the ear). Mathematically, these notes also have the purest ratio relationships with the fundamental. The inverse is also true – notes that first appear further from the fundamental are less closely related to the fundamental; therefore, they are less sympathetic and feel more 'uneasy' or 'unpleasant'. Mathematically, they also have a less pure ratio relationship with the fundamental.

In a piece of music, you will find that notes that correspond to positions earlier in the harmonic series will be used more often than the notes that are more distantly related. When we get deeper into music theory, we will also realize that some of the notes that relate to these closely related harmonics will also serve a functional purpose (both melodically and harmonically) in the way that the music moves/progresses.

To create the set of sympathetic notes, we have to omit some of the steps to avoid notes that clash with each other. Despite what you may be thinking, creating a set of sympathetic notes isn't as simple as only selecting notes that are closely related to the root. Our ears also get an uneasy feeling when two notes that are played simultaneously are too close in pitch to each other. The rule of thumb regarding this is to avoid notes that are closer than three semitones (which we will later call a minor third) away from each other, and this is how our basic scales are built.

This means that we can actually make a number of sets of sympathetic notes from the same root.

3.5.3 Melody

The 'tune' or lead melody is often considered to be the part that is in the foreground of the piece. In popular music, the lead melody (usually a vocal) is the most prominent instrument of a piece. In practice, pieces also have other melodies occurring simultaneously that give texture to the composition. These melodies are often set more in the background and are subordinate to the lead melody. Melodies that play a secondary role in the piece are called countermelodies. Musical breaks in popular music often allow other instruments the opportunity to take the lead when a vocal is not present.

To keep interest, songs will usually include several different melodies or variations. The way that these themes are organized makes up the song's structure, which will be covered later in this section. The main melody in popular music genres is so crucial to the success of the piece that the 'catchiness' of that tune is often deemed a prerequisite.

3.5.4 Scales

While it is possible to write melodies with no formal training or understanding of music theory, it is a huge advantage to know sets of notes that go together well. The harmonic series plays a huge part in our perception of what we find sympathetic, as does the avoidance of intervals that are pitched too close together to be pleasant. But we mustn't ignore the fact that a lot of our 'rules' are based on culture and exposure to particular patterns of notes. To create the Western-friendly patterns that we have become so accustomed to, we take the 12 steps of the chromatic scale and remove some of the notes that we don't find sympathetic with each other.

These sets of notes are called *scales*, and are defined by their starting note and the number of steps within an octave. Most popular music pieces use *diatonic* scales, which means they consist of eight notes (seven distinct notes and an octave). An alternative to the diatonic scale is a *pentatonic* scale, which consists of five notes and an octave, and is often found in blues and rock.

The starting note of a scale is known as the *root* or *tonic*. This is the base of the scale and helps to determine which of the notes aren't going to be sympathetic. The tonic is also the note that will more than likely become your composition's 'home', the note around which you build and resolve the tension. The specific intervals used in the scale are dependent on the emotion the composer is trying to set. The series of intervals used within the scale are called *modes*, which comes from the Latin 'modus' meaning 'method'.

The most common sets of notes in typical Western music arrangements are called major and minor scales. These are technically modes because they define specific interval values but are used so frequently that the term scale is often used interchangeably with mode.

A *major* scale is often considered to be a more 'happy' series of notes, whereas the *minor* scale is considered more 'sad'. The chosen root note and the type of mode used dictates which notes are chosen, which together is called the key of the piece. If the tonic of a piece is C using a minor scale, its key would be classified as C minor. If the tonic is C using a major scale, it would be classified as C major. However, as major keys are the most common, you might also see a C major key listed as simply C. Minor keys, however, will always include the minor extension or sometimes the abbreviation min.

You can memorize the notes in modes by remembering patterns. Thanks to the equal temperament system, the intervals between notes in the scale remain the same regardless of key. This means that you can use the same pattern to work out the notes in every major or minor scale. We can use scale degrees to help memorize these gaps.

Now let's get into the patterns in each major or minor scale. We already know that each of these scales contain eight notes of the 12 musical notes and that the first note in a scale is always the root note; and if you calculate the notes correctly, the eighth note will be the root again but an octave higher. This only leaves six more notes to fill in. To do this, we can remember a simple pattern based on the number of steps (in semitones and tones):

Major Key

In a major key, the intervals between notes are as follows:

1st (Tonic)	2nd (Super-tonic)	3rd (Mediant)	4th (Sub-dominant)	5th (Dominant)	6th (Sub-mediant)	7th (Leading Tone)	8th (Tonic/Octave)
Root	Tone	Tone	Semitone	Tone	Tone	Tone	Semitone

Now, using the chromatic circle pictured in Figure 3.6, let's work out a major scale using the key of G as an example:

1st (Tonic)	2nd (Super-tonic)	3rd (Mediant)	4th (Sub-dominant)	5th (Dominant)	6th (Sub-mediant)	7th (Leading Tone)	8th (Tonic/Octave)
Root	Tone	Tone	Semitone	Tone	Tone	Tone	Semitone
G	A	B	C	D	E	F♯	G

In a diatonic scale, each of the letters of the musical alphabet must be represented. This is so there is no ambiguity between each degree of the scale; therefore, each degree can be quickly identified. This might not be so much of a problem when we are dealing with simple pieces, but it can become a problem when you are dealing with altered chords or more complex melodies.

Therefore, in G major, the seventh degree must be F♯ rather than G♭. If you want to sharpen or flatten any of these notes, you can do so, but you should retain its letter. The sharpening of a note by a semitone is called augmenting; the flattening of a note by a semitone is called diminishing.

If the note you are augmenting/diminishing already has a sharp/flat, you still retain the letter. In these cases, you would represent this note with a double sharp ({doublesharp}) or double flat ({doubleflat}). You won't really see this in popular music, but it's just good to know so it cements the concept that an altered note in a specific key will always retain its letter.

Further Resources   www.audioproductiontips.com/source/resources

 MIDI file of G major as consecutive notes. (17)

Minor Key

Minor keys are not quite as simple as major keys as there are three common variations: *natural minor, harmonic minor* and *melodic minor*. When the name of the variation isn't mentioned, it can be assumed that the key is the natural minor scale.

Natural Minor

As the *natural minor* is the simplest to remember and is the most common in popular music, let's start here.

In a natural minor key, the jumps between notes are as follows:

1st (Tonic)	2nd (Super-tonic)	3rd (Mediant)	4th (Sub-dominant)	5th (Dominant)	6th (Sub-mediant)	7th (Leading Tone)	8th (Tonic/Octave)
Root	Tone	Semitone	Tone	Tone	Semitone	Tone	Tone

Using the chromatic circle, let's work out a natural minor scale, again using the key of G as an example:

1st (Tonic)	2nd (Super-tonic)	3rd (Mediant)	4th (Sub-dominant)	5th (Dominant)	6th (Sub-mediant)	7th (Leading Tone)	8th (Tonic/Octave)
Root	Tone	Semitone	Tone	Tone	Semitone	Tone	Tone
G	A	B♭	C	D	E♭	F	G

Further Resources

www.audioproductiontips.com/source/resources

 MIDI file of G natural minor as consecutive notes. (18)

Relative Major/Minor and Perfect Intervals

At this point, it is worth noting that there is a strong link between major scales and natural minors. If you started on the 6th degree (submediant) of a major scale and worked through eight degrees until you get to an octave, you would actually have a natural minor scale. This means that for every major scale, you have a minor scale with exactly the same notes, just in a different order. The minor scale with equivalent notes to the major scale is called a *relative minor*, and they are as follows:

Major	Relative Minor
C	A
D	B
E	C#
F	D
G	E
A	F#
B	G#

We will be covering the modes that start on other scale degrees shortly.

It is also worth remembering that regardless of whether the key is major or minor, the 1st, 2nd, 4th and 5th degrees of each scale are identical notes.

Scale Degree	C Major	C Minor
1st	C	C
2nd	D	D
3rd	E	D#
4th	F	F
5th	G	G
6th	A	G#
7th	B	A#

The 4th and 5th scale degrees have a high compatibility with other notes from the key (they hardly ever sound out of place). They are known as 'perfect fourths' and 'perfect fifths' because of their simple pitch relationships with the tonic and their high degree of consonance.

Harmonic Minor

The next most common in popular music is the *harmonic minor*. This is identical to the natural minor except that the seventh note is raised a semitone. From a harmonic point of view (chord-building), raising the seventh degree creates extra tension and pull back to the tonic; therefore, the pay-off when a sequence 'goes home' is more significant. We will be learning more about the harmonic minor a little later, but for now it is sufficient to know that the harmonic minor is an altered natural minor scale used to help build a stronger emotional pull back to the tonic.

In a harmonic minor key, the jumps between notes are as follows:

1st (Tonic)	2nd (Super-tonic)	3rd (Mediant)	4th (Sub-dominant)	5th (Dominant)	6th (Sub-mediant)	7th (Leading Tone)	8th (Tonic/Octave)
Root	Tone	Semitone	Tone	Tone	Semitone	Tone + extra semitone	Semitone

Let's work out a harmonic minor scale, again using the key of G as an example:

1st (Tonic)	2nd (Super-tonic)	3rd (Mediant)	4th (Sub-dominant)	5th (Dominant)	6th (Sub-mediant)	7th (Leading Tone)	8th (Tonic/Octave)
Root	Tone	Semitone	Tone	Tone	Semitone	Tone + extra semitone	Semitone
G	A	B♭	C	D	E♭	F♯	G

Melodic Minor

One of the problems with the harmonic minor is that the sixth to seventh note interval is a jump of over a tone, which sounds nasty and is pretty difficult to sing along to. To make a scale more pleasant while retaining the tension from the raised seventh note, the melodic minor was created. The melodic minor scale takes the natural minor and raises both the sixth and the seventh note by a semitone so that there is never a jump of more than a tone:

1st (Tonic)	2nd (Super-tonic)	3rd (Mediant)	4th (Sub-dominant)	5th (Dominant)	6th (Sub-mediant)	7th (Leading Tone)	8th (Tonic/Octave)
Root	Tone	Semitone	Tone	Tone	Tone	Tone	Semitone

1st (Tonic)	2nd (Super-tonic)	3rd (Mediant)	4th (Sub-dominant)	5th (Dominant)	6th (Sub-mediant)	7th (Leading Tone)	8th (Tonic/Octave)
Root	Tone	Semitone	Tone	Tone	Tone	Tone	Semitone
G	A	B♭	C	D	E	F♯	G

When dealing with the melodic minor scale, you can also revert back to the natural minor scale when descending. This is because the tension of the raised seventh degree is not needed when you are not aiming to resolve. In fact, reverting back to the natural minor scale when descending is the most common practice when using the melodic minor scale.

The application of the harmonic and melodic minor scales causes much confusion with songwriters. The reality for popular music is that most of it is written in major or natural minor keys, but when a natural minor key is used, it will occasionally foray into the harmonic or melodic minor to reinforce the tonic.

The rest of this chapter will focus primarily on the major and natural minor scales. However, in the chords section, we will briefly touch upon when and why you might want to borrow from the alternative minor scales.

Further Resources   www.audioproductiontips.com/source/resources

 MIDI file of G harmonic and melodic minor as consecutive notes. (19)

EXERCISES

1. Listen to all of the examples in the scales in this section and write down how each scale makes you feel.

2. Listen to the G major scale example and move the scale to different keys by selecting all the notes in the piano roll and dragging them to different positions. Listen to the similarities and differences when you change the key.

3. Repeat this procedure with all the scale examples, again listening for similarities and differences.

4. With the G major example, create a simple melody by dragging any of the notes randomly (except the first), making sure that only one note plays at a time. How does it sound?

5. Repeat the above, except only using notes that are in the G major scale. How does it sound?

6. Use the interval examples and memorize the sound and the way that each of them make you feel.

7. Move the interval examples to different keys and try to note down any songs that come to mind when you listen to them.

3.5.5 Expressing Intervals by Scale Degree

Now that we know these scales, we can also express intervals by their major and minor scale degrees. For example, the gap between two notes may be called a minor third (three semitones) or a fourth (two tones and a semitone).

The table below shows intervals over one octave. I have also included their frequency ratio value in 12tET and the true harmonics to show how close the 12tET values are to pure harmonics:

Semitones	Interval	12tET Ratio	Perfect Ratios	First Appears in the Harmonic Series
0	Unison	1.0000	1.0000	1
1	Minor Second	1.05946	25:24 = 1.0417	17
2	Major Second	1.12246	9:8 = 1.1250	9
3	Minor Third	1.18921	6:5 = 1.2000	19
4	Major Third	1.25992	5:4 = 1.2500	5
5	Fourth	1.33483	4:3 = 1.3333	21
6	Diminished Fifth/Tritone	1.41421	45:32 = 1.4063	11
7	Fifth	1.49831	3:2 = 1.5000	3
8	Minor Sixth	1.58740	8:5 = 1.6000	13
9	Major Sixth	1.68179	5:3 = 1.6667	27
10	Minor Seventh	1.78180	9:5 = 1.8000	7
11	Major Seventh	1.88775	15:8 = 1.8750	15
12	Octave	2.0000	2.0000	2

Learning the sound of these intervals is very advantageous to not only musicians, but producers too. Getting to grips with them has several advantages:

- You can reverse-engineer tracks faster and more effectively.
- You can fault-find more efficiently.
- When you imagine changes to a melody or progression, you can verbalize that to a musician far more effectively.

A great way to get started is to find examples of these in the phrase of nursery rhymes, popular music or TV themes. Here are some examples. Where the specific phrase is not mentioned, I am referring to the opening notes:

Notice that as we start getting to higher jumps, examples are more obscure? This is particularly true for descending notes, with the descending minor sixth, minor seventh, major seventh and octave being very uncommon.

As well as using these examples, you can also use ear-training programs to help build your familiarity with intervals, scales, modes and chords.

Table 3.3

Semitones	Interval	Ascending	Descending
0	Unison	N/A	N/A
1	Minor Second	Jaws Theme	Für Elise
2	Major Second	Happy Birthday	Mary Had a Little Lamb
3	Minor Third	Smoke on the Water (opening guitar riff)	Hey Jude
4	Major Third	When the Saints Go Marching In	Swing Low Sweet Chariot
5	Fourth	Here Comes the Bride	Oh Come All Ye Faithful
6	Diminished Fifth/Tritone	The Simpsons ('*The Simp*sons')	Purple Haze (opening guitar riff)
7	Fifth	Twinkle, Twinkle Little Star ('Twin*kle twin*kle')	The Flintstones ('*Flintstones*')
8	Minor Sixth	Call Me Maybe ('Hey I just met you and this *is crazy*')	Love Story Theme
9	Major Sixth	Away in a Manger ('*The li*ttle Lord Jesus')	Man in the Mirror ('*I'm star*ting with the . . .'
10	Minor Seventh	Hush Little Baby	Herbie Hancock – Watermelon Man
11	Major Seventh	Pure Imagination (Charlie and the Chocolate Factory) ('In *a world* of pure imagination')	I Love You – Cole Porter
12	Octave	Somewhere over the Rainbow	Willow Weep for Me – Eta James

---

Further Resources   www.audioproductiontips.com/source/resources

 Link to EarMaster ear training program. (20)

3.5.6 Transposition

Transposing is the process of moving a set of notes by a fixed amount to maintain their interval relationships but changing their pitches. Transposition can be used to change the key of an entire piece, to *modulate* (change key within a piece) or to create harmony lines that are copied to work alongside the original pitch. We will discuss harmony soon.

3.5.7 Modes

So far, we have discussed two ways of taking the exact same notes and creating different moods by choosing a different starting position. These were the standard major scale that starts on the 1st degree of a major scale and a natural minor scale beginning on the 6th degree of the major scale. You can, however, make more modes from these same notes by starting on the other scale degrees of the major scale.

The seven modes you can create by doing this are called *church modes*, named after their use by the Roman Catholic Church, who in turn borrowed the concepts from the ancient Greeks. Whichever note you start on, you get a very different feel. Although the major and natural minor modes will be used in the vast majority of popular music, it is still worth knowing about the other modes.

Here is a breakdown of these modes, in relation to the major scale. Warning: These modes have rather confusing Greek names:

Degree	Name	Intervals (T = Tone, S = Semitone)	Notes
1	Ionian (Major)	T-T-S-T-T-T-S	C-D-E-F-G-A-B
2	Dorian	T-S-T-T-T-S-T	D-E-F-G-A-B-C
3	Phrygian	S-T-T-T-S-T-T	E-F-G-A-B-C-D
4	Lydian	T-T-T-S-T-T-S	F-G-A-B-C-D-E
5	Mixolydian	T-T-S-T-T-S-T	G-A-B-C-D-E-F
6	Aeolian (Natural Minor)	T-S-T-T-S-T-T	A-B-C-D-E-F-G
7	Locrian	S-T-T-S-T-T-T	B-C-D-E-F-G-A

> **Further Resources** www.audioproductiontips.com/source/resources
>
> MIDI files of each mode based around C Ionian. (21)

So, you can see that C Ionian, D Dorian, E Phrygian, F Lydian, G Mixolydian, A Aeolian and B Locrian all use the same notes (just starting from a different tonic). When dealing with modes, musicians tend to memorize the Ionian (major) scale but know how to adapt it to make it into each of the other modes. The major scale that contains the same notes as the mode you are using is called the *parent scale*.

When we look at modes this way, the scales look pretty much the same, but let's consider the modes if we were starting on the same tonic every time:

Name	Intervals (T = Tone, S = Semitone)	Notes (Based around C Major)
Ionian (Major)	T-T-S-T-T-T-S	C-D-E-F-G-A-B
Dorian	T-S-T-T-T-S-T	C-D-E♭-F-G-A-B♭
Phrygian	S-T-T-T-S-T-T	C-D♭-E♭-F-G-A♭-B♭
Lydian	T-T-T-S-T-T-S	C-D-E-F♯-G-A-B
Mixolydian	T-T-S-T-T-S-T	C-D-E-F-G-A-B♭

| Aeolian (Natural Minor) | T-S-T-T-S-T-T | C-D-E♭-F-G-A♭-B♭ |
| Locrian | S-T-T-S-T-T-T | C-D♭-E♭-F-G♭-A♭-B♭ |

> **Further Resources**   www.audioproductiontips.com/source/resources
>
> MIDI files of each mode based around the tonic note of C. (22)

Now they start to look and sound very different.

We can simplify this table to express how each scale degree of each mode differs from the major scale (Ionian):

Ionian	1	2	3	4	5	6	7
Dorian	1	2	3♭	4	5	6	7♭
Phrygian	1	2♭	3♭	4	5	6♭	7♭
Lydian	1	2	3	4♯	5	6	7
Mixolydian	1	2	3	4	5	6	7♭
Aeolian	1	2	3♭	4	5	6♭	7♭
Locrian	1	2♭	3♭	4	5♭	6♭	7♭

If you take the time to learn the major scale in each key, you can easily modify it to create the mode that you desire. There is a quick way to remember the sharps/flats in any major or minor key with an illustration called the *circle of fifths*, which we will get to later. For now, though, I want you to do this the long way, as it will help you to build the connections between the notes in your head.

Also, you might be wondering why all but one of these are labelled as flats when we are ascending the scale degrees? And why is the Lydian sharp? The reason for this is that all of the flats indicate that the interval has been taken from a major to a minor by dropping by a semitone. This is the case for all the intervals except the sharp in the Lydian mode, where the fourth is actually raised a semitone from the perfect fourth, and the fifth scale degree of the Locrian mode where the perfect fifth is dropped by a semitone. The fourth degree on the Lydian scale is called an *augmented fourth* and the fifth degree of the Locrian scale is called a *diminished fifth*.

We can organize the modes and tabulate the information to more clearly show the patterns of major/minor intervals (see Table 3.4). The table below orders the modes by brightness with the most uplifting at the top, and the darkest and more melancholic at the bottom.

Notice that the brighter modes consist of more major intervals, and the darkest consist of more minor intervals. This again ties well to the harmonic series. Therefore, modes can also be judged based on their major or minor qualities. This is defined specifically according to whether the third of the scale is a major

Table 3.4

Mode	Starting Note Based on C Ionian	Intervals with Relation to the Tonic						
		First	Second	Third	Fourth	Fifth	Sixth	Seventh
Lydian	F	Perfect	Major	Major	Augmented	Perfect	Major	Major
Ionian	C	Perfect	Major	Major	Perfect	Perfect	Major	Major
Mixolydian	G	Perfect	Major	Major	Perfect	Perfect	Major	Minor
Dorian	D	Perfect	Major	Minor	Perfect	Perfect	Major	Minor
Aeolian	A	Perfect	Major	Minor	Perfect	Perfect	Minor	Minor
Phrygian	E	Perfect	Minor	Minor	Perfect	Perfect	Minor	Minor
Locrian	B	Perfect	Minor	Minor	Perfect	Diminished	Minor	Minor

or minor third. We will learn about why the third is important in the chord section, but for now it is enough to think of the modes as:

Major modes: Lydian, Ionian, Mixolydian

Minor modes: Dorian, Phrygian, Aeolian, Locrian

From here, we can start to think about the emotion created by composing melodies using these modes. As you can see, all of the patterns except Lydian and Locrian still contain perfect 1st, 4th and 5th degrees. These three degrees, and particularly the relationship between 1st and 5th, are often considered the most important to the construction of popular music songs. For this reason, the Locrian mode is almost never used and the Lydian mode is used sparingly.

A word of warning, though: A mode by itself is only a small piece of the puzzle in creating mood, speed, lyrics, instrumentation and effects, and many more factors play an equal or even greater role. To add to this, you don't have to stay in the same mode for a whole song – often, more exotic modes are used for only brief passages. So while it is great to have an idea of the type of moods that are typically created with these modes, you shouldn't pigeonhole them that rigidly.

As we are all so familiar with major and minor scales, we will compare the tonal characteristics of the other modes against their nearest major/minor counterpart. Just like other tonal observations, these are my impression of them – the feeling you get might be different, so you should always experiment and listen to the famous examples to make your own mind up. The modes below are also ordered from brightest to darkest.

Lydian

The Lydian scale is a major scale with an augmented fourth that has a kind of sweet but 'out of this world' sound. Not much popular music uses the Lydian mode, a famous example being the first few bars of 'The Simpsons' theme tune by Danny Elfman. However, many virtuoso guitarists like to use the Lydian mode, and this can be heard on many Steve Vai or Joe Satriani records, most notably parts of Joe Satriani's 'Surfing with the Alien' album.

Mixolydian

The Mixolydian mode is a major scale with a flattened seventh. As the seventh note strongly leads back to the first degree, it is called a *leading tone*. By flattening the seventh, you get a rather bluesy feel, with less of a 'strong' conclusion. The Mixolydian mode is frequently used by blues, funk and jazz guitarist to solo.

Examples of the Mixolydian mode in popular music include 'Sweet Home Alabama' by Lynyrd Skynyrd, 'Hey Jude' by The Beatles (only the outro) and 'Sweet Child O' Mine' by Guns N' Roses (except the solo).

Dorian

Dorian is a natural minor scale but with a major sixth. Because of this, it feels like a middle ground between happy and sad. Compositions using the Dorian mode usually have a slightly melancholic sound, but because of the major sixth it can often sound bittersweet too. Michael Jackson is a famous user of the Dorian mode, and his songs 'Thriller', 'Billie Jean' and 'Earth Song' all use it. The Dorian mode is also used in funk music as an alternative to an overly 'major' sounding progression.

Other Dorian examples are 'Smoke on the Water' by Deep Purple, 'Eleanor Rigby' by The Beatles and 'Scarborough Fair' by Simon & Garfunkel.

Phrygian

The Phrygian scale is a natural minor scale with a flattened second (which makes it a minor second interval). The prevalence of minor intervals in this scale gives it a very Spanish, Eastern and exotic feel. You can also find the Phrygian mode in Latin-infused hip hop. A lot of film composers also sharpen the third to make it a major third, which takes it from Spanish-sounding to an even more Eastern, almost Arabic style.

One of the more surprising uses of Phrygian is in metal, and in particular more progressive metal. This is because of its dark feel such as Tool's 'Forty Six & 2'. Some of Kirk Hammett's Metallica licks suggest Phrygian too, such as 'Wherever I May Roam'. In popular music, the Phrygian is used in 'White Rabbit' by Jefferson Airplane and 'Mr. Man' by Alicia Keys.

Locrian

The Locrian mode is a natural minor with a diminished fifth and a flattened second. It is rather unique because I couldn't think of a single popular music example that uses it and research showed only a single Björk song called 'Army of Me' as an example. The Locrian mode is so unpleasant because the fifth note is diminished; therefore, chords and harmonic structure are very unstable, producing a weird and highly uncomfortable sound, which never really resolves to anything pleasant.

Using the Modes

Notice that each of these modes, except for Locrian, only has one difference from their nearest major or natural minor scale? Remembering the name of the mode, its nearest major/minor and its alteration makes it easier to get used to using modes. It also makes it easier to build chords around these modes.

Mode	Base Scale	Alteration
Lydian	Major	Sharpened fourth (augmented)
Ionian	Major	N/A
Mixolydian	Major	Flattened seventh (minor seventh)
Dorian	Natural Minor	Sharpened sixth (major sixth)
Phrygian	Natural Minor	Flattened second (minor second)
Aeolian	Natural Minor	N/A
Locrian	Natural Minor	Flattened fifth (diminished) and flattened second

One further word of warning when using modes (especially instances where guitarists are using modes to solo): The reason that modes are so hard to grasp initially is because of the fact that they are all relative. If you play a C major scale, you are also playing the same notes as D Dorian, E Phrygian, F Lydian, G Mixolydian, A Aeolian and B Locrian. What establishes the particular mode is the harmony (chords) you are playing over. Without any reference (music in the background), the melody will feel 'ungrounded' and therefore the mode is difficult to distinguish from a major pattern. We will discuss the characteristic chords for each mode when we are discussing harmony. However, there are three tricks to help the listener recognize the mode in your melody more clearly:

1. *Accent the root note of the mode you are going for.* If you wanted to play Mixolydian, you could play the well-known Ionian (major) pattern but start and end phrases on the 5th scale degree. For instance, let's say that I want to play G Mixolydian – I could use the C major scale but focus around G as the tonic.
2. *Repetition.* Using the same example as above, you can also regularly repeat the G note in your licks to imply the mode.
3. *Focus on the unique step.* Each mode has its unique sound. If you were again using G Mixolydian, then you would want to focus a lot of the solo around the flattened seventh. In our example, that is an F note.

3.6 HARMONY AND CHORDS

When we analyse music, we can express it on two planes. There is the horizontal musical space that governs how notes happen over time, and the vertical musical space that governs how notes are stacked simultaneously. The interplay of notes on the horizontal plane creates melody and the stacking of notes on the vertical plane creates harmony. Because of this, we can discuss the interval of a minor third in the context of two consecutive notes in a melody or between two notes found in a chord. Both are equally important – the way that the notes intertwine on both planes makes a huge difference to the emotion created.

The problem is that these two planes are not mutually exclusive. You can create harmony from events on the horizontal plane when two distinct melody lines intertwine and notes fall simultaneously, and who is to say that you can't have as many melody lines as you like all interwoven with each other?

When we think of the origins of music, theorists and historians all conclude that it must have started with the human voice. The oldest melodies were created with a single vocalist. Over time, composers became interested in how multiple singers could interact with each other. At first, experimentation was based around each of these voices being their own distinct entities, which were free to play entirely separate melodies and rhythms. In music theory today, we define a distinct melody as a *voice*, despite the fact that a melody doesn't have to be sung by a human voice. As music developed, composers became aware of the fact that pleasant or unpleasant sounds can be created by voices playing notes simultaneously. The study of how distinct voices interact with each other is called *counterpoint*. Early composers in the West were very much aware of the effects of notes played simultaneously and used it to create emotion. However, their songwriting still focused on thinking in terms of a set of distinct and individual voices. During this time period, special attention had to be given to the way that these voices moved with each other, to avoid unpleasant results. Some of the patterns that emerged from observing which movements gave pleasant results became known as voice-leading principles, with *voice leading* being generally acknowledged as the smooth and harmonious movement of voices. The result of this is an intricate combination of notes that are constantly moving and creating many different harmony events as the voices interact and align with each other.

These days, we tend to think about and define music in a very different way. We are so used to stacking notes on top of each other that we have a definition for them: We call them *chords*. We memorize chords by learning the pattern of these notes, often by visualizing the 'blocks of notes' on your chosen instrument(s). We even distinguish between different patterns by defining them as different chord types. Early composers in the West would not have considered the notes C-E-G simultaneously as a single entity, let alone called it a C major chord. Instead, it would be thought of as a set of voices working together to create a pleasing result.

Because of this, we could imagine the function of a choir of four singers as four-part harmony line (making chords), or as four distinct melody lines. In Chapter 5: Advanced Music Theory, we will discuss voice-leading principles and how you can relate them to our current musical standards.

Rather than having distinct melody lines interwoven with each other (like in classical-era composition), modern pop functions with an arrangement of lead melody and accompaniment. The accompaniment is the supporting cast for the lead melody, designed to bring the best out of the lead instruments. Accompaniment can either be of harmonic (guitar) or rhythmic nature (drums). As well as from chords, on rare occasions a harmony can also be inferred with the use of arpeggios (notes in key played quickly one after the other).

Most of the extra harmonic background within popular music tracks is also provided by chords. Specific chords are chosen because of their relationship to the melody line and the key of the piece. The succession of chords chosen is called a *chord progression*. How chord progressions can be chosen and optimized is coming up shortly.

Sometimes chords progressions and a rhythm bed are written before the lead melody, and the vocal melody is chosen because of its relationship to the chords and the key of the piece. This is a particularly popular route for rock bands and goes to show that there is no sure-fire way of writing a successful song. At other times, lyrics and a lead melody are written before chords are established. In this scenario, the musicians will write around the lead melody. Both methods can produce great results and there is no favoured way of beginning a composition.

Instruments that often play chord progressions include the electric and acoustic guitar, piano, organ and synthesizers. Sometimes accompaniment is also provided by an ensemble of instruments that normally play the melody, such as a violin section.

In popular music, as well as chord-based accompaniment, there is usually a bass instrument (bass guitar, double bass, cello) that plays the bass notes (often the root of the chord) of the harmonic progression.

We are now going to discuss different chord types. Before doing so, I want to assert that you should try to memorize the chord types by their interval relationships rather than the specific notes used for each chord. This way, you can create the same chord type in any key by using the same pattern. After a while, the notes used for each chord will take care of itself.

3.6.1 Triads

Chords are two or more notes in a key that are played simultaneously. The most common chords are triads. *Triads* consist of three distinct notes. The most common and simplest of triads are called major and minor triads. Specifically, major and minor triads consist of the first, third and fifth scale degrees of the root note. Out of the triad, only the third degree is different between major and minor chords.

Triads are the basic building block of popular music and will be far more common than other types of chord. All of the other chord types we will discuss work by building from the triad or by slightly altering them.

Let's take this knowledge and work out both an A major and A minor chord, starting with A major.

A Major Triad

To create an A major chord (often abbreviated as just 'A'), we need to know the notes in the A major scale:

1st (Tonic)	2nd (Super-tonic)	3rd (Mediant)	4th (Sub-dominant)	5th (Dominant)	6th (Sub-mediant)	7th (Leading Tone)	8th (Tonic/Octave)
Root	Tone	Tone	Semitone	Tone	Tone	Tone	Semitone
A	B	C#	D	E	F#	G#	A

Now, to play an A major chord, you simply have to play the first, third and perfect fifth degrees simultaneously, which is: A, C♯, E.

Figure 3.18 A major triad on piano roll

Further Resources www.audioproductiontips.com/source/resources

MIDI file of A major triad. (23)

A Minor Triad

To create an A minor chord (often abbreviated as 'Am'), we need to follow the same rules but using the A minor scale:

1st (Tonic)	2nd (Super-tonic)	3rd (Mediant)	4th (Sub-dominant)	5th (Dominant)	6th (Sub-mediant)	7th (Leading Tone)	8th (Tonic/Octave)
Root	Tone	Semitone	Tone	Tone	Semitone	Tone	Tone
A	B	C	D	E	F	G	A

Again, all we need to do here is play the first, third and perfect fifth notes, which are: A, C, E. Notice that the first and fifth notes are identical in both the major and minor variants? The third degree's natural state is major so when you see a chord labelled as minor, this means that the third is flattened from a major third to a minor third.

Figure 3.19 A minor triad on piano roll

> **Further Resources** www.audioproductiontips.com/source/resources
>
> MIDI file of A minor triad. (24)

3.6.2 Chord Definitions

Before moving on, I want to stress that the notes found within a triad chord are based on the scale degrees of the root note, not the scale degrees of the key the composition is in. The notes found within the key of the piece have no bearing on how a chord is defined. For instance, an A minor chord is the same notes regardless of key.

However, the choice of chords you use in a composition is based on their compatibility with notes found in the key of the composition.

For instance, you wouldn't ordinarily use an A major note in the key of C because the third of an A major chord is C♯, which isn't found in C major (it can be done, but would usually be considered a *modulation*, covered later in the course). However, A minor could be used because all of its notes are found in C major.

1st (Tonic)	2nd (Super-tonic)	3rd (Mediant)	4th (Sub-dominant)	5th (Domi-nant)	6th (Sub-mediant)	7th (Leading Tone)	8th (Tonic/Octave)
C	D	E	F	G	A	B	C

A major triad: A, C♯, E.

A minor triad: A, C, E

When we start discussing larger chords, you will see why it is very important to remember this.

3.6.3 Why Are Major Keys Happy and Minor Keys Sad?

Earlier on, I suggested that major keys are considered happy and minor keys are considered sad. The same guideline is true for chords. The answer to 'why' lies again with the harmonic series.

Most popular music is based around the simple chord triads of the first, third, fifth, which I have just outlined. Let's explore this in the key of A again:

The first, third and fifth intervals of the major triad (A, C♯, E) are found within the first five harmonics of the harmonic series (A, A, E, A, C♯). This means that these intervals have a very strong relationship to the fundamental, or in the case of our major chord, the tonic. This makes the combination of these notes 'stable', 'pleasant' or 'happy'.

The minor triad's (A, C, E) relationship to the harmonic series, though, is much more distant. The first and fifth are the same, so feel stable. The minor third, however, does not appear in the harmonic series until the nineteenth harmonic. Your ear almost expects to hear the major third (C♯), and when that is replaced with a more distantly related note, this makes the listener feel more 'unpleasant', 'tense' or 'sad'.

3.6.4 Other Chords

Now that we are able to work out the notes in major and minor chords, we can begin to look at how to build up more complex chords. It is worth mentioning at this stage that while there are many different types of chord, their application is more limited than with triads. Over time, you will become adept at picking moments where these more exotic chords have the right sort of emotional impact.

Let's start with sixth and seventh chords, which are built by adding one extra note on top of the triad.

Sixth Chords

Sixth chords are also fairly simple. You start with the triad of the chord and add a major sixth degree of the scale to create a four-note chord. Sixth chords are used to add texture or colour to the chord rather than to change a triad's function.

On rare occasions, you may flatten the sixth to a minor sixth degree; these are called flattened sixth chords.

There are three types of sixth chords:

Sixth (A6) – first, major third, perfect fifth, major sixth
Minor sixth (Am6) – first, minor third, perfect fifth, major sixth
Minor flat sixth (Am♭6) – first, minor third, perfect fifth, minor sixth

Using A as our root, that gives us:

A6: A, C♯, E, F♯

Figure 3.20 A6 on piano roll

Am6 (A minor 6th): A, C, E, F♯

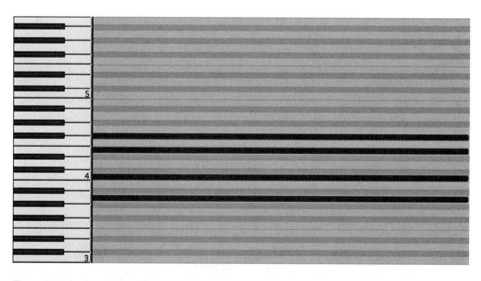

Figure 3.21 Am6 on piano roll

Am♭6 (A minor flattened 6th): A, C, E, F

Figure 3.22 Am♭6 on piano roll

Further Resources www.audioproductiontips.com/source/resources

MIDI files of sixth chords in A. (25)

Each of these styles of sixth chord can conjure different emotions. The context (the other chords that surround it) are very important to the emotion the chord portrays. However, you can get a general gist of the feeling of the chord even in isolation. I personally feel them as:

Sixth chord: Uplifting yet calming

Minor sixth chord: Confused and hesitant

Minor flat sixth: Disappointed and slightly confused

Seventh Chords

Just like sixth chords, sevenths add a fourth note over the top of the triad. You will find sevenths are used much more often than sixth chords and are a common occurrence in popular music. After a triad, they are the most popular form of chord.

Just like sixths, seventh chords can be used to add additional colour to a chord, but in specific locations they can be used for a specific purpose. The addition of a seventh degree can strengthen or change the role of the chord, hence why they are often described as functional chords. The purpose of a functional seventh chord is to create tension that leads very well to the tonic, strengthening the pay-off of the resolution.

Although you will see a variety of seventh chords used, a major triad with a minor seventh note added is the most popular, and is called a *dominant seventh* chord. Although a dominant seventh chord can be used to add colour, they are most often used as functional chords.

Using a minor seventh degree is such a common occurrence that in naming chords, the seventh is assumed to be flattened. Seventh chords in which the seventh is not flat are labelled with the modifier *major* (abbreviated 'maj', or 'M'). The 'maj' modifier applies to the seventh degree only (i.e. you won't see a chord such as 'Cmaj6' because there is no seventh degree in the chord).

Here are the four main variations of seventh chords built on standard triads:

Dominant seventh (e.g. A7) – first, major third, perfect fifth, minor seventh

Major seventh (e.g. Amaj7) – first, major third, perfect fifth, major seventh

Minor seventh (e.g. Am7) – first, minor third, perfect fifth, minor seventh

Minor major seventh (e.g. Amin/maj7) – first, minor third, perfect fifth, major seventh

Using A as our root, that gives us:

A7: A, C♯, E, G

Figure 3.23 A7 on piano roll

Amaj7: A, C#, E, G#

Figure 3.24 AMaj7 on piano roll

Am7: A, C, E, G

Figure 3.25 Am7 on piano roll

Amin/maj7: A, C, E, G#

Figure 3.26 Amin/maj7 on piano roll

> **Further Resources** www.audioproductiontips.com/source/resources
>
> MIDI files of seventh chords in A. (26)

Just like with sixths, each of these variations conveys very different emotions. My personal interpretations are as follows:

Dominant seventh: Tense; wants to resolve

Major seventh: Smooth; hint of tension

Minor seventh: Smooth; slight uplift

Min/maj seventh: Very tense; uncomfortable

Diminished Chords

Earlier on in this section, I mentioned that in some special cases, you can sharpen or flatten the fifth of a chord. When a minor triad has its fifth flattened by a semitone, it is called a *diminished chord*. Only minor triads can be diminished and they occur when a standard minor triad won't fit the notes of the scale.

A diminished – first, minor third, flattened fifth

A diminished chord can be abbreviated as follows: A dimin, Adim, A°, Am♭5

In the key of A minor, this works out to be: A, C, E♭

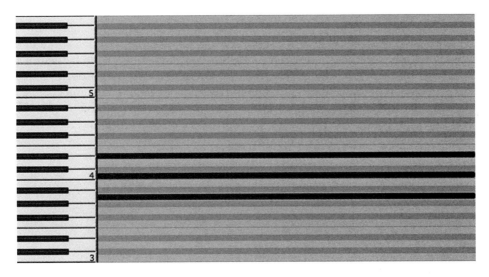

Figure 3.27 Adim on piano roll

> **Further Resources**   www.audioproductiontips.com/source/resources
>
> 🖥 MIDI file of Adim. (27)

Diminished chords sound very tense, eerie or foreboding. In the context of a piece, they make a listener feel like he or she is 'hanging on' and leave them craving for a resolution. This means that diminished chords are often used quickly to connect two more stable chords. They are fairly uncommon in popular music, and particularly in guitar-based music. Hearing a diminished chord out of context, it would be hard to think of a use for them in popular music. However, in context of the right chord sequence, they can work beautifully.

Augmented Chords

You can also create another chord by taking a major triad and raising the fifth by a semitone; this is called an *augmented chord*. Only major chords can be augmented. Augmented chords differ from triads and diminished chords because they don't occur naturally in a major or natural minor scale. However, they can be conceptualized as a chord based on the third degree of a harmonic or melodic minor scale. In practice, though, augmented chords aren't often used that way. They are most likely to function as a way to build tension, like how you would use a dominant seventh chord to strengthen the pay-off of a resolution.

A augmented – first, major third, sharpened fifth

An augmented chord can be abbreviated as follows: Aaug, A+

In the key of A, this works out to be: A, C♯, E♯ (F)

In this particular case, we define the augmented fifth as E♯ rather than F, because there is already an F♯ in A major (A, B, C♯, D, E, F♯, G♯). Any time two notes are equal (e.g. E♯ and F), they are called *enharmonic equivalents*. Remember, we make sure that each letter of the music alphabet is used in order to avoid any ambiguity. Labelling the augmented fifth as F would cause confusion between whether the note was a flattened sixth or an augmented fifth. This might not seem like a big deal, but when musicians are sight-reading this can save quite a bit of effort.

Figure 3.28 Aaug on piano roll

Further Resources www.audioproductiontips.com/source/resources

MIDI file of Aaug. (28)

Augmented chords also sound eerie and very tense, but arguably hold an even greater pull for resolution. The sense of tension is so strong that augmented chords are often found in film scoring, particularly to build tension in horror films.

Suspended Chords

Suspended chords are created by removing the third from a chord and replacing it with either the second or the fourth degree of the scale. A suspended chord can be abbreviated as follows: Asus4, Asus2. The most common form of suspended chord is the sus4, which is often abbreviated as simply 'sus' (i.e. Asus = Asus4). Suspended chords are often used briefly before conforming to a normal major or minor chord (Asus4, A) or as a means to keep the key of the piece ambiguous.

Let's take an A chord and create sus4 and sus2 chords:

Asus4: A, D, E

Figure 3.29 Asus4 on piano roll

Asus2: A, B, E

Figure 3.30 Asus2 on piano roll

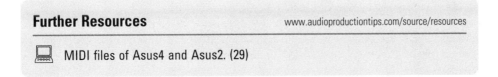

Further Resources www.audioproductiontips.com/source/resources

MIDI files of Asus4 and Asus2. (29)

Because of their ambiguity, suspended chords still sound tense, which makes a listener want a resolution. However, because of the stability of the first and perfect fifth, they sound significantly stronger than diminished or augmented chords.

Altered Chords

As well as the basic chords listed above, sometimes you will want to make a change to a chord in a way that doesn't follow any of the rules above. In these cases, you can state the chord by taking a similar chord and state the change you have made in brackets afterwards. For instance, A7(♭5) would take an A7 and flatten the fifth.

Remember, A7 is: A, C♯, E, G

This would make A7(♭5): A, C♯, E♭, G

Figure 3.31 A7(♭5) on piano roll

Further Resources   www.audioproductiontips.com/source/resources

🖳 MIDI file of A7(♭5). (30)

As well as notating chords with a flattening or sharpening of a note, you can also signify when you have added a specific note.

For instance, A (add9) would take an A major triad and add the 9th degree.

A major: A, C♯, E

A (add9): A, C♯, E, B

Figure 3.32 A(add9) on piano roll

Power Chords

Finally, let's discuss power chords. Power chords are chords that take the triad and omit the third altogether. In classical music, you wouldn't consider a set of two simultaneous notes a chord. However, in popular music, power chords are an exception. Power chords became popular with guitar-based music, particularly

Figure 3.33 A power chord (5th fret low E string) on electric guitar

upon the popularization of distortion. When complex chords are played with heavy distortion, they can often sound 'messy', 'muddy' or 'indistinct', whereas power chords can sound more 'coherent' and, erm, 'powerful'.

Power chords can be played on two strings of a guitar consisting of a first and a fifth or on three strings as: first, fifth, eighth (octave). In A, this would be: A, E or A, E, A. Power chords sound more ambiguous than most chords. While power chords can help to clean up a distorted guitar line, use on more than one or two chords in a phrase can sound 'boring' or 'predictable'. Some genres use power chords more often than full chords (e.g. punk), so this is not a hard-and-fast rule, just another guideline.

Further Resources   www.audioproductiontips.com/source/resources

- MIDI file of A power chord with first, fifth, eighth. (31)
- A power chord on a distorted electric guitar. (32)

EXERCISES

1. Listen to all the examples and write down your own thoughts on how they make you feel.
2. Use the example files to move the whole chords to each different root position; compare and contrast.

3.7 INTRODUCTION TO CHORD PROGRESSIONS

The way that chords progress in a musical arrangement is of particular importance to producers, as the chords give extra texture and emotional meaning to the main melody.

While the scope for drastic alteration of the main melody is usually quite limited, a producer has much more scope to optimize chord progressions that can help craft a stronger emotional tie to the listener.

Remember that a chord progression is the way that chords are used to create musical phrases. It is worth noting that popular music usually has several sections in each song. These differing sections will most likely use different chords or chord orderings. For now, we are going to consider chord progressions over a single section of a composition; we will discuss the way that a piece progresses in the song structure section of the next chapter.

To become proficient at producing songs, understanding how to effectively construct and choose chords is critical.

3.7.1 Chord Numbering

So far, we have outlined how notes of a melody can be chosen from scales, and how you can construct different chord types. But how do you select the right chords and chord types based on the key of the piece? The key of the piece is based on the tonic note and whether you want the piece to have a major or minor feel (e.g. C major).

In the vast majority of pop songs, only notes found in the scale that corresponds with our chosen key will be found in the chords. So this is a great place to start when looking for chords that will fit.

You could do this by getting an instrument and working out all the chords. However, there is a simpler way because the 12tET system makes a pattern of intervals that is the same for every key. Just like with scales, it is simply a case of memorizing the pattern!

Major Chord Numbering

Let's start by discussing the 'long way round' and how you can work out these chords by using an instrument (in this case, a keyboard). Let's use an example in the key of C major, as it only uses white notes on a keyboard. Here is the scale of C major again:

1st (Tonic)	2nd (Supertonic)	3rd (Mediant)	4th (Subdominant)	5th (Dominant)	6th (Submediant)	7th (Leading Tone)	8th (Tonic/Octave)
Root	Tone	Tone	Semitone	Tone	Tone	Tone	Semitone
C	D	E	F	G	A	B	C

If we wanted to create a major triad with a root note on the first scale degree, we would choose the first, third and fifth scale degrees, which give us the notes C, E, G. This makes a C major chord.

If we make a triad from the second scale degree of D, we would use the second, fourth and sixth scale degrees. This gives us D, F, A, otherwise known as a D minor chord.

Let's follow the same pattern all the way through the other scale degrees and we get:

Scale Degree	Notes	Chord Name
1	C, E, G	C major
2	D, F, A	D minor
3	E, G, B	E minor
4	F, A, C	F major
5	G, B, D	G major
6	A, C, E	A minor
7	B, D, F	B dim

Further Resources www.audioproductiontips.com/source/resources

 MIDI file of chords in C major. (33)

Notice that the seventh chord is diminished. This is because B minor is B, D, F♯ and, because the fifth (F♯) is not in the key of C, it has to be flattened to F to remain coherent to the notes in the scale of C major.

This same pattern of major, minor and diminished chords exists in all the major key signatures. Therefore, you can memorize that, when you are in a major key, chords one, four and five are major; two, three and six are minor; and seven is diminished.

To describe the chords within their position in the scale, musicians use Roman numerals. Major chords are listed in upper case, minor chords are listed in lower case and diminished is shown in lower case with the small circle symbol. In a major key, this would make:

I   ii   iii   IV   V   vi   vii°

Minor Chord Numbering

When dealing with natural minor chords, the pattern is a little different. However, as we know that for every major scale (Ionian mode) there is a relative minor (Aeolian mode) with the exact same notes, we can work out the chord types used in minor keys by working with the same pattern except starting from the sixth scale degree. The sixth degree of C major is A, so the relative minor is A minor. If we tabulate this, we get:

Scale Degree	Notes	Chord Name
1	A, C, E	A minor
2	B, D, F	B dim
3	C, E, G	C major
4	D, F, A	D minor
5	E, G, B	E minor
6	F, A, C	F major
7	G, B, D	G major

This gives us chords three, six and seven as major; one, four and five as minor; and two as diminished, or:

i   ii°   III   iv   v   VI   VII

Commit this pattern to memory too.

Further Resources

www.audioproductiontips.com/source/resources

MIDI file of chords in A minor. (34)

When discussing popular chord progressions later, we will primarily use major key examples, although there will be occasional references to minor keys too. A minor scale being used is easily noticeable because of the lower case tonic chord in the Roman numeral system.

3.7.2 Cadences

At points in this chapter, I've had to hint at the concept of tension-resolution and how important it is to creating emotion in music without really going into detail about it. Now that we know about scales and how notes can be stacked to create chords, we can start to think about the journey that popular music goes through. As a chord progression moves, it will travel away from 'home' (which builds tension) before returning home (which resolves the tension). There are numerous routes available when building this tension. You could develop short or long progressions of chords and go about this journey using different chord numbers. But when it comes to resolving the tension, there are fewer options, as certain chords have a stronger pull back home than others. Being able to resolve tension effectively (or deliberately choose not to) at the end of a musical phrase is one of the most important aspects of songwriting. The way that we close a musical phrase is called a *cadence*, and can be defined as the last two chords that conclude a musical idea. Like much of music theory, the term cadence comes from classical music, where its implementation is more definite. In popular music, the most obvious and definite cadence is a stopping point like at the end of a song, but you can also have cadences at the end of a large section, or even between smaller phrases. Therefore, I'd like to define three different categories of cadence based on where you might find them in a popular music track:

Master cadence – a cadence that occurs at the end of a song.
Sectional cadence – a cadence that occurs at the end of a larger section (e.g. going from a verse to a chorus).
Phrasal cadence – a cadence that occurs within a larger section (e.g. halfway through a verse).

Because of the way we often transcribe popular music (lyrics with chords), we can consider cadences as musical punctuation, which usually works in tandem with the lyrical punctuation. A chord progression will move away from home during a vocal line (or set of vocal lines), and return home for the start of the following line. The progression to the tonic strongly emphasizes the tonic as home; when we arrive at the tonic from the IV or, particularly, the V, the progression itself says 'and here we are back at our tonic'. When you add the dominant seventh to the V chord (V7), you get the same identification of the tonic, plus an increased tension that makes the listener feel at ease when it does resolve.

Sectional and master cadences are more definite and easily definable, but phrasal cadences can sometimes be a little bit more ambiguous. Locating phrasal cadences in popular music can be a matter of confusion as pop music is repetitive and cyclical in nature, not to mention the prevalent use of the I, IV and V chords.

Remember, a cadence is defined by the last two chords in the sequence; the fact that many phrases begin on the tonic doesn't mean that resolution is reached when a circle restarts. If anything, the use of the I chord at the beginning of a sequence will give the sense of repeatedly moving away from the tonal centre without fully reaching it. In punctuation terms, it is like having a comma at the end of each line without having a full stop until a phrase ends with the tonic. The final chord of a repeating circle will usually be a V chord, which makes the music feel 'unfinished', and therefore the listener expects another repetition. In the cases of phrasal cadences, I often look to the lyrical grouping for guidance. In some cases the use of I–V, IV–I or V–I is so frequent that it would be unwise to consider every occurrence as a cadence; this is especially valid if they are used in places where the lyrical content has not resolved (e.g. in the middle of a line).

As well as the places in which you can implement a cadence, there are also different types of cadence, which can affect how strongly the listener is pulled back to the tonic. The different types of cadence each have a different feel, so the choice of cadence can have a significant impact on how effective the track is overall. This is particularly true when dealing with sectional cadences. These different types of cadence can be used to create many emotive states, including stability, comfort, melancholy, surprise or a change of gear.

Authentic Cadence

In classical music, the tension built up through the progression is usually resolved with a dominant chord (V chord) resolving to the tonic (I chord). This sequence is often used in popular music too and is called an *authentic cadence*. The master cadence is almost always an authentic cadence, and they are frequently used as a sectional cadence too. Although they have a more definite feel than other cadences, you do occasionally see them used as a phrasal cadence too.

An authentic cadence can be further split into two more types: a perfect or an imperfect authentic cadence.

A *perfect authentic cadence* is where the chords are in root position (meaning the root of the chord is the lowest pitch) and the top voice is the tonic of each chord (in popular music, top voice normally means vocal melody).

An *imperfect authentic cadence* is where there is still a V–I relationship but the chords are not in root position or the top voice is not the tonic of each chord.

An authentic cadence is often made even stronger by making the V chord a dominant seventh. For instance, in the key of C, you would resolve by going from G7 to C. To my ears, a perfect cadence with a standard V–I resolution sounds uplifting and definite. A perfect cadence using a dominant seventh V chord sounds slightly less uplifting but is very smooth and pleasant.

Further Resources www.audioproductiontips.com/source/resources

 MIDI file of perfect cadence. (35)

You will find authentic cadences throughout popular music, with the most obvious placement being at the end of choruses. An example of one such authentic cadence is in 'Stand by Me' by Ben E. King. Where the last line of the chorus has a IV–V–I progression (0:40), the vocal line also resolves to the root, so I would consider it a perfect authentic cadence:

      IV   V   I

Just as long as you stand, stand by me

Plagal Cadence

In popular music, the IV–I is also used as a resolution sequence and is arguably even more common than an authentic cadence. This is called a *plagal cadence*. To my ears, a plagal cadence sounds softer and more melancholic than a V–I resolution. In the key of C major, a plagal cadence would be from F to C. Both sectional and phrasal cadences are often plagal cadences.

Further Resources   www.audioproductiontips.com/source/resources

MIDI file of plagal cadence. (36)

Plagal cadences are also frequent in popular music. An example is 'Hound Dog' by Elvis Presley with a IV–I resolution on the last line of the chorus (0:13):

      F   C

And you ain't no friend of mine

Open Cadence

Sometimes your aim between repeating phrases is *not* to completely resolve. When this is done, you will usually end on a V chord instead of a I chord. In many occurrences, the V chord is frequently preceded by a I chord. In classical music, this type of cadence is called a *half cadence*, as it occurs halfway through a phrase before fully resolving with an authentic cadence. In the case of pop music, this term is a bit misleading as a full resolution may take several repetitions or in many cases never occur; therefore, the term *open cadence* is perhaps more appropriate. In fact, pop songs that don't resolve at all or which only have a master cadence are quite a common occurrence. Most sections in pop music are made up of repeating chord progressions and, because of this, open cadences are the most frequent of all cadences in popular music. To my ears, an open cadence sounds slightly uplifting, but not yet fully resolved; your ears will want more. You will most often see open cadences as a phrasal cadence, occurring at the end of each lyrical line without fully resolving; or as a more formal half cadence, halfway through sections as a way of building suspense before giving the listener the authentic cadence. Sometimes an open cadence can also be implemented as a sectional cadence to transition into a new section and establish a soft 'change of gear'.

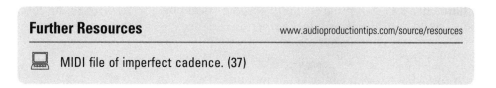

An example of an open cadence is 'Twist and Shout' by The Beatles (or originally The Isley Brothers), which ends each line on the V before repeating the I–IV–V progression.

```
        I       IV      V
```
Well shake it up baby now, (shake it up baby)

```
     I     IV      V
```
Twist and shout. (twist and shout)

Interrupted/Deceptive Cadence

Occasionally composers even like to try to convince a listener that a phrase is going to resolve when it is not. To try this, pick a chord that is normally reserved for creating tension such as a V or vii° and then move to any chord other than the tonic; this is called an *interrupted cadence*. An interrupted cadence that moves to another chord that shares two chord tones with the tonic triad is called a *deceptive cadence* (the vi chord is quite a popular destination). To my ears, an interrupted cadence sounds like a change in direction or 'curveball'. It immediately increases my interest in the music because I feel it is unpredictable. An interrupted/deceptive cadence is often a sectional cadence that is used as a special effect to establish a more jarring change of gear but occasionally can be used as a phrasal cadence halfway through a section.

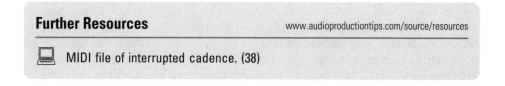

A good example of a deceptive cadence is at the end of 'Ob-La-Di, Ob-La-Da' by The Beatles. All the way through the song, there is a perfect cadence (V–I) at the end of each chorus, except for the very last chorus, which ends at 2:58, where it moves to a vi7 chord as a curveball before giving us the perfect cadence we are waiting for.

3.7.3 Using the Harmonic and Melodic Minor

We have spoken about the way that you make up chords in a major key and a natural minor key. In theory, you could create chords using the scale degrees just like we did with the major and minor scales. This would give us:

Harmonic Minor	i	ii°	III+	iv	V	VI	vii°
Melodic Minor	i	ii	III+	IV	V	vi°	vii°

However, in the context of popular music, a lot of these chords have some rather unstable intervals. The vast majority of pop musicians would find these chords rather unfriendly, but you might see these sorts of chords used in jazz (although it is rare to see a whole song in harmonic or melodic minor; they are more likely to borrow a couple of chords). A borrowed chord is defined as any chord borrowed from a parallel key or mode (i.e. any key/mode with the same tonic).

The harmonic minor and melodic minor are named with their intended functions in mind. The harmonic minor is a consideration when you are building chords. The melodic minor is a consideration when you are trying to build effective melodies, and is especially useful when you are already implementing harmonic minor in the harmony.

Back in the scales section, I mentioned that popular musicians use the harmonic minor to strengthen the resolution. By now, you will realize that I was describing a more powerful cadence. In a natural minor scale, the v–i finish does not have nearly the same sense of pay-off as a V–I. However, by not flattening the seventh degree of the scale like you normally would with a minor scale, a major V chord is created and the resulting V–i resolution is far more gratifying.

So what we tend to see in popular music is musicians composing using natural minor but borrowing the major V from harmonic minor instead of the default minor v chord. What is even more inspiring is that you don't need to do this swap every time you might decide to use the chord; you could decide to use it only when you are using the chord for the purposes of a cadence.

When it comes to writing melodies over the top of a minor progression, you can consider using an un-flattened leading tone too. When doing this, you may sometimes run into problems with an awkward interval between the sixth and the seventh degrees, in which case you can use melodic minor to give a smoother ascension than is possible with the harmonic minor scale. This is particularly useful when you are already borrowing the V chord from harmonic minor.

An example of a song that uses for its cadences a minor scale with a V chord borrowed from harmonic minor is 'House of the Rising Sun' by The Animals:

i	III	IV	VI
Am	C	D	F

I	III	V	V7
Am	C	E	E7

What is interesting about this piece is that there is also a major IV chord used. We could conceptualize this as being borrowed from melodic minor or from the parallel major of A. In the case of this piece, I prefer to think of it as being borrowed from melodic minor.

This goes to show that key-based music doesn't have to be limited to any one scale. The grounding of the key is the tonic and the leading tone that helps to reinforce this tonic further. Any note or chord that supports this tonic can be considered 'in key'. Since the A major and minor scales are based on the same tonic note, they are all elements of the key of A, as would be all the church modes with A as the root. Therefore, when composing, you could technically borrow harmonies from all of these scales and do what 'sounds right'.

However, this doesn't mean that you can start combining these elements willy-nilly; you have to use some discretion. Using too many major scale harmonies in the minor mode can dilute the minor feel of the composition, just as using too many minor scale harmonies in the major mode may dilute the major feel of the composition. The chord that is set in stone is the tonic chord along with its quality of major or minor.

3.7.4 Chord Leading

In theory, you can use any of the diatonic chords in any order to create a solid chord progression. In fact, you can even borrow chords from the parallel major/minor or modes on occasion if that is what sounds right. Culturally and harmonically, though, there are chords that lead better from one another. This concept is called *chord leading*. As a producer, I often find one chord in one of a song's progressions that doesn't 'sit right' with me emotionally. In these cases, I often take a closer look at the progression and suggest a few alternatives. I should stress that unless I have worked with a band extensively, I usually only suggest options and allow the band to choose one that works for them.

Here is a reference table of chords that 'classically' lead well in major keys:

Starting Chord	Leading Chord
I	Any chord
ii	IV, V, vii°
iii	ii, IV, vi
IV	I, iii, V, vii°
V	I
vi	ii, IV, V, I
vii°	I, iii

Let's use this chart starting with the tonic. The chart shows that the I chord can move to any other, so I will move on to ii. From ii, the table shows that you can move to IV, V or vii°. I will choose IV and then back to I. In C, this makes:

I	ii	IV	I
C	Dm	F	C

Sounds pretty nice, eh? Of course, this sort of painting by numbers method of writing music isn't how you would usually write music in the real world, but it is another tool in your box. Let's use the same method to change a chord in the progression we've just made, as this is more likely to be what a producer needs to do.

I decided that in this particular instance, the ii chord isn't working as I'd like. I know that any diatonic chord leads from I, so I need to concentrate on which chords other than ii move to IV, which are iii and vi. In C, these are E minor and A minor. I would usually suggest these chords as options to the band and ask them to try them, and we will make a decision together. In this case, the E minor sounded darker and captured the vibe we were looking for:

I	iii	IV	I
C	Em	F	C

As I previously mentioned, these guidelines are pretty old. These days, popular music has far more leeway, especially with the V chord, which often leads to vi. This same information can also be presented as a flow chart (Figure 3.34).

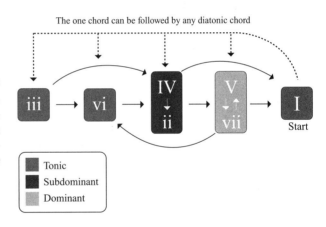

Figure 3.34 Major key chord leading flow chart[1]

The chord leading guidelines for minor keys are very similar to a major key. In the flow chart below, you will notice that the VII chord is an extra step and the dominant chords (V and vii°) are not taken from the natural minor, but are taken from the harmonic minor/parallel major to allow a stronger resolution (authentic cadence) (Figure 3.35).

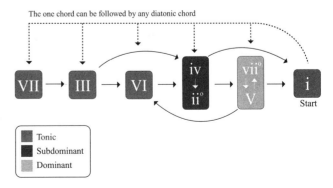

Figure 3.35 Minor key chord leading flow chart[2]

3.7.5 Chord Embellishments

So far, we have only been dealing with regular triads and the occasional diminished chord. In popular music, you will often see chords that are more colourful than a simple triad. To use these effectively, we need to get to grips with selecting appropriate chords. What I mean by this is that all the notes fit the key of the piece. Jazz and experimental musicians frequently go outside of the scale tones or use many modulations; in popular music, you will sometimes see borrowed chords used, but these are the exceptions, not the rule.

Earlier in the chapter, we outlined many of the different types of chords, but as yet we have no context for where and how to use them. As I am primarily a guitarist, I was taught a lot of shapes and chord names rather than sufficiently learning the theory underlying how chords are made up. Because of this, I used my ears and trial and error to work out what I thought worked in the context of the piece I was writing or producing. This worked well but I failed to grasp that picking suitable chord embellishments became a much simpler process when I made sure that all the notes are found in the key's scale.

These days, I've remembered a couple more patterns. However, it is important to understand the principles of why these patterns work. So before I show you the pattern, I want to give a couple of examples.

Let's spice up a IV chord in the key of C. The IV chord is an F major. This contains the notes F, A, C. We can embellish it to make it a four-note chord by adding a sixth or seventh degree as long as the note we add is found in the key of C. For now, we will concentrate on creating a sixth chord. Remember, we already know that our chord is based around a major triad so we can dismiss any chords that are based on the F minor triad. So all we need to do now is work out whether the added note of a minor 6th or major 6th would fit better over the top of our F major when playing in the key of C. To decide this, we will look at the notes found in the F major and F minor scales.

Here are the notes found in F major:

1	2	3	4	5	6	7
F	G	A	B♭	C	D	E

And F minor:

1	2	3	4	5	6	7
F	G	A♭	B♭	C	D♭	E♭

So our sixth intervals are D in F major or D♭ in D minor. To see which of these notes would be the better fit, let's reorder the notes in C major starting from the fourth degree:

1	2	3	4	5	6	7
F	G	A	B	C	D	E

This shows that the sixth we need is a D, meaning an F6 is the correct choice and the Fm♭6 is not appropriate (remember, this is not a minor 6th, which is a minor triad with a major 6th).

When dealing with sixth chords, you only tend to use them when the major 6th interval is the correct choice. This is because a flattened sixth chord can be rather unpleasant due to the close proximity of the fifth and six notes (being only a semitone apart). This means that chords I, ii, IV and V in major keys and chords III, iv, VI and VII in natural minor keys are suitable for sixth chords.

Sevenths are a little different because they are further apart from the fifth, and we therefore see both the major and minor varieties commonly. In the example we used before, the major seventh would again be the obvious choice.

Let's do another example, this time on a V chord (G major) in the key of C. Reordering the scale shows us that the seventh note we need would be F. With a seventh, we would need to use the F note over a G major chord; this would give us a G dominant seventh (G7).

Here is the working out for G7:

Reordered from the fifth degree, this makes:

1	2	3	4	5	6	7
G	A	B	C	D	E	F

This gives us the notes: G, B, D, F.

Notes in G major:

1	2	3	4	5	6	7
G	A	B	C	D	E	F♯

Notes in G minor:

1	2	3	4	5	6	7
G	A	B♭	C	D	E♭	F

Chord types:

G major 7: G, B, D, F♯

G dominant seventh is the only chord type that fits.

What this shows is the correct type of embellishment changes dependent on which chord number you are on. Here is a table you should memorize that shows the correct type of embellishment for sevenths in major keys.

Chord	Seventh Type
I, IV	Major 7th
V	Dominant 7th
ii, iii, vi	Minor 7th
vii°	Minor 7th ♭5

Here is the same table but for natural minor keys.

Chord	Seventh Type
III, VI	Major 7th
VII	Dominant 7th
i, iv, V	Minor 7th
ii°	Minor 7th ♭5

Stacking Thirds

In the examples above, we worked the chords out the long way. Later on, we are going to deal with chords containing five, six or even seven notes. With larger chords, you are far more likely to work them out by stacking intervals. Stacking thirds is the concept of creating chords by adding intervals of a third. Let's look at the interval table again:

Semitones	Tones and Semitones	Interval
0	0	Unison
1	0/1	Minor Second
2	1/0	Major Second
3	1/1	Minor Third
4	2/0	Major Third
5	2/1	Fourth
6	3/0	Diminished Fifth/Tritone
7	3/1	Fifth
8	4/0	Minor Sixth
9	4/1	Major Sixth
10	5/0	Minor Seventh
11	5/1	Major Seventh
12	6/0	Octave

Now let's explain how we can express a chord as a stack of intervals: C major triad is the notes C-E-G. The gap between C and E is two tones, otherwise known as a major third. The gap between E and G is a tone and a half, also known as a minor third. Therefore, we can express a major triad as a stack of a major third followed by a minor third. Most simple chords are a stack of thirds, an exception being a sixth chord, which is a stack of two thirds followed by a second. Table 3.5 shows how some common chords are stacked.

Soon we will discuss some of the most common chord progressions found in popular music. However, before I get to that, I want to outline a system that will help you remember much of the important information for each key that we have discussed so far, including the I, IV and V relationships of each key and the sharps/flats in each key, as well as the diatonic chords found in each key.

Table 3.5

Chord Type	Notes in C	Stack
Major triad	C-E-G	Major third, minor third
Minor triad	C-D♯-G	Minor third, major third
Major sixth	C-E-G-A	Major third, minor third, major second
Minor sixth	C-D♯-G-A	Minor third, major third, major second
Major seventh	C-E-G-B	Major third, minor third, major third
Dominant seventh	C-E-G-B♭	Major third, minor third, minor third
Minor seventh	C-D♯-G-B♭	Minor third, major third, minor third

3.8 THE CIRCLE OF FIFTHS

The *circle of fifths* is a visual representation of the relationships found within the 12 tones of the chromatic scale. It can help you to remember much of the important information that will guide your compositional/production decision-making. I wanted to present this later on in this chapter so that you could spend time working out the notes, chords and interval relationships for yourself.

The circle of fifths illustration is great for those of us with a strong preference for visual memory. The following illustration shows the circular motion in which we get back to the root in 12 steps using intervals of a fifth. Clockwise motion between adjacent blocks shows the interval of a fifth upwards and anti-clockwise shows the interval of a fifth downwards.

One more thing to note about the interval relationships shown on the circle of fifths is that it can also show you intervals of a fourth too. This is because the interval of a fourth is equal to a fifth moving in the opposite direction. For instance, moving anticlockwise around the circle you are also seeing the interval of a fourth upwards. By the same process, moving clockwise shows you either a fourth downwards or a fifth upwards.

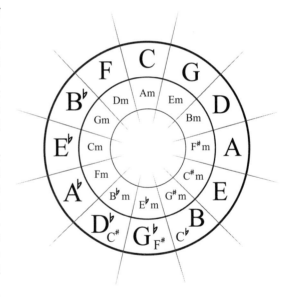

Figure 3.36 Simple circle of fifths

I will illustrate this with G major. As we know that octaves are cyclical, if we count up five degrees of the scale or down four, we end up on the same note D (remember to count the root note as 'one'):

G   A   B   C   D   E   F♯

This means the circle of fifths can show you the first, fourth and fifth relationship of any key.

The circle of fifths also shows you the diatonic chords of any key too. Let's examine this with C major (because it is at the top of the image). Here are the chords that we worked out earlier as being the ones found in C major:

I	ii	iii	IV	V	vi	vii°
C	Dm	Em	F	G	Am	Bm°

Look at the circle of fifths illustration again and you will see that all of the notes are clustered very close to each other. (Figure 3.37.)

Here is a good way to remember the pattern:

> The first six notes in a key are found contained in the boxes either side of the key's root. The final note (the diminished chord) is then found in the minor section of the adjacent box (moving clockwise).

Because the illustration groups the major and relative minors together, the same pattern works for both scale types.

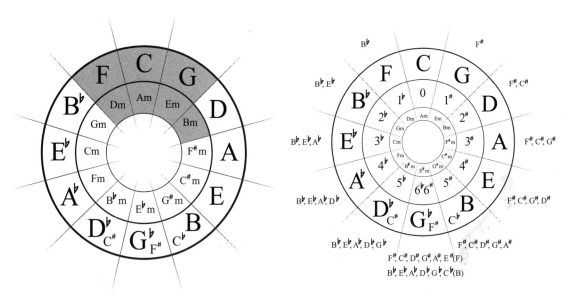

Figure 3.37 Circle of fifths highlighting chords in the key of C major or A minor

Figure 3.38 Circle of fifths with sharps and flats

I also mentioned that the circle of fifths is a great way to remember the sharps and flats found in all the scales too. To do this, we need to look at a slightly more complex version of the illustration (Figure 3.38).

Figure 3.38 shows the sharps or flats in each key. If a sharp or flat is not mentioned, then you assume that it is a natural note (no sharp or flat). For instance, G major contains one sharp (F♯); this means that the notes in the key of G are:

G, A, B, C, D, E, F♯, G

Memorizing the sharps and flats in a key this way is much simpler than trying to remember every single note from each key.

3.9 COMMON CHORD PROGRESSIONS

Chord progressions are formed largely from the composer's imagination rather than from a set of rules imposed during composition. Despite this, more often than not, you will usually find that most pop compositions will follow one of these popular progressions at least one point in the song.

3.9.1 I–IV–V Progressions

This is probably the most common progression as it hinges on the primary chords. Many songs use the I–IV–V progression or simple variations of that theme for the majority of the song, for instance:

- A really simple two chord progression is 'Imagine' by John Lennon, which is I–IV.
- With a slight increase in complexity is 'Blitzkrieg Bop' by The Ramones, using I–IV–V.
- 'Basket Case' by Green Day adds an extra measure of the tonic, being I–IV–V–I.
- Not to mention that 12-bar blues is based around an elongated form of the I–IV–V progression: I–IV–I–V–IV–I.

Further Resources www.audioproductiontips.com/source/resources

 MIDI file of I–IV–V chord progressions in C major. (39)

3.9.2 The '50s Progression

In the '40s and '50s, it became popular to take the I–IV–V and add another chord to make the journey a little more expansive, which led to the I–vi–IV–V progression being named the ''50s progression', which is evident in 'Stand by Me' by Ben E. King.

Another variant of the '50s progression is the I–vi–ii–V, which is used on 'Blue Moon', one of the most covered songs ever, including versions by Elvis Presley, The Marcels, Chris Isaak and Ella Fitzgerald.

> **Further Resources** www.audioproductiontips.com/source/resources
>
> 🖥 MIDI file of '50s progressions in C major. (40)

3.9.3 The Pop Progression

Although the I–IV–V types of progression are very popular, there is another chord progression that is arguably even more popular – so popular it is called the 'pop progression'. This is I–V–vi–IV. Examples of this chord progression are 'Don't Stop Believing' by Journey and 'Let It Be' by The Beatles – or the minor equivalent i–VI– III–VII (vi–IV–I–V) such as Eagle Eye Cherry's 'Save Tonight'.

> **Further Resources** www.audioproductiontips.com/source/resources
>
> 🖥 MIDI files of the pop progression in C major and A minor. (41)

3.9.4 Circle of Fifths Progression

As we discovered in the cadence section, the dominant chord (V) has an extremely strong pull to the tonic and so does the subdominant (IV). We can exploit this phenomenon further by moving in a circle of these perfect intervals back to the root, which gives a very even-sounding and gratifying way to get home to the I chord. The way that you can move these chords in a circular motion even influenced the creation of some of the progressions above, but we will get to that later.

Perfect Circle

First, there is a perfect circle. This is where you circle through 12 chords each by a perfect interval. After this circle is completed, you will have created a chord with each of the 12 chromatic notes as its root. Descending in fifths/ascending in fourths through the circle in C gives the following root notes:

C-F-B♭-E♭-A♭-D♭-G♭-B-E-A-D-G-C

You very rarely see a perfect circle in any genre of music because a 12-chord progression is extremely long. What you are much more likely to see is a *diatonic circle*.

Diatonic Circle

Remember that diatonic means an eight-note scale. So when we talk about a diatonic circle of fifths, we are simply talking about a circular progression that is based on a major or minor scale, rather than a chromatic (12-note) scale that the perfect circle is based on. In a diatonic circle, you will find eight chords until you return to the tonic. Diatonic circles are common in popular music, classical music and jazz.

Let's look at our scale of C major again:

I	ii	iii	IV	V	vi	vii°
C	Dm	Em	F	G	Am	Bm°

So, to do a full circle of fifths, we start by going from the I chord to the V chord, which gives us C–G. Now count five chords more to get the second fifth in our circle, which gives us D minor. Follow this pattern until you get back to C and you get:

C, G, Dm, Am, Em, Bm°, F, C

Although the Bdim to F is actually a diminished fifth movement, it will sound like a V7–IV–I progression.

Let's do the same by going in fourths, which gives us:

C, F, Bm°, Em, Am, Dm, G, C

Notice how the circle in intervals of a fifth ends with a plagal cadence and the circle of fourths ends with a perfect cadence. The type of resolution you require may give a good guide to which circle you might want to include in a section of a song.

Further Resources   www.audioproductiontips.com/source/resources

 MIDI files of a circle of fifths and fourths in C major. (42)

As well as doing full circles, you can also do part circles. To do this, simply go from the I chord to any chord you feel works in context and then follow it through until its natural conclusion back to the root. A little tip for implementing circle progressions is to work backwards from the cadence.

Here is an example of a circle progression. Let's go from chord I to vi and then progress downwards in fifths to get back to the root. This would give us a progression of: I–vi–ii–V–I. Oh wait, this should be familiar! It is the alternative version of the '50s progression we saw earlier.

This type of movement is used in jazz all the time, with jazz progressions often choosing not to start on the tonic, giving you: vi–ii–V–I.

We will now explore a progression that goes from I to IV and then uses a fifth to get to the root. This would give us a progression of: I–IV–I.

As you can see, we can create progressions of many different lengths this way. My only word of caution is that in popular music, you don't often see the use of diminished chords.

Here are two tables showing the chords required to make a circle progression in a major key after the following opening chords:

Circle of fifths:

First Two Chords	Chords in Progression
I–ii	I–ii–vi–iii–vii°–IV–I
I–iii	I–iii–vii°–IV–I
I–IV	I–IV–I
I–V	I–V–ii–vi–iii–vii°–IV–I
I–vi	I–vi–iii–vii°–IV–I
I–vii°	I–vii°–IV–I

Circle of fourths:

First Two Chords	Chords in Progression
I–ii	I–ii–V–I
I–iii	I–iii–vi–ii–V–I
I–IV	I–IV–vii°–iii–vi–ii–V–I
I–V	I–V–I
I–vi	I–vi–ii–V–I
I–vii°	I–vii°–iii–vi–ii–V–I

Notice how the diminished chord is in all but one of the progressions in the circle of fifths, whereas it is only in two progressions in a circle of fourths. For this reason, fourths are very popular with major keys in popular music.

3.10 WHY DO SONGS WITH THE SAME PROGRESSIONS SOUND DIFFERENT?

With so many songs using the same progressions, it is a wonder that they sound different. This is a testament to the effect of other compositional aspects of a song. The most obvious of these factors is that you don't have to stay on each chord for an equal amount of time. Typically, a chord progression in popular music works in phrases of two, four or eight bars. The guideline for the length of time each of these chords is played before changing also works in even-numbered patterns; it is usually half a measure, a full measure or two measures long. This is not surprising given that popular music is very fond of 4/4 time. Four beats to the bar is also mirrored to create four bars to the phrase.

For instance, a I–IV–V pattern that stays on each chord for a full measure will sound very different to a I–IV–V pattern where you stay on the I for one measure and the IV and V for half a measure each.

> **Further Resources**   www.audioproductiontips.com/source/resources
>
> MIDI files of a I–IV–V progression with full measures for each chord, and a I–IV–V progression with a full measure for the first chord and half a measure for IV–V chords. (43)

Although there are many factors or gimmicks that help to give a song its independence, here is a list of arguably the most crucial:

- key;
- other rhythmic detail (syncopation);
- arrangement;
- melody;
- lyrical content;
- chord extensions; and
- chord inversions (this will be covered in Chapter 5: Advanced Music Theory).

EXERCISES

1. Listen to the MIDI progression then move them to different keys.
2. Listen to some of your favourite songs and work out the chord progressions used in the choruses of these songs.
3. Write a chord progression using the I–IV–V, '50s and pop progressions as guides.
4. Write a circle progression in a major key using intervals of a fifth.
5. Write a circle progression in a minor key using intervals of a fourth.
6. Write a chord progression only using the chord leading reference table.
7. Once you have written these progressions, vary the tempo, time signature and virtual instrument selection. How do these changes affect the emotion you feel?
8. Now embellish the progressions using sixth and seventh chords. How does this change the feeling?

NOTES

1. Chord progressions in a major key, accessed June 2014, http://globalguitarnetwork.com/chord-progressions-major-key/.
2. Chord progressions in a minor key, accessed June 2014, http://globalguitarnetwork.com/chord-progressions-in-a-minor-key/.

BIBLIOGRAPHY

Michael Miller, *The Complete Idiot's Guide to Music Theory* (Lifestyle Paperback, 2005).

SONG STRUCTURE, LYRICS AND MELODY

So far, we have covered the musical system and some basic music theory that explains how phrases/sections of a pop song function. The previous chapter was the groundwork that allows me to spend this chapter explaining how a producer might think about a track as a whole, viewing it as the sum of its parts. We will now begin to think of a track in terms of its overall journey as it progresses, rather than viewing it under the microscope of a single progression.

So what better way to start than by discussing the different sections found in a pop song and how they function in relation to each other.

4.1 SONG STRUCTURE

We know that a song has many different elements: melody, rhythm, chords and chord progressions, but in most songs there are a number of sections, each with a different task to achieve.

Typically, each unique section will use differing chord progressions. Sometimes these can be slight alterations; at others, they can be more drastic. In some occurrences, a change in emotion is achieved using trickery such as the layering of instruments, dynamics (loud or quiet, punchy or sustained) or a change in tempo rather than a change in chord progression. The simplest songs use the same chords all the way through and are in fact a series of repetitions of one section. 'Blowin' in the Wind' by Bob Dylan is an example of this. Other songs will have distinct sections, but still use the same chord progression all the way through.

'Creep' by Radiohead is a great example of a song that uses dynamics, layering and vocal melody to differentiate between otherwise very similar sections.

As well as these differences, the various genres of music have established different precedents for how they structure a composition. For example:

Rock, pop and folk, and its derivatives, tend to pick one or two chord progressions and/or melodies (verse and chorus) and repeat them; variety is added by alteration of phrasing and lyrical content.

Electronic dance music (EDM) is similar to rock and pop in the sense that it heavily revolves around repeating sections. However, many types of EDM focus primarily on dynamically building up to the 'drop', which is where the whole arrangement's layering comes together at the point where the main hook (chorus) is starting.

Blues music tends to function over I–IV–V chord progressions with two melodic phrase types played over the top. These are the root, subdominant and dominant chords of the key, which have a strong association with the key. In the key of A, these chords are A, D and E, respectively. Often the chord structure behind the last section contains more complex chord changes, which add tension before the section resolves; this is called a turnaround. Blues is also known for long and expressive instrumental sections, particularly on electric guitar.

An example of the 12-bar blues progression is 'Sweet Home Chicago' by Robert Johnson, which is:

I	I	I	I
IV	IV	I	I
V	IV	I	V

There are many variations on these types of progressions such as quick change, bebop and minor blues.

Jazz music uses the term 'lead' or 'head' to refer its main melody, which is then embellished and altered through improvisation. Jazz is also known to use lots of unorthodox, rich and complex chords.

In Western classical music, composers like to establish a main melody known as a theme and then create variations of that core theme. Western classical music is known for its layering of melody, complex modulation between passages and a large dynamic range.

Obviously, these explanations are *extremely* generalized and simplified. Each of these genres has many nuances and complexities. I've only touched on a few of them in order to give you a basic idea of some of the core elements that give a genre its unique structure. Your job now is to listen to and dissect as much music across as many different genres as you can. Take time reverse-engineering these songs from a compositional point of view, as well as from an audio engineering viewpoint. Later on, I will present some exercises that will help build up these analytical skills. For now, we will break down some of the specific structural sections you will find in pop/rock tracks and also look at the ways you can put them together to create a full song.

4.1.1 Sections in Popular Music

First, let's start with the most common types of section found in popular music today.

Verse

A *verse* gives an insight into the meaning and emotion of the song. There are usually several verses in a song that use similar chords, layering and feel but vary in phrasing and lyrical content. Some songs have no chorus, in which case they contain a refrain.

Refrain

A *refrain* is a set of lines at the beginning and/or end of each verse that have particular importance to the message of the song. The refrain usually consists of the title of the song. A refrain should feel like part of a verse section, rather than being a distinct section. A song will feature either a refrain or a special type of refrain called a chorus.

Chorus

A *chorus* is a special kind of refrain because the melody and/or harmony feel distinctly different to the verse section; therefore, it should be classed as its own section. A chorus is most frequently used directly after a verse and is usually where the main hook of the music is found. The chorus is repeated in several places of a pop song. Choruses are the densest and most dynamically explosive parts of a song. Most of the time, they feature a prominent message that is given extra emphasis by the lead melody; many times, this message is the song's title.

Bridge/Middle Eight

A *bridge* or *middle eight* is an alternative section used to break up an arrangement and give a fresh perspective. It could feature a different groove, tempo, key or a change of instrumentation. It is often called a 'middle eight' in verse-chorus-based songs because it is found roughly two-thirds of the way through the piece and is often eight bars long.

It is well known that listeners get bored of a simple structure like verse-chorus at around the two-minute mark; therefore, the best place for a middle eight is before the listener reaches that point.

Intro

An *intro*, as its name suggests, is an introduction to the musical piece and is usually instrumental (no lyrical content). Introductions most commonly take one of two forms:

1. Based on chord progressions found in one of the other major sections of a composition (usually a verse or chorus). An example of an intro with a verse chord sequence is 'Hotel California' by The Eagles. An example of an intro with a chorus chord sequence is 'Walk This Way' by Aerosmith.
2. Sometimes intros consist of only drums or percussion, like on 'We Will Rock You' by Queen. Other times the intro builds from drums to other instruments like on 'When the Levee Breaks' by Led Zeppelin.

Despite these common forms, there are very few clear rules with intros. Sometimes you will find unique riffs/chords that appear nowhere else, such as in the song 'I Want to Hold Your Hand' by The Beatles or a vocal intro like 'Carry on Wayward Son' by Kansas.

It is rare to find a track without an intro, but there are a few. 'Hey Jude' by The Beatles is probably the most notable example.

Outro/Coda

The *outro* or *coda* is the conclusion of a song. Not every song has a specific outro; some finish on another common part of a song, usually a chorus. Some songs have instrumental outros, particularly guitar solos, like on 'Comfortably Numb' by Pink Floyd or 'Freebird' by Lynyrd Skynyrd. Some songs use a repeating outro section that is unique to the end of the piece. Some outros consist of a repeating chorus, often with ad-libbed parts and a fade out. A good example of a coda is the extended outro of The Beatles song 'Hey Jude' or 'Karma Police' by Radiohead.

Pre-Chorus

A *pre-chorus* is usually a one- or two-phrase section immediately before a chorus to help build anticipation. It usually has a stronger melody than the verses but not as strong as the chorus itself. When the chorus hits, it increases the pay-off. Pre-choruses are often confused with bridges, but due to the fact that a verse-chorus-structured song may also have a middle eight, it is best to define them as pre-choruses.

Solo/Instrumental

A solo is a section of a song where an accompanying instrument temporarily takes the focus. It will often show off the skills of the instrumentalist and often elaborates on a core melody or motif. Solo sections are often found in rock music, where lead guitarists show off their virtuosity. Functionally, a solo section serves a similar purpose to a middle eight (i.e. it breaks repetition and/or increases interest). Just like the bridge, a solo will often be placed after the second chorus. Alternatively, an instrumental break, similar in sound to the intro, could be inserted here. The instrumental is another common place to change groove or key. As previously mentioned, a solo is sometimes also found as part of a track's outro or in combination with the final chorus.

4.1.2 Popular Music Forms

Now that we are aware of the different building blocks we can put together to create a finished song, we will now look at the common forms that songs take. Before we start, these forms are lettered in two ways:

1. By passage, to describe the flow and construction of phrases within a specific section of a song (e.g. a verse of a 12-bar blues track would be AAB).
2. By piece, to describe the grouping of sections throughout a whole track. For instance, a verse-chorus-verse-chorus-bridge-chorus would be ABABCB. These sections are usually eight bars long each.

Because of this, you need to be clear on what type of situation the lettering is describing. In classical music, there are many ways that these forms can interact and move between sections/motifs; they are given rather complicated names such as *ternary* or *rondo*. In popular music, you don't really need to be aware of them, as most tracks will follow one of a few more predictable patterns.

When dealing with these letters, they are assigned by where the phrase/section first appears. This means that a bridge in a verse-chorus-verse-chorus-bridge-chorus structure would be the letter C. In a verse-verse-bridge-verse structure, the letter for the bridge would be B. In terms of musical structure, what makes a verse-bridge (AB) structure different to a verse-chorus (AB) structure is that the chorus carries the refrain of the piece whereas a bridge doesn't.

This kind of analysis covers the repeating sections of a composition so sections that only appear once such as introductions or codas are usually omitted.

AAB Passage Form

The standard 12-bar structure used in blues is based on an AAB pattern. While the other song forms we will be discussing describe the overall structure of the song, in this instance AAB expresses the structure of an individual verse. This means that each section is four bars long. Other blues AAB forms can follow an eight-, 16- or 24-bar form.

An example of an AAB structure in 12-bar blues is 'Sweet Home Chicago' by Robert Johnson.

An example of an AAB structure in 16-bar blues is 'Hoochie Coochie Man' by Muddy Waters. In 'Hoochie Coochie Man', the 16 bars comprise the verse and chorus, each eight bars long.

AAA Piece Form

The simplest structure found in popular music is a one-part song form. It consists of many verses and doesn't contain bridges or choruses. If there is an intro, it will be an instrumental version of a verse. They are most commonly found in folk and country music.

An example of an AAA form is 'Blowin' in the Wind' by Bob Dylan. In this example, the refrain focuses around the lyrics that occur at the end of each verse:

> 'The answer my friend is blowin' in the wind / The answer is blowin' in the wind.'

Another notable example of a one-part form is 'Bridge over Troubled Water' by Simon & Garfunkel, the refrain being:

> 'Like a bridge over troubled water / I will lay me down. / Like a bridge over troubled water / I will lay me down.'

AABA Piece Form

This type of form has two verse sections (A) and a bridge section (B) before transitioning to a final A section. This song form was very common up until the mid-1960s where it began fading in popularity compared to verse-chorus structures.

Just like with an AAA form, the verse carries the main musical context of the song, but this time a bridge section gives it musical and lyrical contrast. Again, just like an AAA form, the verse sections usually contain similar but not identical lyrical content.

An example of an AABA form is 'Somewhere over the Rainbow' by Judy Garland, which has its refrain at the start of each verse.

AABABA Piece Form

A common alternative to the AABA form is to have two bridges and an extra verse, making it AABABA. An example of this form is 'Yesterday' by The Beatles. Incidentally, due to the two bridges, it is often difficult to work out whether the structure is verse-bridge or verse-chorus. My view on this is, if the verse section seems the strongest melodically and features a strong refrain, then it is likely to be a verse-bridge configuration; if the main focus is on the other section, then it is likely to be verse-chorus. In 'Yesterday', we hear the lyric 'Yesterday' at the start and end of verses; the verse is also stronger melodically as well as reinforcing the title. Together, this makes me feel it is an AABABA structure (i.e. verse-bridge rather than verse-chorus).

AB Form

This is the common verse-chorus structure. Unlike the AAA or AABA forms, which focus on the verse, the AB form's musical focus is on the chorus. AB has been the primary form used in popular music composition since the mid to late 1960s. Good examples of the interplay in dynamics and lyrical content between verses and choruses are 'Sex on Fire' by Kings of Leon and 'Livin' on a Prayer' by Bon Jovi.

Like blues progressions, many (if not most) songs based on AB forms are longer than the typical 32-bar length found in AAA, AABA. In AB forms, verses and choruses can be any length, but they are usually an even number of bars long: four, eight, 12, 16 or 24. Songs that are simple verse-chorus structures include 'Roxanne' by The Police and 'Rock and Roll All Nite' by Kiss. However, most verse-chorus-structured songs will also include a C section in the form of a bridge.

ABC Form

If you add a bridge to the AB form, you get an ABC form. With a bridge, the C section is usually after the second chorus or after an instrumental:

verse-chorus-verse-chorus-bridge

verse-chorus-verse-chorus-instrumental-bridge

A great example of a middle eight is the 'There's no love ...' section of '(Everything I Do) I Do It for You' by Bryan Adams, which changes key by dropping a tone. This happens after the second chorus and is immediately before the guitar solo.

Pre-chorus

Another popular way to further enhance a pop composition is to transition between verse and chorus with a pre-chorus. This would make the structure verse (A) – pre-chorus (B) – chorus (C).

A great example of a pre-chorus is in the 'So I start a revolution from my bed ...' section of 'Don't Look Back in Anger' by Oasis.

By analysing your favourite tracks in this way, you will get an idea of which factors are important in creating a successful song, and you will begin to build a map in your head of how elements fit together. Later on, we will do the same process with lyrical content. You may have noted that some of your examples change key within the piece. In the advanced music theory section, we will get into some of the common places to modulate and techniques on how to do it effectively. However, before this, I want to address some concepts regarding the lead melody and lyrical content that may affect how you view the progression of a track.

EXERCISES

1. Pick 10 of your favourite songs and spend some time working out the key of the piece, the type of structure and the chord progressions found in them.
2. Now listen more specifically to each section. Make notes on:

   (a) How long does it take for the lead vocal to come in?
   (b) Is there a breakdown where the arrangement strips right back, or a middle eight where there are chord changes from the verse-chorus structure? If so, after how long does it occur?
   (c) How long does the song last?
   (d) What is the layering of the instruments like, for instance the density of tracks in verse one compared to verse two?
   (e) If it is a verse-chorus structure, how does the arrangement of the final chorus differ to the others?
   (f) Were there any tonal differences in the guitars and/or bass in the verses compared to choruses?
   (g) Are there any key changes or tempo changes in the track? If so, whereabouts are they?

4.2 LYRICS

Now that we have discussed song structure, I want to explore lyrical content. Different lyricists approach writing in different ways. Here are arguably the most obvious ways that writers approach lyric- and melody-writing:

1. They write lyrical ideas when they come to them, as notes or poems. At this stage, this should happen with no concern to how they would be implemented melodically. When it comes to turning it into a song, they will prune the ideas, while playing an instrument, with their band, or along to a pre-written 'beat'.
2. They write fresh lyrical content based on the emotion set by an already-developed rhythmic and harmonic structure.
3. They write the melody and harmony concurrently in a 'jam session'.
4. On rarer occasions, a vocalist may have some lyrics and a basic melody, and the band will work to create rhythmic and harmonic backing for that lead melody.

Regardless of the approach used in the creation of a lead vocal, what is important is that the lyrical, melodic, harmonic and rhythmic content is congruent, thereby creating a strong and unified emotional connection with the listener.

In my experience, the melodic element of a lead vocal line is often written instinctually and the active thought gets put into the lyrical meaning and the rhythmic flow of the words to make them fit the phrase. A singer is unlikely to think about the melody in terms of its notes, like how a guitarist would think about a lead break. Most singers are unlikely to think, 'Let's raise the pitch to hit a B6 at the end of the chorus.' Instead, it is usually far more emotionally driven, such as, 'Let's lift up the vocal at the end of the chorus.' The same can be said for issues with the flow of lyrics: A vocalist might describe a vocal line as 'cluttered' or 'cumbersome', rather than venture into the technicalities of musical rhythm. Therefore, when discussing changes with a vocalist, it would sometimes be unwise to express changes in lead melody or lyrical content in the way that you would speak to an instrumentalist. When you need to express a change to a vocalist, you may do better by trying to sing the alteration to him or her, or play the melody on an instrument. I often use a keyboard to work out a problematic melody line, and use the instrument to test and refine alternative melodies.

Despite the instinctual nature of writing lyrics with a melody, there are benefits to analysing and refining a melody with your knowledge of music theory. Because of this, I will devote a whole section to dealing with the working out and refinement of a lead melody, without concern for the lyrical content.

The rest of this section will explore lyrical structure, ideas and the rhythmic flow of the lyrical content.

4.2.1 Lyrical Content

In my experience, lyrical content is rarely a huge issue for a producer – songwriters and vocalists take a great deal of pride in their lyrics, and therefore the writer will usually have spent a decent amount of time refining them. Some vocalists will occasionally want to rewrite a couple of lines in the studio too, and this is usually a good excuse to help them refine. I like to ask the songwriter for a lyric sheet in pre-production so that I can try to connect emotionally with the lyrical context of the song, as well as check for potential construction issues. The vast majority of the time, lyrical content will have been written from personal experience or a strong pre-existing storyline, so knowing the context of the song before you hear it can help you adjudge the composition more holistically. In my experience, a producer is more likely to have to strip out extra syllables that are unnecessary, fine-tune the rhyming scheme or fix minor grammatical issues than drastically reshape the lyrics. However, I am going to go into more detail as you will be in a better place to critique other people's work if you have experience constructing lyrics yourself. When writing lyrical content, there are many factors to consider:

- Is the lyrical content congruent with the emotional theme established by the music?
- Does the story contained in the lyrics progress with the motion in the music? A good way to think about this is by asking whether the lyrical climax and musical climax occur simultaneously, and work backwards from there.
- Is there a hook? And is it simple enough to be hummable?
- Do the vocal lines flow sensibly? While you don't always have to stick to the same phrasing and number of syllables in a section (verses are particularly lenient), it is always a good thing to establish some sense of repeating rhythm.
- Are all the lyrics intelligible? This can be a diction issue or down to trying to cram too many lyrics into a passage.

- Are the stresses on words in a sensible place? A vocal line will have a rhythm with accented parts just like any musical element. Stresses can be added to important words to state their importance or words can be stressed because of a rhythmic flow. Putting theses stresses in sensible places can greatly affect the believability of the story.
- Are the lyrics grammatically correct? While you don't need to speak the Queen's English in a song, some grammatical concepts such as the correct use of tense and narrative mode (first person, third person) can help the flow of the song. This is not to say that you can't use different tenses or narrative modes in the same song; you just need a clear idea of how the different sections are framed and then be consistent with that plan. For instance, if the first line of the of the first verse is about an event in the present tense, it would usually be unwise to start talking about it in the past tense in the next line. However, you could decide to change to the past tense in the second verse.
- Consider the use of rhyming scheme, repetition, alliteration, similes, metaphors and adjectives. There is a fine line between dull or cliché lyricism and unnecessary complexity leaving the listener feeling disconnected to the context of the song. The best music asserts an emotion on a listener; you will need to use these poetic constructs to create interest, but too much use of them or too obvious usage can lead the listener to either be distracted by them or not feel connected to the piece emotionally.

The answers to these questions depend to some degree on the precedents set in the genre(s) that you are working in. If you have no idea how to answer the questions above, it is time to get listening critically to a lot of music and judge the course of action by other popular material. Despite the nuances of lyrical content being genre-dependent, I can outline some of the key concepts that are more universal. These include how you should approach lyrical content to engage the listener, such as: flow, accents, use of rhyme, alliteration, and also descriptive and metaphorical content.

When we are dealing with concepts to engage the listener, the most important thing to remember is:

Have enough consistency to make the listener feel comfortable but not enough for the listener to get bored.

This might seem rather vague advice, but let me elaborate further. Let's take a verse-chorus-middle eight form. To keep a listener interested, as a songwriter you would want to establish contrast between these sections by altering the lyrical content, melody, harmony or rhythm to make the song feel like a journey. However, by the same token, you would also want to keep the repeats of each section similar in lyrical flow, melody and harmony so that the listener feels familiar with the form that is already established. As a producer, you should be looking out for occasions where a songwriter has not kept consistency between sections or has not sufficiently differentiated between sections in order to keep interest.

Of course, when it comes to music, every rule has exceptions. For instance, some experimental or progressive bands will turn this on its head and want the listener to never feel completely comfortable. A good producer should always know when to throw out the rulebook.

Lyrical Modes

A song can give you a whole breadth of emotion, action or thoughts condensed into a few short minutes. It can be about anything or anyone, from many perspectives. Songwriters should avoid giving the listener mixed messages. The underpinning thought process when writing lyrics should be: What do I want the listener to feel? If you want the listener to feel sad and reminiscent, it would be unwise to use puns or comical rhymes. This incongruence would confuse the listener and make him or her unsure of how he or she should react, and therefore the lyrics would be ineffective. If the intended audience and emotion are always in the back of the writer's mind, they stand a good chance of pulling off the desired emotion(s). When evaluating lyrics as a producer, you should do the same. Most of the time, the purpose of the song will be immediately obvious. If it is not, there is no harm in asking the writer what the song is about and what the intended message is.

When analysing a song's effectiveness as a whole, I like to think about two elements:

- The *fictional mode*. This is a term borrowed from literature to categorize the theme of the story.
- The *narrative mode*. This is the point of view from which the story is told.

Let's break down these modes further.

Fictional Modes

Songs are typically categorized into five modes: history, realism, romance, fabulation and fantasy.[1]

History
Songs in the historical mode depict actual people or real events, and often contain political, social and economic commentary. Examples of historical modes in a song are:

- 'The Hurricane' by Bob Dylan. This is a protest song based on the incarceration of boxer Rubin 'Hurricane' Carter.
- 'Sunday Bloody Sunday' by U2. This is written about the Troubles in Northern Ireland, focusing on the Bloody Sunday incident in Derry where British troops shot and killed unarmed civil rights protesters.

Realism
Songs in the realistic mode observe a situation or tell a story. Songs with realism often reserve their judgement or opinions on the subject matter and rarely overtly tell a listener how he or she should be feeling; they rely on the listener feeling empathy to create emotion. Examples of realism are:

- 'Earth Song' by Michael Jackson. This deals with the environment and animal welfare.

'In the Ghetto' by Elvis Presley. This is about generational poverty concerning a boy who is born to a mother who already has more children than she can feed in Chicago.

Romance
Songs in the romantic mode dominate pop music. It can be about new, lost or jilted love. The romance mode can be about platonic love or even post-break-up animosity. It is any song that shows feelings between individuals. The majority of hit songs are romance songs of some description, and some of the biggest hits have been from movie soundtracks. Here are a couple of examples:

- 'My Heart Will Go On' by Celine Dion
- 'I Don't Want to Miss a Thing' by Aerosmith

Fabulation
Songs in the fabulation mode are skewed or unrealistic but still probable. They often have a romantic backdrop but are usually fun, comedic and not meant to be taken too seriously. Examples are:

- 'If I Had a Million Dollars' by Barenaked Ladies. The song has a hint of romanticism, but the lyrics centre on over-the-top and eccentric purchases they would make if they had a million dollars.
- 'Grenade' by Bruno Mars. The chorus of this song focuses on all the crazy things he would do for a woman he unconditionally loves.
- 'Dude (Looks Like a Lady)' by Aerosmith. A story of an effeminate male (possibly a drag queen) confused with a lady.

Fantasy
Songs in the fantasy mode are out of realms of believability. Just like with the fabulation mode, they are fun and whimsical. Songs in the fantasy mode are not as common in recent times unless written specifically for children or for use with other mediums such as film. However, The Beatles wrote several fantasy-based songs (which are rumoured to be influenced by hallucinogenic drugs). Here are a couple of examples of fantasy songs:

- 'Lucy in the Sky with Diamonds' by The Beatles. This is often considered to be about LSD. However, John Lennon claims it was inspired by his son Julian's nursery school drawing. The title of the drawing also reminded John Lennon of the works of Lewis Carroll, particularly 'Through the Looking Glass', in which Alice floats in a 'boat beneath a sunny sky'.
- 'White Rabbit' by Jefferson Airplane. This classic fantasy-mode song is also rumoured to be about hallucinogens, as well as borrowing imagery from Lewis Carroll.

Narrative Modes
Once you have subject matter and a general idea of the story, you need to think about the point of view the song is coming from. Whose voice is it? Much of the time, the fictional mode chosen will play a part in the perspective of the story;

for instance, a song about love will usually be aimed at the listener, using words such as 'I' or 'you'. There are three points of view that are found in songs: first person, second person and third person.

First Person

A first-person narrative is a point of view where the singer is speaking for and about him or herself ('I'/'me'), or the collective ('we'). Sometimes a song in the first person will also refer to a collective ('you and I'). The majority of pop songs are written in the first person, particularly the romance modes.

When songs are in the first person, the listener will tend to identify him or herself as the 'I' or 'you' in the song. Therefore, you will usually aim to create not only likeable characters that people are happy to be associated with, but also use lyrics that are vague enough for people to relate to events in their own life. When songs are in the collective 'we', the listener likes ideas of togetherness, strength and unity.

Examples of songs in the singular first person ('I') are:

'The Boxer' by Simon & Garfunkel (realistic mode)

'Yesterday' by The Beatles (romance mode)

Examples of songs in the collective first person ('you and I' or 'we') are:

'Someone Like You' by Adele ('you and I', romance mode)

'Waiting on the World to Change' by John Mayer ('we', realistic mode)

Second Person

A second-person narrative is a point of view where the singer is speaking to someone referred to as 'you', or speaking to a more communal 'you' for advice-giving or to impose philosophical ideology.

When writing in the singular 'you', it is important to make the character that is being sung to identifiable. The listener will again try to make him or herself the 'you' or make him or herself the narrator. When writing in the communal 'you', it is important to either tackle a strong ideology that will appeal to many people or conjure images of togetherness and 'unity'.

Examples of songs in the second person with use of an individual 'you' are:

'Candle in the Wind' by Elton John (romance mode)

'You're So Vain' by Carly Simon (romance mode)

Examples of songs in the second person using a communal 'you' are:

'All You Need is Love' by The Beatles (romance mode)

'You'll Never Walk Alone' by Gerry and the Pacemakers (realistic mode)

Third Person
A third-person narrative is a point of view where the singer is speaking about a character as 'he', 'she' or 'they'. The third person can be from a number of perspectives:

- *Bird's-eye view*, which can also be called *camera-eye view*, can be thought of as having an impersonal connection; this implies that the narrator is observing a character or situation without being personally involved or impacting the story.
- *Personal view*. This is where you, the narrator, have a personal connection to the person you are speaking about.

Speaking in the third person is often found in historical or realistic mode songs where the singer is concentrating on building a believable situation.

When writing in the third person, building believable situations and likeable characters is important. When this is done correctly, the listener will have empathy with the character and/or their situation.

Examples of songs in the third person are:

'Eleanor Rigby' by The Beatles (bird's-eye view, realistic)

'The Hurricane' by Bob Dylan (bird's-eye view, historical)

Shifting Points of View
When dealing with narrative perspective, it is important to maintain the same perspective to avoid confusing listeners. The same goes for tense (past, present, future). While it is not extremely common, it is possible to shift points of view or tense mid-song, but you have to lead the listener to it. This can be indicated in two ways:

1. *Sectional shift*. You can shift points of view or tense most easily when the song moves to a new section. For instance, if the verse is in the third person, it is possible to make the chorus second person.
2. *Indicated shift*. If the lyrics suggest the story is leading to another character/group being introduced, or a time period is shifting, it becomes okay to shift perspective.

An excellent example of a shifting point of view is 'Space Oddity' by David Bowie, which alternates between second person and first person, and switches at the start of a new verse. The lyrics directly lead the listener to this perspective shift. In the opening section, the second-person point of view is established by the line, 'Ground Control to Major Tom', followed by the mission control centre issuing orders and instructions before take-off in the singular 'you'. Later in the song, the perspective shifts to Major Tom (1:27) talking about his experiences in space in the first person, 'I'; this is signified by the line, 'This is Major Tom to Ground Control'.

Optimizing Lyrical Flow

The biggest issue I encounter is songwriters trying to cram too much into a phrase. This issue is often caused by the lyricist trying to make the line too descriptive or too fixed in definition, which can often skew the lyrical flow. I don't mean that you need to have ironclad consistency in the number of syllables between lines; some variation in phrasing is in fact interesting and should be accounted for. It is easily possible for the flow of phrases in a verse to feel familiar, even when some of the lines have a different number of syllables. Much of the importance of lyrical flow is not the number of syllables; it is about where accents lie.[2]

In this context, I am not talking about the nuances of regional dialect. I am talking about how specific syllables are accented in both natural speech *and* music. Often, problems can occur when the music's rhythm forces the lyrics to stray from their natural stress patterns during normal speech. This can change the meaning of the line, give the listener an uneasy feeling or make the lyrics sound forced. While the effectiveness of lyrical phrases is often felt rather than intellectualized, to explain common problems we need to be aware of three factors:

1. The *word stress*, which is the common location of the accent in each word. This can change depending on the language being spoken (or sung). In English, the stress will most often fall on the first syllable of the word. When stress is not on the first syllable, it can be in order to distinguish between two otherwise identical sounding words; for example, 'insight' and 'incite' are differentiated in speech by the stress falling on the first syllable in *in*sight and on the second syllable in in*cite*.
2. The *sentence stress*. The stressing of certain words can give context to a sentence or even assert a new meaning. Let's examine the sentence:
'He's the last man for that position.'
This sentence's meaning can be changed by placing the stress on different words. Below are a couple of examples of how it could be construed. I will signify the stressed word in italics and write its implied meaning underneath:
'He's the *last* man for that position.'
Possible meaning: There is no one else available for that position.
'He's the last *man* for that position.'
Possible meaning: He is the only male left in the running for the position.
'He's the last man for *that* position.'
Possible meaning: He is the only option for that position, but other candidates may be suitable for other roles.
3 The *musical stress*, or accents that occur due to the stressing of particular beats. As we have already discussed musical rhythm, the stressing of certain beats musically should make perfect sense, and the lyrical content is no different.

When you are trying to make lyrical patterns match, you need to be aware of all of these variables to create rhythms that give the best flow to the lead vocal. When creating vocal melodies, you don't have to be stuck with one rhythm that you use on every line – implementing different note lengths can help to create interest, align accents or increase/decrease momentum.

Let's give a two-line example that illustrates many of these points:

'Sailing down the British coast,
to see the woman I love most'

If we were to count the syllables, the first line would have seven and the second line would have eight. If our only means of analysis was syllable count, this would probably seem like a pretty awful lyrical couplet and we would most likely decide to lose one syllable from the second line. Doing this would mean the whole line would need to be rewritten, as dropping any syllable causes the phrase to have little meaning, like the following example:

'Sailing down the British coast,
see the woman I love most'

However, it is possible to phrase this couplet in such a way that it flows very well. Let's explain how this is done:

1	&	2	&	3	&	4	&	|1	&	2	&
Sail	ing	down	the	*Brit*	ish	coast,	to	|**See**	the	wo	man
3	&	4	&	|							
I	love	most		|							

As you can see, starting the second line on the '&' before the downbeat creates alignment in accents and a rather fun and interesting flow. There is one potential problem, though: In this example, there is a stress on the word 'I' in the second line. This could allude to the fact he finds himself rather important, but if the song is about his travels, then this is probably fine. However, if the song is about his feelings or specifically a love song, not having a stress on the word 'love' is probably not the best way of phrasing the lyric. To create stress on the word 'love', you might use a triplet and an extra syllable to get the 'love' to fall on the third beat:

1	&	2	&	a	3	&	4
See	the	wo	man	I	*Love*	the	most

When dealing with rhythm, it is important to think about the song form you are creating. If we were to create a whole verse by building on the two lines we've already written, we would normally create a repeatable pattern that we would then apply to all of the verses. When we looked at song structure, we learnt that you can represent structural form by letter to outline passages or larger sections of a piece. For instance, using the two-line example we have already written, you could use the version with the triplet on the second line, and repeat this pattern to create a four-phrase verse with an ABAB pattern.

Let's now think about occasions where you need to remove content – this is usually simpler. It can be as simple as abbreviating or omitting material that isn't essential for the message. For example, if the following line is too crowded:

'the forces that pulled me into you'

this could be changed to:

'the forces that pulled me to you'

This saves one syllable without significantly changing the message.

If that is still too much, then you could even change it to:

'the force that pulled me to you'

We have lost another syllable without affecting the emotional context of the line. If you are unsure how many syllables you should remove, try looking at the example set by other lines in the song by looking at where the accents lie and the rhythm implied rather than just the number of syllables per line.

It is very difficult to give definitive guidelines for when flow is in issue. With a lot of critical listening experience, however, you should be able to feel it. Telltale signs of lyrical flow problems include:

- not being able to make out words clearly;
- not emotionally connecting with the lyrics;
- feeling like the words don't match the music; and
- being distracted from the rest of the music because you are trying to work out the lyrical message.

Some songwriters even like to portray a more conversational feel that may depart from the precedent set by most popular music, so you should judge the effectiveness of the rhythmic flow on a case-by-case basis. However, in most cases, constant changes to the flow will feel unnecessarily disjointed.

Rhyming Schemes

Rhymes are a lyricist's best friend and worst nightmare all rolled into one. When they work, they can be a fantastic tool to create interest; when they don't, they can sound cheesy, cliché or out of place very easily. Again, you should listen to plenty of hits of that genre to work out what is acceptable. For instance, in hip-hop, you are likely to get away with more than you would in a rock track.

Typically, pop and rock music favours very simple rhymes using commonly found single-syllable words, many of which use the most commonly found word in pop music: you. Some of the most popular rhymes are:

Do/You	Be/Me	Me/See	True/You	Go/Know	Through/You
Night/Right	To/You	Day/Way	Free/Me	Say/Way	Too/You

Rhymes usually occur at the end of lines (as they are consciously more memorable there), but there is no reason why you can't use rhymes within a line. Internal rhyming can add to the 'catchiness' of a line, which is a technique that can be used to make a hook even stronger.

Internal rhyming is a strong component in hip-hop music, even from early hip-hop. Multi-syllable words are harder to rhyme without them standing out or sounding try-hard, particularly when the syllable count doesn't match between them. For instance, rhyming 'delicious' with 'promiscuous' would be more difficult to implement. Despite this, more complex rhymes are again a prominent feature in hip-hop. Let's look at a line from 'Rapper's Delight' by Sugarhill Gang, which shows the use of both internal rhyming and rhyming of words with a different syllable count. Rhymed words are shown in bold:

'You **see**, I got more clothes than Muhammad **Ali** and I dress so **viciously**' (1:31)

A more subtle pop example of using rhymes that don't only occur at the end of a line is 'Hey Jude' by The Beatles:

'Hey Jude, don't make it **bad**
Take a **sad** song and make it better
Remember to let her into your **heart**
Then you can **start** to make it better'

When rhyming, you don't have to make every single line rhyme. In fact, a more relaxed rhyming scheme is often advantageous. You will often find specific line numbers within a section rhyming with each other. Therefore, you can use a letter-based form system to signify which lines rhyme. The most common section length is made up of four lines. Let's look at some of the most common rhyming schemes for this length:

AAAA
This is where every line rhymes with each other. For example:

Line one ending with: *case*
Line two ending with: *face*
Line three ending with: *chase*
Line four ending with: *trace*

It is harder to pull off without running out of ideas, or sounding cliché or monotonous.

AABB
This is known as rhyming couplets, when you rhyme the first and second lines, as well as the third and fourth:

Line one ending with: **case**
Line two ending with: **trace**
Line three ending with: *guess*
Line four ending with: *stress*

This is another fairly restrictive form and can easily become tiring.

ABAB

This form groups the first and third lines and the second and fourth:

Line one ending with: **case**
Line two ending with: *guess*
Line three ending with: **trace**
Line four ending with: *stress*

ABBA

This form is a rhyming pair wedged inside another rhyming pair:

Line one ending with: **case**
Line two ending with: *stress*
Line three ending with: *guess*
Line four ending with: **trace**

This form can break the monotony of the more standard rhyming schemes above, but is easier to make things feel disjointed.

XAXA

Now we can start including rhyming schemes where you are intentionally not rhyming lines. The lines that are not rhyming are signified by the letter X.

Line one ending with: make
Line two ending with: *case*
Line three ending with: best
Line four ending with: *trace*

The two non-rhymed lines will allow you some freedom and potentially keep the song from sounding cliché.

AXAA

In this form, only the second line doesn't rhyme:

Line one ending with: c*ase*
Line two ending with: make
Line three ending with: *face*
Line four ending with: *trace*

This form builds a bit of tension in the second line, but it quickly reverts back to normal.

AXXA

In this form, only the first and fourth lines rhyme:

Line one ending with: c*ase*
Line two ending with: make
Line three ending with: say
Line four ending with: *trace*

This form has a looser feel, and lyrical tension is built during the middle section before being resolved at the end.

When it comes to pop/rock/country rhyming schemes, the most important factor is that you stick to the established pattern for each section. If you have set a rhyming pattern in the first verse, you should honour the same pattern in subsequent verses, unless you are deviating for a feeling of uneasiness or for another specific effect. Hip-hop, on the other hand, will deviate from a set scheme more regularly or the scheme is rather irregular and difficult to ascertain in the first place. In many cases, a rapper's artistic aim is to include as many internal rhymes as possible, which helps to break the monotony because the 'beat' is in itself rather monotonous. Beat is in quotes because, in this case, I am referring to a repeating harmonic and percussive structure that a rap is written over, rather than the more traditional definition of a measurement of time in music.

Finally, before moving on, I want to discuss rhyming dictionaries. Many lyricists view them as cheating. The reality is that the more rhymes you have in your song, the more likely you will need to use rhyming dictionaries to avoid predictability. The human brain is great at finding singular-syllabled rhymes but is not the best at pattern-matching more elaborate rhymes, which you might want to implement in one or two places. I've also found that using dictionaries can help take the song's story in a more inspiring direction and also help produce imagery that the listener will appreciate. When I know a client has an aversion to dictionaries, I will rather naughtily find five minutes to check words on my smartphone and pass them off as something that 'came off the top of my head'.

Further Resources www.audioproductiontips.com/source/resources

 Rhyme Zone online rhyming dictionary. (44)

Literary Devices

Being able to make the flow of lyrics interesting rhythmically, and implementing a rhyming scheme, is often not enough to make the lyrical content engaging to the listener. Literary devices are the methods used to portray a message using words.

Unlike other creative mediums, a song has a very short time to create an emotional connection with the end user. When it comes to creating a song that is engaging, songwriters have to use techniques that help to capture the listener's imagination, which in turn will help him or her relate to the context of the song and/or link his or her own life events to the song.

A word or phrase in a song can be classified as either a denotation or connotation. A *denotation* is when a word or phrase is to be taken literally (i.e. the word 'cool' meaning 'at a fairly low temperature'). In contrast, a *connotation* is where a word or phrase invokes a broader emotion or feeling other than its literal meaning (i.e. 'cool' meaning something good, or someone nice, friendly and worthy of respect).

In general, songs will favour use of connotations for three reasons:

- Connotations can give a phrase the same meaning in fewer words than a denotation.
- Connotations create a stronger mental picture for the listener.
- Connotations leave the song open to personal interpretation (which allows people to build their own emotional connection).

A phrase that produces a strong connotation can be considered as a *figure of speech*, because it transcends the literal meaning of the phrase. There are several ways a songwriter can utilize figures of speech to help make the listener use his or her imagination. We will soon get on to how these are defined, but for now let's discuss an example of a figure of speech and illustrate how the three points above create interest.

Let's take 'Moves Like Jagger' by Maroon 5, which is based on Mick Jagger, the lead singer of The Rolling Stones, who is known for his stage show and dance moves. However, Mick Jagger is also known for being a rather prolific Lothario; the writers of the song choose to use this as a way of creating ambiguity and double meaning. Delving into the lyrics would suggest that the song has sexual connotations that the casual listener would likely be unaware of. The use of double meaning, as well as being shorter, more imaginative and catchier than 'I am great at having sex', creates more interest to the listener.

In other cases, song lyrics are even more ambiguous; therefore, true meanings are even more subtle and open to personal interpretation. For instance, 'Every Breath You Take' by The Police is about being obsessed and jealous with a lost lover, rather than a song about true love as many interpret it. 'Summer of '69' by Bryan Adams is about the sexual position, rather than the year 1969.

Now that we know the basic idea behind figures of speech, let's look at some of the most popular literary devices that can help strengthen a listener's bond to a song. Mainstream pop hits are filled with obvious examples of these literary devices.

Metaphor

A *metaphor* is a direct comparison of two things, for instance:

'Baby, you're a firework' (1:00) from 'Firework' by Katy Perry.

Simile

Simile and metaphor are often used interchangeably but there is a subtle difference. A *simile* is a type of metaphor that compares two things by using the words 'like' or 'as'. It is distinct from a metaphor because it is a more indirect and less definitive comparison:

'shine bright like a diamond' (0:02) from 'Diamond' by Rihanna.

Personification
Personification is the attribution of human traits or characteristics to objects, animals or notions:

> 'where the city sleeps' (0:35) from 'Boulevard of Broken Dreams' by Green Day.

Hyperbole
Hyperbole is the use of exaggeration for emphasis or to create effect:

> 'My loneliness is killing me' (0:59) from 'Baby One More Time' by Britney Spears.

Alliteration
Alliteration is the repetition of the same kinds of sounds at the beginning of words or in stressed syllables of a phrase:

> 'Fine, fresh, fierce, we got it unlocked' (1:13) from 'California Gurls' by Katy Perry.

This is by no means a definitive list – more options include: metonymy, euphemism, puns, irony, assonance, oxymoron, onomatopoeia and so on. If you are working with young and developing artists or writing/producing your own music, I urge you to delve further into figures of speech.

Sectional Lyrics

Now that we have discussed a few of the key techniques used by lyricists, let's look at the specific requirements for each of the different song sections commonly found in pop music.

The Hook/Chorus
The usual approach to writing lyrics starts with writing a hook. When discussing melody alone, a hook can be thought of as a strong phrase that compellingly engages the listener and is memorable. Vocal hooks do not necessarily have to give meaning; for instance, the following songs contain strong vocal hooks consisting of 'oohs', 'aahs' and 'doos':

- 'Semi Charmed Life' by Third Eye Blind contains a strong vocal hook of 'doos' (1:09).
- 'In the Meantime' by Spacehog contains a vocal hook of 'oohs' (0:35).
- 'What's Up' by 4 Non Blondes contains a vocal hook of 'heys' (1:29).

That said, the majority of vocal hooks do have meaning and also encapsulate the message of the song in an elegant way. Therefore, the hook is often the title of the song, such as:

- 'I will always love you' by Whitney Houston (0:44).
- 'Sweet home Alabama' by Lynyrd Skynyrd (1:13).

Hooks in popular music are usually reserved for important sections such as the chorus, refrain or coda. The hook is simple, reinforces the message and is usually repeated many times. A hook won't usually develop the sense of story; this is usually the job of the verse. However, a hook can be considered a summary of the meaning of the song. Many songs will have a single hook, but you sometimes find several hooks:

- 'I Just Can't Get You out of My Head' by Kylie Minogue has a 'La, la, la' hook (0:09) and the line 'I just can't get you out of my head' (0:17).
- 'One' by U2 has at least three hooks: I'd consider the lines 'One love, one life' (0:36) and its repetitions as one hook, 'Love is a temple' (2:43) as another and the 'High' (4:10) in the coda as another.

Songwriters often find it easier to start with the hook(s) because:

- The hook will cement your ideas for the message of the song, and writing verse lyrics that develop the story is often easier when you have a general message in mind.
- The hook is usually the strongest, most hummable melody. It isn't common for a verse to outshine the chorus melodically, and working backwards makes it easier to withhold the best melodic lines for the chorus.

When analysing a hook, here are some good guidelines to consider:

1. A hook is simple and effective. Does the key moment of your song encapsulate the message of the song in as few words as possible?
2. A hook is usually the high point of a song. Does the melody match that?
3. Is the melody interesting even without lyrics?
4. Can you repeat the hook? The strongest hooks will usually be ones that you can repeat throughout each chorus of the song and in codas.

The Verse
Now we have considered the hook(s) that will dictate the flow of the chorus, let's look into the process of writing a verse. The job of the verse is to develop the story and give context to the hook, so that not only does the listener expect a hook, but it also makes sense.

Because the verses are about development, this gives them a greater scope. They are often much more lyrically elaborate and rhythmically complex than the chorus/hook. Each verse can be its own micro-story or a writer can build and develop a master story through the verses of a song. Once a theme and/or hook is written, the writer will likely have some ideas for the story, but as a producer you should be able to listen to the verses and feel that the story has developed enough to answer some questions concerning the theme: who, what, where, when, why, which feeling, or how?

Let's break down 'Can't Stand Losing You' by The Police to show how each verse can present new information and progress a storyline:

Verse 1: Indicates a hostile break-up with the guy (the narrator) who is trying to get forgiveness and reconciliation. The last line of the verse indicates that he isn't prepared to carry on with the situation as it is.

Verse 2: Shows explicitly that the woman is angry. This is shown by her actions of 'sending his letters back' and 'scratching his LPs'. The last line confirms that his thoughts of 'not carrying on' are actually darker thoughts.

Verse 3: The narrator's thoughts turn to animosity and desperation with him openly threatening to kill himself. This could be interpreted in a few ways:

1. an action of spite;
2. feeling trapped because he has no other ways of showing his feelings; and
3. they are empty threats in the hope she will take him back.

Middle Eight

As the middle eight is a change of pace or gear, the lyrics should reflect this. A middle eight typically contains new lyrics and a fresh perspective.

For instance, the middle eight of 'Suspicious Minds' by Elvis Presley (1:46) changes time signature from 4/4 to 6/8, the instrumentation drops and the emotion portrayed in the lyrics changes from being centred around the feeling of 'not being able to go on like this' to 'I don't want us to end'.

Coda

By the latter part of the song, the hook might be a little tiresome. By the time a coda comes around, you may want to create an entirely new hook, such as in 'Hey Jude' by The Beatles (3:08), 'Under the Bridge' by Red Hot Chili Peppers (3:14) or 'Karma Police' by Radiohead (2:34).

Alternatively, it is a classic trick to take the main chorus chord progression and ad-lib lyrically or melodically over the top of it in conjunction with a fade-out. An example of this is 'Black' by Pearl Jam (3:20).

EXERCISES

1. Take the favourite songs you analysed the structure of earlier and now focus on their lyrical content.

2. Listen specifically and make notes on the following:

   (a) Does the lyrical content match the music's mood?
   (b) What are the point of view, narrative mode and tense of the song?
   (c) Do the accents line up naturally? Has there been any 'artistic licence' taken to make phrases fit?
   (d) Is there a rhyming scheme implemented? If so, which one(s)?
   (e) What types of lyrical device are used to create interest?
   (f) Where is the hook located and does it contain the song title?
   (g) If there is a middle eight, does it change perspective? If so, how?
   (h) Is there a coda? If so, does the lyrical content change?

4.3 WORKING WITH LEAD MELODY

In this section, I will outline some practical tips on how to work with melody. These concepts will be discussed from two perspectives:

1. When you have a lead melody, but you need to create harmonic accompaniment.
2. When you already have a harmonic structure and are creating or altering a lead melody.

4.3.1 Fitting Harmony to an Existing Melody

Major	Relative Minor	Number of Sharps/Flats
C	Am	0
G	Em	1♯
D	Bm	2♯
A	F♯m	3♯
E	C♯m	4♯
B	G♯m	5♯/7♭
F♯/G♭	E♭/G♯m	6♯/6♭
C♯/D♭	B♭m	7♯/5♭
A♭	Fm	4♭
E♭	Cm	3♭
B♭	G	2♭
F	D	1♭

The first step when trying to write harmonic content is to try to establish the key of the piece. Most of the time, a band will be able to tell you this, as well as any chords they already have. This will narrow down your decision-making process, but sometimes you might be dealing with rappers or untrained vocalists who write to a pre-existing 'beat' and this information may not be known. Because of this, we will assume you have a melody on its own.

When I am deciphering a key, I will usually start by analysing roughly how many sharps/flats I am seeing in the melody. This will help you eliminate some unlikely options for the key signatures. You can see the number of sharps and flats on the extended circle of fifths illustration.

To refresh yourself, here is a quick list of the number of sharps and flats in different keys.

Once you have worked out some potential keys, you can use the specific notes found in the melody and the order that they are utilized to give you further clues to the key. If you have memorized or consulted the circle of fifths, you should be able to work out the notes in any major or natural minor key or adapt the scales to create any harmonic/melodic minor or church mode.

Let's do an example; for now, I'll list all the notes in the major keys to help you in Table 4.1.

Let's assume the melody has a lot of white (piano) notes and a couple of occurrences of F♯; the most likely key signature is G or Em (the relative minor of G). Conceivably, it could also be a D (Bm), where the leading tone of C♯ (2nd in Bm) hasn't been used yet. Other keys with more than two sharps/flats would be much more unlikely.

As well as working out the notes found in the melody, you can use the melody's phrasing, the placement of accents, 'points of rest', the point where the melody 'returns home' or the conclusion of the 'melodic thought' to give you further guidance to the tonic note. For instance, if you hum 'Happy Birthday', you should 'feel' which note is the tonic, and from the use of the other notes we should be able to deduce the scale type (i.e. major, minor or modal). With a new melody, you should be able to come up with an idea of the tonic and work with a couple of options regarding scale type, then find the one that works best.

Table 4.1

Scale	1	2	3	4	5	6	7
A Major	A	B	C#	D	E	F#	G#
B♭ Major	B♭	C	D	E♭	F	G	A
B Major	B	C#	D#	E	F#	G#	A#
C Major	C	D	E	F	G	A	B
C# Major	C#	D#	E#	F#	G#	B# (C)	C#
D♭ Major	D♭	E♭	F	G♭	A♭	B♭	C
D Major	D	E	F#	G	A	B	C#
E♭ Major	E♭	F	G	A♭	B♭	C	D
E Major	E	F#	G#	A	B	C#	D#
F Major	F	G	A	B♭	C	D	E
F# Major	F#	G#	A#	B	C#	D#	E# (F)
G♭ Major	G♭	A♭	B♭	C♭	D♭	E♭	F
G Major	G	A	B	C	D	E	F#
A♭ Major	A♭	B♭	C	D♭	E♭	F	G

Once I am confident of the key, I will usually work out some chords using one or more of the following techniques.

Trial and Error

Use trial and error within the key to work out a suitable bassline to accompany the melody. I will then use the bass note as the root of a chord. So, for instance, if we have a melody in the key of C and the bassline is going C-G-A-F, this would mean the obvious chords to choose would be C, G, Am and F. For the purpose of helping to construct chords, writing a basic bassline will help to reduce your options before working out chord types and qualities.

Closely Inspect the Notes

Use the specific notes contained within measures to further reduce options. Let's take a bar of a song that we have deduced to be in C major, which contains the following notes: D-E-G. There isn't a simple triad that contains all three of these notes; therefore, we might want to look at 6th and 7th chords that use these three notes. A potential choice could be Em7 (E, G, B, D), which works but doesn't feel as uplifting as I'd like. Ordered differently, these exact notes can create a G6 (G, B, D, E) chord and I get a more uplifting feeling.

If in Doubt, Try the I, IV or V Chord

If you are not sure of a chord to put under a section, you will almost certainly be able to place at least one from the I-IV-V sequence. The reason that one of these chords seems to always work is that you find *every* diatonic note of the key within these triads.

Again, let's use C major:

C   D   E   F   G   A   B

I chord: C major (C, E, G)

IV chord: F major (F, A, C)

V chord: G major (G, B, D)

See that every note has been accounted for at least once. Now let's take our bar containing a D, E, G melody from the example before. Looking at the notes, we are unlikely to have success with the IV chord as it contains none of our desired notes. However, the I chord and the V contains two (E, G and D, G, respectively). In theory and practice, both of these chords worked, but they each gave me a very different vibe:

- The C major chord gave me a very rich and 'interested' chord. It would be great for a brief amount of time but I felt it would quickly become tiresome.
- The G major chord underpinned the melody very nicely. Not only did it start the measure strongly with the D note of the melody being the fifth of G major chord, it also ended nicely with the melody joining the chord on G. The E note in the middle of the melody adds texture, which gives the whole arrangement a hint of G6.

4.3.2 Fitting Melody to Existing Harmony

It is often the case that fitting melody to an already-established chord structure comes more naturally to people (myself included). This process is epitomized by rappers and pop stars who like to find pre-programmed tracks (or 'beats') to write lyrical content around. I find that with a chord progression already in place, my imagination starts to automatically find a melody line and all I have to do is follow my emotions and not have to think too hard about it.

The drawback to writing melody to an already established chord structure is that you can easily settle into a melody that is 'boring' or 'forced' compared to when you compose by adding chords to an established melody. When I theorize why this might be, I always come to the same conclusions:

(a) When your mind is free to create a melody, it has to be interesting in itself, without the harmony, but the other way around you have some pre-existing 'interest', that makes you not work as hard with the melody line.
(b) Musicians often use lots of trial and error with chord selection over the top of the melody, which usually results in a more interesting progression. When writing the other way around, musicians often settle for the first melody that pops in their head.
(c) When lyrics or melody are written first, the song already has a strong journey or story to be told. If an emotion is already set lyrically, the chords are easier to match to that.

Of course, these points are huge generalizations, but they are just some potential 'pitfalls' to bear in mind when writing or analysing melody that was written to an already-established harmony. The key point to remember is that you do not need to settle on the first melody line that comes into your head. There is no reason why you can't try a number of options like you would routinely do with chords.

When producing, any pre-existing melodic or lyrical element that I don't like just jumps out at me. (If nothing offends me, however, that doesn't necessarily mean it is perfect – sometimes I just get bored and switch off, which is equally ineffective.) It is like my ear can detect incongruence between the emotion being set with the chords and the one being set with the melody, as well as clashes between lyrical meaning and the emotion set by the music.

Dealing with these issues might be immediately obvious and involve a simple fix such as:

- changing the lyric or note to fit the emotion;
- moving a certain note up a step to fit the chosen scale; and
- reducing rhythmic complexity by removing words or syllables.

When dealing with a melody line that has more complex problems, I like to work out the melody on a keyboard and play around with alterations without being distracted by the lyrical elements. I start by mapping out the starting note in the melody over each chord change. Humming over the top of the progression and matching the piano note to your voice is a good way to get used to doing this.

The starting note for each chord will almost always be part of the chord found in the harmony, and these are called *chord tones*. From there, it is usually fairly simple to join the dots and work out the rest of the current melody. Sometimes these extra notes will be found in the scale but not in the current chord, and these are called *non-chord tones*.

Once the melody is isolated from the lyrical content (and possibly from a pitch-challenged singer), it will be easier to hear problems and/or develop new ideas. Having the songwriter and/or vocalist in the room while doing this should get their creative juices flowing and, between you, you should be able to create something better.

If you don't get that flash of inspiration yet or you have producer's/writer's block, then here are some things to bear in mind:

- *Keep it simple.* Lead melody lines in popular music are 'hummable' and get stuck in people's heads. Creating something overly complicated negates this.
- *Find the hook.* The hook of the song should be immediately obvious and is usually found in the chorus and/or coda. If it isn't, then you probably need to rework it. Again, remember, keep it simple – the hook is usually *very* simple and memorable.
- *Variations.* A pop song is built around familiarity. However, too much similarity is boring. If you have sections with many repetitions, try slightly adapting the melody or rhythm between certain lines. These variations should still align accents, keep to the rhyming scheme, and be similar enough to be recognized as an interesting variation rather than being confused with a new section entirely.

- *'Go home'.* To give that sense of stability, your melodies should feel like they have a tonal centre. This is a note that the melody focuses around. It is usually one of the key's first, third or fifth scale degrees, and a phrase will usually end with your tonal centre (it normally starts there too).
- *Pay attention to cadences.* When you are building a melody, you want to build tension and resolve it at the end of a section. This means that you should end up 'home' (i.e. your tonal centre). Therefore, you can work out the melody for the cadence first and work backwards to make it lead to it emotionally.
- *Think of a phrase in two parts.* In pop songs, you often establish a repeating phrase using an imperfect cadence to set up tension that would be resolved at the end of the section.
- *Think in two, four, eight and sixteen.* When writing melodies, you will usually create sections that are two, four, eight and sixteen bars long. Due to the fact that most popular music is in 4/4 time, melodies and chord changes that are divisible by two are more 'stable' and 'comfortable' than other timings.
- *Stay in key.* When writing melodies, you normally should stay within the key you are in.
- *Establish favoured notes.* When writing melodies, people like stability. Sometimes reducing the notes within a key further creates a stronger emotional connection. Many songwriters write melodies with the *pentatonic scale* ('penta' meaning 'five'). The pentatonic scale uses five notes of the scale: the first, second, third, fifth and sixth degrees (C, D, E, G, A in C major). The fourth and seventh degrees are omitted as they can often create harmonic tension. Sometimes composers use the pentatonic scale for the bulk of the melody and then bring in the fourth and seventh degrees during cadences for the sake of building tension.
- *Give the melody motion.* Melodies often lift you up because they sonically rise in pitch and velocity through the phase.
- *Join the chords.* A melody often leads the chords to their new position. For instance, a chord progression going from C to G might be supplemented with a melody that leads the listener softly to the next chord, for example a melody of C-C-D-B with the B landing on the change to the G chord (B is the third of G major).
- *Keep non-chord tones brief and preferably between two chord tones.* A *passing note* is the presence of a non-chord tone used to connect two chord tones. A passing note has to be used between two chord tones and for a brief time, hence its name. For instance, using a C-Em chord progression and a melody of C-C-D-E, the D is a passing note used briefly to create a smooth transition between the chords.

Later on, we will discuss voice leading, which will help you to further hone the lead melody and backing vocals by thinking about their arrangement holistically. For now, I want to introduce a couple of concepts that will be elaborated on later:

- *Take small steps.* For stability, listeners like small steps of semitones and tones. When melodies make a leap, it is usually for a specific dynamic effect.

- *Make it singable.* Again, singers find it easier to sing small steps, but you also have to consider the vocal range of the client. Keeping the melody within an octave (or just over) is usually enough for even challenged singers to work with. Obviously, if you are working with Mariah Carey, it would be silly to impose such stringent constraints, so judge this on a client-by-client basis.

EXERCISES

1. Take the favourite songs you have already been working on and use an instrument to work out each melody line.
2. Compare the melodies. What are their similarities and differences?

NOTES

1. Sheila Davis, *The Craft of Lyric Writing* (Writer's Digest Books, 1986), 83.
2. Bill Pere, *Songcrafters' Coloring Book: The Essential Guide to Effective and Successful Songwriting* (Creative Songwriting Academy Press, 2009).

5

ADVANCED MUSIC THEORY

Now that we have discussed the Western musical system, basic music theory, song structure and lyrical content, we can begin to get to the music theory that discusses more advanced techniques. These concepts can further enhance a song and build upon what is already a fairly solid structure. A producer implementing some of these techniques in the right places can give a song that extra 10 per cent – an added sense of wow factor.

Before we get into these concepts, it is important to remember that these techniques should be used only where the vision deems them necessary and not just because 'you can'; everything must be in moderation. If a song's structure is strong and conjures a very strong vision, why change it? Much of a producer's job is knowing when to leave stuff alone as well as to add to it.

Much of what you will find in this section can be boiled down to three main concepts: voice leading, reharmonization and modulation:

1. *Voice leading* is the concept that describes how chords move smoothly and effectively to create interest and avoid unpleasant changes and monotony. In this section, we will discuss techniques such as passing chords, chord inversions and pedal points, which are principles derived from voice leading to help intellectualize smooth motion.
2. *Reharmonization* is the technique of taking an existing melody and altering the harmony that accompanies it. A melody will usually be reharmonized to bring extra musical interest or variety in a piece.
3. *Modulation* is the process of changing from one key to another.

5.1 VOICE LEADING

Many inexperienced writers/arrangers think of harmonic content as 'blocks of notes', which bottlenecks you into thinking very one-dimensionally. In reality, chords are much more dynamic and subtle than they may seem. There are countless different options for chord placements, voicings and colours. This paradigm is particularly prominent with guitarists, as beginners are taught to change chord by sliding a barre chord up or down the neck to the desired root pitch.

Even though most popular music functions with a prominent lead melody and a series of subservient backing instruments, this doesn't mean that the voice-leading guidelines used by early Western composers has no bearing in modern music. Fluid, smooth and harmonious flow in the music is still important today; therefore, we need to be aware of the following factors when we create harmony:

1. The pitch order of the notes in a chord significantly affects the feel and emotion created by the music. This is in both the vertical and the horizontal direction.
2. The way that notes flow is not just focused per instrument. A holistic approach can be taken to the whole arrangement to help avoid the instruments all cluttering a similar register.

Despite our remit having significantly changed, there is still a great deal of benefit in learning these age-old voice-leading principles. These guidelines are based around best practice patterns established over a great deal of time. Many of them are unchanged from the choral music (ensemble of singers) of the Baroque period (composers such as Bach).

Before we begin, I've heard a lot of people refer to voice leading as a set of rules. I personally consider them guidelines, because the validity of certain rules is dependent on the application – not to mention the fact that many 'rules' have been broken by successful songwriters/composers in hugely successful songs.

As many of these guidelines were developed in Bach's era, a lot of the observations are based around four-piece choral ensembles. Therefore, some of these practices are concerned with making sure that a choir could actually sing the compositions. When scrutinizing choirs, writers and arrangers started to notice that:

- Leaps (an interval of more than a tone) are harder to sing than steps (steps being either a semitone or tone).
- If voices cross pitches, it makes it difficult to keep track of and sing in pitch.
- If voices consistently sang at the same interval gap away from each other, it was hard to distinguish between them. This is particularly in the cases of octaves and perfect fifths.

It turns out a lot of these principles that were conceived to make a composition easier to sing are advantageous even when a work is performed on instruments that don't have the same constraints, as they aid smooth motion and avoid ambiguity between parts.

Before we examine a more comprehensive list of voice-leading practices, we should explore the following concepts:

- what progressions using four voices typically consist of;
- the classification of which intervals we find pleasing/unpleasing; and
- the way that two adjacent voices can move in relation to each other.

5.1.1 Four-Voiced Progressions

When we are dealing with four-voiced progressions, we typically think of it as one *bassline* and three *upper voices*. In choral arrangements, voices are named and their ranges restricted (from highest pitch to lowest): *soprano, alto, tenor* and *bass*.

Here are the approximate ranges of the voices (C4 = middle C on a piano).

In the days of Bach, when the voices created harmony, it was based on the tones you would find in triads. This makes it a perfect teaching tool for voice leading. Although Bach himself was a bit of a rebel and sometimes threw in a dominant seventh chord (first, major third, perfect fifth, minor seventh), that was pretty much it.

Name	Range
Soprano	C4–A5
Alto	F3–D5
Tenor	C3–G4
Bass	E2–C4

When dealing with these simple four-part structures, you have the three notes of the triad in the upper voices and then the bassline doubling one of these notes. This doubling is often the root note but we will discuss other options for doubling later. When you are dealing with a chord with four notes (such as a dominant seventh), you should use the root in the bassline and the other three notes (major third, perfect fifth and minor seventh) in the upper voices. These days, we use a variety of different chord types in popular music but these guidelines remain similar.

5.1.2 Consonance or Dissonance?

We will begin by identifying how pleasant intervals sound. The term *consonance* is used when the intervals are pleasant to the ear and the term *dissonance* is used when they are unpleasant. Please note that the context also matters; an interval or chord can be considered dissonant because within the piece it gives you an 'uneasy' feeling and makes you want resolution. Therefore, dissonant intervals are used by design and are often used quickly to create tension and to transition between notes. Due to this, consonance is often referred to as a *stable interval* and dissonance is referred to as an *unstable interval.*

Let's first look at some intervals we consider consonant:

- The perfect fifth and the octave are considered *perfect consonances*. These are the most pleasing to our ear. So are notes in unison (but notes in unison aren't technically intervals).

Further Resources   www.audioproductiontips.com/source/resources

MIDI files of perfect consonances. (45)

- The major third and sixth, as well as the minor third and sixth, are *imperfect consonances*. This means they are pleasing but not as much as perfect consonances.

Further Resources   www.audioproductiontips.com/source/resources

MIDI files of imperfect consonances. (46)

- The perfect fourth is dissonant in some cases but consonant in others. Specifically, the perfect fourth is dissonant when it is not supported by a note a perfect fifth or third lower in pitch.

> **Further Resources**   www.audioproductiontips.com/source/resources
>
> 🖥 MIDI file of perfect fourth. (47)

Now let's look at intervals we consider dissonant:

- Major 2nds, minor 7ths and major 9ths can be considered 'soft dissonances'.

> **Further Resources**   www.audioproductiontips.com/source/resources
>
> 🖥 MIDI files of soft dissonances. (48)

- Minor 2nds, major 7ths and minor 9ths are 'sharp dissonances'.

> **Further Resources**   www.audioproductiontips.com/source/resources
>
> 🖥 MIDI files of sharp dissonances. (49)

- The tritone is dissonant.

> **Further Resources**   www.audioproductiontips.com/source/resources
>
> 🖥 MIDI file of a tritone. (50)

5.1.3 Direction of Motion

There are four ways that adjacent voices can move in relation to each other:

Similar Motion

This is where the notes move in the same direction, but the intervals are changing. In other words, both voices move up, or both voices go down, but the interval between them is different.

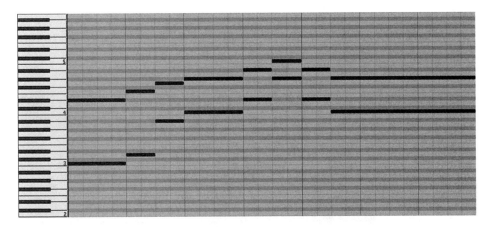

Figure 5.1 Piano roll of notes moving in similar motion

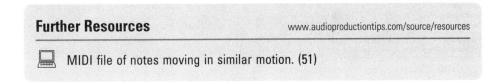

> **Further Resources** www.audioproductiontips.com/source/resources
>
> MIDI file of notes moving in similar motion. (51)

Parallel Motion

The notes move in the same direction, also keeping the same interval between them.

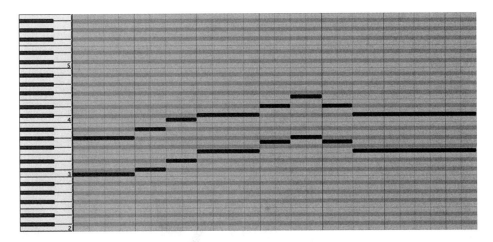

Figure 5.2 Piano roll of notes moving in parallel motion

> **Further Resources** www.audioproductiontips.com/source/resources
>
> MIDI file of notes moving in parallel motion. (52)

Contrary Motion

The notes move in opposite directions. When one of the voices moves up, the other voice moves down.

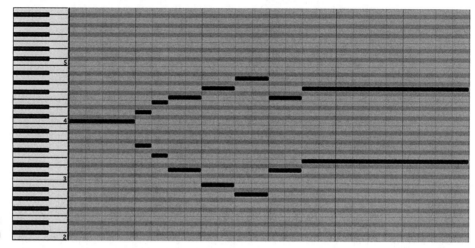

Figure 5.3
Piano roll of notes moving in contrary motion

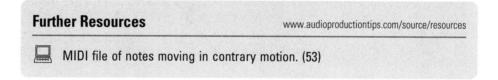

Further Resources www.audioproductiontips.com/source/resources

MIDI file of notes moving in contrary motion. (53)

If the voices always move by the same intervals (in opposite directions), it is called *strict contrary motion*.

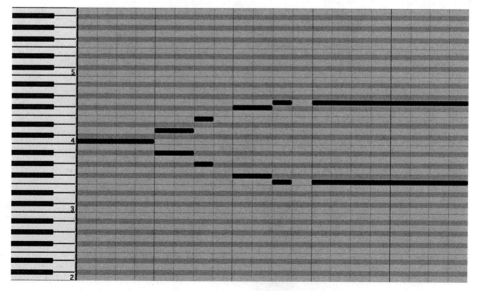

Figure 5.4
Piano roll of notes moving in strict contrary motion

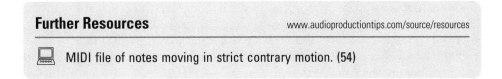

Oblique Motion

One voice moves pitch while the other voice remains at the same pitch.

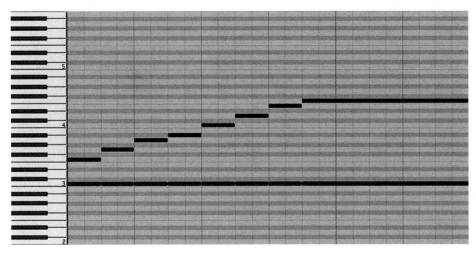

Figure 5.5 Piano roll of notes moving in oblique motion

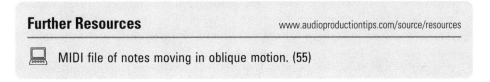

5.1.4 Voice-Leading Guidelines

Without further ado, let's run through the common vocal leading guidelines that I've used to help with popular music arrangement:

- *Pitch range.* Melodic and harmonic writing should be between F2 and G5. This region corresponds to the total pitch ranges for typical male and female voices. Writing within this range is also encouraged when writing purely instrumental pieces.
- *Number of voices.* Harmony should be written using three or more simultaneous voices. The most common harmonic writing employs the four common voice ranges: soprano, alto, tenor and bass.
- *Chord spacing.* When spacing the upper voices, no more than an octave should separate adjacent voices – for instance, no more than an octave between a soprano and alto line and no more than an octave between the alto and tenor.

The bassline is given more of a 'free role' and has no restrictions to the tenor line.

- *Avoid unisons.* No voices should be at the same pitch concurrently.
- *Move only when necessary.* Pitches that occur in consecutive chords should be retained in the same voice.
- *Move to the nearest pitch.* If a part cannot retain the same pitch in the next chord, the part should move to the nearest available pitch.
- *Avoid leaps.* Parts should move in steps wherever possible.
- No crossing. Crossing is where voices interact in such a way that it leaves a lower voice on a higher pitch than a higher voice (and vice versa) – for instance, if an alto voice is higher in pitch than the soprano at a particular point.
- *No overlapping.* As well as crossing voices, the separation between voices is hard to define when voices overlap. This is where the voices do not cross, but they move together in such a way that it leaves the lower voice passing where the upper voice was on the previous chord (or vice versa).
- *No doubling of non-chord tones including the leading note.* When doubling occurs, you should always double one of the notes from the triad. Further guidelines of which of the triad to choose will be discussed shortly.
- *No parallel movement of fifths or octaves.* Parallel movement in fifths and octaves is prohibited because it is not only hard to distinguish between tones, but it is also monotonous and cliché to our ears.

There are also guidelines on how you can use movement to make it easier to adhere to the guidelines above:

- Use contrary or oblique motion when moving between perfect consonances.
- Use any of the three motions when moving from perfect consonance to imperfect consonance.
- Use contrary or oblique motion when moving from imperfect consonance to perfect consonance.
- Use any of the three motions when moving from imperfect consonance to imperfect consonance.

These can be further refined to:

- Use contrary or oblique motion when moving between perfect consonances or when moving from imperfect consonance to perfect consonance.
- Use any of the three motions when moving between imperfect consonances or from perfect consonance to imperfect consonance.

5.1.5 When Is Voice Leading Important for Popular Music?

The principles of voice leading are important in situations where you have individual voices supporting a melody line (e.g. backing vocals). When writing backing vocals, we need to understand these concepts and ranges to not only make harmonies easier to sing but also not to fall into traps to make them boring, predictable and monotonous. In Chapter 14: Tracking Vocals, we will discuss the creation of vocal harmonies using voice leading in more detail.

Instruments that can play chords will also benefit from some forethought about the way that notes move in the most efficient way. For instance, piano parts can sound disjointed when the player is using big leaps in pitch. While playing the piano, it is important to work out ways to move your hands as little as possible. We will soon be discussing chord inversion, which is the way that you voice chords to consider voice-leading principles.

Voice-leading principles aren't always important, though, as there is a lot of layering in pop music. Much of this layering serves to function as reinforcement to the root and fifth relationships found in the chords. As well as this, if you were to strictly follow voice-leading guidelines on every instrument, it would sound very old-fashioned or even just plain wrong. Because of this, people get very confused as to where voice leading is applicable. Remember, they are only guidelines so use them *when it sounds right to you*. Over time, you will get a feel for when you are likely to apply some of these principles and when to leave a part alone.

To apply voice-leading principles on the right parts, let's discuss some common situations where voice leading is not as important in popular music. The most obvious example is genres where barre chords on the electric guitar are used prominently. In a common barre chord, there is more than one doubling of notes (as there are six strings on a guitar and only three notes in the triad). Not only that, but when you move power chords up and down the neck, you are creating parallel octaves and fifths. When you move full barre chords, it is even worse – you get three octaves of the root, and two of the fifth. So why does it sound okay? Isn't this bad voice leading? My answer to this is not a straight answer; it is down to a mixture of factors, including:

- Guitar chords are harmonically rich, especially distorted ones. The blurring of the tones makes our ear perceive it differently to purer tones. Rhythm guitar that is based on barre and power chords is pretty much just reinforcing the root triad and its strongest harmonics (octaves and fifths).
- Guitar music was born from rebellion. It derives a lot of strength and power from such a strong tonal relationship. Punk music, in particular, is famous for its almost exclusive use of power chords; it was all about 'anarchy' anyway, so going against the rules is part of its remit.
- Popular music has the lyrical element that classical music does not have; therefore, there is another dimension to give emotive context. So monotony in the harmonic reinforcement is not so frowned upon.
- Even though guitar chords are often very simple, often the other instrumentation such as strings, backing vocals and lead guitar layering can give the chord structure extra flavours to effectively imply extended chords such as sevenths, elevenths, etc.
- Culturally, we have got used to hearing rock and pop music using parallel fifths and octaves in ways we do not in classical music.

Other instruments such as synthesizer padding parts that are implemented to fill space are usually just harmonic reinforcement of the root or fifth, so are also exempt from voice-leading principles.

On the other hand, voice leading can be important on the electric guitar when using arpeggios or ascending or descending bass notes in altered chords. Therefore, in jazz, funk and fingerpicked music, the implementation of voice-leading principles becomes more frequent.

The bottom line is this, though: As long as there is a lead and/or countermelody present, parallel octaves or fifths in parts designed to reinforce the root don't feel monotonous to our ears. If there is monotony, it is more likely down to other lyrical, instrumentation or arrangement factors.

Note that this doesn't mean you should fill your tracks with reinforcement parts; this can quickly 'muddy' the arrangement. In some genres, a 'wall of sound' is expected; in others, more transparency is optimal. Let your track vision rule. In the next chapter, we will look at where and how you might layer instruments while tracking.

As well as the harmonic elements of voice leading, there are also times where you need to think about how relevant voice leading is in the lead melody. Lead melody lines in popular music will often have slightly bigger leaps in pitch than you would see in classical/choral music. This means, on the one hand, we shouldn't be as worried about leaps in the main melody, but due to this we need to be extra careful not to cross/overlap voices or spread harmony lines too wide.

5.1.6 Analysing a Full Track Using Voice-Leading Principles

Even though there are many layers in popular music, you can also reduce the parts to 'key voices' and apply voice-leading principles to them. For instance, if I have a track with drums, bass, several layers of rhythm guitar, synth pad, lead guitar, vocals and backing vocals, I might deduce:

- The drums are playing only a rhythmic role, not a melodic one, so are therefore not a concern of voice leading.
- The bass guitar is mostly filling the root note of the chords.
- The rhythm guitars are only playing root barre chords, but seem to have been double tracked. So they appear to be there as purely harmonic reinforcement of key chord tones.
- The lead guitar is often playing countermelodies, but in parts of the song it is harmonizing with the lead vocal.
- The backing vocals are harmonizing with the lead vocals, sometimes with a shared lyricism moving in contrary motion and sometimes with 'ooohs' and 'aaahs' in oblique motion.

This analysis would mean that the most important elements of the composition are the bass guitar for the bottom voice, lead guitar, the lead vocals and backing vocals. You should think carefully about how these voices are interacting with each other. This kind of analysis can be great to break a song down to its bare bones and see how strongly the core of the arrangement performs.

With experience, this sort of analysis of parts will become second nature. You will just feel which parts are muddying an arrangement and you will be able to hear potential clashes between two instruments.

5.2 CHORD INVERSIONS AND VOICING

The voice-leading principles I outlined in the last section should have made you think about:

- how chords should move in a fluid way;
- how they shouldn't move in parallel; and
- how notes between chords should only move when they have to.

In pop music, you can apply voice-leading principles to chord-based harmony to get smooth movement between chords. We have discussed that, in some contexts, this is not as important for guitars (especially electrics), but for purer-sounding instruments these principles are still of critical importance. As we have been using MIDI and the piano roll to demonstrate chords, let's use a piano as the example instrument. Its sound is pure enough to make repeated and obvious parallel movement sound formulaic, juvenile or even unpleasant. Its layout and how you operate it also means that you have a lot of scope and power to change the way you place notes to create different emotions. When dealing with chords, you can create sonic variety by inverting the chords or using open and closed chord voicing.

5.2.1 Chord Inversions

A *chord inversion* is the process of using the same notes in a different pitch order to create the desired chord.

Every chord we have built so far had the root note as the lowest pitched note; this is called *root position*. This voicing is quite common but it is not always the case. When the notes in the chord are outside of this order, it is called an *inversion*. To take a chord out of root position, you have to reorder the chord so that the root note is *not* the lowest pitch. Sometimes just changing the inversion of one or two chords in a song can make the difference between an average and an excellent arrangement.

Let's start with a C major chord in root position, which is C-E-G. To put the chord in *first inversion* we simply move the root note to be the highest pitched note by transposing it up an octave; this leaves the E and G notes in the same place but adds the C above the other two notes: E-G-C (Figure 5.6)

Now that we know the first inversion, we can work out the *second inversion*. This time, we take our first inversion chord and move the third to the top by transposing it up an octave. This gives us: G-C-E (Figure 5.7).

With triads, we only have two inversions, because, if we were to move the lowest note of the second inversion (the fifth) to the top, we would end up with the same chord voicing as we started with, only an octave higher: C-E-G.

Further Resources www.audioproductiontips.com/source/resources

 MIDI files of inversions of C major. (56)

Figure 5.6 C major first inversion in piano roll

Figure 5.7 C major second inversion in piano roll

However, chords with more notes have more inversions. For instance, a dominant seventh chord has a minor 7th on top of the triad, meaning we can have a *third inversion* before we end up back in root position.

Before we move on, it is important to mention some potential issues you may run into with chord inversions:

- What matters in a chord inversion is the bass note (i.e. the lowest pitched). The order of the other notes has no implication on the inversion. Imagine you are playing piano with both your hands; in the left hand, you are playing a C2 and a C3, and with the right hand you are playing a C major chord in

first inversion: E3-G3-C4. Despite the fact that the right hand is in first inversion, the chord perceived is still C major root position because the lowest note played underpins the chord.
- Later on in this chapter, we are going to learn about larger extended chords that go beyond an octave. Once you start adding more than four notes in a chord, inverting the chord can actually imply different chords altogether (remember, we only have 12 notes). For this reason, you typically only see chords in first, second or third inversion.

5.2.2 Chord Voicing

As well as the inversion of a chord, you have two ways of voicing the chord:

1. *Closed voicing.* This is where the notes in a chord are the closest possible pitch range. This makes the total span of the chord less than an octave even if you are using sixes and seventh chords. All the chords I've demonstrated so far are closed chords and are created by stacking the triad in intervals of thirds.

Figure 5.8 Closed C major chord in piano roll

2. *Open voicing.* This is where the distance between the notes is greater than an octave. You can create this configuration by 'skipping' the nearest intervals, so that the chord is spread out further. A closed position C major chord at C3 would contain the notes: C3-E3-G3. If we were to use the skip technique to make an open C major chord, we would start on C3 then skip the nearest interval of E3 and go to G3. We would then skip the next occurrence of a chord tone (C4, as we already have a C note) and then finish on E4, giving us: C3-G3-E4. When you have an open chord voicing, you almost always use the root note as the lowest pitch as it helps to stabilize harmony, but on rare occasions you may use the fifth instead.

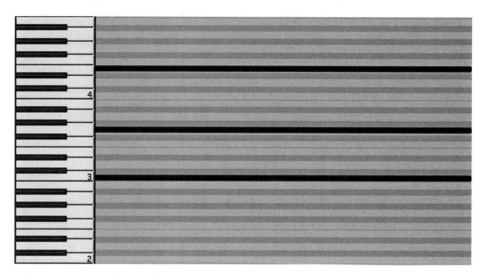

Figure 5.9 Open C major chord in piano roll

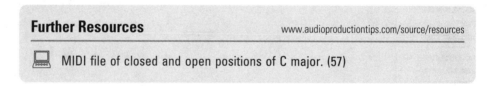

Further Resources www.audioproductiontips.com/source/resources

MIDI file of closed and open positions of C major. (57)

Unsurprisingly, open chords have a lighter sound with greater note definition. Closed chords have a tighter sound but can potentially become 'muddy' sounding. For this reason, you often choose to use open voicings in a lower register and closed voicings in the higher registers. However, remember you can also abuse these techniques for creative effect.

Now that we know chords can be inverted as well as be grouped in an open or closed voicing, we can start to implement these techniques to apply voice-leading principles. In the case of a piano, the general rule is that the less hand movement you can make, the better, and if a finger can stay on the same note, it should. One of the beautiful things about music is that when you know these 'rules', you can start to abuse them and intentionally not follow them for a creative purpose. When a pianist uses larger movements between chords, it is for a specific purpose. You have a few considerations for this:

1. *Implied melody.* When you are playing a series of chords, the movement of the top notes (highest pitched) of each chord will to some degree create an implied melody. Sometimes this implied melody is rather boring, and using alternative inversions and maybe a little more hand movement can make the music more interesting.
2. *Register.* Sometimes even just descending one or two notes lower can darken the emotion you are trying to set (or vice versa). So, sometimes you would actively choose to move the position more to keep the right kind of emotion.

3. *Creating contrast.* When moving between sections of a song, you can create an uplift by moving up to a different register altogether (e.g. moving the whole progression up an octave). You can also do the opposite and create something darker by moving down octaves.

When a piano player uses two hands to play harmony – the left to handle the bassline and the right to handle the upper voices – the note chosen to double in the bassline would usually be as follows:

1. If you are in root position, double the first.
2. If you are in first inversion, double the root or the third. Doubling the third in first inversion is often the best way to avoid parallels.
3. If you are in second inversion, double the fifth.

EXERCISES

1. Program a I-IV-V '50s progression and pop progression in root position using closed voicings. Then take the chords from root position and create smoother motion using chord inversions, bearing in mind voice-leading principles.
2. Compare the original to the new versions. Internalize the differences and similarities.
3. Repeat exercises 1 and 2, this time using open voicings.
4. Compare the results of open and closed voicings.
5. Write a melody line for each of these progressions to go along with these chords.
6. Choose your favourite of the three and note down how it makes you feel. Theorize why the chords and melody may cause this emotional response.
7. Write lyrics to the melody line that match your emotion.
8. Experiment with chord extensions over the piece to further hone the emotion portrayed.

5.3 REHARMONIZATION

Reharmonization can sometimes be a producer's get-out-of-jail-free card. It can be used in many ways to achieve variety or rescue a tired progression. Here are just a few of the common ways an existing melody and chord sequence can be reharmonized:

- The chord structure can be completely changed over the whole song. For instance, many remixers will completely reharmonize a piece to give it a completely different emotional feel to the original track. This is something a producer shouldn't have to do, unless you are involved in genres where a 'sample' vocal or hook is used to create a new composition.

- Film score composers will often reharmonize motifs to build tension, or change the piece between major and minor to portray a different emotion while maintaining comfort in familiarity. The pop equivalent of this is developing new sections such as instrumentals or codas by reharmonizing an existing section (usually a chorus).
- A producer will often reharmonize very specific points to make the arrangement more effective. This might be to replace one chord at a specific part of the song, or a chord that repeats in a specific section; either way, the reharmonization is performed to make the emotive message of the song as a whole more powerful.

Let's have a look at some of the techniques available to reharmonize a composition. Some of these concepts might be best described as changing the chord quality, rather than reharmonizing specifically. What I mean by this is that you retain the same triad but add extra notes to give it a more complex sound, for instance taking a chord from an F to an Fmaj7. True reharmonization is based on replacing chords in the composition; when you swap one chord for another, it is called a *chord substitution*. However, for simplicity's sake, I am including both of these styles of altering the emotion provided in the harmony in a single section.

5.3.1 Chord Extensions

When first looking at what you can do to spice up a chord sequence, it would be wise to think about creating colour without having to change the triad, in which case you could try sixth or seventh chords. However, you can take it a step further and use an extended chord. An *extended chord* is a chord that extends beyond the seventh into a second octave. You start with a seventh chord and then build on top of that by stacking thirds.

If your chord extends into a second octave, you can simply 'carry on counting'. This can be demonstrated with two octaves of A (Table 5.2):

The extended chords only use tones that haven't been used in the chord already, so 8th, 10th, 12th and 14th chords don't exist as they would be doubling the 1st, 3rd, 5th and 7th degrees, respectively. This means that you can only have 9ths, 11ths and 13ths.

Table 5.2

Degree	1st	2nd	3rd	4th	5th	6th	7th	8th (1)	9th (2)	10th (3)	11th (4)	12th (5)	13th (6)	14th (7)
Major	A	B	C#	D	E	F#	G#	A	B	C#	D	E	F#	G#
Minor	A	B	C	D	E	F	G	A	B	C	D	E	F	G

When you are building extended chords, you need to remember the scale degrees with major/minor character. These are:

1. the third, which is assumed major unless stated minor;
2. the seventh, which is minor unless stated as major; and
3. the sixth is assumed to be major unless it is called a flattened sixth, which you will hardly ever see in popular music.

Extended chords will usually be based on a dominant 7th chord when the triad is major, and a minor 7th chord when the triad is minor. Using A as the root, this would make the following chords:

Major: A9, A11, A13

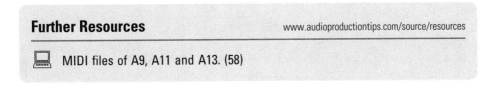

Further Resources   www.audioproductiontips.com/source/resources

MIDI files of A9, A11 and A13. (58)

Minor: Am9, Am11, Am13

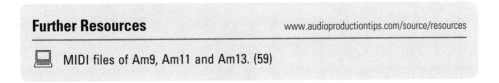

Further Resources   www.audioproductiontips.com/source/resources

MIDI files of Am9, Am11 and Am13. (59)

These large extended chords are often used in genres such as jazz where complexity and unusual tonality is commonplace. They are not seen as often in popular music but that is no reason to not try them out. Maybe you will like them in the context of the track you are working on.

5.3.2 Suspended Chords/Power Chords

Another way I adapt the chord quality is to use suspended chords. Because suspended chords switch the third of the triad to either a second or fourth degree, they sound very ambiguous (neither major or minor). Withholding this from the start of a section, in particular at the start of the song, can cause tension that is resolved when a major or minor chord confirms the piece's key. You can also use suspended chords to create tension and release by using a suspended chord right before the standard major or minor chord:

I	V	V	IV
G	Dsus4	D	C

> **Further Resources** www.audioproductiontips.com/source/resources
>
> 💻 MIDI file of the G-Dsus4-D-C progression. (60)

When using guitars, I sometimes use power chords, which have a similar effect (as they have only firsts and fifths).

5.3.3 Passing Chords

Passing notes or *passing chords* are used to step between two chords, which can aid smooth harmonic movement or briefly create tension. A passing chord can either be diatonic (using a passing chord from the key) or chromatic (not from the harmonized scale).

Passing Bassline

Often the passing chord involves a chromatic stepped movement (either a semitone or tone) of the root of the chord. For instance, descending between C and Am, you could add a passing chord of C/B, which is a C chord with a B in the bass:

I	♭I	vi	V
C	C/B	Am	G

On a piano, that could be as simple as playing a C triad in the right hand, doubling the root with the left. For the C/B chord, you would move the left-hand note to the B while still playing the C triad in the right. End with the Am triad with an A doubled in the bass.

> **Further Resources** www.audioproductiontips.com/source/resources
>
> 💻 MIDI file of C-C/B-Am progression. (61)

Chords

Passing full chords, rather than just ascending/descending basslines, is also common. Depending on the passing chord you choose, you can create brief tension or create a feeling of smooth and pleasant motion. When you are aiming for smooth and pleasant motion, you tend to use passing chords with the same quality as the previous chord that you are joining (e.g. both being seventh chords, or sixths):

I	vii°	vi	V
C	Bdim	Am	G

Chromatic movement can add a bit of tension to the sequence. Passing chords can be thought of as a way of throwing listeners a curveball and temporarily creating tension. For instance, the following progression uses a diminished chord that, due to its flattened 5th, sounds very tense:

I	♭ii°	ii7	V7
C	D♭dim	Dm7	G7

Further Resources www.audioproductiontips.com/source/resources

MIDI file of C-D♭dim-Dm7-G7 progression. (62)

Secondary Dominant Chords

As well as standard passing chords, you can use a secondary dominant chord to create motion and tension between chords in a progression. A *secondary dominant* is the fifth of the chord that you are going to next in the progression. In a major key, there is a secondary dominant chord for every chord except the I, because its secondary dominant is actually the key's primary dominant (i.e. the V chord). Using the vii° chord's secondary dominant is usually inappropriate too as the chord's diminished nature means it doesn't feel like a resolution. In a minor key, these chords would be the i and ii° chords. Again, you can work out the secondary dominant by using the circle of fifths illustration.

For instance, if I was doing a I-ii-V-I progression in C major, I could spice it up with a secondary dominant on the ii followed by a perfect cadence to resolve back to C. This gives us:

C   A   Dm   G   C

This chord can be represented in Roman numerals by a V/ii, meaning the fifth of chord two. This progression would then be: I-V/ii-ii-V-I.

I	V/ii	ii	V	I
C	A	Dm	G	C

Further Resources www.audioproductiontips.com/source/resources

MIDI file of C-A-Dm-G-C progression. (63)

Notice that even though the A major is not in the key of C, it fits; this is because of its V-I relationship with Dm. However, if you stay on A major for too long, it starts to sound wrong, hence why they are used as passing chords. Despite the interest created by the A major, the chord progression still sounds relatively vanilla. To make the secondary dominant more interesting and have a much stronger pull to the next chord, we also apply the same technique we used to increase the tension in perfect cadences, by using a dominant seventh chord. This is so powerful with the secondary dominant, that you rarely see them without it. In Roman numerals on chord two, this would be expressed as V7/ii. Using dominant seventh chords throughout the progression would give us: I-V7/ii-ii-V7-I.

I	V7/ii	ii	V7	I
C	A7	Dm	G7	C

Secondary dominants give a similar effect to a perfect cadence but within a chord progression.

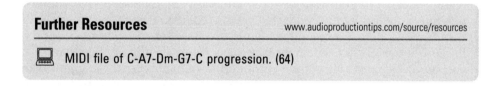

Further Resources www.audioproductiontips.com/source/resources

MIDI file of C-A7-Dm-G7-C progression. (64)

The topic of passing chords is huge. Jazz, blues and funk genres use passing chords so frequently, it is almost a prerequisite. In many cases, these genres use even more colourful passing chords with the extensions past an octave.

5.3.4 Chord Simplification

Although far more rare, you may actually want to make a progression simpler. *Chord simplification* is the opposite of chord extension. It may be removing a chord change altogether; it may be stripping a ninth chord back to a seventh or even a seventh back to a triad. This technique can work well when you want to develop contrast between repetitions or to develop a build in momentum. For instance, I could take a progression that uses a major 7th chord on every repetition and strip them out on all but the last chord of the entire section, so that the section feels like it is about to go somewhere different and therefore heightens that particular cadence:

I	V	vi	IV	
C	G	Am	F	x3
C	G	Am	Fmaj7	

> **Further Resources** www.audioproductiontips.com/source/resources
>
> 💻 MIDI files of this chord simplification. (65)

Equally well, I could decide to do the opposite and keep the Fmaj7 chord in on all but the last chord and create a differing cadence that way:

I	V	vi	IV7	
C	G	Am	Fmaj7	x3
C	G	Am	F	

> **Further Resources** www.audioproductiontips.com/source/resources
>
> 💻 MIDI file of this chord simplification. (66)

5.3.5 Diatonic Substitution

As we have discussed already, the I, IV and V chords are the primary chords in any key; therefore, it probably isn't a surprise when I mention that the ii, iii, vi and vii° are called *secondary chords*. All these secondary chords can be thought of as having a relationship with one of these primary chords. Due to this, you can directly substitute any of the chords within the same family with one other, as long as the melody permits it (i.e. it doesn't infer a dissonant interval). Here is the relationship between the primary chords and their secondary chords:

Diatonic Chord Family	Major	Minor
Tonic	I, iii, vi	i, III
Subdominant	IV, ii	iv, VI, ii°
Dominant	V, vii°	v, VII

As well as knowing which chords substitute with one another, it is important to know that substitutions work due to the common notes they share. If we look at the notes found in tonic family chords in C major and their diatonic sevenths (i.e. the type of seventh that fits the key), we will be able to see a strong correlation:

Chord Family	Triad Notes	Seventh Notes
Tonic	C (I) – C, E, G	Cmaj7 – C, E, G, B
	Em (iii) – E, G, B	Em7 – E, G, B, D
	Am (vi) – A, C, E	Am7 – A, C, E, G

What you can see here is that there is a strong correlation between these notes and their replacements. In the case of the Em, they are actually the upper three notes of a Cmaj7. This is a particularly good replacement because our minds are so used to hearing the way that chords are stacked that we can often remove a bass note entirely and to some extent our brain compensates for it.

When you substitute a primary chord with any of the secondary members of its family, you are changing its type from a major-style chord to a minor-style chord, which gives the piece a very different feel and starts to give a minor feel to the melody line. This is great to use when you need contrast. If you pay particular interest to the lyrical content of the song, you can use a diatonic substitution any time the lyrics and music are not congruent with each other.

Let's take the I-V-vi-IV progression from before and use diatonic substitution to replace the I chord on just the last repetition, to make:

I	V	vi	IV7	
C	G	Am	Fmaj7	x3
iii	V	vi	IV	
Em	G	Am	F	

This makes the last repetition uplift prior to the chorus and gives the section a sense of motion.

> **Further Resources** www.audioproductiontips.com/source/resources
>
> 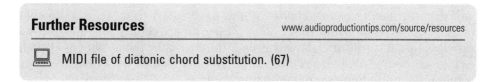 MIDI file of diatonic chord substitution. (67)

5.3.6 Relative Major/Minor Substitution

As well as diatonic substitutions, another common type of chord substitution is a relative major/minor substitution. This chord substitution is as simple as it sounds – you replace a chord with its relative major or minor equivalent. In practice, you can find this chord quickly by taking the major chord's tonic and dropping it a minor third interval (three semitones) and then make a minor chord there. From a minor to a major, it is the opposite way – go up a minor third and make it a major chord. Let's do a similar substitution as above, changing the I chord for its relative minor, a vi chord:

I	V	vi	IV7	
C	G	Am	Fmaj7	x3
vi	V	vi	IV	
Am	G	Am	F	

> **Further Resources** www.audioproductiontips.com/source/resources
>
> MIDI file of this relative minor substitution. (68)

5.3.7 Tritone Substitution

The tritone interval is the interval halfway between the octave, and it can be expressed as either an augmented fourth or diminished fifth interval. You find intervals of a tritone apart in many chords, although they are particularly important in dominant seventh chords (between the third and flattened seventh). A tritone substitution is where you take a dominant seventh chord and replace it with another dominant seventh chord with the same character notes. When I talk about character notes, I am talking about the specific intervals that distinguish a dominant seventh chord, which is – yes, you guessed it – that tritone interval between major third and minor seventh. To find the correct chord to substitute, all you need to do is to find the dominant seventh chord that has the same two character notes, just in reverse. Let's do an example in C major using a typical ii-V-I cadence that is often found in jazz-infused tracks:

I	ii	V7	I
C	Dm	G7	C

With the following chord progression, we could replace the G7 with another chord a tritone apart. Doing this would give us a chord with the same character notes, but the opposite way around (i.e. the note that was the major third is now the minor seventh, and vice versa). In the case of G7, this would give us D♭7. Let's examine the notes between the two chords to prove this:

Degree:	1st	3rd	5th	7th
G7:	G	**B**	D	**F**
D♭7:	D♭	**F**	A♭	**B**

This would make the chord progression:

I	ii	subV7/I	I

This Roman numeral means the substitution for the dominant seventh of the I chord.

C	Dm	D♭7	C

> **Further Resources** www.audioproductiontips.com/source/resources
>
> MIDI file of tritone substitution. (69)

5.3.8 Applications of Reharmonization

It is easy to see reharmonization one-dimensionally. However, it has way more flexibility than to just directly replace one chord for another. You can mix and match the different types of reharmonization (e.g. use diatonic replacement on one chord, a chord extension on another and a tritone replacement on a third chord).

As well as this, you can embellish an arrangement by using reharmonization to add more chords to the progression. For instance, let's add more chords to the simple I, V, vi, IV progression from earlier to give a more interesting sense of harmonic motion:

	I		V		vi		IV
Original:	C		G		Am		F
Adapted:	C	Am	G	G7	Em		F

> **Further Resources** — www.audioproductiontips.com/source/resources
>
> 🖳 MIDI file of progression with multiple types of reharmonization. (70)

In this example, I added a relative minor chord between the I and V chords, altered the colour with a seventh chord between the V and vi chords, and used a direct diatonic substitution instead of the vi chord. You could use the adapted progression throughout, or choose to use it to create a coda, instrumental or any new phrasing to a song. It doesn't have to stop there. I could've added some chromatic passing chords, circle of fifths or secondary dominant chords to make the motion even smoother – or even used suspended chords to create tension-resolution between chords.

The ways that you can apply these techniques are almost endless, and if you bear in mind voice-leading principles (choosing appropriate chord inversions), I'm sure you will be blown away by the results they can produce.

The methods you decide to use and how often you deem it necessary to reharmonize should be based on your vision and also what the melody allows you to do. If you create a super-cool reharmonization that the melody doesn't fit with, you could even alter the melody slightly.

There are no hard-and-fast rules here – just allow your ears and the band's thoughts and feelings to dictate the best course of action. These techniques will give you options to try, but you should never implement based on theoretical knowledge alone. If you have perfect pitch and can visualize the changes perfectly, that's great, but the vast majority of us won't know for sure until we try it. Experience and time just makes it more likely we will be right!

5.4 MODULATION

Before we move on to the factors affecting the arrangement and orchestration of a piece of music, I want to discuss using modal chord progressions – and one more compositional ace up our sleeves: modulation. As we have already outlined, modulation is the changing of key within a musical piece. I've saved it until now to discuss because the knowledge of diatonic chord substitution and secondary dominants can be extremely useful to smoothly modulate to a new key. For instance, you can use diatonic chord substitution to find a common chord between each key to help make the modulation smoother and you can use secondary dominants as a way to enforce the new key. Before we start looking at specific modulation methods, let's examine some of the terminology used. When you decide to modulate, you need to find some common ground between the keys. There are two ways in which you can find commonality:

1. *Related keys*. The two keys share a number of notes.
2. *Parallel keys*. The two keys share the same tonal centre.

Let's look at related keys first. If a key has any notes in common with another, the keys are considered to be related. The more notes in common, the more closely they are related and thus the easier it is to modulate between them. If two keys have six or more notes in common, they're called *closely related keys*. If two keys share fewer than six common notes, they are called *distantly related keys*.

To find out which keys are closely related to each other, you can use the circle of fifths. In the key of C, the closely related keys are those adjacent to the key of C on the circle of fifths. These related keys are C, F and G, plus their relative minors: Am, Dm and Em, respectively. As you already know, the relative minor (Am) shares all of the notes with C, and each of the other four keys (Dm, Em, F, G) share six notes.

As well as related keys, each key will have one parallel key; the one with the same shared tonal centre. If we were to use C major as our example again the parallel key would be C minor. The parallel keys will share only four notes with each other; despite this, modulating between the two is fairly smooth due to the shared tonic.

There are far more related and parallel keys if you also consider modes other than Ionian and Aeolian. However, this is a little out of the scope of this section.

Now that we have outlined the possible commonalities, let's discuss common ways modulation is used in popular music. The two most frequently used methods of modulation in popular music are called *direct modulation* and *pivot modulation*.

5.4.1 Direct Modulation

Direct modulation, or *phrase modulation*, is the most common type in popular music. A direct modulation is where you change key suddenly with no preparation. Direct modulation usually happens at the end of one section and start of another. You can directly modulate in any interval you please. However, it usually happens in a stepped motion (semitone or tone). Other popular types of direct modulation are switching between major and relative minor (i.e. C to Am) and switching to the parallel major and minor (C to Cm).

Let's take a I-IV-V chord progression in C major and use a direct modulation up a tone into D major:

I	IV	V	I	IV	V
C	F	G	D	G	A

Notice that after the V chord, we modulate to D major with no warning, making I, IV, V the chords D, G and A, respectively.

Popular music often favours direct modulation as it provides a more brash and obvious change of key centre and is used as a change of gear in an arrangement.

It should be noted that both parallel major/minor modulation and relative major and minor modulation sound a lot smoother than brashly moving in a stepwise motion. This is due to the fact that the parallel key shares the same tonic and the relative major/minor share the same chords.

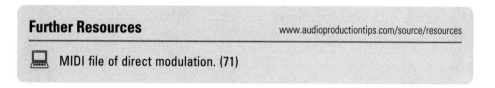

Further Resources www.audioproductiontips.com/source/resources

MIDI file of direct modulation. (71)

Here are some examples of direct modulation in popular music:

Relative Major/Minor

'Comfortably Numb' by Pink Floyd – Relative minor (Bm) in the verse and major (D) in the chorus.

'One' by U2 – Relative minor (Am) in the verse and major (C) in the chorus.

Parallel Key

'Perfect Day' by Lou Reed – Verses in minor key (B♭m) and choruses in major key (B♭).

'I'm Still Standing' by Elton John – The song's intro is in a minor key (Bm), the verses are major (B) and the choruses are again minor (Bm).

Ascending

'My Generation' by The Who – This takes the same riff and modulates it to new keys. 'My Generation' changes key three times in the final half of the song from its original key of G into A after the bass and guitar solo (1:20), then B♭ (1:50), before ending in C (2:25).

'Mandy' by Barry Manilow – In many pop ballads and show tunes, the final chorus of the song modulates up a step along with extra instrumentation and layering.

This technique is often considered too cliché for modern pop productions and is rather unaffectionately called the truck driver's gear-change. 'Mandy' moves from B♭ to C (2:46) for the final chorus.

Descending

'(Everything I Do) I Do It for You' by Bryan Adams – Direct modulation can be used to create an emotional shift in a middle eight, too, like in '(Everything I Do) I Do It for You' by Bryan Adams. This goes from C♯ to B (2:15). Key changes will most often move up in key rather than descend, but in a middle eight descents are a little more common, particularly when the lyrical content is more sombre.

'Penny Lane' by The Beatles – Just to show there are no rules, 'Penny Lane' by The Beatles actually drops a tone from C in the chorus. Don't try to play along with Penny Lane, though, as it is not in true concert pitch due to the engineer's speeding/slowing down of tape machines to be able to use parts of two takes.

5.4.2 Pivot Modulation

A direct modulation is brash, obvious and simple. Modulating using a pivot is a lot more subtle and the motion is therefore much smoother. In popular music, a key change is often used as a dramatic effect or as an obvious change of gear, so, compared to direct modulation, pivot modulation is used much less often. However, where you don't want the contrast to be so stark, it can be useful. You will also come across pivot modulation regularly in classical music and film scoring.

A *pivot chord modulation* or *common chord modulation* works by moving from one key to another via a chord that is present in both keys. After the pivot chord is introduced, you need to firmly establish the new key; this is usually done using a cadence (V-I or V7-I of the new key).

'No One but You (Only the Good Die Young)' by Queen is a good example of pivot modulation between the keys of C major and D major. This occurs between the verse and chorus, and first happens at 0:55:

I	V				
		IV	V7	I	iii	IV
C	G	A7	D	F#m7	G
No one but	yooooouuuuu		One	by	one

In this example, the G chord acts as a pivot between C and D, which is the V chord in C and a IV chord in D. The new key is established by the V7-I cadence of A7-D.

Further Resources www.audioproductiontips.com/source/resources

 MIDI file of pivot chord modulation. (72)

Although implementing the secondary dominant after the pivot chord is very popular, you don't have to use it. Here is a good example:

'Walk the Line' by Johnny Cash – This song is a series of verses containing I-IV-V chords. At the end of each verse, Cash holds either a IV or V chord for four bars, which acts as the pivot to a new key. Without the use of a secondary dominant after the pivot chord, this key change sounds more abrasive. Johnny Cash hums the note before each new verse to help establish himself in the new key.

For instance, the first verse progression is as follows:

I	V	I	V	I	IV	I	V	I	IV
									I
F	C7	F	C7	F	B♭	F	C7	F	B♭

As you can see, the B♭ acts as both a IV chord in the original key and is then established as the I chord in the new key. The next verse is then B♭-E♭-F7 which again modulates using the IV chord (E♭). Verse 3 is E♭-B♭-A♭7 but this time modulates on the V chord (B♭). Verse 4 is therefore back on B♭-E♭-F7 and, like verse 3, it modulates using the V chord (F). This makes the final verse F-B♭-C7.

In jazz and classical music, you will also sometimes find *common tone modulation*, which is where modulation occurs through two chords with a shared note (which is often also in the melody). This is not commonly used in pop music, so it is not analysed here.

Before we finish talking about modulation, I wanted to quickly point out that while a modulation is a useful change of gear in a song, not every song needs it. In fact, I'd say that the majority of songs you hear on the radio today don't have a key change. That is not to say that these songs don't have a change of gear or curveball; it's that a key change isn't your only option to achieve it. Tempo, time signature, a change of chord progression, the depth of layering and many more factors have an effect on the song's flow. So before you assume you need a modulation, think about whether that is really what you need to do.

5.5 WORKING WITH MODES

Before moving on to techniques regarding arrangement and orchestration, I want to quickly discuss a final advanced topic: working with chord progressions using the church modes.

When we discussed chord progressions in major and minor keys, we reviewed the Roman numeral system; the same process would work for applying chords to the church modes, which would give us (see Table 5.3):

Please note that the ♭ symbol at the start of a numeral for each scale degree helps show which chords are flattened compared to the ones you would find in Ionian (major scale).

Table 5.3

Mode	Scale Degree Numeral	Chords in C
Ionian (major)	I, ii, iii, IV, V, vi, viiº	C, Dm, Em, F, G, Am, Bmº
Dorian	i, ii, ♭III, IV, v, viº, ♭VII	Cm, Dm, E♭, F, Gm, Amº, B♭
Phrygian	i, ♭II, ♭III, iv, vº, ♭VI, ♭vii	Cm, D♭, E♭, Fm, Gº, A♭, B♭m
Lydian	I, II, iii, ♯IVº, V, vi, vii	C, D, Em, F♯º, G, Am, Bm
Mixolydian	I, ii, iiiº, IV, v, vi, ♭VII	C, D, Eº, F, Gm, Am, B♭
Aeolian (natural minor)	i, iiº, ♭III, iv, v, ♭VI, ♭VII	Cm, Dmº, E♭, Fm, Gm, A, B
Locrian	iº, ♭II, ♭iii, iv, ♭V, ♭VI, ♭vii	Cmº, D♭, E♭m, Fm, G♭, A♭, B♭m

However, we don't really think of chord progressions outside of Ionian and Aeolian in this way for three reasons:

1. This is a lot of information to have to remember, especially given that we already know there will be a major key that shares the exact same notes/chords (parent scale), but will just have a different tonal centre (tonic).
2. Some of the modes have very few options to resolve effectively back to their tonic. So in that sense, you might only find yourself using two or three of the chords in some modes and using repeating phrases to reinforce the tonal centre. Phrygian, for example, is very limited in what you can do with it and Locrian is almost impossible to use. Ionian (major) and Aeolian (natural minor) are so popular because they allow so much variation and give songwriters a lot of options.
3. As the modal chords are the same as the parent scale but in a different order, it is *very easy* for the modal chord progression to be confused with its parent scale. It is also possible to confuse the most closely related major or natural minor scale because it shares the same tonic note and several chords.

Let's examine this last point further with an example using the D Dorian mode. When we are looking to make progressions with a mode, we need to ask ourselves two questions:

(a) Which chords will be characteristic of the mode, and thus impart the mode's unique 'sound'?
(b) Which sixth and seventh chord extensions can we use to establish the mode even more and help clear any ambiguity?

To work out the answers to these questions, we need to study the D Dorian scale and any others that we are trying to clear any ambiguity with. D Dorian could be heard as C Ionian (parent scale), D major or D minor (because they share the same tonic note) if we are not judicious in our choice of chords, progressions and resolutions.

D Dorian chords:

i	ii	III	IV	v	vi°	VII
Dm	Em	F	G	Am	Bm°	C

C Ionian:

I	ii	iii	IV	V	vi	vii°
C	Dm	Em	F	G	Am	Bm°

D major chords:

I	ii	iii	IV	V	vi	vii°
D	Em	F#m	G	A	Bm	C#

D minor chords:

i	ii°	III	iv	v	VI	VII
Dm	Em°	F	Gm	Am	B♭	C

When we are picking chords, we know that the tonic will always be an important chord, but the other chords that give the mode its character will be the ones that establish the unique raised sixth degree. The triad chords that use that degree are two, four and six; therefore, these chords help to give D Dorian its 'sound'. With the sixth chord being diminished, it won't be used often, which leaves us with chords two and four. Now to further clear ambiguity, let's try to embellish the chords with sixths and sevenths that further affirm the modal character.

Let's take chord ii, which is Em. Reordering D Dorian starting from E would give us the target notes of C (sixth) and D (seventh).

Let's look at the degrees of E minor scale:

E minor:	E	F#	G	A	B	C#	D

This shows that working a C into an Em chord would require flattening the sixth, so that is a no-no in popular music. However, the minor seventh of E is D, so we could use an Em7 chord. If you follow the same process for the other character chord, chord four, you get a choice of either G6 or G7. I like the way that G7 sounds as it emphasizes the 'minorness' of D Dorian, because of the minor third interval (F). However, without careful use, G7 could cause confusion with the parent scale of C major as it is its dominant seventh chord, making G6 a safer bet. With the tonic chord Dm, chord six emphasizes the characteristic raised sixth interval of the mode so a Dm6 is the obvious choice. This would leave Dm6 (i6), ii7 (Em7) and IV6 (G6) as chords that strongly assert D Dorian.

Just like when we examined sixth and seventh chords in Chapter 3: Basic Music Theory, the way we analysed above is the long way around. A quicker way to remember suitable sixth and seventh chords in a mode is to remember them from the parent scale. For instance, we know that D Dorian's parent key is C major, which gives us the following options as character chords:

Suitable sixth chords: C6, Dm6, F6, G6 (as they use the major 6th interval)

Suitable seventh chords: Cmaj7, Dm7, Em7, Fmaj7, G7, Am7 (flat fifth chord omitted as it is deemed too dissonant)

As well as this method, another common way of emphasizing a mode is to leave the tonic in the bass of the chords. This way, you have to be less worried about the chords sounding too much like C major while still giving it a Dorian feel:

Dm, Am/D, Em/D, Dm

Let's now look at the rest of the modes and their unique chords that distinguish them from the parent scale, and the most related scale with the same tonic note:

Mode Chords	Scale Type	Unique Note	Characteristic
Ionian (Major)	Major	N/A	I, IV, V
Dorian	Minor	Minor with ♯6 (major sixth)	i, IV
Phrygian	Minor	Minor with ♭2	i, ♭II
Lydian	Major	Major with ♯4	I, II
Mixolydian	Major	Major with ♭7	I, IV and ♭VII
Aeolian (Minor)	Minor	N/A	i, iv, v

I haven't included Locrian here, as so few songs (contemporary or otherwise) use it. That doesn't mean you shouldn't at least try it; see what you can come up with. By adapting the techniques for extensions we used for Dorian, you can create progressions that focus around any of these modes.

BIBLIOGRAPHY

Edward Aldwell, Carl Schachter and Allen Cadwallader, *Harmony and Voice Leading* (Brooks/Cole, 2010), Chapters 5–6.
Randy Felts, *Reharmonization Techniques* (Berklee Press, 2002).

ARRANGEMENT AND ORCHESTRATION

In this chapter, we will deal with the ways a producer might take a client's work and adapt it so that it is the best incarnation of itself. Of course, this is a rather subjective process and is best done in agreement or participation with the artist. The process of optimizing a track will combine different elements of composition, arrangement and orchestration. However, for simplicity's sake, where there might be crossover, I will refer to any of these changes as an alteration of a song's arrangement.

A producer arranging a track might suggest many types of adjustment, including:

- Changing the style. Although not likely to occur too often, sometimes a song will benefit from a rethink in style. For instance, I have on occasion suggested that a song might be more effective if it was arranged for acoustic instruments rather than as a full electric version. You could also take a song with a more traditional '60s/'70s instrumentation and make it more modern using cutting-edge synthesizers.
- Embellishing or simplifying the current layering by adding or removing parts (e.g. synth parts, string sections, vocal harmonies, etc.).
- Adding or altering structural elements such as by shortening the intro or adding a coda or middle eight.
- Strengthening the core compositional elements: rhythm, harmony, melody or lyrical content. This can include:
  - *Simplification* if the parts are cluttered, too complex or too difficult for the instrumentalists to play.
  - *Embellishment*. If parts are too simple, they can be enriched by using chord extensions, chord substitutions or more complicated rhythms to create a stronger hook.
  - *Adapting* the lyrical content to improve poeticism or rhythmic flow.

While the specific decisions you make on each of these topics will be different for every track that you produce, this chapter will cover some key concepts that you will need to understand and apply in order to analyse how a given track could best be arranged. To conclude, I will also discuss some of the most common 'mistakes' I find with the bands that I produce. I work in rock/metal and indie so these 'mistakes' will be based on observations in those genres. Obviously, the term 'mistake' is also rather subjective – a mistake to me might be a selling point to someone else, so you always need to think about the end listener and the feelings of the band. Nevertheless, listening out for these potential problems will help you to start analysing tracks for yourselves.

Over the last few chapters, we have been discussing production from the perspective of composition in order to refine the song's basic melody/harmony and structural form. However, this is only one piece of a huge jigsaw. In reality, at least as much importance should be given to the arrangement and orchestration. Proper arrangement and orchestration not only aids the way that a track progresses so that the listener's interest is kept, but it also does a lot to outline the genre of the track. Much of the merit of choosing a particular producer is his or her vast knowledge of a specific genre. Therefore, if you want to produce music in an unfamiliar genre, one of the first things you should do is listen to as much music in that genre as you can!

To illustrate just how much the arrangement affects the feeling of genre in a track, listen to these famous covers and their originals:

'All along the Watchtower' by Bob Dylan and the cover by The Jimi Hendrix Experience.

'Tainted Love' by Gloria Jones and covers by Soft Cell and Marilyn Manson

'Mad World' by Tears for Fears and the cover by Gary Jules

'Heartbeats' by The Knife and the cover by José González.

'Speak to Me/Breathe' by Pink Floyd and the cover by Easy Star All-Stars

My point is simple: The raw song form (i.e. structure, lyrical content, harmony and melody) creates a strong foundation that can *easily* be reworked into many genres. Put simply, the arrangement and rhythmic content imposes genre much more strongly than the core melodic, harmonic and lyrical structure. For instance, a change of tempo and instrumentation could be the main differences between a power ballad and a club dance track like in 'Heaven' by Bryan Adams (power ballad) and the DJ Sammy (dance) version. That is not to say that the core harmony doesn't have any effect – it does. Sometimes moving from one genre to another will involve some alteration of the core structure (usually the harmony). For instance, taking a pop song and making it a jazz track is likely to take a lot of reharmonization to make the chords more unique and colourful.

Later on in this chapter, we will discuss some of the key orchestration criteria for different genres, as well as some tools to help keep the listener's interest. Before that, however, we will look at arrangement from the angle of optimizing the current orchestration to create balance.

6.1 THE FREQUENCY AND TIME DOMAIN

The mix engineer of a project is going to have a much more pleasant time if he or she receives a track that conveys, to a good extent, an inherent balance/mix due to judicious arrangement and orchestration choices having been made earlier in the production chain. There are two main considerations when trying to get a balanced arrangement:

- The frequency domain – the frequency content found in a particular instrument or set of instruments.
- The time domain – how the amplitude of a sound (or a collection of sounds) changes over time.

Please note that, in actual fact, the frequency domain and the time domain are not two separate problems, they are part of the same system.[1]

When dealing with the frequency domain, an engineer can use EQ to adjust the frequency response of a signal. In the time domain, we are dealing with the dynamics of sound and therefore the primary tool is compression/limiting. In this section, however, we will explore the more organic art of manipulating how we capture audio at source to achieve these results, rather than simply relying on processing audio post-recording (i.e. during the mix phase).

All of the subjective terms you hear from engineers such as 'space', 'transparency', 'clarity', 'punch' and 'balance' are interrelated and can be traced back to the frequency and/or time domain. When arranging, you will be making decisions on how the parts fit together to create the most pleasant result. It is important to establish that the most pleasant end result might actually make a specific instrument sound 'bad' or 'odd' in isolation. If too much focus is being given to every single instrument, the result is a mess. The most important aspect to keep in the back of your mind while engineering, producing or mixing is that every instrument used in a track should serve a *specific purpose appropriate to the track*. Not every element is of equal importance, and therefore you will decide how much focus each instrument deserves in the context of the particular track you are recording.

Take time in pre-production to consider the role of each instrument in the arrangement and plan its tone accordingly. Making these decisions will affect your choice of instrument, mic, mic placement or even the sort of room you choose to record in.

6.1.1 Introduction to the Frequency Domain

When making arrangement and orchestration decisions, they often involve making sure that a certain frequency range is captured. To be able to do this effectively, we need to outline how certain frequency areas or *bands* are defined. Splitting the auditory spectrum into bands allows us to describe aspects of the sound. Figure 6.1 is a chart commonly found in audio engineering circles and shows the frequency ranges we commonly associate with the various frequency bands, as well as some more specific and somewhat subjective terms you hear engineers use.

Note that you should proceed with caution when using the subjective terms as they are often open to the individual's interpretation.

As well as the bands shown in Figure 6.1, I often like to envision a band, not found on the standard chart, called 'low mid', which falls between 150 Hz and 400 Hz. I like to define this particular band of frequencies because the majority of instruments in a rock/pop mix are capable of producing fundamentals

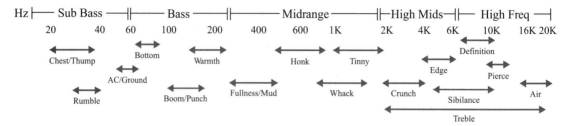

Figure 6.1 Frequency bands and subjective terms[2]

in this range. This means that it is extremely easy to end up with cluttering or 'muddiness' if you do not pay particular attention to this area of frequencies.

The frequency ranges that a specific instrument occupies within a mix can differ dramatically because of genre precedents and the producers intended vision for the track. For instance, an acoustic guitar in a folk trio will need to be treated differently to an acoustic guitar in a pop mix. In the folk trio, it is likely to provide most of the 'weight' of the entire production; therefore, it would need to be present in the top end, rich in the mid range, but also full and 'warm' in the bottom end (i.e. have a larger frequency range). In a dense pop-rock arrangement, however, you are likely to need much less low end and mids from the guitar, because bass guitar and synths are likely to cover that area. This would result in you wanting to engineer the acoustic guitar so that it is rather 'toppy', which acts in the arrangement more like a trebly percussive drive rather than the cornerstone of the entire arrangement.

Obviously, these two applications require vastly contrasting tones; therefore, the instrument should be captured differently to align with the differing vision. You have countless ways of achieving this, including:

- choosing a different guitar with the required tone;
- asking the player to use a different plectrum (or fingers) to get a different tone;
- moving the player in the room;
- changing mic or mic placement; and
- changing the register of the acoustic guitar to best fit the need in the arrangement.

Remember, all of these organic methods are much more likely to get more natural results than resorting to EQ!

When you start to engineer to your vision rather than for neutrality, you will start to hear how other top-level engineers are doing the same. By doing this and listening critically to many different genres, you will come to realize quite how much variation there is in frequency distribution between them. Every instrument will have a frequency range where it will operate in a finished mix, but the full operating range that the instrument is capable of producing will almost always exceed this range. You will hear that engineers are being selective in how they fit the jigsaw puzzle together. The most crucial part of arrangement is that there is enough frequency distribution to fill out the entire audible spectrum in a satisfying and balanced way. However, it doesn't necessarily matter which instrument is producing each frequency range as long as the total mix is balanced relative to genre expectations.

For instance, the electric guitar in a rock track might take up a huge amount of space in a metal track but in a classic rock or punk track the low mid and mid range space is usually shared much more evenly with the bass guitar. I think you would be surprised at the amount of variation in the frequency distribution even within a small subset of guitar-driven genres. Let's examine some of the key attributes of a few common subgenres of rock music to illustrate my point:

- **'60s rock** such as 'I Wanna Hold Your Hand' by The Beatles. This has much more the sound of a band in a room, relying more on overheads in the drum sound and rather 'unhyped' (more natural sounding) and more distant mics. Bass was very 'round' in the bottom end with very little high frequency. In many early '60s tracks, they are actually live takes.
- **Punk** such as 'God Save the Queen' by The Sex Pistols. This has more close-mic'd-sounding drums than '60s drums without being too hyped, and low-frequency bass sounds with a little bit more of a presence peak to compete with gritty distorted mid-range-heavy guitars.
- **Classic rock** such as 'Black in Black' by AC/DC. This created depth and space by hyping the drum kit and 'scooping the mid range' more than earlier rock records. The kick drum has more 'sub thump' and the snare drum has more 'crack'. The guitars are thick and full from double tracking, rather than the more high-gain and 'fizzy' guitar sounds in punk.
- **Thrash metal** such as 'Master of Puppets' by Metallica. This had lots of layered tracks of bass-heavy electric guitar sounds that occupied a lot of the spectrum that the bass guitar typically used. To counter this, these records gave very little focus to the bass (i.e. it was very low in the mix). As well as this, the drum sound was very 'present' and top-heavy. All of the 'beef' in the track was in the rhythm guitar.
- **Nu-metal and pop-punk** such as 'Nookie' by Limp Bizkit. This developed from where thrash left off, using chugging rhythm guitars to carry the weight in the low mids and a higher-frequency or distorted bass sound to cut through the saturated bass-laden rhythm guitars. Drums were starting to become more clinical with triggering and high-frequency EQ. Often snare drums were cranked higher to achieve a 'popping' type of sound.
- **Indie rock** such as 'Cigarettes & Alcohol' by Oasis. This sounds rather more 'raw' and like a live band than many more 'perfect' recordings of the time. The focus is on real drums, bass and guitar recorded together where possible. However, to get an aggressive sound, there is a lot of focus on the high mid frequencies on most of the instruments. It also features extreme brickwall-style limiting.
- **Post-millennium alt rock** such as 'Crushcrushcrush' by Paramore. Low-frequency bass guitar makes a comeback. Drum punch is also back in the low end, although triggering to achieve a larger-than-life drum consistency is becoming the norm.

Being able to create the correct soundscape for the genre will hinge on your understanding of the specific types of performance, instrument, amplifier and effects required in the first place – the mix engineer will not be able to fix this. For instance:

- If you have a punk track, no amount of compression can compensate for a lack of palm-muting.
- Without a suitably designed amplifier and guitar, a metal sound with high gain settings and lots of low mids will probably sound awful.

If the above explanations of guitar tone are going over your head, don't worry – later chapters are designed to help you make decisions when picking equipment.

If you don't optimize the performance and equipment for the style, you are only ever going to be playing catch-up, and the reality is that, despite extra time and effort, it probably won't sound as good.

6.1.2 Sound Envelope

It would be easy to attribute tone just to the frequency domain, but sounds are also interacting in the time domain. For instance, part of the signature sound of a high-gain guitar chord is the way that it is likely to extend over more time than a clean guitar chord. The way that a sound's amplitude changes over time is called the sound's *envelope*. A sound's envelope is divided into four sections:

1. *Attack.* This is the length of time the sound takes to rise to its peak level from the initial generation of sound.
2. *Decay.* This is the length of time it takes for the sound to drop to a more stabilized level.
3. *Sustain.* This is the length of time that the sound stays at this more stable level.
4. *Release.* This is the length of time the sound takes to decay to silence once the sustained signal level starts to drop.

The abbreviation ADSR is used on many samplers/synthesizers for settings that allow you to adjust the envelope of a sound. The envelope of a sound is much easier to comprehend when you can picture it (Figure 6.2).

Different types of instruments will have a very different envelope. For instance, the waveform of a snare drum will look significantly different to that of a distorted rhythm guitar. When it comes to explaining a sound's envelope more loosely, engineers will often talk in terms of punchy sounds or sustained sounds.

The term *punch* is often used to describe a sound that has:

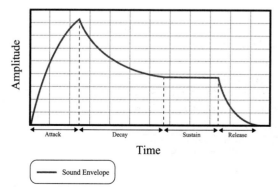

Figure 6.2 Sound envelope

- a sharp transient attack;
- quick decay;
- short sustain; and
- sharp release.

Because of their sharp initial transient and short sustain and release, percussive sounds are considered punchy (Figure 6.3).

The term *sustain* is often used to describe a sound with a prolonged length, especially if the attack is not as pronounced. Because of this, synth pads, bowed strings and even distorted electric guitar chords can be considered more of a sustained sound (Figure 6.4).

Again, proceed with caution, as musicians will throw around more subjective terms to describe time domain response. Some of these can also be related to or be confused with frequency-based observations, such as 'thick' or 'fat'.

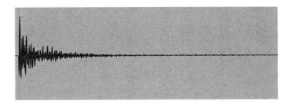

Figure 6.3 Snare drum hit waveform

Figure 6.4 Distorted guitar chord waveform

As well as the natural envelope of the sound, other factors will also have an impact on the way that a source's sound changes over time. For instance, consider an acoustic guitar in a reflective room, where the room's acoustics cause an alteration to the natural sound envelope of the guitar. The added factor of the influence of the room has a knock-on effect that you need to account for when deciding how you will record the guitar:

1. The instrument has too long a sustain and release due to a reflective room.
2. You don't have access to any absorption panels so you decide to move the mic closer, which then makes the instrument have a different frequency response.
3. It now sounds too 'boxy' (i.e. too much low mid, due to proximity effect), so the mic is moved nearer to the neck and facing off-axis to the sound-hole. Alternatively, the pick-up pattern and/or mic can be changed to correct this.

At this stage, understanding all the terminology I used in the example above is not important. What is important is that you understand that by experimenting with equipment, rooms and placements, you will be altering the sound in both the frequency and time domains. Many engineers are so focused on the frequency domain that they forget about the time domain. Remember that the combination of player, equipment, placement and room is affecting both domains! Sometimes the best result will be the best compromise between the two, especially where live-room environment, equipment or playing style is less than perfect. Knowing the point at which it is as good as it gets is an art form in itself.

6.2 PRACTICAL APPLICATIONS OF THE FREQUENCY AND TIME DOMAIN

Now that we have discussed frequency bands and the time domain, I want to clarify the term 'focus' that I used earlier. Focus simply means that you should have a clear vision of how the instruments are going to fit together in terms of not only their frequency range and time domain response, but also:

- how 'upfront' you intend a particular instrument to be in the final mix; and
- how all of the rhythms interact as a whole.

It may be counter-intuitive at first, but if you capture every source like it is a lead instrument, and balance, EQ and compress it so that it feels 'upfront', you actually end up with an arrangement that *lacks* focus. Let's go back to the acoustic guitar example in the frequency domain section. In the pop arrangement, the sound is focused in the top registers of its range. In the folk trio, it was focused on its whole range. By restricting the instrument's range in the mix, you are making it take up less headroom and giving the other instruments space. This might not be enough though. Another facet is that you can manipulate the acoustic guitar's level and envelope to give an even greater illusion of space. It doesn't even stop there: reverb can also go some way to push an instrument 'out of focus'. Creating 'space' and 'depth' out of the raw ingredients really is key.

A good analogy is to think of a song like a recipe for a meal. Let's imagine a recipe for spaghetti bolognaise. There are three core ingredients you cannot do without: the bolognaise tomato sauce, spaghetti and meat (usually beef or a veggie substitute). The defining feature is arguably the style of sauce, but without the spaghetti or meat it wouldn't be nourishing. Both the spaghetti, which adds 'weight', and the meat, which adds 'grittiness', are pivotal to the meal, but if either element overpowers the sauce it would still constitute a culinary failure. If you were to liken this to rock music, the sauce is like the lead vocals, the spaghetti is much like the fundamentals of the drum kit and the meat is like the backline (bass and electric guitar). After placing these elements in correct proportions in your recipe, you have a myriad of other ingredient options to further add to the flavour, such as salt and pepper, basil, onions, garlic, chilli and parmesan. These ingredients will all be used in much smaller proportions and could be likened to other instrumentation in the rock track such as percussion, synths, piano, backing vocals, etc. In our analogy, smaller proportions means less overall level, or restriction of frequency range or of the sound's envelope. While there are many similarities between everybody's versions of spaghetti bolognaise, every person's incarnation is slightly different. The same goes for a song's arrangement. However, as long as the core ingredients are accounted for in the correct proportions and the rest of the elements are used more subtly, you would recognize it as spaghetti bolognaise. My version of spaghetti bolognaise varies from my parents' because I like spicy food so it is heavier on garlic and chilli. Some people would like that more than others, but both would still recognize it. Going back to music, if you are able to engineer according to the expectations of the genre of music with your own ideas and personal preferences, you are on the right track as a producer. When visualizing frequency ranges and the envelope of the accompaniment, it is important to think about every instrument critically and consider the role of this instrument in the bigger picture. Is this instrument like the sauce or is it more like a pinch of salt? Even the room choice, mic selection and placement can be accounted for in the recipe analogy. This is more like the heat, length of time cooked and the order ingredients are added, all of which affect the resulting flavour.

In musical terms, thinking about all these factors should result in the following questions:

1. Do I have the correct player and instrument and/or amplifier?
2. Does this instrument sound like my vision as it is set up in the live room?
3. How does it feel with these other instruments?
4. Does the mic choice and position do justice to my vision?

Often you will have predicted and prevented some of these problems during pre-production. However, you won't always catch everything.

When you eat a meal, you don't taste each ingredient in isolation (in fact, that is a particularly nasty way to taste some of the more extreme elements) – you taste it all together. Listening to music is the same. This means that you should avoid listening to instruments in isolation too much (i.e. in solo). By listening in isolation, it is easy to get caught up trying to make each track as full, present and larger than life as possible. Remember, all that matters is the way they interact and sound *together*. When you are making decisions, you should hear them in the context of the entire arrangement. If all the instruments are not tracked yet, you should at least visualize how they fit with the other parts. As you get more experienced, you will become better at that. By the way, I am not saying that soloing a channel doesn't have its uses – it is a great way to identify or fine-tune a specific problem. However, you must have *always* outlined the need to correct an issue before listening to instruments in isolation.

Remember, throughout this process, you will be making decisions and optimizing. Don't be afraid to do so. Be confident, follow your vision and avoid neutrality. You should *always* try to get closest to the tone you require at source and just keep practising.

If you are particularly worried about a specific decision, you can set up a 'fail-safe mic' or 'dry signal'. This mic is positioned to capture the instrument's entire frequency range, and as natural an envelope, in as pleasing a way as possible. Once the fail-safe mic is checked, it should be muted and forgotten about, so that other decisions aren't based on it. With guitars, you could split the signal before hitting the amp and record a 'dry' version to re-amp in case of emergency.

In the next few sections, we will outline some of the ways to hone your musical recipe to create space and interest.

Due to the number of variables, genres and personal taste at play regarding arrangement, I can't possibly tell you how to arrange a song. All I can do is outline some of the elements to look out for and give some practical examples that, as an audio engineer/producer, you are likely to run into on a regular basis.

6.2.1 De-Muddying the Low Frequencies

When discussing frequency bands, I mentioned that the low mid frequencies are very important to the clarity and transparency of an arrangement. Over the years, many engineers have told me, 'Get the bass right and the rest should fall into place'. This is particularly wise advice, and when you are optimizing an arrangement you should start by considering the low mid band. Sub-bass frequencies are usually less of an issue as far fewer instruments are active in this range, but note that, without proper planning, you can build up clutter there too. Often high-pass EQ filters are used to clear up these regions on channels other than the kick drum, bass guitar and bass synths.

Because the low and low mid frequency ranges create the most 'muddiness' or 'clutter', you will need to decide which instruments are going to do the heavy lifting in these ranges and then make sure that the other instruments leave space for it.

I find that assigning which instrument plays the lead role in the low end is often dependent on the element that is most valuable to the rhythm. This trend also means that there are some patterns to be found based on the song's genre, such as:

- Rock – bass guitar or possibly a saturated rhythm guitar.
- Jazz – upright bass.
- Dance – kick drum.
- Hip-hop – kick or synth.

Of course, these are rather sweeping generalizations. In practice, you should make the decision of to which instrument you assign the primary focus in the low end on a case-by-case basis. This is based on the instrument that is contributing most to the rhythm (i.e. the instrument that seems to drive the track in the low end).

Let's move forward by assuming that we have a pretty standard modern rock track and the bass guitar is the obvious contender for the lead instrument in the low mid frequencies. This would mean other instruments with low-frequency content such as the kick drum and electric guitars should not encroach on its territory. Some overlap is inevitable, but each of these instruments consistently sitting on top of each other is a definite no-no. If the bass guitar is focused on the low mid, it would be sensible to focus the kick drum at frequencies below that. A tight 'thump' down in the 40–100 Hz range might be exactly what the track needs to complement the bass and fill out the bottom end. When auditioning microphones and/or positions for the kick drum mic, something that scoops the low mids a little would mean less hard work for the mixer later on and a more natural end result. This should all be flagging up the principles from earlier: organic tone adjustment is better than resorting to EQ or compression.

So far, we have been discussing the frequency domain, but we also need to consider the time domain. The time domain should be considered in two ways:

- Faster tempos have less time for the low frequencies to sustain.
- Having a number of bass instruments with sustain will clutter the low end.

Instruments that fill the bottom end of the frequency spectrum eat up a lot of 'space' and are also less forgiving rhythmically. All of the instruments in the arrangement should breathe with the tempo of the track, but this is most important with bass instruments. Again, this is best started at source, but mix engineers are also known to manipulate the envelope of sounds much further in post-production.

Let's start by discussing tempo. If the song has a fast tempo, then the sustain of the bass notes can't be too long otherwise notes will 'smear' and start to feel like they are joined together, and thus lose definition. Typically, you will want a sound to be in the release phase of its envelope (or even silent) by the time the next note strikes. This can either be 'forced' by the musician muting or by altering the envelope by other means. Shortly, we will discuss rhythm from a musician's perspective but for now I want to briefly discuss altering the envelope from an engineering perspective.

For example, a drummer isn't easily able to mute the kick drum mid-beat. So the fast tempo and frequent use of double kick drum pedal in thrash metal could eat huge headroom and clutter the arrangement. This then explains why the kick drum in thrash metal is often engineered to have a very short envelope and sound 'top-heavy' and 'clicky'.

Let's think about the time domain in the context of a less extreme example by going back to the interaction of bass guitar and kick drum in a modern rock track. For us to have that 'fat' and 'thick' sounding bass guitar, we need to create space in the low frequencies for the bass guitar to breathe. We already know that the kick drum should not encroach on the bass guitar's territory so we give it emphasis on even lower frequencies than the bass, but we also need to think about how the attack and sustain of both instruments affect their balance in the time domain. With drums in modern productions, engineers are often looking to exaggerate their punch (sharp attack, short sustain). Making sure that the kick drum has a sharp attack and short sustain would allow the sustain of the bass to come through and take control of that frequency range. In this particular example, you could also sacrifice some of the initial attack of the bass guitar to give the kick drum its own space as long as it doesn't come at the expense of note definition. Mixers often exaggerate this in post-production by utilizing compressors with a side-chain function to temporarily duck the level of the bass guitar during a kick drum hit using super-quick attack and release settings.

This approach of having the bass occupy the low mids with a lot of sustain and the kick thumping below that with a shorter envelope is my normal go-to when recording live instruments for modern productions. However, it should be stressed that it is *not* the only approach – decisions should be taken on a *case-by-case basis*. The point is not for me to tell you how to do it, but to show you how to put thought into what you're looking for and make you question the decisions you make. For instance, there have been times where I've done the opposite with the kick and the bass, by resorting to a more 'woolly' mid range heavy kick drum sound (with a little more sustain) with the bass guitar handling the lower register for a more 'subby' sound. This results in a more 'earthy', bluesy sound rather than a more 'hyped' and punchy rock or metal type of sound.

Now that we have defined space for the kick drum and bass guitar in our modern rock arrangement, we can start thinking about how to set up the electric guitar tone. If this were a metal track, we might have started with the guitar, but in a modern rock track we will be looking for enough bottom end to gel the instruments but not enough to clutter it. This means that the electric guitar tone will be most prominent in the mid range and upper mids. To be able to correctly judge the amount of bottom end, you need to monitor the sound in context of the other bass instruments, and in particular our lead low end instrument: the bass guitar. If you were to monitor the guitar in isolation, you would almost certainly make very different decisions. From here, you can utilize a variety of techniques to help manipulate the low mids:

1. Ask the player to play harder or mute between notes to alter the envelope.
2. Use a compressor (either as a pedal, in the DAW or outboard hardware).
3. Reduce the bass control on the amp to create space.
4. Reduce the gain on the amp to aid definition.
5. Move the microphone further from the amp (to reduce proximity effect).
6. Use a high-pass filter (that cuts low end below a specific frequency) on the pre-amp or in the DAW to clear space.

What is important here is to understand that you have several options available to achieve your goal. Adapting the tone to reflect the other instrumentation should result in a lot more transparency, clarity and power in that frequency range, which gives you the ability to crank the guitars higher in the mix than would have been possible before. Recognizing the fact that every instrument does not have to handle everything or even sound good in isolation is pivotally important to achieving a clear and transparent production.

The same processes we have expressed apply to any sounds that might be competing in the same register. However, as we will shortly discover, you have a bit more leeway when you get to the upper mid band and above.

6.2.2 The Importance of the Mid Range

After the bass and low mid are handled, a clearer musical arrangement is a lot easier to establish. Let's look back at our chart (Figure 6.1) and you can see that the mid range can be described roughly as between 250 Hz and 2 kHz. This is where a lot of the definition, power and realism in your music come from. When you consider that a lot of consumers are listening to music on laptops, tablets and phone speakers in which the frequency response of the speaker system is rather limited in the bottom and extreme top, you can see why the mid range becomes very important. In Chapter 9: The Critical Listening Environment, we will discuss in detail how to pick a variety of speakers so that you can judge the entire audio spectrum critically, but for now all I want to mention is that many control rooms have a dedicated set of monitors that focus in this area (such as NS-10s or Auratone 5C). By listening to your favourite songs on these monitors, you will begin to understand how important this frequency band is and how your arrangement should fit together in this area. What you will typically find while doing this is that pretty much every instrument in the mix can still be heard clearly and in the correct balance, despite the fact that these types of speakers typically deliver a somewhat skewed frequency response. Of course, it might not be as brilliant or exciting but everything is audible.

We have already discussed the way that the low mids (150–400 Hz), which take up part of the mid range, can muddy the arrangement, so now we need to consider the sonic attributes of the rest of this frequency band and how they can affect your arrangement decisions. Once you get above the low mid region (approximately 400–600 Hz), rather than feeling muddy, a build-up in this area will usually be called 'boxy'. Slightly higher, between approximately 500 Hz and 1,000 Hz, is often called 'honky'. When there is too much of these regions, the result literally sounds as though the music is being played from inside a small box or small tube. The final area of the mid range, between 1 kHz and 2 kHz, can sound 'tinny' when a build-up occurs.

Some thought needs to be given to the envelope of sounds in the mid range but it is nowhere near as difficult to make instruments fit as it is in the bass and low mids. However, I often find myself in a position where there is a serious build-up in mid range frequency areas even in a sparse arrangement. This is usually caused by reflections from the room boundaries of the live room itself tainting the sound that is being captured. While this is not technically an arrangement issue, it can colour the sound significantly enough in the frequency and time domain to make it worth mentioning here.

This problem is caused by a lack of space and is therefore regularly encountered by people with budgetary restrictions (i.e. most novice engineers or home recordists). In a small room, the floor, ceiling and the sidewalls will all be close to the performer *and* the microphones. Reflections from nearby surfaces create significant problems that affect both the frequency and time domain response. Acoustic interference from the room can cause a multitude of different sonic drawbacks, including 'hollowness', 'boxiness' and lack of definition. In the past, I ran into this most often in the boxy region. Because of acoustic interference, it is extremely difficult to get an 'open' sounding account of the instrument in a small room, so positioning the instrument in a suitable area of the room and the subsequent microphone positioning becomes even more critical (and it's still critical even in a well-treated room).

When using such small environments, I will usually resort to extremely close mic'ing using some portable acoustic panels, duvets or blankets to isolate the microphones from room reflections. As well as this, it is often necessary to resort to some EQ to help correct issues that you wouldn't encounter in a more perfect environment. This processing work can sometimes be a delicate balancing act because it is very easy to destroy your mid range this way. The rule of thumb is that the more imperfect the environment, the more likely you are to have to cut some boxy mid range content with an EQ. If you are able to treat the live room environment, I *highly* recommend that you do so. We will learn more about some of the fundamentals of room acoustics in Chapter 9: The Critical Listening Environment. While the aim of that chapter is to guide the creation of a suitable critical listening environment, you can also adapt the same techniques to create a workable live room environment.

When dealing with the mid range and beyond, you do actually have another tool available when creating space in the arrangement: the option of panning (i.e. the movement of sounds across the stereo field), which allows for a greater separation between instruments. Bass instruments such as the kick drum and bass guitar tend to be panned in the centre of the stereo field for two reasons:

1. More power is achieved by having your low-frequency instruments coming out of two speakers instead of one.
2. A precedent was set by early recordings having to utilize bass instruments in the centre due to manufacturing issues – when music is in vinyl form, anything loud and low made the needle jump out of the groove if it was only on one side.

The final decision on placement will usually be left to the mix engineer, but putting the arrangement into provisional panning positions can help to hear how the instruments are more likely to fit together in the finished product. When provisionally spreading the instruments, try to use the extremes and centre as much as possible. Once you have spread out the instruments in the stereo field, you should get a good idea of how competing instruments can be separated in this way rather than resorting to more destructive processing.

6.2.3 Working with Upper Mids

Now we are getting to the range where the human ear is most sensitive. The frequency range of the upper mids is typically between 2 kHz and 6 kHz. It is the range of frequencies that are often boosted by the mix engineer to help bring an instrument more 'in your face'. However, if overcooked, frequencies in the upper mids can quickly become very fatiguing. Even though we are getting fairly high in frequency, the upper harmonics of many bassier instruments live in this range. For instance, between 2 kHz and 4 kHz is where the attack of percussive instruments is usually found (the sound of the beater or drumstick on the skin), as well as being a hotbed for distorted guitar and a very important area for speech recognition. The range slightly higher than this at 4–6 kHz is called the presence region. Boosting this range makes many instruments (including vocals) sound closer and more detailed.

Just like with the other frequency bands, time should be spent to make sure that there aren't too many instruments sharing this frequency band. If the drums are getting a nice percussive attack at 2.5 kHz, it might be best to consider making the guitars cut in another region. The area where you decide to give emphasis to competing instruments such as the attack of the drums and grit of the guitar will be based on two factors:

1. *The areas where the instrument naturally sounds good.* Dependent on many variables, the instrument will have frequency areas that sound smoother and more advantageous to emphasize than others. For instance, on some tracking sessions, I have used a mic right next to the skin inside the kick drum to emphasize the attack of the beater. Depending on the tuning of the skin, the beater itself and the way the drum is hit, sometimes the perfect presence

point is at 4KHz; other times it is nearer 5 kHz. Sometimes having the kick drum attack in that area doesn't sound right for the genre. Each of the above variables will make a difference to which mic you choose for the job.
2. *The other instruments*. Remember, everything has to sound good together! The area that you decide to emphasize might make the vocals less intelligible. When I decide on how I make the instruments fit together, I typically give the priority to what makes the instruments nearer the front of the mix sound best. This means in most arrangements, the vocal will trump everything else. For most vocalists, a large amount of intelligibility can be found around 3 kHz. Of course, this will vary dependent on vocalist, mic choice/placement and environment.

If you have done that due diligence and considered your vision when placing and auditioning mics, you won't usually find many issues with the upper mids. Because we are entering into higher frequencies, the envelope of the sound is also a little less important as the sound decays quicker and sounds are absorbed by their environment much more easily. However, if you have an untreated and reflective room, or have chosen an unsuitably bright mic for the instrument and/or placed it badly, you can run into issues in this range. So it is by no means certain that this range will take care of itself. This is particularly apt considering the current trend of a lot of cheaper mass-produced mics entering the marketplace. Many audio professionals lambast the type of shrill and fatiguing top end produced by these mics, which should give you the impetus to try before you buy any piece of gear!

Despite the lack of problems in arrangement, mixing the high mids is actually one of the biggest causes of error. It is quite easy to get caught in the trap of making too many boosts in the upper mids to make everything sound more detailed. So, if you like to EQ a bit on the way in, be wary of going too far. Again, this is another case of making sure that you are making decisions based on the arrangement of the whole rather than tracks in isolation. Fatiguing upper mids can also occur from overzealous use of compression and limiting to try to make everything as loud as possible.

6.2.4 Utilizing the Highest and Lowest Frequencies

When you are listening to bigger studio monitors, they will have a flatter and larger frequency response than most of the environments that a casual listener will have access to. On one hand, this can be interpreted as meaning that this area is not as important. On the other hand, there is plenty of scope to make music sound even more amazing than is fathomable on basic speaker systems. As a tracking, mix or mastering engineer, you should be concerned with making sure that not only does the music sound great, but also that it plays back as brilliantly and cohesively as possible on every conceivable playback device.

Let's start with the treble frequencies, ranging from 6 kHz to 20 kHz. This band of frequencies is responsible for the 'brilliance', 'sparkle' and 'air'. Basically, it is just the upper harmonics of treble instruments that live in this range. Boosting here will give a larger-than-life clarity to instruments. However, it can also sound very brittle if you are not careful. Not many instruments live up in the treble frequencies, and your main concerns in this register will usually be cymbals and the vocals. Acoustic guitars and some less important elements of the electric guitar can live into the treble band too. Because this range contains mainly harmonics, it isn't so much a job of arranging as it is more like 'pruning' the soundscape. Listening to the genre you are producing will give you an idea on the sort of mics to use to capture the correct treble response. Many more retro-sounding genres such as Motown used old-school recording gear that exhibits a treble roll-off, so some of the character of the genre comes from that. Of course, the character that comes from analogue hardware is not limited to treble roll-off, and hardware non-linearity and distortion characteristics can have a whole range of other benefits *and* disadvantages, but that is a debate for another day!

In the extreme of the bottom end, we have sub-bass. This is the range between 0 Hz and 80 Hz, but between 0 Hz and 40 Hz there is next to nothing worth keeping. While the lowest fundamental of a piano is technically 27.5 Hz, it has been proven that you are actually hearing more of the low harmonics than the fundamental itself on most audio that is recorded. On most audio that is recorded, the 0–40 Hz region will contain nothing but noise and rumble.

This means that the 40–80 Hz range is the only usable region of sub-bass. You tend to 'feel' sub-bass more than you hear it, and a larger-than-life power can be attributed to sub. However, a huge drawback to sub-bass is that smaller speaker systems can rarely produce sound in the lower end of this range – this even goes for studio-grade nearfield monitors. This range is difficult to hear at all at lower volume levels (because of the equal loudness contours). This makes this range particularly difficult to judge properly.

Being able to properly audition the extremes of our hearing range is one of the reasons why professional facilities with treated rooms and large full-range monitoring can be advantageous. (If you are keeping score, that is likely to be 1–0 to the mastering engineer.) Hopefully, you have remembered the fact that we are going to discuss monitor choice and room acoustics in Chapter 9: The Critical Listening Environment.

Working heavily with this range of frequencies is not a pivotal concern for the tracking engineer, and much of the life is brought into these ranges during post-production. However, the mix engineer will thank you if you at least give him or her a bit of thought as quality tracking allows him or her to process these regions much more effectively.

For instance, mix engineers often like to boost the extreme top end of the vocals and cymbals with a high end EQ to create 'air' and 'sizzle', respectively. By doing this, these instruments can come alive and sparkle and make your recording sound much more expensive. However, there are many problems that can hamper the mix engineer's progress:

- Your mic has captured very little top end.
- The cymbals are cheap and nasty-sounding, or the singer's technique is off.
- The mic sounds brittle.

A common cause of harshness in the treble range is vocal sibilance, which happens during consonant syllables (such as *s*, *t* and *z*), also known as a vocalist being 'essy'. This sibilance band is usually found around and just above the presence region – about 5–8 kHz. This problem is caused by the singer rather than the gear itself, but it can be exaggerated by microphone choice, placement and technique. In Chapter 14: Tracking Vocals, we will learn some techniques to help reduce sibilance.

Being able to exaggerate extreme bottom end is important too, but the sub-bass end can be difficult with standard microphones. Specialist mics such as the Yamaha Subkick can be used to capture sub-bass. When dealing with sub-bass, engineers have an ace up their sleeve: it is actually commonplace for mix engineers to use a sub-frequency sine wave triggered with a bass instrument to create the illusion of extra sub-bass. An alternative to this that some mix engineers prefer to utilize is sub-harmonic synthesizers to artificially increase the sub-bass response.

When dealing with the extremes, it is important to think about the yin-yang relationship between these frequency ranges. Picture a mix that is sounding a little flat. Now imagine we spice things up by adding some top end EQ to our imaginary cymbals and vocal. This might sound fantastic in isolation but in the context of the rest of the mix it skews the frequency response, making the mix seem lopsided and unnatural. A natural instinct might be to reduce the amount of 'air' you have just added. This might be the right decision, but in reality the processing you did might have been correct and you just needed to balance that processing by adding more sub-bass content. This same principle works between other 'opposite' frequencies such as the relationship between low mids and upper mids.

This is a concept that is important to remember while arranging. If you think there is too much focus on the upper mids, is it because there really is too much focus on those frequencies or could it actually be the fact that you don't have enough low mids? Listening to similar commercial releases, and experience, will give you an answer to that.

6.2.5 Creating Punch

As we learnt previously, the term punch is a general term that relates to an instrument's sound envelope. For an arrangement to have punch, there must be a significant amount of difference between the peak energy of a track and the sustained portion of the waveform. To give any arrangement more punch, you need to exaggerate this difference; this can be done organically by adding/removing instrumentation or by the balancing of percussive instruments compared to instruments with a stronger harmonic purpose.

The term 'punchy' is used to describe the use of sharp dynamics that are sustained long enough to momentarily raise the average perceived volume level above the standard volume level. To create more punch in modern productions, it might be as simple as giving a lot of focus (and presence) to the kick and snare drums. There are several ways in which you can create punch:

- Mic closer to the source (while being aware of proximity effect).
- Use microphones that have a faster transient response.
- Use compression to exaggerate initial transients.

- Add other percussive sounds aligned with the kick and snare such as a handclap on top of a snare or a low-frequency sine wave on top of a kick.
- Use drum samples to reinforce natural sounds.

If your track fails to have punch, it is usually down to two issues:

(a) **Not enough percussive sounds.** To create punch, you need sharp transients, particularly those with low-frequency content: kick drum, snares, handclaps, cowbell or wood block add punch. To a certain extent, even other sounds with sharp transients can create a level of drive or momentum that some would say constitutes punch (i.e. acoustic guitar).

(b) **Too heavily compressed/limited.** The nemesis of punch is compression. If compression is reducing the initial transient peak of percussive sounds too much, then you will suck all the life out of those types of sounds. Like I've mentioned above, compression can also be used to create *more* punch if the settings on the compressor leave the initial transient alone and compress the sustain of the signal. A compressor with a low threshold, heavy ratio, slow attack (50 ms or more) and quick release will actually exaggerate rather than compress your dynamics. This is why you typically see these longer attack times on percussive instruments.

Punch works hand in hand with transparency. If you have a cluttered or muddy arrangement, instruments are going to have a hard time 'cutting through', meaning that you are going to have to use a lot of mix trickery to create punch in the mix. Therefore, it is always better to avoid that problem by first fleshing out the arrangement.

6.3 SELECTING APPROPRIATE INSTRUMENTATION

When you are selecting instruments or orchestrating for the production, you will often have some restrictions imposed by the instrumentalists of the band/artist. For instance, there is little point changing the instrumentation to be heavily synth-based if the artist is a rock band in which there is no keyboard player. On the other hand, a manufactured pop band that will be singing to backing tracks and/or a live session band might not have such stringent restrictions. The practicality of live performance will be an issue to the majority of artists, particularly at the lower end of the market (unsigned artists, etc.).

Despite practical restrictions, you will often have some scope to change the orchestration and add additional parts to the arrangement. When deciding on adding new parts, it is important to think systematically about three main factors:

1. What areas are we lacking? For example, is it a particular frequency range, percussive element, a time domain issue or simply a rhythmic foundation that there is a shortage of?
2. Can we solve this by altering the implementation that is already there? For example, mics, instruments, placements or the instrumentation already being used?
3. If not, which potential instruments fulfil the sonic needs *and* are suitable for this particular song?

If you think you can solve the issue with what you already have, then try that first!

Deciding to add instrumentation should always come about because there is actually a need for it, not because it is 'just there'. If everything seems balanced and full, then it probably is. Granted, a good mix engineer will make decisions on superfluous instrumentation so if you have a couple of different ideas you may want to track them. However, you can waste a lot of time on peripheral instrumentation that probably won't make the cut. As you get more experienced, you will get better at judging this and avoid falling into this trap as often. If you are starting to lose perspective and are not sure if the track you are producing is balanced well enough or meets the criteria of the genre, take a break. When you come back, you should listen to comparable commercial releases for guidance.

6.3.1 Instrument Registers

Now that we have outlined the thought process behind whether or not to add extra instrumentation, let's have a look at the frequency ranges of some common instruments. This is best done with an illustration (Figure 6.5).

Remember, your aim is to balance the frequency *and* the time domain. So selecting an instrument that is comfortable in the range is important, as is making sure that the instrument is capable of filling the gap in the time domain. For instance, if you wanted a thick top-heavy instrument to fill out a large amount of space, you are much more likely to choose an acoustic guitar rather than a flute (yes, that really was the first instrument that came to mind).

6.3.2 Genre Precedents

Now that we know the reasons why you might decide to add instruments and the registers they will typically operate in, let's look at how the genre of music might affect the decision-making process. Throughout the last few chapters, I've been harping on about genre precedents a lot. When you get used to a genre of music, you will start to develop a knack for the types of instruments that are commonly associated with it. For instance, you are unlikely to hear a mandolin in a heavy metal track. This is a rather crude, comical and exaggerated example, but having a basic knowledge of many genres can help you bridge genres or arrange a song for a new genre entirely.

Before we begin, I quickly want to state that I will be discussing genres of music that became popular to the general public from the 1940s onwards. It would be rather an impossible task to include everything, but here are some key attributes of the many genres and subgenres that have contributed to the charts over the last half a century or so.

Jazz/Easy Listening

Jazz is unique because of its prominent use of experimental techniques such as improvisation, polyrhythms, syncopation and 'swing'. Jazz is an extremely broad genre and can encompass many styles and attributes. However, from the perspective of the development of popular music, it was the way that the big band/swing era influenced early/mid-twentieth-century popular music.

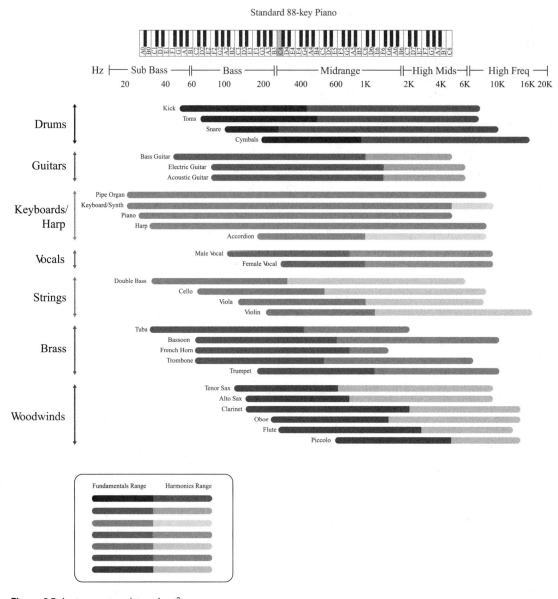

Figure 6.5 Instrument register chart[3]

During the 1930s and 1940s, jazz orchestras and swing music dominated the charts, particularly in the USA, but this style of music was conceived in the late 1920s. During this era, the band leader/arranger and/or prominent soloists from within the group would gain eminence. Examples of this are Glenn Miller, Cab Calloway and Duke Ellington.

These jazz orchestras contained a standard rhythm section of drums, double bass or bass guitar, guitar, piano and sometimes percussion, along with a large brass and woodwind section consisting of trumpets, trombones, saxophones, clarinets and flutes.

By the end of World War II, jazz-style or studio-style orchestral backing was primarily being used as accompaniment to vocalists. A studio orchestra is similar in instrumentation to a jazz orchestra except with the addition of a symphonic string section. Typically, vocalists would be male performers nicknamed 'crooners'. Examples of renowned crooners are Frank Sinatra, Dean Martin and Tony Bennett.

This style of music came to be known as easy listening, which encompassed many styles of 'middle of the road' music during the 1950s–1970s (i.e. radio-friendly and rather unobtrusive music).

Rhythm and Blues

Rhythm and blues (R&B or RnB) is a genre of popular music that originated in the 1940s. The term was originally used by the recording industry to describe recordings marketed to the African-American community in the USA.

As time has passed, the term RnB has evolved in meaning. In the 1950s, the term was often used to label electric blues records. From the mid-1950s, the term started to shift into musical works that were influenced by not only electric blues, but gospel and soul music too. Going into the '70s, RnB was being used for soul and funk. In the 1980s, a fusion of mainstream popular music and RnB was developed, and became known as 'contemporary RnB'. Table 6.1 describes some of the common traits and typical instruments found in the sub-genres of Rhythm and Blues.

Reggae/Ska/Rocksteady

Originating from Jamaica in the middle of the twentieth century, all three are primarily focused around dance. The word reggae can sometimes be used to express the entire Jamaican music scene but reggae is technically a style in itself.

Ska was developed in the 1950s and was the first of the Jamaican styles to be conceived. It combined elements of traditional Jamaican music and borrowed concepts from rhythm and blues and American jazz. At ska's core is:

- Comparatively high tempos compared to other Jamaican styles (92–132 bpm).
- Repetitive drum rhythms often with the snare drum being played side-stick (the drum being hit on the rim by the side of the stick rather than hitting the skin itself) or with rim-shots (hitting the rim of the shell and skin at the same time to give a more defined accent).
- A walking bass line.
- Guitars or organs playing on the offbeat (called a *skank*).

In common time, a skank rhythm would stress the two and four or, if playing in double time, this would be the offbeat 'ands' of a quaver (eight-note) rhythm. In many forms of Jamaican music, the guitar would play on the two and four and the organ would accent the 'ands' in double time. The following rhythm illustrates this with bold and italics indicating the notes played on each instrument:

```
Guitar
1     *2*     3      *4*
Organ
1     *and*   2      *and*   3      *and*   4      *and*   1
```

Chapter 6 ARRANGEMENT AND ORCHESTRATION

Table 6.1

Subgenre	Description	Notable Artists	Typical Instruments
Electric blues	Blues originated from the African-American communities in the Deep South of the USA towards the end of the nineteenth century. Early blues compositions were typically melancholic and contained themes of injustice, spiritualism and group unity. The blues is famous for its 12-bar forms and its derivatives. By the middle of the twentieth century, it was starting to cross over from acoustic instruments to amplified instruments, which gave a new lease of life to the genre and added to its intensity. Electric blues inspired new generations and different cultures, which led to the creation of rock and roll and the music of the British Invasion (The Beatles and The Rolling Stones).	B.B. King, Muddy Watters, Buddy Guy	Drums, bass guitar or upright bass, guitar, piano, vocals, harmonica and less frequently horns (mainly saxophone)
Gospel	Gospel music is a type of worship music popularized by the African-American community. Gospel music is often characterized by dominant vocals with strong use of harmony. Gospel music is often lyrically overt with its referencing of Christianity.	Yolanda Adams, Fred Hammond, John P. Kee	Drums, tambourine, bass guitar, guitar, piano, organ, vocals, full choirs
Soul	Soul music combines elements of gospel music, electric blues and jazz. Soul is synonymously linked with the Motown record label.	Aretha Franklin, Marvin Gaye, Otis Redding	Drums, bass guitar, electric guitar, piano, organ, keyboards, clavinet, horn section, string section, vocals
Funk	Funk originated in the late 1960s as African-American musicians created a more rhythmic and danceable form of music. Its roots are still a mixture of soul music, jazz and RnB. But funk music often de-emphasizes melody and harmony and emphasizes a rhythmic groove. Funk music also brought drums and electric bass to the foreground of the musical arrangement. Funk songs are less reliant on chord progression and can even be improvised over a single chord.	James Brown, Parliament, Sly and the Family Stone	Drums, bass guitar, electric guitar, keyboards/synths, Hammond organ, clavinet, horns, congas
Disco	For a brief time in the late 1970s, disco music was extremely popular. It is a derivative of soul and funk but with a more regular bass beat. Disco was also famed for its incorporation of the then-modern technology of synthesizers and drum machines.	Donna Summer, Kool and the Gang, The Bee Gees	Drums/drum machine, percussion, bass guitar, keyboards/synthesizer, electric guitar, horn section, string section
Contemporary RnB	Today, the term RnB is mainly used to describe African-American music since the decline of funk and soul. Modern RnB is known for its polished production, use of modern technology such as dance-orientated sampled/drum machine rhythms and a use of smooth, lush style of vocal arrangement.	Whitney Houston, Boys II Men, Lauryn Hill	Drums/drum machine, keyboards/synthesizers, bass guitars/bass synth, piano, vocals, large vocal harmonies

Ska was also known for its prominent use of horns such as trumpet, trombone and saxophone.

Ska music has proved quite popular historically and developed further from its Jamaican roots in both a second and third wave. The second wave arose in England in the late '70s, and mixed ska with some elements of punk rock, creating a style called 2-tone, which is typified by The Specials, The Selecter and Madness. The third wave in the 1990s in the USA saw the punk rock electric guitar rhythms being mixed more prominently with ska to create a sound that was often heavier than previous incarnations. It is often linked with the skater punk movement and extreme sports. Bands such as Less Than Jake, Reel Big Fish and Rancid are noted for their ska punk style.

Rocksteady developed from ska and has many similar traits, including skanks, but is typically slower in tempo than ska (80–100 bpm). The slower tempo allowed musicians to play 'around the beat' more and the bassline often had a syncopated feel. As well as this, drum patterns were often significantly different to that of the ska era. Drum patterns in rocksteady recordings often utilized a 'one-drop' drum beat, which features little accent on the one beat and a heavily accented second and fourth beat by hitting the kick drum and the snare together. Due to political and economic issues in Jamaica, as well as other factors, rocksteady's popularity was short-lived, but some of the rhythmic influence, and in particular the one-drop, was utilized regularly in reggae.

Reggae is typically slower than ska but faster than rocksteady, and just like the other Jamaican genres, it still has accented offbeats. There are three common drum patterns in reggae: the one-drop, steppers and rockers. *Steppers* emphasize the kick drum on every beat of the bar, which gives the music a more driving and military-like feel. *Rockers* also emphasize every crotchet (quarter note) but are more flexible and can include syncopation. Trying to define the differences between steppers and rockers is often a talking point between reggae musicians.

To most listeners, the distinction between ska, rocksteady and reggae might be fairly subtle, especially as a lot of reggae artists will compose and perform songs across all of these subgenres. Each of these subcategories tends to feature similar instrumentation: drums, bass guitar, electric guitar, organ, trumpet, trombone, saxophone, vocals and piano (more prominently in early works). Examples of prominent Jamaican musical artists performing these styles are Bob Marley and the Wailers, Desmond Dekker and Lee 'Scratch' Perry.

Here are some examples of ska, rocksteady and reggae:

Genre	Track
Ska	'Pressure Drop' by Toots and the Maytals
Rocksteady	'Rocksteady' by Alton Ellis
Reggae	'One Love' by Bob Marley
Reggae	'Israelites' by Desmond Dekker & the Aces
Ska (2-tone) – second wave	'Too Much Too Young' by The Specials
Ska (ska punk) – third wave	'All My Best Friends Are Metalheads' by Less Than Jake

Acoustic/Contemporary Folk

Acoustic music, as the name suggests, revolves around stringed acoustic instruments. Defining other characteristics can be difficult because there has become a trend of artists who primarily perform in electrified or digital mediums to then perform acoustic versions of their songs. Examples of this are MTV Unplugged and Radio 1 Live Lounge (UK).

The dictionary definition of folk is also fairly vague:

> 'Music originating among the common people of a nation or region and spread about or passed down orally, often with considerable variation.'

Contemporary folk is therefore also difficult to define in musical terms. However, my interpretation of contemporary folk is any music that was originally written and performed with acoustic instruments that couldn't be classified as being traditional. Many popular contemporary folk musicians could be considered a folk-rock fusion, such as Bob Dylan, Simon & Garfunkel and Mumford & Sons.

Instruments typically found in contemporary folk are kick drum, double bass or bass guitar, acoustic guitar, banjo, mandolin, ukulele, harmonica, organ, piano, fiddle, vocal, backing vocals and flute.

Country/Easy Listening

Country is a genre of American popular music that originated in the Southern United States in the 1920s and is derived from folk music. Country songs are very often ballads or can sometimes be dance tunes. Country is similar to pop music in terms of core instrumentation structure and form; it is set apart from pop music by its unique vocal style and use of vocal harmonies. More traditional forms of country will also focus heavily around acoustic stringed instrumentation such as banjos, acoustic and pedal steel guitars, fiddles and sometimes a reed instrument such as a harmonica.

More traditional folk artists include Dolly Parton, Garth Brooks and Willie Nelson. More mainstream crossover country artists include Shania Twain, Taylor Swift and Keith Urban.

Rock

Rock music developed from 1950s rock and roll popularized by the rebellious 'white youth' of America, which in turn took its roots from the genres popular with the African-American population such as electric blues and soul mixed with the more mainstream swing and country. As rock and roll evolved, it became less reliant on swing and shuffle rhythms and was reclassified simply as rock music. Rock music became extremely popular and spawned many subgenres. Despite all the variations, the core instruments remained similar with a backbone of drums, bass guitar, electric guitar and vocals, although piano and organ were sometimes incorporated too.

As we illustrated some of the differing sonic attributes of many styles of rock music earlier in the chapter, I don't want to tread over the same ground. However, here is a list of some of the subgenres and notable artists so that you can judge for yourself:

Subgenre	Notable Artists
Rock and Roll	Chuck Berry, Little Richard, Elvis Presley
Surf Rock	The Beach Boys, Dick Dale, The Shadows
Classic Rock	The Beatles, The Rolling Stones, The Who
Progressive Rock	Pink Floyd, Rush, Yes
Hard Rock	AC/DC, Black Sabbath, Led Zeppelin
Punk Rock	The Sex Pistols, The Ramones, The Clash
Heavy Metal	Metallica, Van Halen, Pantera
Grunge	Nirvana, Alice in Chains, Soundgarden
Alternative Rock	Pearl Jam, Smashing Pumpkins, Foo Fighters
Indie Rock	Oasis, Blur, The Libertines
Experimental Rock	Radiohead, Frank Zappa, The Flaming Lips
Funk Rock	Red Hot Chili Peppers, Rage Against the Machine, Faith No More

Pop

'Pop' originally derived from an abbreviation of 'popular music', but has evolved to become a genre in its own right. These terms are still used interchangeably today so it is important to distinguish between pop the genre and 'popular music', which can take any style.

As a genre, pop music is diverse and often borrows rhythmic/harmonic/melodic elements and instrumentation choices from other musical styles, including rock, country, hip-hop and dance. Despite the eclectic mix of influences, pop music is generally categorized by its basic structures (often verse-chorus), strong melody and catchy hooks. The vast majority of pop music is designed to be danced to. Modern pop instrumentation tends to focus on standard rock instruments as well as predominant use of synthesizers, samplers, looping and heavy use of effects. Notable pop artists are Michael Jackson, Madonna and Britney Spears.

Hip-Hop

Hip-hop is a musical genre that is primarily known for its use of rapping. In the early days of rap around the late '70s and turn of the '80s, the funk, soul and disco roots were very apparent, which can be heard on records by the early pioneers of the genre such as The Sugarhill Gang, and Grandmaster Flash and The Furious Five. During this era, songs were often created by DJs using turntables to sample, scratch and create breaks to 'update classic tracks'. However, as drum machine and digital technology advanced, hip-hop music started to develop a style of its own and diversify further. Run DMC's classic 'It's Like That' is an early example of hip-hop music produced using completely synthetic accompaniment. During the '80s, it also became commonplace to use digital looping/sampling prominently. A great example of this is 'Rhymin' & Stealin'' by The Beastie Boys, which used the drum groove of Led Zeppelin's

Table 6.2

Subgenre	Synopsis	Further Subgenres	Notable Artists	Example Track
House	House music is known for its repetitive 4/4 rhythms, offbeat hi-hat, and pulsing kick drum and synth bass. It often features a drop where all the instrumentation 'comes together'.	Deep House, Progressive House, Electro House, Tech House, Tribal House	David Guetta, Benny Benassi, Deadmau5	'Don't You Worry Child' by Swedish House Mafia
Techno	Much like house, techno is known for its repetitive nature and is often instrumental music. The rhythmic structure is quite often common time (4/4) and the kick drum will often mark each crotchet (quarter note). The two and four are often marked with the snare and/or clap. Open hi-hats are also integrated on every second (eighth note). The tempo tends to vary between 120 and 150 bpm.	Acid Techno, Ambient Techno, Minimal Techno, Tech House	Juan Atkins, Carl Cox, Richie Hawtin	'Kernkraft 400' by Zombie Nation
Trance	Trance music refers to the state of 'lessened consciousness' of the same name. The music achieves this by using layering and repetitive rhythms to create build and release. Typically, a trance song will have a mid-song breakdown similar to a middle eight in rock and pop music, with the instrumentation in this section being broken down without percussive rhythm tracks.	Progressive Trance, Tech Trance, Vocal Trance, Goa/Psytrance	Armin Van Buuren, Tiësto, Paul Oakenfold	'Adagio for Strings' by Tiësto
UK Garage	UK Garage is well known for its distinctive drum patterns that feature heavy syncopation in the hi-hats and kick drums. It is slower and more sensual than many other EDM genres at around 130 bpm, although many faster subgenres of garage were also conceived. Garage tracks also commonly feature manipulated vocal samples – a trait that is also popular in other EDM genres that followed, such as jungle/drum and bass and dub step.	Speed Garage, 2-Step Garage, Breakstep, Future Garage	Artful Dodger, So Solid Crew, Shanks & Bigfoot	'Re-Rewind' by Artful Dodger feat. Craig David
Jungle/Drum and Bass	Jungle and drum and bass are similar 'dark' styles of EDM that originated from the rave scene. An extremely high tempo (160–180 bpm) and prominence of the drum and bass line gives DnB its name. Jungle and drum and bass features heavy use of syncopated rhythms, as well as the sampling and manipulation of Jamaican dub and reggae music. However, the genre has expanded to include influences from many other genres.	Drumstep, Liquid Funk, Techstep, Neurofunk, Jump-Up	Pendulum, Chase & Status, Sub Focus	'Tarantula' by Pendulum
Dubstep	Dubstep is another 'darker' style of EDM that originated from DJs influenced by jungle and incorporating aspects of drum and bass into 2-step garage. Dubstep features syncopated drum rhythms with a clap or snare usually inserted every third beat in a bar. Wobble in the bass line is highly prominent in .	Future Garage, Post-Dubstep, Brostep, Trap	Skrillex, Flux Pavilion, Caspa	'Bass Cannon' by Flux Pavilion

Table 6.2—continued

Subgenre	Synopsis	Further Subgenres	Notable Artists	Example Track
	dubstep, which is produced by using a low-frequency oscillator, which can manipulate many parameters such as volume or distortion. This is often called 'wub'. The tempo of dubstep is normally between 138 and 142 bpm.	Future Garage, Post-Dubstep, Brostep, Trap	Skrillex, Flux Pavilion, Caspa	'Bass Cannon' by Flux Pavilion
Grime	Grime is a development of UK garage and drum and bass fused with rap vocals and other hip-hop traits.	Grindie, Electro-Grime	Dizzee Rascal, Wiley, Lethal Bizzle	'Fix Up, Look Sharp' by Dizzee Rascal
Breakbeat	Breakbeat is a fairly fast form of EDM known for its use of breakbeats (who would have thought?). The most popular form of breakbeat is called big beat, and it is famed for its use of loud, compressed and often distorted breakbeats and extensive looping from obscure '60s/'70s rock, pop, funk and jazz tracks.	Acid Breaks, Big Beat, Breakcore, Broken Beat, Progressive Breaks	The Prodigy, Chemical Brothers, Fatboy Slim	'Push the Tempo' by Fatboy Slim
Ambient	Ambient tracks are usually slower and more subdued, atmospheric, unintrusive and sometimes more playful than typical EDM. Often they have minimal drums and percussion and, where there is, it is often drenched in reverb. They also tend to be lengthy and therefore rather unsuitable as a commercial form of music. As ambient developed, a subcategory that is the antithesis of ambient, called dark ambient, was developed.	Dark Ambient	Aphex Twin, Squarepusher, The Orb	'Pulsewidth' by Aphex Twin

'When the Levee Breaks' and the guitar riff of Black Sabbath's 'Sweet Leaf'. By the late 1980s/early 1990s, rap had broken into the mainstream with the likes of Public Enemy and NWA giving it a more politically charged edge or highlighting 'gangsta' culture. Hip-hop became so popular that it started to cross over into pop circles with novelty-style tracks such as 'U Can't Touch This' by MC Hammer and 'Ice Ice Baby' by Vanilla Ice.

Such was the crossover appeal of hip-hop that, by the late 1990s, pop tracks often included rap breaks; this trend has continued through to the modern day.

Electronic Dance Music

Along with hip-hop, electronic dance music, or 'EDM', benefitted from the digital music boom. EDM is electronic music produced for dance-based environments such as nightclubs. EDM typically uses drum loops and/or samples to build a 'beat' that is then layered with a multitude of synthesizers, samples, pre-made loops and sometimes live instrumentation. Just like rock music, EDM can be categorized into many subgenres. EDM can sometimes be difficult to categorize as many DJs will have influences and compositions spanning several genres. Table 6.2 gives a breakdown of a few of the most popular subgenres and a brief synopsis of each.

continued

6.4 RHYTHMIC ARRANGEMENT

When it comes to creating an interesting, concise and transparent arrangement, the choice of rhythmic patterns is critically important.

6.4.1 The Rhythm Section

Every bit of melody or harmony will have inherent rhythm but in this section I am talking specifically about the underlying rhythm or pulse provided by what is commonly called the *rhythm section*. If the rhythm section 'grooves' and 'gels', you will find that you need less instrumentation to create interest and a whole lot of effort will be saved come mixdown to make everything fit. In this section, I will be talking about the way that the drums and bass interact to create the solid foundation for the rest of the arrangement to be built upon. The rhythmic connection between pretty much every hit song ever is that the rhythm bed of drums and bass holds a *steady, reliable groove* that allows the lead melody space to shine. The same principle applies to the vast majority of Western genres of music (progressive rock, jazz and some types of metal being notable exceptions). The rhythm section can also include the addition of a simple chordal part such as a rhythm guitar or keyboard.

Despite the similarities, different genres will utilize the rhythm section differently. In some styles of folk, the drums are further back in the arrangement; in genres such as rock, they punch right through almost as loud as the lead vocal. The bass can be intricate and forward in funk and soul styles or very simple and a background influence like in much of '80s pop. However, in all cases, the rhythm section provides the groundwork that everything is built on and defines the amount of space that the other instruments have.

So, what makes a strong rhythm bed so special? And why should you pay special attention to it in your arrangement? Many of the world's top rhythm players will talk about that magical gelling effect that happens when the rhythm section plays cohesively as being 'in the pocket'. The 'pocket' is a rather esoteric and hard-to-define term because the creation of emotion through music is very subjective. A solid drum and bass groove makes people want to move, whether it be nodding their head, tapping their feet or dancing. It would be easy to define playing in the pocket as: good timing while playing a pattern that works in sympathy with the melodic and harmonic structure of the song. Looking at it from an engineering perspective, it is more than that. There is a good reason why the top session drummers and bass players are highly sought after: to some degree, they mix themselves. It's all about the way that they control the dynamics of their instruments to the tempo and genre of the music; they just seem to understand the sensibility of the music being played. Think of a drummer playing a rock beat and then the same player moving to a blues beat with the same tempo. The drummer could be playing a quaver-based (eighth note) rhythm in both, and even a similar basic pattern, but the grooves will still feel distinct. If you listened to these beats again and took notes, you might find that the ferocity that the drums are hit, ghosting of snare hits and also the type of cymbal that is used to supplement the kick and snare separates the two styles.

The actual rhythmic patterns can be similar or even identical. The same can apply to the bass player. His or her use of fingers or plectrum, and the way that he or she mutes the strings between notes can change. This should illustrate quite how much arrangement can matter and all of this is before we have considered instrument choice or tone control. My equation for a successful rhythm section is:

> Good timing + sensible pattern chosen in sympathy with the song
> + suitable dynamics

Again, this is judged on a case-by-case basis according to the precedents of the genre. A good rhythm section will adjust their dynamics to breathe with the tempo and create the right 'sonic space' and leave enough room for the rest of the musical ensemble to embellish when appropriate. No wonder I've heard many of the top session cats say, 'Playing in the pocket is as much about knowing what not to play as it is knowing what to play.' The musicians should be playing their instruments (and setting them up) such that the envelope of the sound is releasing in sympathy with the pulse of the music and not fighting against it. For instance, listen to a series of slow-tempo rock songs. How many have really tightly tuned and high-pitched kicks and snare drums? My guess is not many because it is likely to sound odd as there is not enough sustain to the sound to fill enough space. On the other hand, a faster-tempo song is much more likely to have a higher-pitched tuning, which is further dampened to help control the sustain.

6.4.2 Avoiding Clutter with Competing Rhythms

Now let's think of a situation where rhythms are cluttering the arrangement. This is particularly important in the bottom end as it occupies more headroom and your ear is not as good at differentiating between two closely pitched bass notes. In classical orchestration, it is often taught that it's best to avoid tight spacing of chord tones in the lower registers.

So let's take a song that has a bass guitar playing quavers (eighth notes) but in which you want to incorporate a string arrangement. To do this, you might want to include a cello to fill out the lower registers of the string arrangement. When writing the cello part, it would usually be unwise to have the cello playing either:

- A note too close to the pitch that the bass guitar is playing.
- A 'quick' rhythm that competes with the rhythm of the bass guitar.

Therefore, a long slow-bowed cello that plays minims (half notes) or semibreves (whole notes) is likely to fit better. The de-cluttering of fast or competing rhythms is particularly important in the bass and low mid bands of the auditory spectrum, in higher frequency bands panning can often be used to aid separation between competing rhythms.

6.4.3 Keeping Interest

When we discussed rhythm in Chapter 3: Basic Music Theory, we established that rhythm should:

- drive the song forward; and
- keep the listener's interest.

Keeping the listener's interest rhythmically is one of the hallmarks of a great arrangement. It is one of the major factors contributing to how you create progression or contrast through a track. Let's go back to the example of 'Rock with You' by Michael Jackson, which we used when we examined simple rhythms. In this song, a lot of momentum is built and interest is maintained by the hi-hat moving to the semiquavers (sixteenth notes) rhythm in the chorus compared to the quavers (eighth notes) rhythm in the verses. While the momentum in the track is often built heavily around the drum groove, momentum can also be built using other instruments. For instance, in many rock songs momentum is built using the rhythmic pattern of the electric guitars. In verses, guitars often consist of long sustained chords or drones and then the chorus uses more intricate or faster-paced patterns. A simple example of this is 'Big Balls' by AC/DC, where the lead guitar sustains long chords in the verses (semibreves/whole notes) and moves to a more snappy (crotchet/quarter note) rhythm in the chorus. U2 also use a similar trick in many of their songs, including 'With or Without You', 'Beautiful Day' and 'Pride (In the Name of Love)'. In U2's case, the guitar often plays brief arpeggios or swells in verses before a much more complete guitar arpeggio or barre chords in choruses.

6.5 MAKING A SONG HAVE A JOURNEY

Back in Chapter 2: Project Management and Pre-Production, I explained how a good song should be a journey; it should take you somewhere. As popular music progressed with music technology, music producers started to utilize the extra power at their fingertips in a creative way. Songs released in the early '60s were a very good representation of what a band or artist could achieve completely live. A pop song in the early '60s would be successful primarily because the core of the song was interesting, catchy and the musicianship was good. But by today's standards, the restrictions of simply replicating a live performance are not as commercially acceptable. Even songs that appear to be more faithful to live performance are often subject to much audio trickery. Our attention span, considering the simple soundscape provided by a three- or four-piece band, probably explains why song lengths in the '60s were short by modern standards (two to three minutes). As time has passed and we have gotten extremely used to the 'larger than life', 'perfect world' recordings, utilizing and exploiting modern recording/editing and multitracking techniques has become the norm. Song lengths have increased as there is more 'trickery' to prolong interest; it also means that the layering of instrumentation needs to build with the track.

To put it simply, you need to make sure there is a sufficient pay-off at crucial points of the journey, and to create pay-off you need contrast. In an 'average' pop track, this would usually mean that:

1. Each sequential verse is denser than the last.
2. The choruses are denser and have a stronger hook.
3. There is a point around 2 minutes where there is a change of gear.
4. The final chorus or coda is the most dense and exciting.

When arranging, you can work backwards to create the 'journey', making sure that only the final chorus or coda 'throws in the kitchen sink' by stripping away layers in other sections. Making the song feel like a journey is important regardless of genre or arrangement. Even a simple song with only an acoustic guitar and a vocal should still feel like it goes somewhere!

Here are some tracks that I use as reference when I want to figure out how to create an effective musical journey:

'You Shook Me All Night Long' by AC/DC – This is a great example of how to get a masterful arrangement out of a simple five-piece rock band. All of the members of the band know exactly how to create space for each other. Several key points are:

- The drums only play kick, snare and hat in the verses and save crashes for the rhythmic build-up and the choruses.
- The bass doesn't come in at all until the build-up right before the first chorus.
- After being in the intro, the rhythm guitar drops out until halfway through the first verse.
- Gang vocals are present fairly low in the mix in the chorus to add extra weight.
- During the first half of the solo break, the rhythm guitar and bass play rhythmic stabs to give the solo more space before building up in the second half.
- The drums crescendo with the vocal right at the end of the track before stopping dead.

'Go Your Own Way' by Fleetwood Mac – One of my favourite Fleetwood Mac songs, the song structure is straightforward but for me the genius about this song is the way that momentum is built in the choruses. While the verses are up-tempo, it manages to sound more subdued by the fact that the bass guitar is more static, the acoustic guitar is more palm-muted and the drums focus on the toms. By the time the chorus drops, there is more hi-hat and cymbals (certainly in the latter choruses), the acoustic rhythm opens up with no muting and the shaker/brushed cymbal drives with a semiquaver rhythm (sixteenth notes), which pushes the momentum significantly. The introduction of a lead electric guitar as a point of primary interest from only the solo onwards gives the arrangement somewhere to go and the coda utilizes the electric guitar riffing to carry the emotion with the backing vocals taking the backseat, as well as a classic fade-out.

'Just Dance' by Lady Gaga – This is the perfect example of how to use modern production techniques to keep excitement in a vocal performance. There are shadow vocals, various delay lengths, 'glitch' vocals and the obligatory rap section.

There is also a rather unexpected middle eight towards the end of the track that features trippy vocal delays and a low-pass-filtered synth.

'Hot n Cold' by Katy Perry – In the modern pop domain, I think some of Katy Perry's tracks are arranged incredibly. Starting with an '80s-style drum sample and synth before the guitars take over in the chorus with lots of low mid weight, by the end of the track the guitars and synths are both blaring with plenty of extra vocal production too. Many 'cheeky' pop lyrics sound very cheesy to me, but the lyrics here toe that line perfectly.

'Here Comes the Sun' by The Beatles – How could I not include a track by The Beatles? They really were the kings of creating an interesting arrangement. This one is a Harrison tune. I like how, in parts, the lead vocal line and the acoustic guitars mirror and wind around each other. By '60s standards, the instrumentation is very full and layered, yet everything has its own place because so many of the layers double the notes of the lead vocal.

'Take the Power Back' by Rage Against the Machine – So many songs from the debut album, including this one, show how heavy rock can have 'space' and 'groove'. The grooves are so infectious that many of these tracks could go on forever and no one would care. This album is also testament to tone selection and engineering because each instrument sounds natural (certainly by today's standards) yet can hold up to any modern production in terms of production values and sonics.

'Paranoid Android' by Radiohead – This choice was a toss-up between this song and 'Bohemian Rhapsody' by Queen for the way that the core structure is unusual. 'Paranoid Android' masterfully develops over four seemingly contrasting sections rather than any sort of conventional structure. The way that the instruments intertwine and weave up and down in the arrangement/mix is phenomenal. Despite such an unusual structure, the song is cohesive and if you didn't think about it you might not notice that there doesn't seem to be a chorus.

'The Boxer' by Simon & Garfunkel – This song illustrates how you can break from the normal instrument and percussion choices to create something unique and beautiful. The track starts with the signature picked acoustic guitar(s) and tight vocal harmonies, before developing by:

- Having a less-than-standard kick drum rhythm-panned hard-right.
- Bass harmonica in a way that almost sounds like a kazoo or accordion.
- A snare drum that is heavily manipulated to sound like a gunshot. Rather than artificial reverb and tape echo, it is actually from the drums being set up in front of an elevator shaft in the CBS building in New York.
- A piccolo trumpet solo (which I speculate to having been given a slower-geared attack that changes the timbre).
- A few little Dobro (resonating guitar) licks.
- String arrangements that creep in and build in layering and intensity throughout the coda.
- Baritone voices at the end thickening the arrangement.

'Ain't No Sunshine' by Bill Withers – For me, this is the definitive reference point on how to still have space in a record despite a large string section.

This soul classic knows the importance of the lead vocal performance, and while the string section is prominent, it never overpowers the emotion in Bill's performance.

'Skyfall' by Adele – I could have chosen a number of Bond themes such as 'Live and Let Die' or 'Nobody Does It Better' as examples of drama in a pop track. Bond themes are perfect examples of the way you can intertwine a full orchestra with popular music instruments in a way that maximizes pay-off.

On more complex songs, I often outline how I want a song to progress and the emotional meaning of each section using a 'song map'. This can be a table or a diagram, and its purpose is for you to be able to plan and visualize the way that a song is going to build. Later in this chapter, you will see the song map I made for 'Take Her Away'.

6.6 KEEPING INTEREST WITH EXTRA PRODUCTION

Back in the '60s, songs had to be interesting, well written and catchy to succeed; there were few quirks and gimmicks involved. These days, with the amount of technology and media available at our fingertips, things have changed. Modern productions need to not only stir your emotions, but also compete against other factors such as:

(a) listeners using laptop and phone speakers;
(b) music being listened to while performing other tasks;
(c) media oversensitivity of the younger generations; and
(d) poor tracks filled with production gimmicks.

Each genre will have different precedents established that will affect the way that you decide to arrange the track. Pop music typically tries to compete with all of these factors, leading to commercial pop tracks having 'flourishes' of production every five to 10 seconds.

With rock music, you have more leeway, but you should consider adding at least one flourish to each section of the song. A flourish could be:

- a variation in rhythm or melody, or a change of expression in the lead vocal;
- an alternative riff or alteration of rhythm;
- a cool little reverb, delay or modulation effect;
- the manipulation of the stereo field;
- use of distortion, extreme compression or creative use of EQ filters for special effect;
- adding a new instrument or taking one away; and
- adding vocal harmonies.

Basically, anything that makes your ears prick up. I'm beginning to sound like a broken record, but listening critically to reference material to judge what is right for the genre will help give you inspiration.

Be wary, though: adding flourishes does not mean that a dense arrangement is required. The majority of flourishes should occur on instruments that are already present and the variation of emotion is most often achieved through the lead vocal. Many of the most successful records are actually relatively small arrangements; you just have to listen to some of the biggest-selling records of 2012–2013 in the UK for evidence of this:

Robin Thicke feat. Pharrell – 'Blurred Lines'

Miley Cyrus – 'Wrecking Ball'

Gotye – 'Somebody That I Used to Know'

The common denominator in all of these is an interesting and killer lead vocal. Get listening to your favourite records and work out how they take you on a journey and keep you interested. Here are some of the more obvious flourishes that you could use to keep a listener's interest while maintaining the original song structure.

6.6.1 Double Tracking

The first way that you could make sections of a track have more impact is simply by double tracking elements of the arrangement. For instance, it is commonplace to have single tracks of electric guitar lines in verses and then double track or even quadruple track them for choruses.

6.6.2 Panning

Panning (i.e. the movement of sounds between speakers) can also play a huge part in making an arrangement progress. A trick I often utilize is to keep lead instruments (guitars, keyboards, strings) in the arrangements either central or biased towards one side during the verses, then pan them hard-left and hard-right for the choruses. This helps to give an explosive feeling to the music. While this is a trick for mixing, I often record parts with this in mind.

6.6.3 Adding/Removing Instruments

Giving a listener something different is an important part of your arrangement. If the second verse seems musically identical to the first, the listener will probably lose attention. Consider using an additional instrument or part in verse two, and the same can be said for each chorus. This does not mean that songs should just get denser and denser ad infinitum; you still need to be wary of not cluttering your arrangement. In these cases, you may consider taking one instrument away as you are adding a new one in.

While adding instrumentation to an arrangement can help a song to progress, a continual increase in instrumentation can become jaded and predictable; sometimes you need to surprise the listener! You can do this by reducing the instrumentation and/or softening the dynamics. Employing a drop-chorus before the final chorus is a good example of how to reignite a listener's interest via a reduction in complexity. You could also use this treatment with bridges, pre-choruses or for later verses.

6.6.4 Special Effects

A lot of excitement can be added through the use of a few little production quirks. This is one occasion where you may get great inspiration from listening to genres that you are not used to producing or even listening to, and note down any special effects you hear. You will be surprised by the wide variety of places you can use these special effects, and having plenty of techniques in your arsenal is advisable. Sometimes these production quirks can even add interest to an otherwise direct and dense production. Here are just a few examples of some quirky special effects being used in productions:

- *Mid-long delays* are most commonly used on vocals on certain words to give the listener something new. These are often introduced in verses where typically the melody is not as constant. Sometimes delays are only introduced later in the song. Sometimes they are introduced early but used on different phrases in each section. An example of the use of mid-long delays as a special effect is 'High Hopes' by Pink Floyd on the line 'sleepwalking back again' (2:20).
- *Vocoder.* This is often used as a special effect such as on 'Get Lucky' by Daft Punk (2:20) and 'Intergalactic' by The Beastie Boys, where a vocoder is used on the main hook to give it an 'out of this world' sound.
- *Autotune.* In many modern rap, dance and RnB tracks, you can hear autotuners being abused to create special effects, for example 'Buy U a Drank' by T-Pain or 'Believe' by Cher.
- *Distortion.* As well as the use of distortion as a guitar effect, you can also utilize distortion as a special effect by pushing analogue equipment past its limits, running other instruments through guitar effects or using specially designed distortion/saturation plug-ins. Distortion is often used on vocals in rock music for special effect. For example, 'White Limo' by Foo Fighters seems to use distortion mixed with a radio effect type of EQ filter on the lead vocal.
- *Percussion effects.* Many RnB and pop songs (particularly in the '90s) were stylized with the sound of tuned chimes, sweeps and more abstract percussion (wood blocks, xylophones and claps), for example 'Sometimes' by Britney Spears and 'You Are Not Alone' by Michael Jackson.
- *AM radio/phone treatment.* A drastic EQ that gets rid of all bass and treble and leaves you only with mid range. Usually used in intros, bridges and pre-choruses. It can also be used to create prosody (see below) for example 'Telephone Line' by ELO and 'Wish You Were Here' by Pink Floyd.
- *Samples* can be used to help set a scene and create prosody (see below). For example, the bells at the start of 'Hells Bells' by AC/DC or the cash register in 'Money' by Pink Floyd. Many modern arrangements use samples or well-known licks from other tracks to form a major part of a song's hook, for example 'Gold Digger' by Kanye West.

6.6.5 Prosody

Prosody in music production occurs when the musical arrangement and lyrical content align in a way that helps to further emphasize the lyrical content.[4]

A simple example of prosody would be a vocalist's melody line rising in pitch as he or she sings the word 'higher'. In this example, elements of the instrumentation could also rise in pitch to give the effect of prosody. You can use prosody with special effects to create particularly strong production flourishes. For example, say a lyric in the song you are working on is 'This will go on forever', and there is a gap at the end of this line. You could use a long vocal delay to further emphasize the word 'forever'. Conversely, if the line were 'It has got to be the end', you could get the musicians to abruptly stop for a beat or two, adding a rhythmic pause after the word 'end'. Given the right genre, you could further enhance this with the sound of a record player slowing down.

6.7 COMMON ARRANGEMENT MISTAKES

From my years in production, I've noticed a lot of similarities in the way that artists' compositions fail to interest me. We will be talking mainly about composition and structure here, with a sprinkling of arrangement. As I produce mainly rock/metal/indie bands, I usually find that a lot of the extra instrumentation and production quirks are formulated in pre-production and during the recording session, meaning that you, as the producer/engineer, will have a lot of control over the direction of those ideas. If you are working in more synthesized or sample-heavy genres, you may also have to keep a careful eye on quality of samples, software instrument choice and production quirks that the artists start with. After some experience, the arrangement/production 'issues' I have been describing in this chapter should jump out at you. Until then, here are some specific weaknesses to look out for.

6.7.1 Structural Changes

Probably the most commonly found issue is that the song's structure makes the track too long or boring, or lacks momentum to keep interest. Here are some factors that can contribute to this:

Cut the Intro

The intro is too long. A lot of bands have tracks that have 40–60-second intros without good reason. This happens a lot when bands come up with lyrical content as the last step in their songwriting process, or when musicians like to solo. An intro should give a taste of what is to come – usually a snippet of the main chorus melody or riff. In most cases, an intro should leave the listener wanting more, not leave him or her thinking, 'When are the vocals going to come in?' Yes, for genres such as prog rock and prog metal, expectations are different, but for most rock tracks you should be thinking about ways to get the vocal in before the 30-second mark. If you are going for a commercial and radio-friendly sound, the 10–20-second mark is advisable.

Too Many Verse-Chorus Repetitions

A modern commercial single is usually 2–3 minutes long. A lot of the bands I've recorded in the last few years seem to write songs 3:30–4:30 minutes long.

For instance, if you've cut 20 seconds out of an intro of a song that was 4:30 minutes, you could still probably cut one whole verse-chorus repetition. Of course, it depends on the individual song, and this applies less to album tracks and for bands who are of a less commercial nature, but even so, I still advise you to closely scrutinize track length. Sometimes radio edits and cut-downs should be proposed early on, too.

Song Structure Is Too Simple

I've lost count of the amount of songs that have the following simple structure:

> long intro – verse 1 – chorus 1 – verse 2 – chorus 2 – long outro

Remember that the listener will usually get bored around the two-minute mark, so a change of pace around this time will usually help the overall effectiveness of the track. If you are adding a new section, you may also need to shorten other areas of the song to accommodate this.

Potential new sections could be pre-choruses, middle eights, solos/instrumentals or a coda. When you have this issue, it could be best for the band to write a new section on their own. Some bands will be more open than others to the producer playing an active role in the composition of the track, so play this one carefully. Even if the band want to write new parts themselves, they will usually be open to suggesting the placement, the length and even the general 'vibe' of the new music.

Phrase Lengths

Most tracks will have even phrases in each verse or chorus (usually fours or eights) and alternate between A and B parts. Even though you might have established a pattern, that doesn't mean you can't avoid the norm by having a double chorus at the end or two consecutive verses at the start. More unique options such as a half-chorus at the end can also create a very different feel and end with some suspense.

In alternative/progressive genres, you might find threes and sixes. On even rarer occasions, you might find that verses and choruses within the same song will have different lengths; this is usually found in those bands really trying to avoid the mainstream.

Remember, even though these are the most common structural alterations you can make, please don't let that stop you or the band experimenting with other options.

Too Many/Too Long Instrumentals

Some bands, particularly hard rock and indie bands, tend to have too many instrumental/solo sections, which is okay if it is classic rock, blues or other suitable genres. However, modern productions seldom have such lengthy instrumental sections. If there is a riff-heavy intro, solo section and outro, you may try to convince the band to drop a section or try to put vocals on some of these parts.

6.7.2 Tempo Suggestions

As well as structural changes, you are likely to come across occasions where the speed of the track is a little off. Most commonly, you will find that a song will feel a little 'pedestrian' or 'ploddy'. You can try increasing the tempo a little to add more urgency. This might be for the whole track, or, for songs that have a rigid structure and arrangement, you could add steady tempo increases throughout the track. This might be just a few bpm for the final chorus, or it could be a subtle 1–2 bpm increase between each section. This sort of gradual increase is almost completely unnoticeable to the listener, but it provides a sense of increasing intensity. This kind of increase reflects how many bands perform live, naturally speeding up during a performance when there isn't a metronome to restrain them.

Occasionally, there is the opposite problem. When a track is too fast, you will start to feel as though the musicians aren't comfortable playing it, and you may even start to hear playing errors more often than in other tracks.

6.7.3 Avoid Too Much Repetition

You could also have a song that has quite a complex structure but nevertheless sounds formulaic. If every instrument in a song's arrangement is simply repeating the same patterns for each phrase of a verse or chorus, the song will sound formulaic and lack interest. While too many differing phrases can make a song sound disconcerting, too few and a song can be predictable. My advice is that if a verse is four phrases long, you should add a variation of some description at least once in each verse to keep the listener's attention. A variation might be a snare roll or tom fill. It could be a guitar lick, vocal harmony, slight change in guitar arpeggio, an alternative melody or even a different accent on beats. In fact, it could be almost anything.

6.7.4 Masking

Songs can sound lacklustre despite a decent structure, tempo and interesting variations. The most common cause is when too many instruments are playing in the same register or a specific frequency range isn't accounted for, creating a 'sonic hole'. This happened just recently on a project I was not only producing, but playing on too. To help give me a fresh perspective, I hired a producer friend of mine to act as an 'executive producer' on the project. He listened to a few tracks and hit the nail on the head first time: 'The bass guitar is playing high up the fretboard and is being masked by the guitars.'

At this point, I thought of a couple of ways that this could be corrected. First, I could have asked the bass player to play the same part an octave lower. I chose not to do this because the register of the bass felt okay for the genre and there was a pleasing 'gelling' effect with the guitar as the bass was lightly distorted. This combination also gave a thickness in the upper mid range that I liked. However, we had a space in the low mids where the bass guitar usually resides. To deal with this, we composed a few synth bass lines that operated in the sub-bass through to the low mids. Often this synth was a direct copy of what the bass was playing but an octave lower. While this was appropriate for most of the songs on the album, there were songs where it didn't work. Where a synth was inappropriate, we used an octaver on the bass to achieve a similar effect.

Of course, EQ was an option too, but it is almost always a poor substitute compared to arranging the music so that sounds are filling the frequency spectrum in a way that removes the need for such dramatic EQ to 'fill space'. Don't misunderstand me, though; sometimes a little masking is unavoidable and tasteful use of EQ will definitely help clean up the track and make instruments sit in their own space. For instance, when recording electric guitars, there is often extraneous low-end information picked up by the microphone (rumble, boom) that doesn't really have a sonic benefit to its tone. Provided you don't go too far, removing these frequencies with an EQ (high-pass filter) will help give more definition to the kick drum and bass guitar without negatively affecting the sound of the guitar.

6.7.5 Uninteresting Hook/Lead Vocal

The lead vocal will almost always be the focal point. It should carry the song and provide interest even if the accompaniment is fairly plain. Too static a melody line or a vocal that is too 'choppy' can mean that an otherwise suitable composition seems dull or doesn't quite hit the heights it should. To test if the vocal is interesting enough, you might want to play it in isolation and see if it remains at all interesting. If it doesn't, you have some work to do. Sometimes the melody is simply lacklustre; at other times, a vocalist might be prone to having too many long gaps in his or her melodies. You might consider getting the vocalist to add extra variation or content to the melody or to include some 'shadowing' or harmony backing vocals.

6.7.6 Lack of Progression

This is the biggest and also the vaguest of these 'mistakes', and many of the issues described above can result in a lack of progression. Ultimately, the biggest pitfall in a musical arrangement is a lack of pay-off when the chorus drops or a lack of development in repeating sections in latter parts of the song. Either of these would cause a musical journey that doesn't resonate with the listener. I've found this particularly troublesome with smaller ensembles who stick with a more common song structure. I often find that once you reach the second or third chorus, you feel like you have heard it all. While there is nothing wrong with a tried-and-tested song structure, it does mean that an interesting hook and a strong core is essential. Having contrast between verse and chorus helps too.

Utilizing extra production and use of layering will help to maximize pay-off but it usually needs to happen in tandem with variation in the core instruments. It may be that the last chorus employs a more aggressive rhythm in the lead vocal, or the lead guitar arpeggio in the coda could be an octave higher than it is in the chorus, or the hi-hat could move from quavers (eighth notes) to semiquavers (sixteenth notes) in the chorus. There is almost no end to the number of options at your disposal and there are likely to be several ways to combat a lack of song development in any given situation; therefore, it is sometimes difficult to choose which path to explore. Given time, it is always best to experiment, but a good rule of thumb is to listen to the instrumentation in pre-production and pick instruments that haven't developed strongly as the track unfolds.

EXERCISES

1. Take some time to think about the genres of music that you would ideally like to work in.
2. Listen to as many critically acclaimed albums in these genres as you can and make notes on their make-up. This includes composition, structure, layering, arrangement and orchestration.
3. Pick a few of your favourite songs and think about the frequency domain and time domain ranges of instruments in these tracks.
4. Load these songs into your DAW and use an EQ or multi-band compressor to isolate and listen to the content in the following bands: 0–80 Hz (sub), 80–150 Hz (bass), 150–400 Hz (low mid), 400 Hz–2 KHz (mid range), 2–6 KHz (upper mids), 6–10 KHz (treble), 10 KHz+ (air). What instruments can you hear in each range, how dominant are they, and what are their attributes? Remember to think about the frequency and time domain.
5. Now listen to some classic tracks/albums for which there is a lot of documentation and compare what you are hearing to what is being explained in the documentation. You are doing this so that you can take inspiration from their approaches. SoundOnSound's 'Classic Tracks' and VH1's 'Classic Albums' series are particularly good for this.

6.8 'TAKE HER AWAY' COMPOSITION AND ARRANGEMENT

This section explains my thoughts and the notes I made when I first went into pre-production with Luna Kiss.

6.8.1 Core Structure and Journey

One of the first things I noticed was how solid the core structure of the song was. While I often suggest tweaks to the core structure to help keep the listener's interest, I didn't have to in this instance. Its length of over seven minutes didn't bother me because it seemed apt for the progressive rock style of the instrumentation. Besides, when you take out the experimental section after the bridge (which I call the breakdown), you actually have a rather more standard core structure:

Intro – Verse 1 – Chorus 1 – Instrumental 1 (Clean Guitar Solo) – Verse 2 – Chorus 2 – Bridge – Instrumental 2 (Distorted Guitar Solo) – Chorus – Outro

It is also worth noting that chorus 2 is double the length of the other two choruses.

Another key observation was the fact that the arrangement needed to be explosive between the chorus and the verse. I also visualized the breakdown needing to gradually build in intensity. This led me to want to make the verses more mono, light and clean, and the choruses more widely panned, heavy and distorted. In the breakdown, I wanted to build the amount of distortion in the guitar tones. The breakdown of the track and the distorted guitar solo that follows the bridge is all about incremental progression and can be thought of as consisting of three emotional states as it builds:

1. Confusion/chaos. Not particularly aggressive but a little 'off-putting'.
2. Calm before the storm/tranquillity that gradually gets more aggressive and twisted.
3. Aggressive and up-tempo, distorted and confrontational.

When an arrangement is quite a complex musical journey such as this, I often make an 'arrangement map' to help visualize how the journey unfolds.

This can come in the way of a table or a more graphical form. Both methods have their merits and sometimes I do both. The table form allows you to write some key information about each section, and a graph allows you to better visualize how the intensity changes between sections and gives you a great impression of the 'bigger picture'. Below is the arrangement map I made in pre-production for 'Take Her Away'. As I had a lot of notes on dynamics, I decided to do both the table (Table 6.5) and graphical form (Figure 6.6).

6.8.2 Rhythm, Tempo and Metre

Rhythmically, the use of metre and tempo in 'Take Her Away' is ingenious. It reminds me of the masters of progressive songwriting, Pink Floyd – or, in certain ways, like more modern progressive writing for a mainstream rock audience such as Karnivool or Biffy Clyro. Luna Kiss managed to make rather complex accenting seem natural and unforced. To the casual listener, they might not even notice how out-of-the-ordinary it is. Before the session, the band provided me with a tempo and metre map that they had programmed in Logic before the session. They sequenced all accents into many different time signature changes so that they followed the drum accents. Listening to the song, you could argue that what is occurring rhythmically is more like very advanced syncopation (shifting of accents) rather than a distinct change to the pulse/metre. However, firmly establishing changing accent patterns via alterations in the metre provided several benefits:

1. For helping us to keep track. When you are shifting around the accents so much, it can be easy to get lost on where you are in the track.
2. For the musicians' tightness. Having accents in the click in similar places to the instrumentation should highlight mistakes and help to lock the musicians into the groove.
3. For myself and my assistant to get to grips with the unusual accents more quickly.

Table 6.5

Section	Notes
Intro	Similar chords to the chorus, heavy rhythm and lead solo. With maximum width and intensity.
Verse 1	Lighter guitars, more mono. Bass clean and fingered, drums dynamically lighter. Long vocal delays between phrases (low cut to sound whispery?).
Chorus 1	Heavier guitars, wider pannings, slightly distorted bass with plectrum, drums dynamically harder hit.
Instrumental Break	Soft reverb/delay heavy solo guitar.
Verse 2	Retain very similar dynamic and layered feel, back to more mono.
Chorus 2	See Chorus 1.
Breakdown Pt. 1	Crazy samples with lots of panning variation, guitar lead and static clean guitar arpeggio. Volume rides throughout section?
Breakdown Pt. 2	Clean guitar arpeggios with clean lead guitar. Add lead harmony lines? Lead guitar is automated or controlled via pedals to get more distorted throughout the section. As this is happening, the arpeggio starts to be replaced with chords and stabs from the rhythm guitar. Just before third section hits, there is an 'eerie' pause.
Breakdown Pt. 3	Lead guitar solos in overly distorted tone – such distortion that it starts to sound square-waved and compressed. Rhythm guitar is doing distorted chords and stabs.
Chorus 3	The song goes back into a chorus this time. However, there is 1 bpm increase in tempo. Group vocals to sound out of tune (to help sound more like a crowd?).
Outro	See Intro.

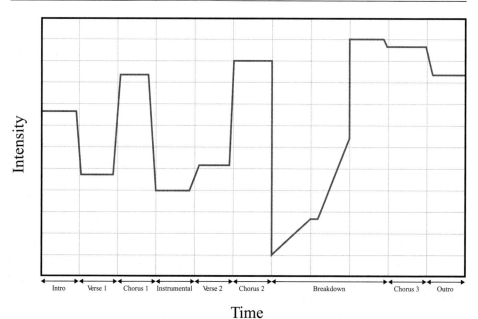

Figure 6.6 Arrangement graph for 'Take Her Away'

4. Having the accented beats on the click aligning to the key points of the drums saved a lot of potential head-scratching when editing, particularly on the drums.

The band did a fine job with the time signature and tempo map, which is probably not surprising given the fact that both guitarists have been classically trained. They programmed the tempo to be:

- 60 bpm for the intro/outro, choruses, breakdown and instrumental 2.
- 57 bpm for the verses and instrumental 1.

The tempo fluctuations between sections were incremental/decremental over the course of a few bars rather than being an instantaneous jump. I only made a few tiny adjustments to the tempo to further aid the song's journey:

1. A slight tempo increase was added for the intro and outro to give it more intensity and separation from the choruses. It went from 60 bpm to 62 bpm.
2. The last chorus is 61 bpm rather than 60 bpm to help subtly increase intensity. Increasing the tempo by 1 bpm can make a huge difference to the feeling of momentum without becoming noticeably faster. The increase happens slowly during the guitar solo so that it seems smoother and is more like how a band might be getting slightly faster because of adrenaline.
3. I increased instrumental 1 and verse 2 by 1 bpm for similar reasons to the above. I wanted the second verse to have slightly more momentum than the first.

The drum patterns were well chosen and expressive, had great use of toms and the rhythms seemed to build well. My main concerns that I noted were on making sure the timing was accurate, as any slip-ups in such a complex rhythmic arrangement could make the listener disconnected from the music. I also noted that I'd have to make sure that the expression and subtlety of the quieter sections were maintained.

6.8.3 Melody and Harmony

The melodies and harmonies used throughout the song were very strong. No changes were suggested to the guitarists or bass player. It was noted that the distorted guitar solo at the end of the breakdown could be improved. The part hadn't been fully written yet so we decided that we would experiment with a few options while tracking and probably make a composite track, which involves taking the best parts of several takes and combining them into one finished part.

The song is in the key of Am and the progressions are based around the following chords with riffs interspersed between them, although there are several different variations on these themes coloured by the use of differing chord extensions:

Intro/outro riffs – Am, B7 and E7

Verse/soft guitar instrumental – Am, E7, Dm7, G♯ aug5

Chorus – Am and G♯ aug5

Bridge – Am, Dm, E7

Breakdown – Am, C and D

All of these base chords are in the key of Am except B7 and G♯major aug5. The B7 is fairly easy to explain as it is the secondary dominant of E. B7 pulls to E(7); E7 pulls to A(m). The G♯ aug5 chord is a lot more exotic and jarring and on the surface of things much more confusing. However, substitute this chord for an E major and you get a rather friendlier V-i cadence. Both G♯major aug5 and E major share two common notes, E and A♭, which explains why the two chords share some common tonality. Therefore, I consider this chord a rather jarring and unexpected way to make you want to resolve back to the i chord.

6.8.4 Lead Melody and Lyrics

My notes regarding altering the vocals were minimal as both the melody and lyrical content seemed superbly constructed and the message fits the emotion given by the accompaniment. I did comment that the space between lines in the verses allowed an opportunity for different types of delays as a special effect.

The lyrics are based on a relationship where the girl has turned crazy as she has become obsessed over her man and the obsession is eating her away. The first part of the song describes her fading away into a cruel obsessive personality, while the second part describes the juxtaposition of how she used to be and how she is now. The chorus is a cry for help as the male tries to constantly flee from her. Here are the complete lyrics to the song:

Intro

Verse 1
Take a look at yourself
Are you the girl you wanted to be?
With those hazy green eyes
You were no good to me
From the marks on your neck I could tell
Your body has wasted away to its shell
Your bruises still young left your body to fault
Melting away to the heart.

Chorus 1
Take her away, she was trying to follow me home
Throw out the key, she won't leave me alone.

Instrumental 1 (clean guitar solo)

Verse 2
The glimmer that wilts in your eye
Once shone like the stars moving soft through the sky
A heart made as one's now a heart made of stone

Close to the touch of your skin.
From your lips there had once sat a smile
Shining like coins that have grasped at the sun
Now sits a girl perched arrest on that tear
I'm drifting away from the gaze of her fear.

Chorus 2
Take her away, she was trying to follow me home
Throw out the key, she won't leave me alone
Take her away, she was trying to follow me home
Throw out the key, she won't leave me alone.

Bridge
Lately she sits awake tryin' to figure out what she's become
Lately she sits awake tryin' to work it out, sort it out, figure out, what she's become.

Breakdown

Instrumental 2 (distorted solo)

Chorus 3
Take her away, she was trying to follow me home
Throw out the key, she won't leave me alone.

Outro

Before moving on, it is important to realize that a lot of music production is about taste. Developing your own personal preferences and to some degree the 'hallmarks' of your production style will help you progress. The following exercises will help you start to think about your own tastes.

EXERCISES

1. Listen to 'Take Her Away' and, if you play an instrument, play along as best as you can.

2. Analyse the effectiveness of the structure and the production. Write a list of positives and negatives.

3. Write down anything that you might have done differently and why.

4. Now make a list of mainstream songs that you think are successful productions, as well as a list of mainstream songs that are not as well produced in your opinion. Make sure that you list your reasons why.

5. Write down 10 specific aspects of the successful songs that you have chosen and where you might utilize these aspects in your own productions (e.g. the varying lengths of quirky vocal delays found in many early Lady Gaga tracks I might like to utilize in one of my own pop productions).

NOTES

1. Many frequencies will exist in a musical instrument's timbre, but the rate of decay for individual frequencies isn't identical. The reason for this is down to many factors:

   - The physical decay of differing frequencies in air. Low frequencies are not as easily absorbed by air as high frequencies; therefore, they travel further and take longer to decay. This means that the timbre of an instrument will typically start out brighter and it will get progressively duller as it decays.
   - The reflectivity and materials in the environment. Often, high frequencies are more easily reflected, but low frequencies often pass through objects.
   - The mechanical properties of the instrument themselves can cause certain frequencies to build up or reduce in amplitude.
   - Electrical properties of the instrument and the proceeding recording chain.

   The reason I am choosing to discuss these two attributes separately is so that you begin to critically listen to how sounds are interacting both in terms of frequency range and how a sound is changing over time. Inexperienced engineers often overlook the way in which a group of sounds mesh over time, or worse, they attempt to correct these problems with static processing, such as EQ (meaning that, over time, everything that passes through is being processed identically).

2. Audio frequency charts, *Sound on Sound Magazine,* January 2012.
3. Audio frequency charts, *Sound on Sound Magazine,* January 2012.
4. Bill Pere, *Songcrafters' Coloring Book: The Essential Guide to Effective and Successful Songwriting* (Creative Songwriting Academy Press, 2009).

DEMYSTIFYING RECORDING LEVELS

Although we have outlined exercises to develop your critical listening skills, explored pre-production and discussed compositional and arrangement concepts, we're not quite ready to jump into the recording process just yet. I know what you're thinking: 'I'm bored – just get to the bit with all the knobs and buttons!' While getting creative with production is why you are reading this, taking this time to really explore the foundations of the craft that are so often overlooked will save you a whole heap of pain later on and will allow you to concentrate on the 'good stuff'. So now it is time to recap on some fundamentals, bust some myths and outline some recording practices that will help you record anything!

7.1 ANALOGUE AND DIGITAL AUDIO

Now I hope that all of you reading this will know the differences between analogue and digital audio, but let's just make sure. The process of changing a physical sound wave into an analogous electrical signal using a transducer (such as a microphone) is called *analogue audio*. In an analogue system, the instantaneous electrical level is directly proportional to the instantaneous air pressure captured by the transducer. The analogue signal is then amplified and can be stored on an analogue medium such as tape (which works by magnetization) or converted further into a series of discrete mathematical numbers. Once it is converted to a numeric form, it is called *digital audio*. Digital audio works by storing the amplitude of the analogous signal many thousand times a second. These 'snapshots' are called samples and the number of times per second a sample is taken is called the *sampling rate*. The amount of detail that each of these samples can store are its *resolution* or *bit depth*. Once the signal has been converted to a series of numbers, it can easily be stored or processed using a computer.

Both analogue and digital systems have inherent strengths as well as weaknesses, so it would be unfair to call either definitively superior.

Analogue pros

- An analogue signal is an extremely efficient way of representing sound, but its accuracy can be hampered by the recording chain and storage medium.

- Many producers feel that the subtle imperfections of analogue recording methods sound warmer and more pleasing to the ear.
- By pushing the signal level and overdriving, the analogue signal path can cause saturation, which might, in certain circumstances, deliver even more pleasing results than digital methods.

Analogue cons

- Recordings are susceptible to degradation.
- Copies of the original recording are noisier and more distorted.
- Editing is more difficult and time-consuming.
- Noise from the storage medium (tape hiss) and recording device become a part of the recording.

Digital pros

- Digital editing is more powerful and generally easier.
- Duplicates are exact copies that do not degrade.
- When recording at high enough bit depths, the dynamic range (range between noise floor and clipping point) of digital audio exceeds the human hearing ability.

Digital cons

- Overdriving a digital signal results in unmusical and harsh-sounding distortion.
- With lower bit depths, conversion between analogue and digital must be done carefully to avoid loss of fidelity.
- Digital recordings have been called sonically 'colder' or 'more sterile' than fully analogue recordings. While this is a matter of taste, quality of equipment and processing methods, I have found merit to analogue recording techniques.

Because of the distinct advantages during editing and storage, digital audio has become the prevalent recording medium over the last few decades, so much so that many of today's aspiring engineers know a lot about sampling rates or bit depths but very little of the fundamental electronics terminology that would have been staple knowledge in the 1960s–1980s. Despite the popularity of digital audio, knowledge of basic electronic principles is still relevant in a modern recording environment, not least because the signal will be undoubtedly converted to or from an analogous form at some point (even if it is just to drive your speakers). As well as this, you will often encounter analogue gear and digital models of the same classic gear. Knowing basic electronic principles will also help you to avoid breaking equipment, for example by incorrectly matching amplifiers to speaker cabinets.

Luckily for you, Appendix A covers the rudiments of simple electronic circuitry such as voltage (volts, V), power (watts, W), current (amps, A) and impedance/resistance (ohms, Ω).

7.2 INTRODUCTION TO GAIN STAGING AND METERING

Before we get into gain staging, let's quickly define some terms that will be used throughout this section. First, on a system that captures and reproduces audio, there will always be two operational extremes. The *noise floor* defines the point at which a signal is so low that hiss and other system noise begin to have a detrimental effect on the quality of capture. The other extreme is the *clipping point*, which is where a signal level is so high that it cannot be accommodated by the system; therefore, the signal is distorted in such a way that it is detrimental. Two other terms are derived from these to help describe the quality of audio capture:

1. *Signal to noise ratio* – the level of noise in a recording in relation to usable signal. It can also be written SNR or S/N.
2. *Headroom* – this is the range between your operating level and the system's clipping point.

Gain structure (also known as *gain staging*) is a hot topic in the audiophile community. Bad puns aside, it is basic, key knowledge that is often overlooked entirely. Put simply, gain staging is the process of setting your levels throughout the recording chain. The art of gain staging is to record the signal 'hot' (loud) enough to avoid excess levels of unwanted noise in the recording but also give enough headroom to avoid the audio distorting in unpleasant ways. Over the years, our metering needs have changed with technological advancements, and because of this there are several common metering scales still being used in the audio industry, which has led to some confusion. The period of transition when digital audio was first introduced was particularly troublesome. What is worse is that these scales are completely different and sometimes even incompatible with each other. Demystifying these types of meters is crucial in the modern recording studio, where you are likely to see:

- analogue peak meters;
- VU (volume unit) meters;
- analogue dBu meters;
- digital peak meters; and
- digital RMS ('root mean square')

Before we start to explore the different types of metering, it is important to outline the biggest fundamental difference between analogue and digital metering. On a digital system, zero is the *maximum operating level* – go any further and undesirable distortion will be introduced. In an analogue scale, zero is the *optimum operating level* – levels can go beyond this without distortion, and even when the point of distortion is reached it might even be desirable. As we move through this chapter, we will elaborate further on the whys and hows, but understanding this fundamental difference will help you get to grips with gain staging much more quickly.

Before we outline the common types of audio metering, there is one more caveat. Remember, from Chapter 1: Production Philosophies, Your Ears and

Critical Listening, the decibel is a unitless ratio, used to express a change in power. So to calculate absolute values, we need to have a reference value. The same goes for all signal levels in audio equipment. Regardless of whether the scale is analogue or digital, it is only representing the strength of a signal based around a set reference; what changes between metering scales is one of two factors:

- The reference type *and* value used to measure the signal strength (e.g. 0.775 volts RMS, 1 milliwatt or 0 dBFS).
- Whether the meter is measuring peak or average (RMS, root mean square) signal level.

7.3 ANALOGUE METERS

Some of this next segment might get a bit maths-heavy. I have purposely omitted the working out as what is important is not the calculations themselves but remembering the relationships between scales.

7.3.1 The VU Meter

The VU (volume unit) meter was the first commonly found metering system in analogue pro-audio equipment. Its origins go all the way back to a time when all audio equipment had an input impedance or resistance of 600 Ω. This was due to the fact that when audio was transmitted over large distances, the length of wires could approach the electrical wavelength of the signal, causing reflective noise in the system. Reflection is caused by unmatched impedances, so, to maximize power transfer, a standard impedance of 600 Ω was used.

It was decided that the VU meter should read zero when there was an optimal balance between clean signal (i.e. minimal distortion) and low noise. This equated to a power level of 1 milliwatt (mW).

Figure 7.1 Example of a VU meter on a UREI 1176 classic compressor (picture courtesy of Metropolis)

The VU meter (Figure 7.1) was designed to give an indication of average volume rather than peak levels. It was designed like this because it is similar to how your ears perceive loudness. Due to the slow response time of the needle (approximately 300 ms), they are unable to show peak values. You can find VU meters on much of the classic outboard gear that is still widely in use today.

Because the VU meter is based around an optimal level of performance between headroom and SNR, optimal recording levels were calculated on older large-format mixing consoles by setting gain levels to meter roughly 0 VU (volume unit) on the loudest instruments.

Since we know the resistance (600 Ω) and the power level (1 mW), using Ohm's law, the voltage needed to zero the VU meter can be calculated as 0.774596669 (rounded to 0.775) Vrms (volts, root mean square).

7.3.2 The dBu Meter

Since then, the pro-audio industry has moved away from having a standard resistance in equipment, but the same signal strength of 0.775 Vrms remained a reference value for professional audio equipment. A new scale was created to reflect this, known as 'dBu' (decibels unloaded), or, more specifically, '0 dBu'. These days, however, many pieces of equipment employ the dBu scale of reference using LED displays. As the dBu scale's reference is volts RMS, it is also an averaging meter and does not show peak values.

7.3.3 The PPM Meter

To combat this, another type of analogue meter was developed called a PPM (peak programme meter) and these were developed by the BBC to show peak values in the analogue domain.

Hybrids of these two types of meters were also developed and are called 'peak-hold meters'. They show both averaged signal level and peak signals by displaying the average value of the signal as a solid bar and the peak value as a floating point at the top of the meter. This peak mark usually has a temporary or permanent 'hold' function to give us time to recognize the peak values.

7.3.4 The dBV Meter

The meters above will cover the majority of the analogue metering in pro-audio applications. However, it is worth noting that since then, a second separate operating level was devised for semi-pro or consumer audio equipment. This is based around an operating level of 1.0 Vrms, which is known as 'dBV'.

7.3.5 Analogue Operating Levels

We can use the power ratios I outlined in Appendix A to work out changes in voltage between signal levels. For example, when we see a figure of +3 dBu, it means the voltage gain has produced a power gain of +3 dB over 0 dBu (which is 0.775 Vrms). We work out the voltage gain ratio from the power gain to find the new voltage level of: 1.095 Vrms.

For +3 dBV, we would use the reference value V, 1.0 Vrms, finding the new voltage to be 1.41 Vrms.

These days, pro-audio equipment operates at +4 dBu, which translates as a nominal voltage of 1.227 Vrms. This level is called 'line level'. Without a 600 Ω load, 0 VU is equal to +4 dBu.

Consumer audio gear operates at a nominal power of −10 dBV, meaning a nominal operating voltage of 0.316 Vrms. Often you will find switches on pro-audio equipment that can switch between +4 dBu and −10 dBV, which makes devices more compatible with each other.

Here (*right*) is a quick breakdown of the standard operating levels and their value in volts RMS.

Unit	Level in Vrms
+4 dBu	1.227 Vrms
0 dBV	1.0 Vrms
0 dBu	0.775 Vrms
−10 dBV	0.316 Vrms

As well as VU, dBu and dBV, there are many other analogue audio signal measurements, such as:

- Vpp (peak to peak voltage);
- dBr (decibels relative to reference level);
- mW (milli-watt); and
- dBm (decibel relative to a milli-watt).

This may be confusing, but the reason for so many scales is the fact that different applications have chosen different types and values as a reference point. In reality, even as an experienced engineer, I only know what about half of these actually mean. In practical terms, you won't come across them often, if ever.

7.3.6 Analogue Scale Headroom

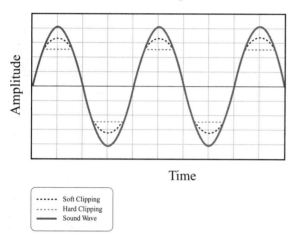

Figure 7.2 Analogue clipping example – soft and hard

Because analogue meters are centred on optimal recording levels, analogue scale meters have headroom above 0 before they reach their point of distortion. This point is often referred to as clipping. The point at which clipping starts to become audible depends on the quality of the equipment. Some high-end gear even adds subjectively pleasing artefacts to a signal when it is softly distorting (see Figure 7.2). Because of this, clipping wasn't the primary concern when it came to metering in the analogue domain.

It is a common rule of thumb that any peaks rising up to 8 dB above 0 VU (+4 dBu) are acceptable. Peaks beyond +12 dBu start to audibly distort (soft clipping), and the viability of using such levels is dependent upon the musical context and the quality of equipment used. Undesirable hard clipping can start to occur around +18 dBu or +24 dBu on high-end systems.

7.4 DIGITAL METERS

The invention of digital audio and the CD in the early 1980s caused a revolution in the pro-audio industry. Suddenly, recorded music was much more easily transferred, copied, edited and reproduced. Unlike analogue recording systems, digital signal processing produces no additional noise when changing the amplification of the signal or when going through a string of digital plug-ins (although it would amplify noise that is already there). Despite the practicality of digital audio, there are significant drawbacks: when digital audio clips, it degrades in a far harsher, more undesirable way. Analogue processing and its soft clipping properties create a 'warmer' sound that is subjectively more musical and pleasing to the ear (there will be more on this later).

The fact that digital clipping is so detrimental to sound quality meant that all current metering solutions were ineffective. It was no longer appropriate to have a meter that showed average volume, but instead a peak meter was needed to show the moment a signal clips. This led to the development of a new digital metering system called dBFS. FS stands for 'full scale', where 0 dBFS is the highest figure possible in the digital domain. Anything that goes beyond that level is a clip. Unlike analogue meters, a digital signal can only be represented by negative figures rising up to its maximum level (0 dBFS). In order to analyse digital average levels in your DAW, you need to use a dBFS RMS meter.

One other point to mention about digital signals is that once a signal reaches peak value, it will simply truncate (cut off) the wave. This is the main reason why digital clipping sounds so intrusive and far more audible. As well as truncation, another form of distortion called aliasing can also occur when a digital signal is clipped, which introduces artefacts. Aliasing will be explained shortly, but for now it is sufficient to say that when a digital system is overloaded, harmonic artefacts caused by aliasing might be introduced. These harmonics are, at worst, unrelated to the fundamental and, at best, less related, and are therefore dissonant/unpleasant to the ear. This is in contrast to an analogue signal: when an analogue signal overloads, it produces distortion that introduces more even-order harmonics, which are more heavily related to the fundamental and are therefore more pleasing to the ear. Therefore, analogue distortion is often referred to as sounding more 'musical'. This section is not meant to scare you away from digital recording; in fact, recent improvements in audio technology and metering, combined with sensible gain staging, means that truncation and aliasing artefacts shouldn't be a concern as long as you operate the system correctly (i.e. within its limits).

Let's now look at the parameters involved in creating digital audio.

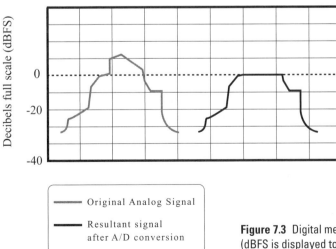

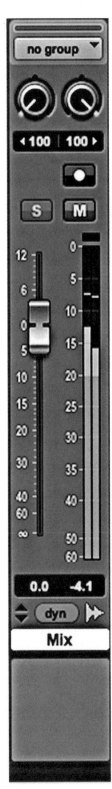

Figure 7.4 Example of a clipped digital signal

Figure 7.3 Digital meter in Pro Tools (dBFS is displayed to the right of the meter with its analogue equivalent peak programme meter on the left)

7.4.1 Conversion to and from Digital Audio

To create a digital audio signal, you first must convert a physical waveform into electrical energy (i.e. into an analogue audio signal). You then need an analogue-to-digital converter (ADC). This device takes a measurement of the electrical voltage of the analogue signal and represents it as a binary number (made up of zeroes and ones) so that it can be sent to a computer. By taking samples thousands of times per second, you can get a very accurate representation of the original audio signal (Figure 7.5).

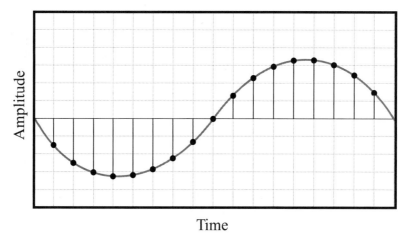

Figure 7.5 Sampling rate diagram (analogue signal is shown in red, with its digital samples shown as 'lollipops')

The same process applies in reverse when playing back a digital audio signal. First, you need to convert the digital information back into an analogue signal with a digital-to-analogue converter (DAC), then turn the electrical energy back into sound energy with a speaker.

There are two factors that determine the quality of a digital recording:

Sample rate
The rate at which the samples are captured or played back, measured in hertz (Hz), or samples per second. Common sample rates used are 44.1 KHz (44,100 samples per second), 48 KHz, 88.2 KHz and 96 KHz.

Higher sampling rates allow a digital recording to accurately record higher frequencies of sound. According to the *Shannon-Nyquist theorem*, the sampling rate you choose should be at least twice the highest frequency you want to reproduce.[1] The highest frequency a system can accurately reproduce is called the *Nyquist frequency*, which is half of the sampling rate. If a frequency that is higher than the Nyquist frequency is calculated by a ADC or DAC, it is misrepresented and causes the phenomenon called *aliasing*.

To explain the effects of aliasing, let's imagine we are recording a sine wave (i.e. a single frequency) that is steadily increasing in pitch using a sampling rate set at 48 KHz. When the sine wave is at lower frequencies, the waveform is sampled at many points in its cycle. As the sine wave's frequency increases, less points in its cycle are represented. Once you reach the Nyquist frequency (in our case, 24 KHz), each cycle is only represented by two samples. If the sine wave's frequency rises past this point, the waveform will not be expressed accurately by the system. The result of inadequate sampling is that frequencies above the Nyquist frequency are misrepresented in a way that makes them the equivalent of a lower frequency. The way that the signal is misrepresented is, in fact, highly predictable: it appears to 'reflect' around the Nyquist frequency. Using a 48 KHz system, this would make a 25 KHz sine wave indistinguishable from a 23 KHz sine wave, a 30 KHz sine wave indistinguishable from the 18 KHz and so on. Once you consider the implications of aliasing on a signal that contains more than one frequency, you can see how detrimental this can be to sound quality. To avoid this, a low-pass filter called an *anti-aliasing filter* is used to block out the frequencies above our audible range, therefore eliminating aliasing artefacts.[2]

The maximum hearing range for a human being is 20 Hz to 20,000 Hz, so any sampling rate above 40,000 Hz (40 KHz) would create an audibly perfect representation of any audible sound. The reason that the lowest standard sampling rate of 44.1 KHz (4,100 samples *above* 40 KHz) was chosen is due to slope of the anti-aliasing filter and the need to keep the entire audible range unaffected by the filter.

However, despite the anti-aliasing filter, it should be reiterated that aliasing will still occur when the signal has digitally clipped, as the distortion is introduced after the low-pass filtering stage.

Bit depth
This is the number of binary digits used to make the digital representation of each sample. The bigger the bit depth, the more accurate each sample is to its original analogue form.

There are two types of bit depth: fixed or floating-point. These types define how numbers are represented in binary form. A fixed-point bit depth has a set number of digits after the decimal point (or sometimes before), which means that fixed-point bit depths have a rigid minimum and maximum value, but between these ranges it is very accurate. A floating-point bit depth is able to move the decimal point dependant on the number of significant digits, which means that they have a far greater range. When a bit depth is listed, a fixed-point system is assumed unless stated. A floating-point system is usually indicated by the word float listed in brackets after the bit depth (i.e. 32-bit (float)). Common bit depths are 16 bit, 24 bit and 32 bit (float).

Higher bit depths give a greater dynamic range and also provide a better SNR ratio, but come at the cost of using far more space on storage media. In a fixed-point system, each extra bit gives 6 dB of extra usable dynamic range. Usable dynamic range is the difference in level between the system's clipping point and the level at which you hit the noise-floor. Therefore, theoretically, the dynamic range of fixed-point processing is 96 dB for 16-bit audio, and 144 dB for 24-bit. In reality, the figure for converters is slightly less than the theoretical values. Our own hearing has a dynamic range of approximately 120 dB from the threshold of hearing to the start of physical discomfort; this means that 24-bit can easily

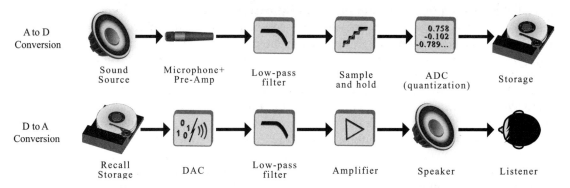

Figure 7.6
Digital audio process flow chart

cover the whole dynamic range of our hearing without clipping or hitting the noise floor. A 32-bit float system has a theoretical dynamic range of 1,680 dB, which is rather mismatched for the dynamic range of our hearing, but as we will soon find out it can have an advantage in digital signal processing.

7.4.2 Implementing a Suitable Sampling Rate and Bit Depth

When it comes to choosing a sampling rate and bit depth, there is some contrast in opinion between audio professionals regarding the quality benefits between settings. A rule of thumb is to use the highest sample rate and bit depth your computer can handle comfortably. This rule of thumb is good up to a point; when deciding settings, you should be concerned with the compromise between audio quality and computer performance.

Sampling Rate

We all know that computer speeds are improving all the time and it is no longer the case that you need the latest Pro Tools HDX rig to get the very best audio quality and professional-sounding mixes. You can make very professional-sounding mixes with a sampling rate of 44.1 KHz and if that is all your computer can handle, then you are not going to lose out an awful lot. In fact, it is true that any waveform can be completely recreated by using a sampling rate of twice the highest frequency you are wishing to reproduce (Shannon-Nyquist theorem). Therefore, recording at 44.1 KHz or 96 KHz sounds identical. This may seem counter-intuitive but you just have to take it as fact. However, there is a slight advantage to using 96 KHz over 44.1 KHz in the digital processing that comes after you have recorded. To avoid overcomplicating matters, it simply makes the maths involved in non-linear plug-ins slightly more accurate. In fact, when using lower sample rates, many of these plug-ins actually up-sample within themselves, do its processing and then down-sample back down to the session's sample rate. It doesn't make a big difference, but if you've got the power, why not use it? I still currently mix at 48 KHz as I find it to be the most efficient sample rate for the speed of my computer and sizes of the mixes I am completing, and I know many top-level guys with various HD rigs and DSP chips who still do the same.

Please note that I do not advocate the use of 192 KHz sample rates, as any improvements are inaudible at best or actually less accurate at worst over

96 KHz.[3] Also, the CPU power and storage facilities needed to process this information start to dramatically drain system resources and slow down workflow in even the best systems.

Bit Depth

Whenever audio is converted between analogue and digital, and vice versa, fixed-point systems are used. This means that any converter, interface or soundcard will be fixed-point. Modern soundcards can handle at least 24-bit recording and computer hard drives are big enough to handle their file sizes, so there are no real benefits to recording at 16-bit any more.

Once the signal is in the digital domain, your DAW can represent and store these 24-bit samples in a floating-point format. Such formatting will not increase the fidelity of the audio in any way (as the source was still a 24-bit resolution). However, there is an advantage; implementing 32-bit (float) within your DAW means that you cannot irreparably damage the audio by clipping a signal within the DAW (as the range of numbers is so huge). There is a catch, though; before the audio signal can be converted back to analogue again (for use on speakers, or to use analogue hardware), it has to formatted back into a fixed-point system. The DAW does this automatically but it means that the signal level must be back within the range of a fixed-point system to avoid clipping.

If you practise proper gain staging, the difference between 24-bit and 32-bit (float) won't matter! But if you can afford the 50 per cent bigger file sizes of 32-bit float, a safety net can't hurt.

To summarize, you should really be looking to record at a minimum of 44.1 KHz and 24-bit to get professional results.

7.5 GAIN STAGING

The biggest problem that inexperienced engineers face when recording to a DAW is that they record everything too hot.[4] There are a number of possible explanations for this – it could be confusion/hangover from the analogue days of 'set the levels to zero'; or to avoid the noise floor when recording at 16-bit; or just human instinct to think that 'the little bars should be near the top'. In your DAW, you should leave a sensible amount of headroom to 0 dBFS on every channel.

Here are several issues that will potentially cause problems if you don't leave enough internal headroom even when working all-digitally:

- You run the risk of clipping internally through plug-ins. Imagine that you are peaking at –3 dBFS on a snare drum track and you add a +3 dB high shelf to add more sparkle to the sound. You start to run a real risk of clipping the digital signal within the DAW. You can deal with this by turning down either the input or output of the plug-in down to compensate for the increase in signal. Alternatively, you could just leave more headroom initially.
- If you are using analogue-modelled plug-ins, they have been modelled to work best at a level equivalent of 0 VU, just like the originals. It is not as much that a plug-in will 'expect' a certain input level; it is just that many of these analogue models have the same varying knee behaviours and non-linearity just like the

original units, thus making them sound better being run at the same level as in the analogue domain. Each plug-in will be calibrated to a particular level, which you can find in the product's manual. *Waves CLA* series classic compressors, for example, are modelled with −18 dBFS RMS equalling 0 VU.
- If you are running levels high, it means that your levels are going out to any auxiliary subgroups and master fader equally hot, and you will need to lower the faders significantly to prevent the signal clipping when the channels are summed together.

Because of these factors, I actively try to use similar headroom levels when mixing ITB (in the box) as you would do when using analogue or hybrid systems. It must be stated, however, that working with headroom in your DAW does not mean that quality of audio is adversely affected. The mixer in your DAW is clean right up to the point of clipping (and beyond, with a floating-point bit depth); therefore, the benefits are down to practicality and improved performance from analogue hardware and emulations of such equipment. There are no adverse effects in audio quality in the DAW as long as:

- you aren't clipping on input;
- you aren't clipping in plug-ins;
- you are managing the levels into analogue-modelled gear effectively;
- you leave the master fader at unity (because its insert points are post-fader); and
- you aren't clipping going back into the analogue domain from your DA converter.

So how do you run similar levels in a DAW compared to the analogue domain? Turn it down! Every channel in your DAW should be running at peak values between −6 dBFS and −12 dBFS. Where you set the peak values in this range also depends on the nature of the instrument, which we will get to soon. If in any doubt about how hot to record an instrument, always lean towards recording at the lower end of this spectrum as you can always boost it in the digital domain later.

Even more experienced engineers can get confused with recording levels. I often see professional engineers recording all of their tracks at −18 dBFS into their DAW. While this is not going to do any harm, it is also not likely to get the best out of any analogue equipment prior to the AD conversion. This is because the recording level is likely to be lower than what you would set in the analogue domain.

So why have professional engineers made this 'mistake'? To answer this, we need to take one more look at comparisons between metering scales. When engineers started to integrate both analogue and digital processes in hybrid set-ups, manufacturers suggested recommended digital levels for engineers to calibrate their analogue VU meters to. In Britain, the calibration process involves sending an uninterrupted sine wave test tone at 1 KHz out of their system at −18 dBFS and setting the analogue equipment's meter to equal 0 VU (in the United States, it has been recommended at −20 dBFS).

This calibration process caused much confusion with engineers when it came to implementing recording levels. What engineers extrapolated from the calibration process was that they should record everything into the DAW at −18 dBFS too as this would equal 0 VU. This is wrong! Comparing 0 VU and −18 dBFS is like comparing apples to oranges, as VU is an averaging level and dBFS is a peak level. The level calibration process is based on a continuous sine wave and, because a sine wave is a pure tone consisting of only one frequency and the tone never stops, the calibration tone has an *identical* dBFS peak and RMS level. In practice, the RMS and peak levels of instruments won't be equal – the disparity between them will be based on the transient nature and decay of the instrument. Your decision on where you place the peak level between −6 and −12 dBFS should be based on:

- the transient nature of the instrument; and
- the density of the mix.

For instance, a snappy snare drum transient with a short decay is going to have a higher peak value but a much lower RMS level. On the other hand, a heavily distorted electric guitar is likely to have a much smaller disparity between its peak and RMS level. In practice, this means that you are likely to want to have the snare drum peaking a little higher in level so that it feels more balanced with the guitar without having to adjust your faders. If your track's layering is going to be very dense with keyboards and guitar parts, I often give these instruments a little less level on input so that I am not running into problems overloading plug-ins later when it comes to the mixing process.

When recording in the digital domain, we need to see any clips in real time, so monitoring peak levels is still pivotally important, but for balancing purposes you also need to be aware of the RMS signal in the same way as you would in the analogue domain. You can use an RMS meter if you wish, but over time you will get used to setting recording levels by only looking at a peak meter and by balancing instruments.

7.6 HYBRID SYSTEM CALIBRATION

If you are working in a hybrid system, where you have analogue equipment (console, analogue summing mixer, or other hardware) along with your DAW, it is important to use the calibration method mentioned above. However, you can only do it this way if your DACs have an output level trim, which is considered a premium function in converters, and generally only the most expensive will have that option.

If you don't have this option available to you, the simplest and easiest way is to consult your interface or converter's technical specs. For instance, I own a Prism Sound Orpheus, and after looking at the specifications I can see that the analogue inputs and outputs are +18 dBu at 0 dBFS. Therefore, to output at line level, I need the signal being sent to my analogue hardware to be at −14 dBFS RMS in my DAW (Table 7.1).

This means that I have 14 dB of headroom from line level to peak. This headroom is less than on some of the flagship converters, which are usually set between −14 dBFS and −18 dBFS. So on the Orpheus, we need an output signal level of −14 dB RMS to the analogue gear to equal 0 VU. Ideally, we would like

Table 7.1

0 dBFS	=	+18 dBu
−1 dBFS	=	+17 dBu
−2 dBFS	=	+16 dBu
−3 dBFS	=	+15 dBu
−4 dBFS	=	+14 dBu
−5 dBFS	=	+13 dBu
−6 dBFS	=	+12 dBu
−7 dBFS	=	+11 dBu
−8 dBFS	=	+10 dBu
−9 dBFS	=	+9 dBu
−10 dBFS	=	+8 dBu
−11 dBFS	=	+7 dBu
−12 dBFS	=	+6 dBu

to work at the UK or US standard levels, which are −18 dB RMS for the UK and −20 dB RMS for the US.

In order to do this, we would need to adjust the interface's settings to make the system reach line level at −18 dBFS instead of −14 dBFS. Unfortunately, this is a rather premium feature for a DAC and is not offered on the Orpheus. So I just have to remember when mixing that I have to output at −14 dBFS RMS to outboard equipment. In practical terms, this means I have to be slightly more careful with peak values in the DAW and channels may require more compression or limiting to avoid clipping while retaining an RMS of −14 dBFS. This is because peak values will be proportionally closer to 0 dBFS with a system calibrated to be line level at −14 dBFS than one calibrated to be −18 dBFS or −20 dBFS.

If you cannot get hold of your interface's tech spec or it doesn't mention the output sensitivity, then you can use a digital multi-meter (DMM) to work it out. All we need to know is that 0 VU = 1.227 AC volts RMS on the output of the DAC:

1. Insert the black patch lead that comes with the DMM into the COM port, and the red patch lead into the VΩ port, then set the multi-meter to read AC voltage (~).
2. Set up an auxiliary channel to send a 1 KHz sine wave tone out of your specified output at −25 dBFS RMS.
3. Insert a lead into the chosen output and on the other side of the cable place the multi-meter's probes on the + and − connectors. These are pins 2 and 3 of an XLR connector or the tip and ring of a TRS connector.
4. Keep gradually increasing the output level of the sine wave until your DMM reads 1.227 AC volts RMS (if it only displays two significant figures, use 1.23).
5. Once it reads 1.227 AC volts RMS, read the level in dBFS that you are sending out of the output. This level will be 0 VU for your audio interface/converter.

Once you know how your system is calibrated, you should have no problem running some of your key instrument faders in the DAW at near unity, and even after summing it should also leave plenty of headroom on the master fader.

7.7 CONSIDERATIONS OF WORKING WITH HYBRID/ANALOGUE SYSTEMS

Even those without experience recording with tape in a fully analogue system will probably be aware of the argument that analogue processes sound warmer and subjectively better than digital ones. Such is the reverence of analogue equipment that the world's top plug-in manufacturers are making emulations of the very best tape machines, consoles, EQs, compressors and anything analogue they can get their hands on!

But why is this happening? Digital recording is a definite improvement on analogue systems in terms of data storage, portability, ease of use, editing facilities, flexibility and workflow, but analogue systems had innate faults and tonal colour caused by the altering of the harmonic structure that can subjectively make the end result more pleasing to the ear. These faults, including the distortion introduced by overdriving tubes, tape saturation, the top-end loss in tape-machines, tonal colour of classic hardware and countless other small imperfections, can help bring a sense of cohesion and 'glue' to a multitrack recording.

This does *not* mean that analogue is better than digital. In fact, unless you have the budget for the top-level consoles or hardware and the maintenance budget to match, it usually isn't. I am not saying there aren't bargains out there, but try equipment out before you buy it. In my experience, on a smaller budget you get comparable (if not better) results by opting to buy plug-ins that emulate top-end analogue technology rather than buying cheaper hardware.

Working digitally is not an excuse for why you are not getting warm, punchy and pleasing results while recording and mixing entirely in the digital domain, or 'in the box' (ITB). There are plenty of techniques you can use to make the result of a digital recording sound more 'analogue'. Many top-level mixers such as Tchad Blake, Andrew Scheps and Dave Pensado have mixed commercial-grade results entirely ITB. I'd go so far as to say that the difference you hear in commercial recordings is actually 90 per cent the skill of the engineer, the quality of the rooms used to record/mix in, the quality of performers and instrument build, and only 10 per cent is the 'sheen' added by the top-of-the-line pro-audio equipment. There will be a deeper discussion of this topic in Chapter 10: Decision-Making and General Recording Techniques.

If you do have the budget to buy analogue hardware or even want to set up a classic analogue-only system based on a tape machine, you will need to think a little bit more about signal-to-noise ratio. Fully analogue recording processes were notorious for developing electronic hardware-based SNR issues. A few standard operational procedures help to reduce the causes of a low SNR:

- Use amplifier volumes and mic positioning to maximize level without having to crank the pre-amp's gain.
- Minimize the amount of equipment that the audio passes through. Only make the recording chain as long as it needs to be. This includes tape-bouncing only when necessary.
- Avoid reducing gain in one processor only to increase it again in another.
- Regularly maintain analogue audio gear.

7.8 MONITORING LEVELS AND THE K-SYSTEM

In Chapter 1: Production Philosophies, Your Ears and Critical Listening, we discussed the importance of reasonable monitoring levels. Now that we have learnt more about metering, we can start to calibrate monitoring levels in a more fail-safe way.

In the music industry, it has become standard practice for producers to set their own monitoring levels. In the film and television industry, however, significant standards have been set to help unify the levels of audio. This level is calibrated so that 0 VU equals 83 dB SPL.

Renowned mastering engineer Bob Katz sought to change this with the K-System, which involves a three-tiered monitoring gain and metering system, which you change dependent on application.[5] While the K-System is not universally implemented by audio professionals, it is a great starting point for the inexperienced to make sure they are monitoring correctly. It is designed to:

1. Allow you to easily return to a calibrated listening level that is a compromise between the point at which your ears have the flattest frequency response (Fletcher-Munson), and also a comfortable listening level that can be used for extended periods of time without ear fatigue.
2. Optimize headroom in your mixes and to make sure that dynamic range is not excessively reduced, which can result in loss of excitement and impact.

The three tiers are called K-20, K-14 and K-12. The proposed applications for each of these are as follows:

- K-20 is for music with a high dynamic range, such as the large theatre film mixes, classical music, etc. (20 dB of headroom).
- K-14 is for the mixing of standard pop-rock music, etc. (14 dB of headroom).
- K-12 is for programmes dedicated for TV or radio broadcast (12 dB of headroom).

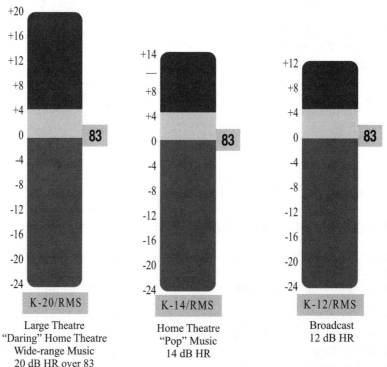

Figure 7.7
K-20, K-14 and K-12 meters

To explain Figure 7.7, the top of each meter is always your system's clipping point and 0 is always your reference listening level of 83 dB SPL. What changes is the amount of headroom that the system has between 0 and the clipping point. So working in K-12 will have a generally higher average level in your DAW than working in K-14 or K-20. This means that more tightly controlled dynamics are required to stay closer to 0 while also not exceeding the limits of the meter (which would result in clipping). This generally means that individual channels have to be compressed/limited and to a certain extent more aggressively EQ'd to control the dynamics.

To properly integrate the K-System into your workflow, you need to be aware of three factors:

1. You *must* use the standardized monitor level outlined by the K-System. This states that each monitor should be set to a level of 83 dB with an SPL meter, which should be set to C weighting with a slow attack.
2. You *must* use the K-Meter on the mix bus to ensure the sections of your track are falling within the required levels suggested by the K-System. Luckily, many of the most popular limiters and spectrum analysers come with K-Metering.
3. The K-System was developed to help govern the overall dynamic range of a mix. The K-Meter was only designed to meter an entire mix, and therefore is not relevant when determining correct levels on individual tracks within a mix (proper gain staging determines this).

7.8.1 Monitor Calibration for the K-System

Here are the steps to correctly set up your monitors for the K-System:

1. Turn your sound card's output to 0 dBFS (and any monitoring controller you might have).
2. Output pink noise from your DAW at −20 dBFS RMS.
3. With the pink noise outputting to your monitors, each speaker needs to be turned up to 83 dBSPL using an SPL meter at the listening position. You need to make sure that you do each speaker separately, and that only the speaker you are setting is turned on. You also need to make sure that the SPL meter is set to C weighting (which has a flatter bass response) and that the attack time is set to slow.

Now that you have done this, you can operate at K-20. If we then choose to operate at K-14, our unity point (0) is 6dB higher than when we were operating in K-20. To maintain our ideal monitoring level of 83 dB, we need to trim 6 dB from our monitor output. To operate at K-12, we need to trim 8 dB.

7.8.2 Using the K-Meter

To implement the K-System, you need to insert a plug-in that has a K-Metering feature on the last insert slot of the mix bus. Both PSP's Xenon and FabFilter's Pro-L brickwall limiters have this capability. To work out how to use the meter, let's examine the K-14 scale some more.

Figure 7.8
K-14 meter

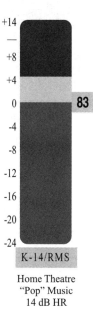

The K-meter is split into three operating ranges, the green, amber and red zones. Your aim is to operate in the green zone (less than 0) during the quieter sections of a song, the amber zone (0 to +4) for louder sections such as choruses, and the red zone (+4 and above) for the loudest sections (typically last choruses/solos/outros). Values over +8 should be seldom used, as this starts to leave the mastering engineer too little headroom to work with. These limits are guidelines and should be based on personal taste and the style of song. For instance, a particularly muscular heavy metal track may be in the red zone more often than a pop track. Similarly, a stripped-down acoustic track may always operate in the green or amber zones.

EXERCISES

1. Intentionally clip analogue and digital systems to learn how each type of distortion sounds.
2. Use your sound card's documentation or a multi-meter to work out the RMS signal level required to output at 0 VU.
3. Play around with different sample rates and bit depths and see if you can tell the difference.
4. Open a completed mix. Are you guilty of recording too hot? Have you peaked within plug-ins? Have you left enough headroom for a mastering engineer?
5. Set up the K-System in your project studio, remembering that the K-System is about monitor calibration *and* metering.

NOTES

1. The reason most ITB mixes don't sound as good as analog mixes, accessed January 2013, www.gearslutz.com/board/so-much-gear-so-little-time/463010-reason-most-itb-mixes-dona-t-sound-good-analog-mixes-restored.html.
2. Hugh Robjohns, What is 'aliasing' and what's the cause of it?, accessed January 2013, www.soundonsound.com/sos/jan06/articles/qa0106_2.htm.
3. Shannon-Nyquist sampling theorum, accessed January 2013, http://en.wikipedia.org/wiki/Nyquist%E2%80%93Shannon_sampling_theorem.
4. Dan Lavry, The optimal sample rate for quality audio, white paper accessed January 2013, www.lavryengineering.com/pdfs/lavry-white-paper-the_optimal_sample_rate_for_quality_audio.pdf.
5. Bob Katz, *Mastering Audio: The Art and the Science* (Focal Press, 2007).

MICROPHONE, PRE-AMP AND LIVE ROOM CHOICE

To be able to make decisions about the recording chain, it is advisable to know the basic operation of the core parts (i.e. the microphone, pre-amp and room selection). Later chapters will cover general placement guides and the specifics of capturing each instrument. For now, I want to concentrate on the technical operation of these variables.

8.1 BUYING EQUIPMENT

When it comes to purchasing equipment, it is always worth testing it out beforehand as it is really about your ears, taste and the program material you are going to use it on, to see whether it is worth your investment. If you have a decent amount of money to spend, a retailer is usually able to take a deposit and send you some demo units to try.

When you are buying your first set-up, you will need a few different pieces of equipment, so you may have to compromise on quality here and there. My advice is to put priority on the monitoring and acoustics first and foremost. Once you have a fully functioning set-up, you should make your upgrades count. If you are really serious about making music production your career, avoid the temptation for intermediate upgrades – save for longer and buy premium equipment. Not only will this mean you'll get longevity from your purchases, but the equipment holds its value and can be easily resold. A used £1,400 Neumann U87 can be sold for the same amount you bought it for, and if you are lucky it may even go up in value. However, the same amount of cheaper used mics will usually depreciate much more. In the early days of my career, I lost count of the amount of times I bought equipment only to want to upgrade it six months later.

8.2 MICROPHONES

When choosing a microphone for a specific purpose, it is not just its frequency response that is important in making a decision. How quickly the microphone responds to a change in pressure (this is called *transient response*), and the different directions that microphones can pick up sound in their environment (this is called their *polar pattern*), are also important.

8.2.1 Microphone Types

All studio mics operate on the same basic principle: sound energy moves a diaphragm, and this motion is then converted to an electrical signal. However, their mechanisms can be divided into several categories. Regardless of the type of microphone, lighter diaphragms require less energy to move than heavier diaphragms. Therefore, lighter diaphragms react more quickly to sudden changes of sound energy than heavier diaphragms. This means they typically have a faster transient response. Heavier diaphragms are less sensitive but have lower self-noise. They will also typically appear to be 'warmer' or more bass-heavy because the diaphragm will have a lower frequency extension and not react as well to higher frequencies.

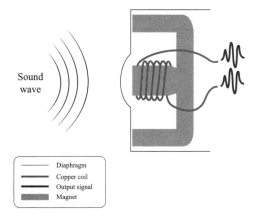

Figure 8.1
Dynamic mic mechanism[1]

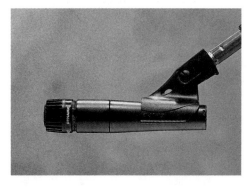

Figure 8.2
Shure SM57 dynamic instrument mic (picture courtesy of Metropolis)

Dynamic

A dynamic microphone works via electromagnetic induction. A cone is attached to a coil of wire that moves in the field of a magnet. When the diaphragm vibrates, the coil moves in the magnetic field, producing changes in current that directly relate to the changes in air pressure. However, dynamic mics tend to have the heaviest diaphragms and do not respond in a linear fashion to all audio frequencies, and therefore can be considered less accurate.

Dynamic mics are very good for close-mic'd sources. They are cheap, robust, durable and can traditionally withstand very high sound pressure levels. They are particularly good on sources with heavy bass and mid range content. Dynamic mics suffer from a lack of 'air' in the top end and also bass response when not close to the sound source. There is commonly a presence peak between 2 and 5 KHz.

Ribbon

Just like dynamic mics, ribbon microphones employ electromagnetism to convert sound energy to electrical energy. When the long thin strip of conductive foil moves within the magnetic field, it generates a voltage. The foil's lower weight compared to a moving coil like in a dynamic mic gives it a smoother and higher frequency response. However, despite the efficient transient response, they tend to roll off higher frequencies (usually from around 10 KHz and above).

Ribbon mics are great for applications where you want a smooth high frequency roll-off and a warm low end. Even though ribbon technology has improved recently, they are still typically the most fragile mics – be particularly careful using them on high-SPL sources such as bass drum, snare drum or high-gain guitar cabs. Apart from a handful of newer hybrid models, ribbons do not need phantom power and using it can damage them. They need to be used with a high quality pre-amp as they generally require a lot of gain. They are often used as drum overheads to compensate for 'thrashiness' from cymbals, and also on electric guitars (often in combination with a dynamic mic).

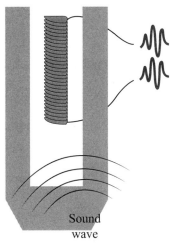

Figure 8.3
Ribbon mic mechanism[2]

Figure 8.4
Royer 121 ribbon mic (picture courtesy of Metropolis)

Condenser

In a condenser mic, the diaphragm acts as one plate of a capacitor, and the vibrations produce changes in the distance between the plates and therefore a change in electrical energy. Condenser mics typically have the lightest diaphragms of the main microphone types, meaning that they have the fastest transient response. They also typically have the broadest and flattest frequency response, therefore making them the most accurate. Because their diaphragms are light, they are usually more fragile than dynamic mics.

There are two types of condenser microphones, called 'true condensers' and electrets. A true condenser needs power to charge the back-plate of the diaphragm. The back-plate of an electret is already charged and remains charged for the life of the microphone. Despite the fact that an electret microphone has a charged back-plate, both varieties still need some sort of voltage to run (because of their FET impedance matching circuit requiring power). This power is usually supplied by +48 v phantom power on the pre-amp. However, older varieties often run from batteries. Electret mics are usually cheaper than true condensers but suffer from a gradual but minimal reduction in charge of the back-plate.

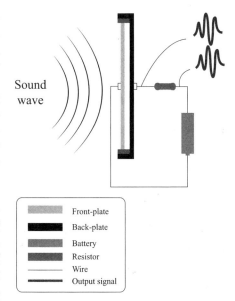

Figure 8.5 Condenser mic mechanism[3]

Condenser mics are perfect for instruments with a lot of top end content such as cymbals, acoustic guitars and studio vocals. They are frequently used on instruments (including drum shells) when a more natural or airy sound is required. They are common in genres such as jazz or country.

Tube

Tube mics are condenser mics with a tube pre-amplifier stage. They are characterized by an extra warmth compared to a condenser. These shouldn't be used instead of a condenser when sound sources are especially quiet or mic'd from a distance due to their lower signal-to-noise ratio.

Figure 8.6
Neumann U87 large diaphragm condenser mic (picture courtesy of Metropolis)

Boundary

A boundary microphone, or PZM (pressure zone microphone), is a special type of condenser mic, mounted facing a metal plate. It works by utilizing a phenomenon that occurs near a reflection point: the source signal and reflection combine to create twice the pressure of the source alone. A major advantage of using PZMs is that, because they capture the first reflection point, you get less of the room's tone. This is particularly good when you want a depth to the drum tone that can be lacking when recording in a small tracking room (when normal microphone set-ups are giving you a comb-filtered sound).

Figure 8.7
Telefunken AK47 tube condenser mic (picture courtesy of Metropolis)

Figure 8.8
Realistic PZM boundary mic (picture courtesy of Metropolis)

8.2.2 Polar Patterns

The polar pattern of a microphone is how well the microphone picks up sound from different angles or directions. Some microphones have a set polar pattern, while others can be switchable. Engineers use different polar patterns to help reduce spill between instruments or in some cases to affect frequency response.

The most common types of polar patterns are: omnidirectional, cardioid, super-/hyper-cardioid and figure of eight.

Omnidirectional

The omnidirectional microphone has equal sensitivity at all angles, meaning it picks up sound from all directions. This is advantageous when you want the microphone to pick up ambience. An omni mic is a disadvantage when you are trying to avoid spill or are trying to isolate sound in a live event (to reduce possibility of feedback). Condenser and tube mics often have switches to make them omnidirectional (Figure 8.9).

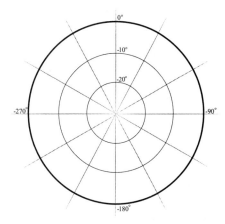

Figure 8.9 Omnidirectional mic pickup pattern

Cardioid

A cardioid microphone is most sensitive to sound at the front, and rejects sound from the back extremely well. This means that a cardioid mic is very directional and is great for reducing bleed and in live sound scenarios, reducing feedback. Dynamic mics are usually cardioid and most condensers will either be cardioid or are easily changed (such as a multi-pattern switch or interchangeable capsules) (Figure 8.10).

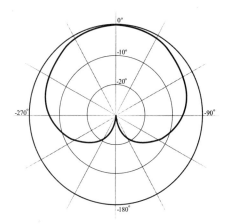

Figure 8.10 Cardioid mic pickup pattern

Super-/Hyper-Cardioid

A hyper-cardioid microphone is similar to a standard cardioid, but with a slightly larger figure-of-eight contribution leading to a tighter front sensitivity compared to a standard cardioid, but with a small amount of rear sensitivity.

Super- and hyper-cardioids are useful when reducing bleed is of paramount importance, like vocals on-stage or times where there is a lot of spill between snare and hi-hat (Figure 8.11).

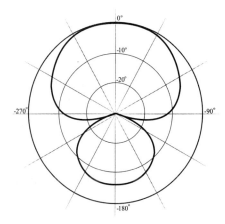

Figure 8.11 Hyper-cardioid mic pickup pattern

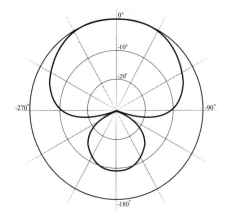

Figure 8.12 Super-cardioid mic pickup pattern

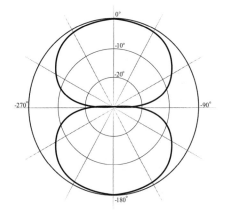

Figure 8.13 Figure-of-eight mic pickup pattern

A super-cardioid microphone is similar to a hyper-cardioid, except there is a larger front pickup and less on the rear. Hyper- and super-cardioid mics are usually dynamic mics.

Figure of Eight

A microphone with a figure of eight (or figure eight) polar pattern picks up the sound from the front and rear but not on the sides (90-degree angle). Ribbon mics have a figure-of-eight polar pattern and large diaphragm condensers often have a figure-of-eight switch.

Because figure-of eight-mics have such great rejection on the sides, they are often used on singer-songwriters who are playing acoustic guitar and singing at the same time. By using figure-of-eight mics on both sources, you can considerably reduce spill compared even to cardioid patterns.

8.2.3 Proximity Effect

Proximity effect is an increase in the bass response of a microphone, which occurs when the microphone is in close proximity to the source. Proximity effect is caused by microphone porting, which is an opening that allows air into the microphone only in certain positions and blocks air from hitting the diaphragm from any other direction. Porting creates directional polar patterns, so omni microphones do not exhibit the proximity effect. The physics behind this can get a bit complicated so a full explanation will not be beneficial; what is important is that the amount of bass boost is dependent on the microphone design and distance from the source. Figure-of-eight microphones exhibit the most proximity effect, and cardioids and their derivatives exhibit moderate proximity effect; omnidirectional microphones exhibit no proximity effect.

Proximity effect can be both a strength and weakness in engineering situations. It is often utilized to create a 'fatter' or 'warmer' sounding source, but the proximity effect can often accumulate on all of the sources to create a massive low-end build-up and a 'muddy mess' that would take lots of corrective EQ to sort out.

8.2.4 On-/Off-Axis

When you are recording you can place a microphone so that its pickup direction is pointing straight on to the source, or you could place it at an angle. Pointing the microphone directly at the source will get a more accurate representation

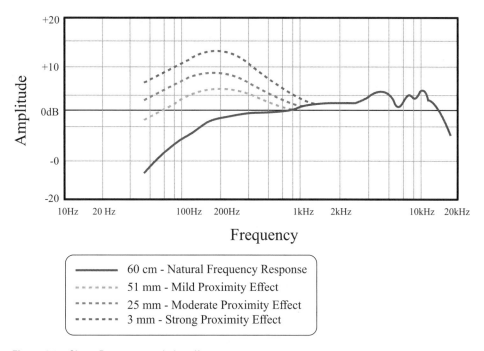

Figure 8.14 Shure Beta 57a proximity effect

of the source because the mic's frequency response is usually the flattest there, but this is not always what you want.

When the microphone is placed at an angle to the sound source, the microphone's frequency response will change, and a lot of the time this change is not subtle. You should experiment with hearing how all of your mics sound off-axis and internalize this so that you can make informed decisions on how you capture a source; when you do so, I think you will be surprised at the level of variation. The most commonly found and obvious change when a mic is off-axis is that the top end is significantly rolled off, leaving a darker, more 'warm' tone. Turning a mic off-axis can have many benefits, including:

1. Making an instrument fit better in the context of the track without having to resort to signal processing (EQ, compression, etc.).
2. To avoid overly plosive (b and p) or sibilant (s, t, z) sounds because air isn't directly hitting the diaphragm.
3. To get a larger-than-life representation of the source, you could place the mic off-axis and then use a smooth-sounding EQ to add in a little 'air' (which are the ultra-high frequencies that give the track that sparkle).

In effect, turning the mic off-axis is like a form of natural EQ, and the best part is that the amount of natural EQ that is applied is based on how far off-axis the microphone is. Generally speaking, engineers will experiment with placements between completely on-axis and at an angle of 45 degrees away from the source.

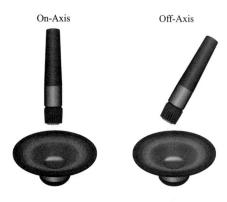

Figure 8.15 Microphone on- and off-axis

Remember, if you can solve a problem with mic choice and placement rather than EQ and compression, you should probably do it. In Chapter 10: Decision-Making and General Recording Techniques, we will talk about mic choice and placement in more detail.

8.3 MICROPHONE PRE-AMP

In combination with the microphone, pre-amp selection can have a significant impact on the overall tonality and quality of your signal. I would wager that many of you are most familiar with the stock pre-amps in your audio interface, but you may have read about dedicated 'high-end' pre-amps or used them in another studio. My belief is that while the choice of pre-amp is important, there are more important issues to get right first. Only when I have matched the right musician with the right instrument, and placed him or her in the best spot of the best room and selected and placed microphones suitably, do I consider which pre-amp to use.

Mic pre-amps come in many different 'flavours'. In the last couple of decades, it has become a bit of fad to own a series of different pre-amps. Through the years, bands have made records entirely on a studio's house console without using extra pre-amps. This shows that having a variety of pre-amp options is a nice luxury, but is not of critical importance. However, for the home recordist, it is good to invest in a quality mic pre-amp (preferably two channels with some in-built compression). This can give your recordings that extra wow factor without having to remortgage your property.

8.3.1 Microphone Pre-Amp Types

Pre-amps can be classified into two main categories – 'colour pre-amps' and 'transparent pre-amps'. A colour pre-amp is built with components that add a tonal character to the signal. The purpose of transparent pre-amps, however, is to add as little tonal character as possible. In rock and pop music, you are often aiming for everything to sound 'larger than life', and in this respect coloured pre-amps can be very useful. Coloured pre-amps will usually contain tubes and/or transformers. Transparent pre-amps are usually solid-state and transformerless. The major advantage of transparent pre-amps is that they more accurately replicate what you naturally hear with your ears. By and large, they are also quieter and cheaper to produce.

Tube versus Solid-State

Microphone pre-amps are based on either tube or solid-state technology. Many producers like to mix and match dependent on the source, bearing in mind the type of instrument they are recording, the type of signal they are receiving and/or the genre of the record.

Some advantages of tube pre-amps are:

- Warmth: If you are looking to impart some 'vintage warmth' or are finding a vocal performance 'thin', whatever microphone and placement you choose then a tube pre-amp may be the way to go.
- Distorts more 'musically': When a tube pre-amp starts to run into distortion, it produces even-order harmonics.
- Natural compression: To a certain extent, tube pre-amplifiers, when driven, add a little bit of 'musical' compression.

Some advantages of solid-state pre-amps are:

- Less distortion until they reach clipping point; this means they are great for adding clean gain.
- They sound great on instruments with percussive attack as they have a quicker transient response and retain more high frequency information.
- Generally they are more efficient, produce less heat and require less maintenance than tube-based equipment.

Tranformer versus Transformerless

Just because a pre-amp is solid-state, it doesn't necessarily mean that it cannot sound 'coloured'. In fact, some of the most popular and colourful pre-amps are actually solid-state (take a bow, Rupert Neve!). The famous Neve 1073's rich bottom end is largely attributed to the transformer circuitry. Similarly, API pre-amps – famed for their punch – incorporate transformers. Where you are looking for a true-to-life pre-amp, transformers are usually omitted from the design. Transformers are still used not only because of their tonal benefits, but also because a transformer bridges the input and output section via magnetic coupling – and therefore no hardwiring. This gives the pre-amp's circuit more electronic isolation, and the unit is therefore less noisy and produces less hum when connected to other gear than transformerless designs.

Figure 8.16 Neve 1073 solid-state transformer-based pre-amp (picture courtesy of Metropolis)

Hybrid and Modelling Pre-Amps

Recently, pre-amp design has advanced so that even more tonal flexibility is available. This can be done in two ways, via *hybrid* and *modelling* designs.

A hybrid design can use various mechanisms to allow choice between tube and solid-state technology. For instance, the summit audio 2BA-221 uses solid-state circuitry on the input stage with an output gain control that drives a tube.

A modelling pre-amp usually uses a transparent solid-state front-end with digital processing to emulate the sound of classic pre-amps.

Class

When people read 'class A' on the pre-amp's box or marketing materials, it is easy to assume that it refers to a magical quality assurance board for electrical products. However, sadly, such a thing does not exist. All it really means is that it is a certain type of circuit design where current to the amplifier flows all the time. This is great for accuracy but not for efficiency. In class B designs, current is only flowing half of the time, meaning that only the peak or trough is amplified, which creates larger amounts of distortion but is much more efficient. In class AB, two drivers are used, one that does the peak and the other doing the trough, to make sure that both sides of the wave are represented. In general, Class AB pre-amps are a middle ground between class A and B, creating more distortion than a class A pre-amp but less than a class B, and they are also between the class A and B in terms of efficiency.

While there are no hard-and-fast rules to which pre-amp class is 'the best', you will find that most designs are class A or class AB. Generally, a class A pre-amp is accurate and has low distortion, which is a very positive attribute for transparent pro-audio applications, but their power-hungry nature and the fact that they run hotter are negatives. Another potential disadvantage is that manufacturers tend to make their enclosures larger on class A designs to help heat dissipation. As well as class A pre-amps, class AB pre's are fairly popular in pro-audio applications and often provide subjectively pleasing distortion characteristics that in some applications might be just what is needed. The class of the pre-amp is not the only factor in how they sound so I tend to judge pre-amps not by their class, but by testing them; in my opinion, the term class A can be used as a marketing buzzword to try to sell more products.

Impedance

A lot of people don't realize that the impedance of mic pre-amps can have a significant tonal effect. Impedance can affect the level of distortion, output volume and the amount of high frequency information. These days, the impedance in most modern pre-amps is set so high you can mix and match pretty much any mic and pre-amp. However, trying to combine different variations of mics and pre-amps can give interesting results. Some pre-amps even have variable impedance so that you can utilize this EQ-like effect. The difference is particularly noticeable on ribbon mics where impedances are typically very low. The rule of thumb here is that the input impedance of the pre-amp needs to be at least five times the output impedance of the microphone, with ratios of 10:1 being deemed optimal. If you have variable impedance, as you move from low impedance to high, you should expect to hear an increase in volume and a less dark and mid-heavy tone with more 'airy' high-frequency content; it may also give more room tone and appear to 'flatten out' the frequency response of the microphone. The advantage of using variable impedance is that it means you have some tone control without having to patch anything else into your chain.

Choosing a Mic Pre-Amp

At the end of the day, a lot of the nuances between mic pre-amps are subtle and are largely determined by personal taste. If you only have your interface's stock pre-amps, you shouldn't lose too much sleep. If you do have a larger budget or are considering a specific studio based on their equipment, my typical preferences are:

- For drums and guitars, given a range of pre-amps at my disposal, I will quite often favour solid-state designs with a subtle transformer-based colour; API pre-amps could be my preferred choice here.
- For vocals and bass guitar, I often opt for a tube pre-amp or a more coloured transformer design such as a Neve.
- If I was going for a vintage sound, I might consider a Neve or tube pre-amp throughout.

8.3.2 Line Inputs/Hi-Z/DI Boxes

Most pre-amps these days have alternative input(s) usually on 1/4" jack inputs, as well as the XLR input(s) you will find for microphones. This gives your pre-amp the ability to accept other strengths of signal. The operation of these inputs is different and shouldn't be confused as it can lead to distortion, noise or other negative effects.

Line Input

A *line input* is designed to work with signals that have already been amplified to around +4 dBu. This is the signal strength that mixers, audio interfaces, etc. use to pass audio signals around. Microphones and instruments (excluding many keyboards and synths) do not directly output line-level signals so a pre-amp is needed to boost this signal up to line level.

Hi-Z Input

In the pro-audio world, the letter 'Z' is used to represent impedance. *Hi-Z* inputs are used when the output of a device that you wish to capture has a high impedance; this is typically the case on acoustic, electric guitars and basses. To ease confusion, these inputs are sometimes called instrument inputs.

DI Boxes

If you don't have a hi-Z input, you will want to convert the high-impedance signal into a low-impedance mic-level signal so that it can be amplified by your mic pre-amp. The job of a *direct injection (DI) box* is to do just that. Changing the hi-Z signal into a lo-Z signal also allows you to run a long cable without adding noise or losing signal quality. High impedance signals such as an electric guitar or bass are more prone to noise, therefore keeping cable lengths short (under 8 metres) is preferable. Another advantage of a DI box is that it allows you to output directly from a multitude of instruments through the DI box and into a studio's XLR wall box. DI boxes tend to come in one of two varieties: *passive* and *active*.

A *passive DI box* uses a transformer and needs no external power. Passive DI boxes are great for signals that have already had some sort of amplification, such as the built-in battery-powered pre-amps of active basses or acoustic guitars, or the output level of electronic keyboards.

Figure 8.17 BSS AR133 active DI box (picture courtesy of Metropolis)

An *active DI box*, on the other hand, has a pre-amp built into it and is great for higher-impedance sources such as standard electromagnetic pickups on electric guitars and basses, Rhodes pianos and magnetic acoustic guitar pickups.

These days, the majority of producers will prefer to record DI guitars and keyboards in the control room, meaning that the hi-Z inputs build into pre-amps are usually of a high enough quality and convenient enough to not warrant the use of a DI box. However, you will see them used when much of the recording process is happening in the live room. In studio environments, you will also see a variant of a DI box called a *re-amp box*, which is used to convert a lo-Z signal back into a hi-Z signal. We will get to this in Chapter 12: Guitar Tracking.

8.4 LIVE ROOMS

Many of you reading this will not have access to several types of dedicated live rooms to record in, let alone a selection of them in one studio. In fact, when recording 'Take Her Away', I had a single live room to record the whole track in. This live room was also not the greatest for tracking (recording), particularly for drums, so don't be disheartened by the space you have – a lot of the time, you will be forced to work within your means. In Chapter 11: Tracking Drums, I will discuss dealing with recording drums in small environments, but for now I want to share some options with you that you should consider if you have a larger budget.

As with all recordings, engineers make choices determined by the genre of music, the budget and also the skill of the band. As I record a lot of rock music, my preference is to find a room that will suit the drum sound I am after. Having a good-sounding room will make the processing of the audio come mixdown a lot easier and more natural. Instruments that rely on ambiance such as drums, acoustic guitars and strings will benefit hugely from a specifically designed live room. This is usually an environment that has ambience from a high ceiling and non-equal dimensions, as well as acoustic treatment to tame problematic reflections. When budgets are tighter, I will sometimes choose to record drums in a studio with a great-sounding drum room and do all the rest of the tracking in a less-well-equipped studio.

Having a wooden floor is a plus, and with some specialist subgenres of rock music, it is also advantageous to have brick/stone walls too. A nice and controlled reverb time across all frequencies is vitally important, so I would look for specific room design details, for instance:

- Diffusion and absorption panels to help control reverb.
- Specifically designed angles and ceiling designs.
- Enough 'gobos' (movable absorption/diffusion panels) to change room tone.

It is also worth remembering that not all drum rooms are developed to be completely flat across all frequencies. I've read about drum rooms with wood used on the walls to help dampen some of the top end from cymbals while retaining some more reverb decay in the bottom end. I've also used the main room at Circle Studios, Birmingham, England, which has a specific reverb tail in the top end to aid a nice 'crack' on snares. They also have rotatable wall panels that have diffusers on their reverse side to increase the reverb time of the room without adjusting the room's tonality. What really matters here is that there is controlled reverb. A procedure I use for listening to a room's acoustics is the age-old 'clap' test. I literally clap while walking around the whole room. I am looking for several things here:

- I want to hear some ambience: a nice amount of reverb that is either flat across all frequencies, or has a slight high end roll-off. I often look for something around the 0.6–1.2-second mark of RT60 time (time it takes for reverberation to drop by 60 dB). I think that cathedral and church reverb times are too long, except in special cases such as choral work.
- I listen out for any problem areas in the room. Does the bass response change in any corners? Is there flutter echo anywhere? Is there 'boxiness'?
- I am also listening for 'sweet spots' to set up instruments. As with microphone placement, there are usually positions in a room where they will sound best. Even the world's best rooms have sweet spots!

Here are a few specialized room types that are worth considering for recording drums in, with a generalized description of their acoustic characteristics. To a certain degree, every live room sounds different.

8.4.1 Stone Rooms

These are probably the least flexible of all live rooms. However, they have a great deal of attack and a huge aggressive sound. If you are after that big '80s pop sound like Phil Collins, then a stone room is the way to go. Stone rooms should be large and made up of lots of different sizes, shapes and types of stone. This gives them diffusive properties and allows the reverb tail to be more controllable. Stone rooms generally have only one wall made out of stone. The smaller the room, the harder it becomes to control the ambience, so be wary of this when selecting a stone room. A good rule of thumb is to position the kit on the opposite side of the room to the stone wall, at least 10 feet away from the wall; this helps prevent troublesome early reflections from the stone.

Figure 8.18 Circle Studios Stone Room, Birmingham, UK (picture courtesy of Trevor Gibson, www.circlestudios.co.uk)

8.4.2 Wooden Rooms

Wooden rooms offer a good alternative to stone rooms and are much more flexible. Like stone rooms, wooden rooms can be used to produce a large reverb tail. However, dependent on the acoustic properties of the wood used, the reverb tail tends to be darker with some top-end absorption. This helps tame cymbal reflections that can become abrasive when mixing. They are also easier and cheaper to build than stone rooms, and can incorporate a mix of different woods, absorption panels and diffusion.

Figure 8.19 Circle Studios Wood Room, Birmingham, UK (photography by Iain Gibson, www.igibson-photography.com)

8.4.3 Drywall Rooms

When your budget is tight, you will typically use a standard drywall room. It is important to examine the thickness of the drywall and the other acoustic panels used. You are looking for 5/8-inch thickness, and preferably double-layered. Tapping on the walls should give you an idea of thickness – if the drywall is too thin, it could amplify sounds in the room. In terms of room design and treatment, you are looking for rooms with unusual shapes, non-parallel walls, diffusion, an uneven ceiling and/or ceiling absorption and bass trapping (sometimes hidden within walls). Remember, the best test for your environment is the clap test.

8.4.4 Don't Forget the Vision

Remember that, at all times, you should consider the vision you formulated in pre-production. If you want the instruments (particularly the drums) to be hyped, aggressive and reverb-heavy, a stone room is probably best. If you want a larger-than-life reverb tail that is subtler and warmer, a wooden room may be for you. If you want drier drums, a standard drywall room with plenty of absorption and diffusion will suffice. For a rock or indie production, I would lean towards a wooden room or even a stone room, and for an RnB or funk track I would probably favour a drier environment and use digital ambience if required.

EXERCISES

For some of these, you may need to hire a studio or rent equipment to get the opportunity to experiment. While this might work out to be expensive, it is definitely worth it for your development. Maybe get together with other like-minded individuals and split the cost and compare opinions.

1. Get hold of as many microphones as you can (get at least one of each type) and experiment recording as many instruments as you can. Make sure that you use a nice and neutral pre-amp and keep the mic reasonably close to the source, and on-axis. Write down your conclusions on the tonality of the mics; this includes frequency response and transient response.

2. Zoom into each waveform. Does the visual representation of the wave recording match what you would expect from your conclusions?

3. Experiment with each mic at varying distances from a source, listening out for proximity effect and room ambience. Write down your thoughts on how distance affects each mic.

4. Experiment with recording a good quality acoustic guitar with a single at various positions in the room to hear the effects it has. Write down how different distances and positioning affects its timbre and ambience.

5. Experiment with polar patterns. First, on a single source to see how much proximity effect is exhibited between patterns. Second, try recording several sources live in a single room and experiment with how different polar patterns can reduce spill and ambience.

6. Now experiment with taking the mics off-axis. Capture a source at varying positions between completely on-axis and 45 degrees off-axis. Write down your thoughts.

7. Once you feel that you are getting to grips with the differences between mics, experiment with changing mic pre-amps. Write down your thoughts.

8. Book time in a stone room, wood room and drywall room, set up a drum kit with overheads, a mono room mic and a close mic'd kick, and compare the results in each environment.

NOTES

1. Microphones, accessed June 2014, http://hyperphysics.phy-astr.gsu.edu/hbase/audio/mic.html.
2. Microphone polar patterns, accessed June 2014, http://en.wikipedia.org/wiki/Microphone#Microphone_polar_patterns.
3. Beta 57A(r) Supercardioid Dynamic Instrument Microphone Product Specifications, accessed June 2014, http://cdn.shure.com/specification_sheet/upload/122/us_pro_beta57a_specsheet.pdf.pdf.

THE CRITICAL LISTENING ENVIRONMENT

The subjects of room acoustics and monitoring are complex enough to write an entire book about. While we don't have scope to cover them in full detail, a basic understanding of these subjects is of significant importance to anyone wishing to make music production their profession. Acting on the advice here will allow you to judge every decision you make more accurately and with more certainty than you have ever been able to before.

Here are a few questions and decisions you have to make in a recording session:

- Where shall I place the microphones on a guitar cab?
- Which mics and pre-amps should I use?
- What EQ should I use on a kick drum?
- How much reverb should I add to the lead vocal?

As I mentioned in Chapter 1: Production Philosophies, Your Ears and Critical Listening, you make decisions by judging the sound that you hear. If your speakers and your room are not telling you the truth, how can you possibly make the best, most informed decisions while producing, mixing and mastering?

When deciding what sort of room to use for your studio, there's a delicate balance to address: the acoustic quality of your environment versus the practicality and cost of the space. The solution you arrive at for your room should directly affect your choice of monitoring.

9.1 ROOM ACOUSTIC PROPERTIES

Most of you reading this will already have some sort of home studio set-up. You will probably have also noticed how the tracks you have mixed in your home studio sound different in your car, your bandmate's bedroom and the PA in your local music venue. A critical listening environment will quickly make your mixes sound better in all of these environments and you won't need to revise your mix 12+ times.

Having a critical listening environment that has a lot of uncontrolled reflections greatly undermines our ability to correctly judge our mixes. The majority of what we are hearing should be the direct sound from the speakers and not a wash of reflections coming from every room boundary. Hearing too much of this 'room tone' causes issues such as time-smeared transients, imposed ambience, amplification of certain frequencies and near destruction of others. This even varies hugely according to listening position and monitor position.

The majority of problems experienced by project studio owners and hobbyists are caused by room acoustic issues. It's somewhat ironic, then, that room treatment is often a low priority for people looking for that 'magic pill' to create good mixes. Yes, bass traps are not as sexy as the latest plug-ins or the most heralded interface made by your favourite manufacturer. But I have found that correctly applied room treatment will be the best investment you will ever make as a music producer, mixer or mastering engineer.

Maybe people don't know what a good room sounds like? I would have certainly put myself in that category in my early days. It was only once I had moved into a post-production room designed by a professional acoustician that I realized what I had been missing. 'Tuning' this room and buying the construction materials cost me approximately £2,200, which is less than my audio interface alone, yet it really made more improvement than all my other gear combined. If you have less of a budget, a significant improvement can be achieved for £500, which is a similar price to a mid-level interface. Got the picture? No? Okay, well I would go as far as saying I would rather mix in an acoustically treated room with no mixing desk, outboard and only my DAW's stock plug-ins than mix in a small untreated environment!

For the purposes of this chapter, I am going to focus solely on a control room environment. This isn't a course about building recording rooms, and if you're just starting out it's more likely that you will be recording and mixing in a single control room environment. When you come to record drums and other complex acoustic sources (pianos/acoustic guitars, etc.), I strongly recommend you hire a professional sound-isolated and -treated facility. Control rooms are typically more 'dead' sounding (meaning having less room ambience) than the majority of live recording rooms.

Acoustic design and materials exist to alter the sound of your room so that it serves its intended purpose. For example, an ideal control room environment is sonically flat with a small reverb tail to help make it as neutral as possible. A live room, on the other hand, may have to serve many purposes, and you might find several areas treated differently within one room, including adjustable screens to help isolate areas where necessary. You can have 'wet' areas with higher ceilings and more reflective surfaces such as stone, wood and concrete.

These can create ideal environments for recording drums and acoustic instruments. You can also have 'dead' areas full of absorbent materials, which are often great for vocals and amplified instruments. There is a great amount of science involved in creating these environments, and acousticians are paid vast amounts of money to design bespoke rooms for all sorts of applications.

Acoustic treatment should not be confused with soundproofing, which has the sole purpose of isolating sound from entering or escaping a room.

Figure 9.1 shows my critical listening environment at my post-production facility.

Over the course of this section, you will learn the principal causes of acoustic problems, techniques that will help you correct them, the types of material

Figure 9.1 Acoustically treated critical listening room

used to treat acoustic problems and the importance of the positioning of such treatment. Because of the complexity of these variables, wherever possible, it is best to hire a trained acoustician to design the treatment of your room. It is typically harder to achieve a flat reverb time with lower frequencies as they are harder to absorb. More specialist materials and complex bass trapping designs are therefore required.

Nevertheless, it is possible to get good results much more cheaply by using common building materials, simple mathematics, spectral analysis programs and some good old-fashioned trial and error.

So, where to start when creating a suitable critical listening environment? Before you even think about treating a room, you need to think about the size and dimensions of the room.

9.1.1 Room Modes

The first thing to do when choosing a room to set up as a critical listening environment is to measure the dimensions. The dimensions of a room will give you a basic idea of how hard you will have to work to make the space usable. A good rule of thumb to bear in mind is that smaller rooms and rooms with even dimensions (square rooms) are usually more problematic than larger rooms with more uneven dimensions. On the other hand, if you are using a room that is very large, you will have a different set of problems to deal with, such as having to heavily reduce reverb time.

Measuring the room will also give you a good indication of potential problem areas called *room modes*. Room modes or room resonances are the frequencies

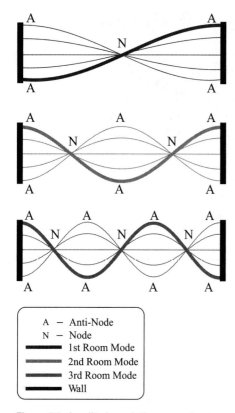

Figure 9.2 Amplitude variation according to position within a standing wave

at which your room will either reinforce (amplify) sound or act upon it destructively (decrease in amplitude) in different areas of the room. Because of this, larger rooms will suffer *much* less from room modes as the sound has sufficient time to develop and dissipate. Room modes are caused by the interaction of the sound reflected between walls – and not the interaction between the direct sound and its reflections, which is called *comb filtering* and will be covered later.

Room modes are caused by a phenomenon called *standing waves*. A standing wave is a frequency resonance caused when two waves travelling in opposite directions interfere with one another. Depending upon the position where two waveforms meet within a standing wave, they will interact in different ways (Figure 9.2). At certain positions, the interference will vary between complete constructive interference and complete destructive interference, and at other positions the interference will be continuously 100 per cent destructive (complete cancellation). Destructive areas are called nodes and constructive areas are called anti-nodes.

Above 300–400 Hz, room modes become so closely spaced that the negative effects are negligible. Furthermore, above 300 Hz, other acoustic properties have a greater effect than modal frequencies. Below 300 Hz, they become more widely spaced and distinct which increases their negative effects. Therefore, close attention should be paid to the room modes below 400 Hz. In positions where low-frequency standing waves occur constructively, the effect is often termed 'one-note bass', due to the persistent perception of a single, low-frequency anti-node.

For instance, if you had a room that was 4 m in length, it would have its first room mode at 43 Hz and then at every multiple of 43 Hz thereafter. You can work out the first modal frequencies for your room by using the following formula:

first modal frequency = $c/2L$

where L is the length between two walls and c is the speed of sound (343 m/s at room temperature).

The severity of the effect of standing waves changes significantly depending upon listening and monitor positions. For instance, a monitor placed in the anti-node of a frequency will excite that frequency fully (causing an increase in amplitude of that frequency). A monitor positioned in a node of a frequency will mean the speaker is unable to excite that frequency (resulting in no standing wave at that frequency). The same physical phenomena occur in relation to listening position too. Because of this, it is very important to identify the best listening position in the room early on. This can be done with experimentation.

Later in this chapter, we will discuss the best theoretical positions for your listening position and speaker placement.

Remember that each of the speakers' placements and the boundaries of the room itself create complex interactions, which result in the frequency response in a particular listening position. This means that while a single reflection and its result are easy to comprehend, the combined result of all of these reflections and their interactions with both the direct sound and many other reflections means these results cannot easily be explained in a simple model, let alone calculated. What I am saying here is that you should trust your ears.

A mode caused from one dimension of reflection is called an *axial mode*. However, in every room, there are three dimensions: length, width and height. This means there are three series of axial modes per room and it also means different types of modal frequencies develop between walls (i.e. two or three dimensions). Modes that take into account two dimensions are called *tangential modes* and are half of the strength of axial modes. Finally, there are *oblique* modes, which cover all three dimensions. These are a quarter of the strength of axial modes. The three room modes are shown in Figure 9.3.

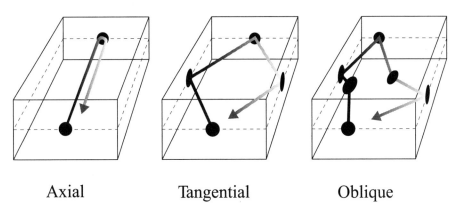

Axial   Tangential   Oblique

Figure 9.3 Types of room modes[1]

Further Resources   www.audioproductiontips.com/source/resources

 As long as your room boundaries are parallel with each other, you can use this room mode calculator to work out the potential room modes you will have to deal with in your chosen room. (73)

If you have the luxury of choosing from several rooms or are building one from scratch but don't have the budget to consult with an acoustician, then there are lots of preferred 'golden' room ratios that can help obtain a better modal performance even before you have started treating the room. Some of these ratios include (see *right*):

Height	Width	Length
1	1.14	1.39
1	1.28	1.54
1	1.6	2.33

Height	Width	Length
2.4	2.7	3.3
2.4	3.0	3.7
2.4	3.8	5.6

Here (*left*) are examples of the above, in metres, based on a standard UK ceiling height of 2.4 m:

However, it must be noted that these golden ratios are *not* the most important factor when deciding on a room. Overall size is an equal (or greater) part of the puzzle, and therefore I would not usually sacrifice moving into a significantly smaller room because it had a better ratio. Typically, any room in a standard home will be smaller than the optimal sizes for a critical listening room; therefore, there will usually be some kind of compromise. If you were to run a commercial facility with no budgetary restrictions, a ground area of 30–50 square metres is a good rule of thumb. Ceiling height is also important and I'd recommend it being at least 3 m, and if you can get a total volume of around 120 cubic metres, it is recommended.

As well as room modes, other acoustical problems such as comb filtering, flutter echo and stereo imaging deficiencies can affect the accuracy of a critical listening environment. We will discuss these before moving on to how to treat them.

9.1.2 Comb Filtering

Comb filtering is a phenomenon that causes peaks and nulls across the frequency range of a sound due to a series of constructive and destructive phase interferences (Figure 9.4). This happens when a direct sound combines with a delayed version of itself. In the case of a room, this is caused when sound reflected from the room boundaries combines with the direct sound from the speakers. Using an example of a simple sine wave and only one boundary, it is easy to see how the reflected sound interacts with the direct sound to cause constructive or destructive interference. If the reflected sine wave arrives at the listening position in phase with the direct sound it causes constructive interference; if the reflected sine wave arrives out of phase with the direct sound, the result is destructive interference.

In Figure 9.4, you can see two examples of a sine wave being reflected off of the same boundary. The frequency of the sine wave is different in the two examples and we can see how the resultant time delay would affect the phase relationship of the two frequencies when they arrive at the listening position.

For any frequency that is encountering constructive interference, the same effect will be found at any multiple of this frequency. For instance, if the time delay between the direct sound and the reflected sound arriving at the listening position is 1 ms (one thousandth of a second), the

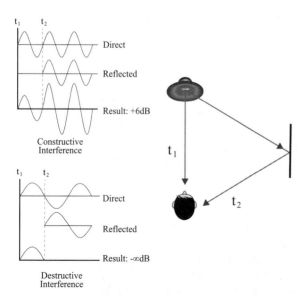

Figure 9.4 Constructive and destructive interference[2]

lowest frequency that would arrive fully in phase would be 1 kHz (as it will have completed one full cycle). As stated above, the next frequency that would arrive in phase would be at 2 kHz, then at 3 kHz, and so on.

Using the same example, the lowest frequency that would arrive completely out of phase would be 500 Hz (because it has completed half a cycle). At this frequency, the reflected sound arrives at the listening position only half a cycle behind the direct sound causing complete destructive interference. The same phenomenon occurs whenever the reflected wave and the direct sound meet at the listening position with the direct sound halfway through its cycle. These points are halfway between each of the in-phase frequencies mentioned above. After 500 Hz, the next frequencies affected would be 1.5 kHz, 2.5 kHz and so on.

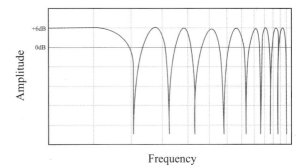

Figure 9.5 Comb-filtering distributions[3]

This simple example, which exhibits complete phase interference, does not fully represent the complexity of a real-world situation. When we extend the above example from a simple sine wave to a broadband signal (one that carries many frequencies), you can see the peaks and troughs caused by constructive and destructive interference occurring across the frequency spectrum. (Figure 9.5.)

As well as this, you have many other attributes that affect how strongly comb filtering affects sound. In a real-world environment, not only would you have many boundaries of different angles and distances from the source, but also various types of surfaces that would reflect and absorb varying amounts of each frequency. So, just as with room modes, the effects of these phenomena are incredibly complex and result in an extremely variant frequency response within an untreated room. Due to this, computer programs are extremely useful to be able to visualize where comb filtering and other acoustic problems are occurring.

Before moving on, I want to quickly mention one misconception that often confuses people about phase. People often talk about a signal being 'out of phase', but a broadband signal can never be completely out of phase (i.e. complete cancellation). This is because each frequency has its own wavelength, so when one particular frequency in the broadband signal is experiencing complete destructive interference, others will not be. The idea of a signal being completely in or out of phase only applies to the interaction of two pure sine waves at the same frequency. This concept is particularly important when listening for phase relationship problems on multi-mic'd sound sources. This will be elaborated on in the next chapter.

9.1.3 Flutter Echo

Flutter echo or slap echo is a type of one-dimensional (axial) reflection that has a distinctive high-pitched bright 'ringy' sound. It is caused by quick repeated reflections between two parallel surfaces (Figure 9.6). A flutter echo occurs for very short, percussive sounds only.

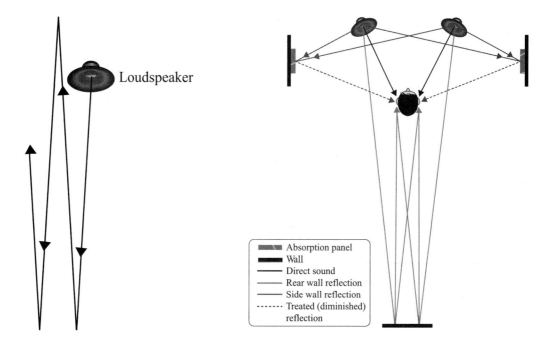

Figure 9.6 Flutter echo example[4]

Figure 9.7 Early reflection causes – the untreated rear wall reflections (blue) can give the illusion that the sound also originated from the opposite speaker; however, the side wall reflections have been treated with absorption panels, diminishing this effect[5]

9.1.4 Stereo-Imaging Deficiencies

As well as comb filtering, too many early reflections can cause the stereo imaging in the listening position to be affected. This occurs because these early reflections reflect off a boundary and into the opposite ear than the one expected. This affects stereo imaging when a sound hits a side or rear wall and reaches the listening position so quickly that the ear cannot distinguish between the original signal and the reflections (less than 20 ms). This can trick your mind into believing that the sound came also from the speaker located on that side of the room (Figure 9.7). Your mind should still be able to detect when a sound is fully left, right or in the centre of the stereo field, but less extreme panning is less defined. The result of this is a shifting of the stereo field more towards the centre.

9.2 MONITOR PLACEMENT AND LISTENING POSITION

Now that we have discussed the types of issues you are likely to encounter, it is time to discuss the ways that you can minimize their effects. When analysing a space, you should always start by finding the best listening position then decide on treatment from there. Speaker placement alone can make a significant improvement in the low end response of your critical listening environment and also dictate where to put first reflection treatment. Best of all, it's free!

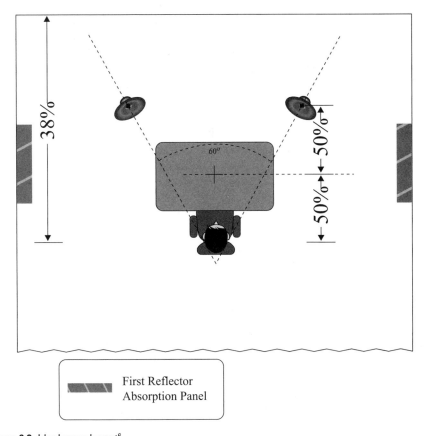

Figure 9.8 Ideal room layout[6]

While every room is different, there are some good rules of thumb that will help to find the best monitor placement and critical listening position a lot quicker. Remember, though, these are just guidelines, and experimentation to fine-tune is also advised.

Here are the steps to help you find the best listening and monitor positions for your environment (Figure 9.8):

- Speakers should be placed so that they face down the longest dimension of the room.
- Bass response is technically worst at halfway points between any walls. This is true whether the dimension is length, width or height. Avoid listening and monitoring positions in these places. However, from a stereo field perspective, having a listening position in the centre of the room width-wise is important. Sometimes it is acceptable to shift your speakers width-wise slightly to avoid the worst effects. Just be careful not to negatively affect your stereo imaging.
- Theoretically, the best listening position occurs 38 per cent down the length of your room. You can work out monitor positioning based on this, although I would always recommend using room analysis software to work the flattest possible listening position for your room.

- Your speaker placement should form an equilateral triangle between the speakers and an imaginary point slightly behind the listening position.
- Monitor tweeters should be placed at ear height – except when that leaves the woofer evenly spaced between the floor and ceiling, in which case offsetting it slightly would be beneficial (see second point, above). When doing this, use your measurement software and/or your ears to choose the optimum monitor position that produces the flattest frequency response.
- Speakers should not be placed directly against any walls. (This should not be confused with placing them directly against bass traps with air gaps.)

9.3 ROOM TREATMENT

Now that we have outlined some of the key acoustical problems and decided on the listening position and speaker placement, it is time to start outlining some of the materials, tools and tricks of the trade to help you get the most out of a small critical listening room.

Remember that the smaller the room, the more problem frequencies you will typically encounter. This is because sound does not decay sufficiently before hitting walls and other boundaries and reflecting back into the room. Therefore, rather than using acoustic treatment targeted to certain frequencies, it is better to treat as many frequencies as possible. This is called *broadband treatment* and it is what we will be focusing almost entirely on. Rooms that are smaller than 50 cubic metres will need to have significant broadband treatment to help absorb mid-to-high frequencies and there will also be a need for bass trapping where space permits. Rooms larger than 120 cubic metres are generally optimal to convert into a critical listening environment. This is because there is enough distance for the sound to decay sufficiently and cause minimal problems. Another advantage comes from there being enough space for acousticians to incorporate more complex designs such as angled walls.

There are generally three phases to consider when you are treating a room yourself:

1. *Bass trapping.* This is absorption treatment placed in the corners of rooms with the purpose of dampening low-frequency sound energy in order to create a flatter low-frequency response in the room.
2. *Creation of a reflection-free zone (RFZ) at the listening position.* This is the placement of absorbent material in any position where early reflections (also called first reflections) could interfere with the high-frequency response and also alter the stereo image you perceive. This happens when the sound from your monitors arrives at your ears through different paths – either directly from the speaker, or a delayed sound caused by reflection(s) off a nearby wall.
3. *Controlling ambience.* After treating the corners with bass traps and creating a reflection-free zone, you might still have to reduce the overall reverb time further. Treatment in areas that aren't first reflection points may be necessary.

Before we move into how to approach these three phases, let's quickly outline the types of treatment that are commonly used by engineers treating their critical listening space.

9.3.1 Types of Acoustic Treatment

When dealing with smaller environments, there are typically two ways an engineer might control acoustical energy:

1. *Absorption.* This type of treatment (Figure 9.9) 'soaks up' the sound and turns the sound energy into heat through friction. Most acoustic treatment products available work by absorption, including foam products and Rockwool-based insulation products. The vast majority of acoustic treatment will be absorption, including bass trapping and first reflection treatment.
2. *Diffusion.* This works by scattering or dispersing sound waves unevenly, thereby reducing any potential standing waves and echoes, thus creating a better listening environment and stereo image. Diffusers are most commonly made from wood (Figure 9.10) but can also be made from plastic or sometimes even polystyrene. Diffusion is used when you want to reduce acoustical problems without reducing the reverb time of the room.

Figure 9.9 Foam absorption panel

Figure 9.10 Skyline diffuser

9.3.2 Bass Trapping

Dealing with the low end in any small room is probably the most difficult task when tuning a room. This is because the wavelengths of low-pitched sounds are longer and have more energy that has to be absorbed. This means that bass traps are bigger, more unsightly and heavier than acoustic treatment produced for mid-to-high frequencies. In the smallest rooms, this can often lead to a catch-22 situation where accurately flattening out the low end takes up more space than you actually have in the room.

Bass accumulates most in the corners of rooms and, wherever possible, it is best to have absorption panels on all four corners; absorption panels designed at combating bass frequencies are called *bass traps*. If possible, it is best to make them yourself so that you can tailor them to your space and budgetary requirements. Where space and budget is a concern, foam corner traps can make a small improvement, but are usually useless at dealing with sub-bass frequencies.

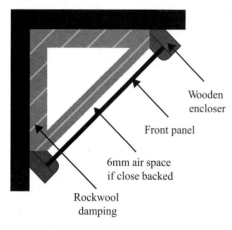

Figure 9.11 Basic Rockwool corner trap with air gap

To make bass traps yourself, you have to construct a wooden frame placed across the corner, creating the shape of a triangular prism (Figure 9.11). This is then filled with Rockwool with a density of around 60 kg/m³ or fibre-glass at 45 kg/m³ for panels up to about 15 cm. Deeper panels or completely filled corners benefit from less dense material, around half this density. The frame is then covered with a fabric that is porous to air (mainly for aesthetic reasons).

Completely filling the corners is the most effective method. However, it requires more material so it's more expensive. If you don't have a big enough budget to completely fill the corners, having an air gap with hanging strips of Rockwool is recommended to make the bass trap more efficient. Or, third, a simple absorption panel placed across the corner will still significantly improve your bass response. Sometimes it's also advisable to face corner traps with reflective material to stop them killing too much of the high mid ranges and above.

Don't use reflective facing on early reflection panels, though! Remember to also treat wall-to-floor and wall-to-ceiling corners and tri-corners.

The basic rule of thumb with bass trapping in a small-to-medium room is that you just fit as much as you can feasibly fit (or can afford to fit) into the room. It is that simple!

9.3.3 Creating a Reflection-Free Zone

Figure 9.12 Simple panel absorber

Just like with bass trapping, absorption is generally the most commonly used method for dealing with early reflections from the sidewalls and ceilings. This can again come in the form of a Rockwool-based panel (Figure 9.12), or out of acoustic foam.

It is also possible to DIY flat rectangular Rockwool absorbers using the same materials as with the corner bass traps. In the case of absorption panels, it is recommended that you leave an air gap between the panel and the wall that is the same depth as the absorption material. Absorption becomes less efficient the closer the panel gets to a boundary, and therefore the efficiency of the panel is improved by being placed away from the wall.

Again, using foam for absorption treatment is in general a far less efficient solution when compared to acoustic panels made from Rockwool or similar synthetics. If you decide to use foam absorption for early reflections, then it is really important to choose the foam correctly. Before buying any foam, pay attention to the noise reduction coefficient (NRC) table provided in the product's spec sheet. This table shows the amount of absorption in different frequency bands. A value of 0 indicates complete reflection and a value of 1 indicates complete absorption.

This is important for two reasons:

1. To make sure that you are buying reliable products.
2. To hone its efficiency. Although foam absorbers generally have a high level of absorption at all mid-to-high frequencies, you can use the information on the NRC table to further improve the room's response. For instance, if your room has specific problems at 4 kHz and 6.3 kHz, then pick a foam that has a high absorption value at those frequencies.

Once you have decided on the type of absorption material, you need to work out where to place it. While there are methods to work out the placement of first reflection points with a bit of mathematics, for most people it is easier to do this with the help of a friend and a mirror. To find the areas that you need to cover with absorption traps, simply sit in the listening position and get your friend to hold the mirror against the sidewall and move it around. Any position where you can see either monitor is a place that needs treating. Just to be safe, it is also best to cover a *slightly* larger area than needed. You can repeat the same steps to find the first reflection points on the ceiling. When dealing with absorption panels on the ceiling, you need to hang them a small distance from the ceiling. This can be done using furniture hooks and chain, fishing wire or cable ties.

9.3.4 Controlling Room Ambience

After treating the corners with bass traps and creating a reflection-free zone, you will have alleviated a lot of acoustical problems from the room. However, there will usually be a need for more treatment to control the reverb tail. A good rule of thumb in small- to mid-sized rooms is to cover 40–50 per cent of the free wall and ceiling space; the sidewalls should also have staggered treatment. This means that the pattern that is used on one sidewall should be inverted for the opposite wall. This is to avoid potential problems with flutter echo.

If you want to get a bit more in-depth, room measurement software can be extremely useful to measure a room's reverb time. We will shortly discuss such software. While there are guidelines regarding reverb time when creating a great critical listening room environment, there is also a lot of scope for variation due to taste and current production trends. For instance, in the mid-1980s, there was a control room trend to deaden (cut reverb times) much more than had previously been advised. Because of these control room environments, mix engineers tended to add a greater amount of reverb and delay to tracks. This, combined with the advances in synthesized sound and sampling, led to the unique sound of '80s pop productions.

Environments where a room has no 'sound' are called *anechoic chambers*; this is where sound is completely absorbed without reflecting off any surfaces. Rooms that are this dead are 'eerie' and uncomfortable, and actually some room sound is important to not only help give a sense of how your mix will translate into other environments, but also to help increase the perceived loudness of your monitors. The key to room acoustics is that these room reflections need to be small and controlled, particularly those from the side of the room nearest the listening position.

The measurement for controlling reverb time is called RT60, and this is the amount of time it takes for the sound to decrease by 60 dB. A control room with an RT60 of between 0.2 and 0.5 seconds at the side of the room nearest the listening position is thought to be optimum; this could be described as very 'dead' sounding. This RT60 time is actually very similar to an average living room. However, most living rooms would be a less than perfect critical listening room because the RT60 needs to be as uniform across all frequencies as possible. If you have a mid-sized critical listening room or larger, it is advisable to create a more lively area on the side of the room furthest from the listening position.

In a smaller environment, you need more absorption, thus leaving very little space for diffusion. If you are in a mid- to large-sized room, once the RT60 time is brought down to reasonable levels with absorption, it is very common to find that it isn't flat enough in the mid to high frequencies. This is where diffusion becomes important. Diffusion doesn't reduce the sound energy in the room; it simply spreads it more evenly, thus helping to stop standing waves developing. Diffusers should be positioned in places that are not first reflection points. Although you can find diffusers on the sidewalls and ceiling, they are generally used on the rear wall, particularly at ear level.

9.4 ROOM ANALYSIS SOFTWARE

Unlike with bass trapping and early reflection treatment, it can be very difficult to learn how to judge the correct amounts of additional absorption versus diffusion to evenly reduce the room's RT60. This really is where a trained acoustician is worth every penny.

Fortunately, if you cannot afford an acoustician, then this is where Fuzz Measure, Room EQ Wizard or any one of the many other acoustic measurement applications will be extremely useful.

> **Further Resources**   www.audioproductiontips.com/source/resources
>
> Fuzz Measure and Room EQ Wizard are programs used to help analyse the acoustics of your room. We will be using these later to help test for problems in a room. (74)
>
> While it's beyond the scope of this course to show you how to use specific software, here is a set of YouTube videos made by the support staff of Fuzz Measure, showing you how to set up the program. (75)

Before you start making measurements to analyse, there are a few simple rules of thumb to be considered:

1. Make sure that you measure the whole human hearing spectrum (20 Hz–20 kHz).
2. Take the majority of measurements from the listening position at ear level. You can use other measurements to help find problems, but the listening position is your primary concern.
3. Ensure that both speakers and your subwoofer (if applicable) are all turned on when making measurements.
4. Correctly calibrate your speakers when making measurements
5. Avoid clipping your interface when making measurements, but make sure that the signal is as hot as it can be, as signal-to-noise ratio is key when making measurements. This can be done by turning up the gain on your interface.
6. Take several measurements at each position to ensure there are no data anomalies occurring.
7. Use a specially designed measurement mic.

Remember, it is essential to take measurements before and after acoustic treatment so you can track your progress. You should also remeasure at every critical stage of the process. This is to ensure that all the treatment you originally planned to make is still necessary. For instance, bass trapping a room will have an effect on the mid-to-high frequency response, and remeasuring could potentially prevent you over-deadening your room.

Once you have made some measurements, you can choose to display the information in a series of graphs, which can help you assess the weaknesses in different areas of your room.

9.4.1 Frequency Response Graph

The first is a standard frequency response line graph (Figure 9.13). This plots frequency on the X axis and amplitude on the Y axis. This type of graph is particularly useful to spot comb filtering in the mid range and treble.

Another useful trick when dealing with mid-to-high frequencies is to average out the values using a smoothing filter so that the graph is easier to read. The graph in Figure 9.14 is smoothed to one-third of an octave.

Although these commonly cover the full human hearing spectrum, it is also useful to scale the graph to cover only the low-frequency information to help with judging the results of bass trapping (Figure 9.15). A 1:24 or higher resolution is recommended, especially when looking at the modal range, 400 Hz and below.

9.4.2 Waterfall Plot

One important factor that is not shown at all on a frequency response graph is time. A waterfall plot (Figure 9.16) shows not only the frequency and amplitude (like on a frequency response graph), but also the decay time on a third axis.

Figure 9.13
Critical listening room frequency response graph

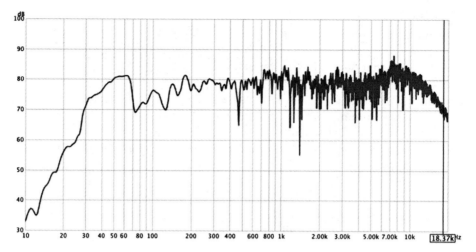

Figure 9.14
Frequency response graph smoothed to one-third of an octave

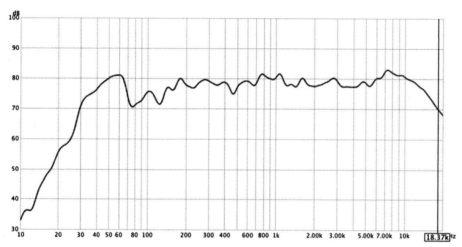

Figure 9.15
Frequency response graph limited to 400 Hz

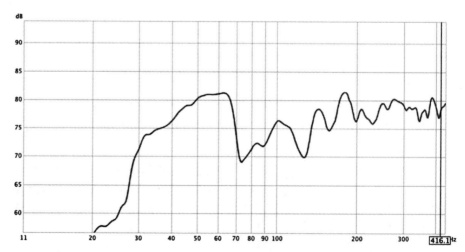

Chapter 9 THE CRITICAL LISTENING ENVIRONMENT

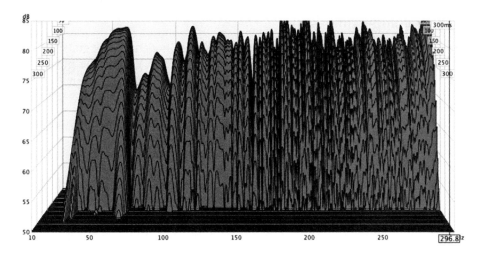

Figure 9.16 Critical listening room waterfall plot

Waterfall plots are extremely useful to be able to analyse modal ringing – this is particularly important in your low-end analysis.

9.4.3 RT60 Graphs

As well as waterfall plots, an RT60 graph (Figure 9.17) is also useful to help identify time-related issues. It is a particularly easy graph to read as it smoothes out decay into octave bands. This graph is great to quickly see if you have an even RT60 time across the whole frequency range.

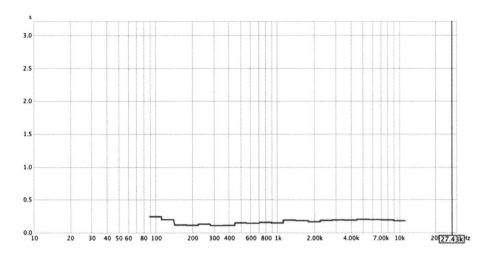

Figure 9.17 Critical listening room RT60 graph

9.4.4 Envelope Time Curve

The last graph we are going to cover is an envelope time curve line graph (Figure 9.18). It plots *time* on the X axis and *amplitude* on the Y axis. The purpose of this graph is to find out if specific room boundaries are the root cause of significant reflection problems. Typically, you are looking to correct any reflection that isn't at least 10–15 dB lower than the original source. For instance, if your direct sound pressure level (SPL) is 80 dB, you want to correct any reflection above 65 dB.

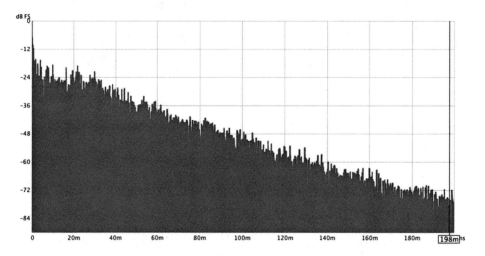

Figure 9.18 Critical listening room envelope time curve

You can work out the possible causes of problematic reflections by working out the distance the sound would have travelled when the spike on the graph occurred.

You can work out the distance based on the fact that sound travels at 343 metres per second at room temperature.

So for example, if a spike occurred at 15 ms on the graph, you would have a total travelling distance of 5.145 metres:

0.343 m × 15 ms = 5.145 m

Remember that the sound has to travel to a boundary and reflect back to the measurement microphone (at the listening position) within that time. In other words, it is not just the time taken to travel to the boundary.

Usually in very small rooms (particularly dreaded square rooms!) this graphical analysis is less useful because room boundaries are closer together, making it harder to ascertain which room boundary could have caused the problematic reflections.

It is also important to check on the software whether the envelope time curve is normalized. In this context, normalizing the graph only shows the reflections and not the direct sound.

9.4.5 Correcting Specific Frequency Issues

It is likely that in your quest to treat your own room, you will come across problems you were not expecting, or that you will want to target specific problem frequencies after you have used broadband treatment. Troubleshooting these issues is particularly difficult. There are many great books, websites and forums where people give information or have created handy guides to make DIY treatment. Also, on some of these forums, people will be more than willing to give advice and critique your work.

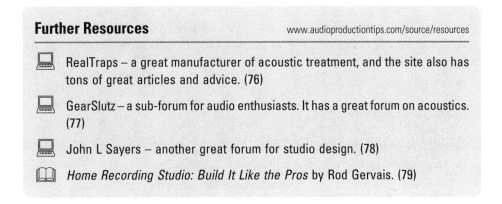

Further Resources www.audioproductiontips.com/source/resources

- RealTraps – a great manufacturer of acoustic treatment, and the site also has tons of great articles and advice. (76)
- GearSlutz – a sub-forum for audio enthusiasts. It has a great forum on acoustics. (77)
- John L Sayers – another great forum for studio design. (78)
- *Home Recording Studio: Build It Like the Pros* by Rod Gervais. (79)

9.5 USING EQ TO CORRECT ACOUSTIC ISSUES

Before we move on to looking at monitors themselves, I'd like to quickly cover why you should never rely on EQ correction to compensate for room acoustic issues. Having made your measurements, you may look at your frequency response graph and think, 'Well, can't I just EQ it to be flatter?' The short answer is no. Here's the long answer!

EQ'ing your room is not categorically evil, but it usually causes more problems than it is worth. You can reduce and improve your room using subtractive EQ on some of the worst peaks. This is why it is common to see live sound engineers using a 31-band graphic EQ on their master output before beginning a sound check. This helps to reduce the worst resonances in the room, thus the engineer needs to use less severe EQ on individual channels and it also helps to avoid feedback. However, given the right materials, space and time, it would be a far better alternative to treat your room. It is not only in recording studios where you'll see acoustic treatment in action. Visiting a symphony hall, top-end concert venue and many theatres, you will see a range of acoustic treatment to help tame modal frequencies.

In a control room environment where the room's sole purpose is to sound great, acoustic treatment has major advantages over EQ:

- Room EQ can do next to nothing to improve the nulls caused from cancellation. In fact, with the severity of the additive EQ needed, you are more likely to start blowing up amplifiers and monitors than create an audible improvement.

- In a small room, the frequency balance can change dramatically depending on your position in the room. If you had EQ'd the room to be flat based on your primary monitoring position, as soon as you step outside of that area, you'll find the EQ will be doing more damage than good.
- A lot of the problems caused by low frequencies happen because of the length of decay time. EQ cannot improve or correct this. There are some advanced software programs that claim to help with this, but in my experience these improvements are far less drastic than bass trapping and also suffer the same problems as EQ when moving around your room.

To conclude, I personally avoid using EQ to help improve the response of rooms. However, in a particularly bad environment, you may wish to EQ the room (subtly), but it should, in my opinion, only be considered after you have treated the room to the best of your ability.

Suggested Treatment Plan

- Go to a well-treated commercial studio (or more if possible) and listen to some of your favourite tracks in their control room.
- Empty your mix room except your monitors and computer system (remove any current acoustic treatment you may have). Listen to your reference tracks, making notes about the tonal quality of the sound in the empty room.
- Move the monitors around to find the best speaker and listening position (bearing in mind the guidelines set out in this chapter).
- Make a set of acoustic measurements from the ideal mix position with the room still empty using the advice in the monitor placement section.
- Build DIY bass traps and repeat the measurements.
- Now add first reflection absorption and collect more measurements.
- Listen to the same tracks again and compare differences in tonality.
- Now add any extra absorption and diffusion you deem necessary to reduce and flatten your room's RT60.
- Take some final measurements, and compare graphs.
- Listen to the same tracks for a final time to listen for more tonal changes.

9.6 MONITORING

Now that we have discussed acoustic treatment, we can discuss the other half of the critical listening environment: the speakers themselves. The acoustic environment and monitor choice and placement are important to making tonal decisions throughout the recording process, and mix engineers will also consider how those choices will sound to the casual listener. However, I'd like to assert that understanding the systems that the song is ultimately going to end up on isn't just the jurisdiction of the mix engineer. These considerations can also affect the arrangement and the way you layer instruments during tracking.

Professional mix engineers typically switch between multiple types of monitor while mixing, to help judge how music will sound on different sizes and qualities of system. You can do this using multiple outputs from your interface/converters or use a monitor controller.

I also like to have multiple monitors to switch between while engineering/producing. Not only does this help me to hear the song more like the end listener might hear it, but also:

1. Certain speakers focus specific frequency bands more prominently so I can hear sonic 'holes' in the production more clearly.
2. Switching to a bigger, more powerful speaker temporarily can give the musician that extra 'buzz' to get the best out of their performance.

Let's dissect some of the different types of speaker you are likely to find in a commercial studio:

9.6.1 Mains

In the 1960s and much of the 1970s, most top-level recording studios focused their monitoring around large full-range speakers, otherwise known as 'mains' or 'bigs'. These speakers would usually have dedicated drivers for bass, midrange and high frequencies. These speakers were usually very accurate, expensive and extended through the full range of the human hearing spectrum. Mains are capable of crystal-clear sound at high SPLs and are usually hung from the walls

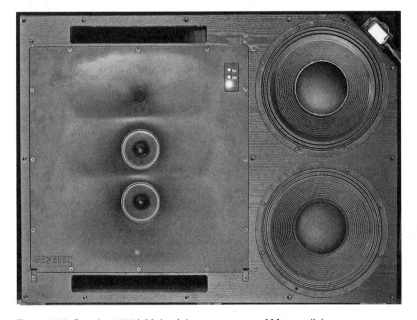

Figure 9.19 Genelec 1035A Mains (picture courtesy of Metropolis)

with chains or are specifically catered for by acousticians and built into the walls. While mains are still very popular in major studios today, there were huge drawbacks to solely using them:

- *Listening distance.* Because of their size and number of speakers used, the dispersion of each speaker means that you need to be several metres away from them before you get a balanced and accurate sound.
- *Acoustic issues.* Due to the fact that they can produce higher SPLs and are physically larger, it is more difficult to place them without the room causing interference in terms of stereo imaging and also frequency response. Therefore, mains need to be in large and well-treated control rooms.
- *Translation problems.* Because mains were higher calibre than the average listeners' systems, it was found that many mixes didn't translate well to smaller speakers. This is even more of an issue now there is a precedent for using laptop and mobile phone speakers.
- *Fatigue.* Due to the distance from the listener and the speakers' power, it is easy to run them too loud and cause ear fatigue quickly. Exposure day in, day out at these SPLs can even cause hearing damage.

For these reasons, even studios with mains now tend to use them sporadically compared to their nearfield monitors, usually reserving them for checking low-end high-SPL performance and impressing clients.

9.6.2 Nearfields

By the 1980s, smaller two-way speaker designs were dominant in the pro-audio industry. Due to the listening distance being considerably closer than mains (usually one to two metres), they got the name 'nearfields'. One driver was used for the bass and midrange and another for high frequencies. Nearfields to this day are the main monitoring choice for most mix engineers, and even when other speaker options are present most mixers spend around 90 per cent of the mixing process using them. They are usually placed on the meter bridge of the console or placed on stands just behind. In the early days of nearfields, they were usually passive (meaning that the amplifier was separate to the speaker enclosure). These days, however, the majority of nearfields are active. Nearfields suffer less from the limitations of mains, but their slight lack of low-end extension compared to mains makes both speaker types necessary for a top-level mixer with the budget to implement both.

One particular model of nearfield became incredibly popular in the late 1970s, and by the mid-1980s was in just about every top studio in the world. In fact,

Figure 9.20 Yamaha NS10M Pro nearfield monitors (picture courtesy of Metropolis)

it is still very popular today even though it has been discontinued. Yes, you already guessed it; it is the Yamaha NS10, pictured in Figure 9.20. This nearfield monitor was released as a consumer hi-fi product in 1978 and it was panned both critically and in terms of sales. So why was a poor and unpopular hi-fi speaker such a great studio monitor?

Without going into too much detail, hi-fi enthusiasts didn't like it because the bass frequency response of the NS10 rolled off higher than is preferable and thus the overall frequency response was not very flat. These characteristics made it a very 'boring' and even 'bad-sounding' speaker for hi-fi purposes. However, its closed-box design (meaning air cannot escape from the enclosure) and the special type of curl-and-join paper speaker cone meant that it produced incredible time domain response and distortion characteristics. These characteristics, it turns out, are at least equally as important as frequency response in studio monitors. With such incredible time domain response and low distortion, it meant that if the results sounded good on an NS10, then it would sound good anywhere, as long as you compensated for the 'sound' of its frequency response. As most studios also had mains, this wasn't usually a problem.

9.6.2 Dangers of Ported Nearfield Designs

The NS10 closed-box design leads me perfectly to a debate that is gaining ever more attention in recent years, thanks to Mike Senior's 'Mixing Secrets for the Small Studio' and Philip Newell, Julias Newell and Dr Keith Holland's research paper on why the NS10 was so popular.[7][8] This debate regards the effect porting has on small speakers.

A ported speaker has a hole that allows air to leave the enclosure, which causes a greater resonance from the enclosure itself; this in turn can increase the speaker's bass response. A huge percentage of modern budget nearfields are ported. On paper, this might sound like the perfect remedy to improve the problematic sub-100 Hz response caused by the size of the enclosure and the speakers in nearfields. Unfortunately, though, this increase in response comes at a heavy cost:

- *Bass ringing.* Using a ported design alters the bass response in the time domain and bass notes ring far longer than with closed-box designs. See the waterfall plots for the NS10 and a typical ported speaker in Figures 9.21 and 9.22.
- *Resonant side effects.* The ringing of the port can overshadow the real fundamental frequency, which can cause frequencies to be difficult to differentiate from each other. This makes EQ decisions harder and potentially incorrect.
- *Distortion.* As well as the ringing and masking characteristics, a port can also cause distortion into the midrange of bass instruments. This could cause you to scoop the midrange of bass instruments more than is necessary.
- *Difficulty judging sub-bass levels.* A port increases the overall bass level but the sub-bass rolls off much more steeply than closed-box designs. This means that you are likely to want to EQ the sub (20–50 Hz) regions of the kick drum more heavily.

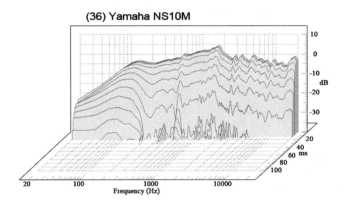

Figure 9.21 Yamaha NS10M waterfall plot[9]

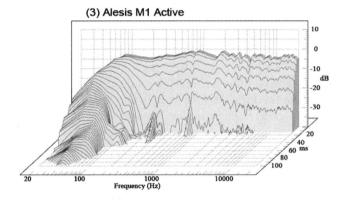

Figure 9.22 Typical ported nearfield (Alesis M1 active) waterfall plot[10]

Figure 9.23 Auratone 5C single-driver speakers (picture courtesy of Auratone)

Sometimes porting is done specifically to improve specs or allow manufacturers to cut costs. For instance, the port can make the frequency response look flatter on a simple frequency response graph. More importantly, though, the effects of porting 'muddy up' the time domain response meaning that manufacturers can cut costs on tweeters knowing that artefacts and 'brittleness' are often masked by the bass ringing.

Just because it has a port doesn't mean it should always be avoided. Porting can sometimes be slightly advantageous in a well-designed unit. However, the less money you have to spend, and the smaller the enclosure, the more wary you need to be of ported designs.

9.6.3 Auratones/Grot-Boxes

To further aid translation, many engineers like to also use single-driver speakers to help emphasize the important midrange frequencies. These speakers are often called *grot-boxes* and can be used as a singular mono set-up or in a standard stereo pair. Just as with the NS10, one model of grot-box became very popular: the Auratone 5C. Despite their unappealing sound, some engineers choose to use them even more than their nearfields. 5Cs are specifically useful to see how instruments that dominate the bass and treble frequencies such as kick drum, bass guitar and cymbals, sound on smaller, more mid-heavy systems. They are also great for highlighting crucial elements of the mix that fuller-range speakers are more flattering of, such as midrange masking between instruments and relative levels between vocals and guitars. With playback devices getting smaller and smaller in line with modern technological advancements, there are many engineers who are choosing to use grot-boxes with an even smaller range, including cheap iPod docks and phone speakers.

9.6.4 Other Monitoring Options

Although we have discussed the three most common types of monitoring options found in a commercial studio, there are three other options that can also be considered.

Headphones

While doing a whole mix on headphones isn't recommended, it is important to check a mix on headphones. It is also important to choose the right kind of headphones for tracking and mixing purposes. As many people listen to music on the go using earbuds, it is important to see how your music will sound in these circumstances. One of the big issues with headphone listening is that the left ear only hears the left side of the stereo mix and the right side is only heard in the right ear; in other words, there is no crosstalk between the sides of the stereo field. On loudspeakers, this is not the case – you hear the direct sound coming from the opposite speaker a little after and at a slightly reduced volume; then there are the reflections from the room's boundaries arriving at each ear. In practice, the lack of crosstalk and room reflections mean that hard-panned sounds can be easily sound unnatural and crucial central sounds can feel like they are coming from inside your head, which is why the use of spatial effects such as reverb and delay is important. Another reason why checking your productions on headphones is important is due to the fact that subtleties and tiny details are often easier to hear. These include unwanted click bleed, hiss, pumping and distortion. Finally, if you don't have an ideal critical listening room, making some tweaks on headphones can go a long way to improving how your mix translates to different listening environments.

Studio headphones are usually of the large, over-the-ear variety rather than earbuds, and are commonly found in two varieties: open-back and closed-back. Open-back headphones have grilles that expose the rear of the drivers, which gives them a more natural or 'airy' feel. However, the drawback is that because they are exposed, they aren't suitable for tracking as there is too much spill. Closed-back, on the other hand, are great for tracking, as spill is reduced

Figure 9.24 Sennheiser HD595 open-back studio headphones

dramatically, but they can feel fatiguing when used for long periods and often suffer more severe bass-resonant side effects compared to open-back designs. One issue that headphones are renowned for, though, is having problems producing a realistic 'bottom end' like you would get out of loudspeakers. Unfortunately, this means that for a reliable bottom end, you are likely to have to opt for a higher-end (and most likely open-back) headphone.

Subwoofer

Another foreseeable issue with your monitoring environment is that you could have problems judging ultra-low end but not have a big enough room or budget to implement a set of mains. In this case, a subwoofer could be the answer, but in extreme cases it could even make your bass response issue worse. Some engineers like subwoofers; others don't like them, so it is very much a personal preference. The critical factor is again your acoustic environment. If you have a poor bass response in your room, then any subwoofer is very unlikely to produce satisfactory results.

Provided you have resolved poor room acoustic performance, you will want to know the types of subwoofer to buy and also how to set them up.

First, subwoofers typically cover only the sub-bass region of 20–80 Hz (although many can be used up to about 150 Hz if required). On the subwoofer itself, you select a frequency at which it passes all higher frequencies to your nearfields; this frequency is called the *crossover frequency*. In the pro-audio world, there are two schools of thought regarding setting the sub and its crossover frequency:

1. Leave the nearfields without any crossover frequency and just use the sub on a separate output, using the crossover frequency to control what frequency the sub produces up to.
2. Set nearfield and sub on the same output by first going into the sub then out into the nearfields. Set the crossover frequency to where the bass performance of the nearfield starts to become efficient.

I tend to prefer to use the nearfields without any frequency restriction and like to turn the sub on and off using my monitor controller; therefore, option 1 is my preference. In most cases, the bass performance of a nearfield monitor starts to kick in between 50 and 70 Hz but crossover frequency should be set on a case-by-case basis, listening to several tracks while critically analysing bass performance – or, better still, using test tone and measurement equipment.

Most studios only use one subwoofer. This is because frequencies below 100 Hz exhibit very omnidirectional characteristics (sound travels from the source in all directions). This, combined with the nature of room reflections when indoors, means our hearing system cannot extrapolate the directionality for these low-frequency sounds.

As well as the correct setting of the crossover frequency where required, you need to make sure that the sub is at the appropriate volume level, it is in the optimum position, and it is phase-matched to your nearfields (both in the case of polarity and time alignment). Below are instructions for setting a subwoofer:

1. First, place the subwoofer in the listening position and set appropriate volume levels between the nearfield monitors and the sub. If you are using a crossover frequency, approximate it and leave the phase at 0. It is fairly easy to set levels of the sub too high at first, so be critical when you are choosing.
2. Next, test the subwoofer in different positions and pick the best bass response. Sometimes the loudest overall bass is not the best position, so you will want to try to analyse the trade-off between extra bass response and unwanted resonance (i.e. ringing in the time domain). If you are taking room measurements, waterfall plots are extremely useful. Please note that not every place in the room will be appropriate to try. A general guideline, though, is that you will want to keep your subwoofer close to your nearfields (usually no more than 1.2 metres away). This is so there won't be problematic time alignment issues.
3. After the best position is chosen, sweep the crossover frequency and listen for the position where the bass response is clean but full. If you are using the sub on a separate output, you need to make sure that the sub and nearfield aren't overlapping in bass frequencies too much, causing a muddy or cluttered bass response. Bear in mind the 50–70 Hz approximations for nearfield monitors; with speakers such as the NS10 and other smaller bass-/mid-range drivers, you might need to set it higher.
4. Next, adjust the polarity and analyse whether the change is better or worse. The telltale sign that there is a problematic phase relationship is if there is a boosting or attenuation of level at frequencies close to the crossover frequency. Some subs have a simple phase switch and others have a knob that rotates the phase; the latter is a little bit more difficult to set effectively but can be even more accurate.

Mid-Field Monitors

In recent years, there has become a large demand for speakers that are in between the sizes of mains and nearfields, which can produce full-range sound but are also sized for reasonable listening levels and use in medium-sized rooms. Because of this, smaller three-way loudspeakers were designed for this purpose. While nearfields are designed for use 1–2 metres from the listening position, mid-field monitors are optimal between 2 and 4 metres.

If your room is too small for mains but you still want a full-range speaker set-up without having to resort to a sub, mid-fields may be for you.

9.7 MONITORING CONSIDERATIONS

Now that we have been through the monitoring options, you may be clearer on the options available but still be a little unsure on how to effectively pick monitoring equipment for your critical listening environment. In this case, your first consideration should be picking the correct-sized speakers for your room. If you have a small room, there is no point spending your money on a set of mains as their efficiency will be smeared by room tone and you might go deaf pretty quickly. Generally, these are also very expensive, so unless you have a substantial budget they are probably best to steer clear of.

In terms of nearfields, most critical listening rooms would suit 6- or 8-inch drivers and, as discussed, it is usually best to avoid ported models. Those just starting out must take time to get used to nearfield monitors; the good ones tend to sound clear and precise without flattering the music. They might not be the most immediately brilliant-sounding speakers, but you need to learn to love them for what they are: tools to help you analyse your mixes. Your monitors should show the flaws in the production and make you work hard to make it sound good – that is what people loved about NS10s and it remains true today. This does not necessarily mean that the monitor has to sound bad; it just has to be honest.

Figure 9.25 Unity Audio Rock nearfield monitors (picture courtesy of Metropolis)

When I've shopped for nearfields in the past, I look for monitors that have a fast transient response. This is how faithfully a monitor can reproduce transients. I look for clean transients in the bottom end (i.e. like that heard in many closed-box designs) and clean and punchy transients in the top end where the results of sharp transients such as snare drum hits are clean and clear. A lot of the overall timbre of the sound comes from the nature of the transient. A monitor that smears transient response will ultimately alter our perception of the sound more than most other factors. My current monitoring set-up focuses around a pair of Unity Audio Rock monitors, which I find to have an incredibly quick transient response. They are closed-box, have a great frequency response and are very faithful to the nature of what you have actually recorded.

I also regularly use grot-boxes and headphones to check for potential translation and bleed issues. Just like many other producers, I find that because the frequency response is so good on the nearfields, I often like to listen to more 'mid-heavy' speakers and have a monitor controller that allows me to switch between these monitors and check my productions/mixes in mono quickly (which helps discover phase and balance problems). I also own a subwoofer to help check ultra-sub-bass; just like most nearfields, the Rocks have a little bit more bass roll-off than I'd like. However, I only use it sparingly.

I find being able to switch between several different types of monitoring system in a well-treated room allows me to be confident in the decisions I make and I do not have to check in many different surroundings. The *only* advantage I find in listening in alternative environments these days is that it allows me to at least try to listen to the music I've made as a consumer and not as a professional; this can have its own benefits in understanding the 'bigger picture'.

The only other advice I'd like to give on monitoring is that even once you have spent time learning the sound of your specific nearfield monitors, there is still a surprisingly large difference in tonality (even between top-end models). Pick monitors that appeal to your style and taste and *learn that specific model*

– each monitor is different and will have its own learning curve. Remember to always do a few test productions/mixes on new speakers before doing paid work, and allow time for new speakers to settle in. Manufacturers can say it takes up to 30 hours for them to be working optimally.

EXERCISES

1. Go to a local pro-audio dealer who stocks a range of monitors and listen to commercial music that you are very familiar with (make sure the volumes are matched).
2. Listen to music in a car, on a cheap iPod dock, phone or similar, on headphones, through nearfield monitors and full-range speakers (if you have access to some). Critically listen to the time domain, transient and frequency response of each speaker.

NOTES

1. What are room modes, accessed November 2012, www.gikacoustics.com/what-are-room-modes/.
2. Acoustics 101, accessed November 2012, www.audioholics.com/room-acoustics/acoustics-101-course-by-john-dahl-of-thx/acoustics-101-course-by-john-dahl-of-thx-page-2.
3. Acoustics 101, accessed November 2012, www.audioholics.com/room-acoustics/acoustics-101-course-by-john-dahl-of-thx/acoustics-101-course-by-john-dahl-of-thx-page-2.
4. Acoustics 101, accessed November 2012, www.audioholics.com/room-acoustics/acoustics-101-course-by-john-dahl-of-thx/acoustics-101-course-by-john-dahl-of-thx-page-2.
5. Creating a refection free zone, accessed November 2012, http://realtraps.com/rfz.htm.
6. How to setup a listening room, accessed November 2012, http://realtraps.com/art_room-setup.htm.
7. Mike Senior, *Mixing Secrets for the Small Studio* (Focal Press, 2011), 5–9.
8. Phillip R. Newell, Keith R. Holland and Julius P. Newell, *The Yamaha NS10. Twenty Years A Reference Monitor. Why?*
9. Phillip R. Newell, Keith R. Holland and Julius P. Newell, *The Yamaha NS10. Twenty Years A Reference Monitor. Why?*
10. Phillip R. Newell, Keith R. Holland and Julius P. Newell, *The Yamaha NS10. Twenty Years A Reference Monitor. Why?*

DECISION-MAKING AND GENERAL RECORDING TECHNIQUES

When a record producer is working his or her magic, he or she is constantly making important decisions quickly. To make correct decisions mid-session, the producer needs to be able to trust his or her instincts. The theory and techniques we have been learning so far are strong foundations that will allow you to think about audio engineering and record production in the right way. This chapter will take this further and tie up some other key points that relate to decision-making, as well as provide a couple of generalized techniques that will help you record anything.

Before treading new ground, let's quickly put into perspective what has been covered so far in this course. By now, the bigger picture should be starting to click in your mind and you should be beginning to see the way that all of the techniques we have been learning are connected to each other. Putting the key points into a few paragraphs will help you see the bigger picture:

Understanding human hearing. We have learnt that human hearing isn't flat, and at different volumes we will perceive sounds very differently. This means that you need to pay careful attention to your listening volume and be mindful of the effects it can have on your decision-making. You should also be aware of how the human ear fatigues and recognize when you need a break.

Trusting your critical listening environment – you need to get to know the environment that you are producing in. You should feel confident that the monitors and room are telling you exactly what you need to know. This is why I must emphasize that optimizing the critical listening environment is your number-one priority when investing in home studio equipment. Even if the space isn't very acoustically neutral, or if the monitors are not particularly suited to your needs, you should still know how the environment sounds and understand its strengths and weaknesses. When I am producing in a new environment, I will play a disc of reference material and make a mental note of the positive and negative aspects of the room (e.g. it has a lumpy bass response in the low mids, is very reflective in the upper mids if the window is uncovered, or the ported monitors seem to exaggerate bass response). Taking some time to do this will help you to make decisions that translate well into other acoustic environments and on other monitoring systems.

Preparation. You will produce better music if you know the strengths and weaknesses of the band members' technical abilities, the music and their personalities ahead of the session. Preparation time for the band to get used to any structural alterations will ease stress and allow for studio time to be used efficiently.

Organisation. You need to be able to manage a session so that a reasonable amount of time is set for every task, and keep on top of the planning so that you don't forget anything. Also, as fatigue sets in at the end of a long session, you will be thankful that you labelled the track sheet and marked the completion of parts on a whiteboard. At the end of a long day, it is very easy to forget which pieces of outboard equipment are connected where and which parts you still have left to record! Spreadsheets, notepads or whiteboards at the ready and don't forget to label channels and keep session tidy in your DAW!

Understanding composition and arrangement. A great production starts with a great song. The structure and the arrangement of the song should be optimized where necessary before you start to record anything.

Know the genre. To produce a track, you need a vision, and to create a vision that is suitable, you need to know what the band are aiming for and who their influences are. Have knowledge of which instruments are suitable and their ranges and tonality in the particular genre, as well as the type of layering and density of tones you are likely to come across. You should be aware of the other production and compositional precedents set in the genre and also when to break the rules. Getting to grips with this is a case of listening critically to as much suitable material as possible before the session.

Knowing the equipment. We might know what we want to hear, but we also need to know how to make this a reality. Most clients will not have the budget to spend hours deliberating on the correct mic choice, placement and pre-amp to capture the perfect representation of the band's tone. Therefore, you should:

- Know how the room you are recording in sounds, and have some idea of potential sweet spots. This includes consulting your vision as to whether you want a certain part to be bright or dark, alive or dead, etc.
- Understand the tone of the microphones you have at your disposal, both on- and off-axis, and the strength of their proximity effect.
- Know the sonic qualities of any outboard equipment that you may be recording through.

Correct gain structure and DAW settings. You should be aware of how to implement correct gain structure so that you are keeping your signal clean with an acceptable signal-to-noise ratio. Of course, if you are aiming for something more gritty and distorted, you might choose to implement gain structure differently (but be sure that this is what you want since distortion can't be removed once it has been recorded). However, this is definitely a case of knowing the rules before breaking them. Understanding gain structure also means understanding metering. You should be able to tell which meters are peak or averaging, which reference they are based on and where you might find them. You should also understand that a peak meter cannot be directly compared to an averaging meter.

Getting to grips with all of these elements will take time (a lot of it), experimentation and experience, but once you are there you will start making inspired decisions quickly and authoritatively mid-session. However, I feel it necessary to explain a few miscellaneous tips and techniques that will further help to create a strong foundation of audio engineering and production knowledge before we move on to discussing individual instruments in depth.

10.1 LISTENING TO A SOURCE

I know how easy it is for novice (and even mid-level) engineers to get caught up in a control room crammed full of inspiring yet daunting equipment. It may look hideously complicated, but don't lose sight of the fact that beyond all the knobs, buttons and faders, the principles are the same simple principles you learnt from the beginning. Take a deep breath and remember that as long as you are used to listening critically, your ears will tell you the answer to every question you need to ask.

Don't get caught up with the gear itself. Just because you might have all of the equipment imaginable, it doesn't mean that you should use it all. Picture a restaurant: they might have dozens of spices and herbs, but they are probably only using a couple in each dish. They just need to have them all so that they can cater for the whole range of dishes they have on their menu. Large studios are the same: they have to cater for every eventuality. Your vision is what dictates your decisions, not the amount of gear available, the price bracket or brand; they mean nothing if your ears are telling you that it is not what the track needs. This point echoes one of the key mantras of this course: get the sound as close to your vision as early in the chain as you can. *Get the sound right at source!*

Take nothing for granted. Let your ears guide you, not your preconceptions. If you just listen to what you are recording and experiment until the sound fits your vision, it is hard to go too far wrong (as long as you can trust your monitoring). Of course, in a session, you will rarely have time to try every mic, pre-amp and gain setting for every instrument you record. This is not only tedious, but counterproductive for 'creative flow', for both the production team and the artists. This is why learning the sonic attributes of your equipment (or popular equipment in general if you record in many studios) is important. Just remember,

EXERCISES

1. Take some time to listen to each of your mics and pre-amps, capturing as many different sources as possible. If you don't own many, try to borrow/rent some or try to sweet-talk a local recording studio into giving you a little time to experiment.

2. Armed with this new insight, listen critically to some commercial releases and try to pick out the chain of recording equipment they might have used on a particular instrument.

3. When recording, actively try to get the sound in your head from as small a recording chain as possible.

no matter how well you know the gear, you should *always* ask yourself: is this the sound I am after?

10.2 HOW TO PROPERLY CONDUCT A/B COMPARISONS

One of the biggest problems I see in novice engineers/producers is their inability to properly conduct A/B comparisons. An A/B comparison is a test where you are comparing two identical sources except for one variation to help discern which is subjectively better (i.e. comparing the same source with two different compressors). This can be split into two different sections: volume matching and decision speed.

10.2.1 Volume Matching

To make an informed decision on any parameter, you need to level-match playback volume. You might be trying to decide between two outboard compressors, two different settings on a plug-in or between two takes to judge which is better; this principle still matters. This doesn't mean matching the level with peak meters on your DAW – we already know that our ears are inclined to perceive the averaging volume level. This does not mean that you should level-match to a VU or dBu meter either – the Fletcher-Munson curves show how our hearing isn't flat across all frequencies. It means that before you evaluate, you should first match volume using your ears. Of course, averaging level can help you get it in the ballpark quickly, and to a certain extent peak meters can do the same – but your ears trump whatever your eyes are telling you! So why is it important that the levels are matched? It's because louder sounds better. What this means for engineers is that it is pretty much impossible to make critical A/B judgements unless you are hearing the material at the same volume.

10.2.2 Decision Speed

Another problem I have noticed with some engineers is that they labour for too long over making decisions, particularly insignificant ones. Watching them work can be painful at times, and there are many practical reasons why making decisions so slowly is a potential problem, including:

- Your ears very quickly get used to a particular sound. Therefore, you lose objectivity very quickly.
- You are second-guessing every decision rather than trusting your ears, which can sometimes leave you delaying decisions.
- You can dwell on decisions that do not matter in the bigger picture.

Regardless of the reason, working slowly is likely to interrupt your workflow and stop you from working 'instinctively'. The reality is that most decisions will not make or break your production. For instance, choosing between a 5 ms attack time or a 15 ms attack might seem like it makes a considerable difference in isolation but in the context of an entire mix it probably won't matter that much.

This is not to say that you don't need to worry about compressor settings; it just emphasizes the point that the bigger, broader decisions will matter much more. An example of one such bigger decision would be, 'Is the bass guitar going to carry a lot of weight or should I feature the Hammond organ bassline more?' As long as you don't lose sight of your overall vision and you trust your instincts, you will produce the best results you are capable of at that point in time.

EXERCISES

1. Ask a friend to help set up a blind test between settings with some instances *subtly* louder and see which ones you like best.

2. On your next recording session, try to make all of your decisions quickly and decisively.

10.3 MICROPHONE PLACEMENT AND SELECTION PROCEDURE

Just a word of warning to the gear-head within you (admit it, we all have it lurking somewhere inside!): mic placement is as important as mic choice! To elaborate, a well-placed budget microphone in a good room will sound better than a poorly placed expensive microphone in a bad room. Of course, this doesn't mean that you don't need to buy/rent/borrow expensive microphones to further improve your recordings, but it does mean that the price of your gear should not be your prime focus.

One of the biggest mistakes people make when recording is to get stuck in habits. I myself am guilty of this, especially when time is precious. Using a 'cookie-cutter' approach to microphone selection should be avoided, although I do recommend having favourites. If you don't have an idea of the general tone and characteristics of the various microphone types explained in Chapter 8: Microphone, Pre-Amp and Live Room Choice, it's time to experiment! As described in one of the previous exercises, just spend a few days playing around with as many mics and placements as possible on as many instruments as possible. Drums are particularly difficult and important to get to grips with. While in future chapters I am going to suggest some potential microphone positions and discuss my personal favourites for each job, you should always experiment for yourself. Making your own opinions and judging for yourself is an important part of the learning process.

To help you record any instrument, I have a tried-and-trusted approach to mic choice and placement that helps me achieve my vision in the fastest way possible. First, I am assuming that source factors such as the room, placement in the room, the instrument, amplifiers, effects and the player's style are suitable. If not, you have some work to do before considering microphones.

After the source is correct, I will walk around the instrument/amplifier, listening for areas that sound the best, literally moving and bending with one ear to the source. Where it sounds good to that ear is where you should initially place the mic. I then consider the overall track and anticipate the tone I will want from the instrument in the mix. Remember, not every instrument should be a focal point, and there will be a natural range of frequencies and type of transient attack and sound decay that is required. I pick a microphone I think would suit the task. Once you have found a sweet spot and selected a mic, take a listen in the control room. You won't always get it right first time, so listen and then hone. Don't be afraid to try other mics if you didn't get it right first time. When you are satisfied with your choices, it is time to check for bleed and phase-correlation problems. We will be covering phase in depth shortly, but the gist of it is that you should try to find sweet spots where the bottom end is present but tight. During this process, you should also be trying to find a balance where the tone is still optimum but where the mic's polar pattern rejects sound from other sources (e.g. maximizing hi-hat rejection on the top snare mic). Where bleed is a serious issue with your chosen microphone, you may decide to reconsider. When auditioning mics and placements this way, it is extremely useful to be able to listen from the control room while an assistant moves the mic around in the live room, following your directions.

The advice here is great for finding sweet spots, but sometimes you will want to use several microphones to capture a stereo spread of the instrument(s). There are several different placement techniques to get a stereo representation of a source, and these will be discussed when we consider drum overhead placement.

For now, let's picture us mic'ing a drum kit for a soulful track that is a bit of a throwback to the late '60s and '70s. In this scenario, I would immediately think of a retro-sounding drum kit in a Motown style. After listening to some reference material and potentially researching the techniques of that era, I might decide to strip back the number of mics and reach for a Coles 4038 (a classic ribbon mic with large high-frequency roll-off) as a single mono overhead, a single kick drum mic (probably an AKG D12/D112), and use a simple dynamic snare mic on the bottom of the snare such as an SM57. This stripped-down approach and lack of condenser mics would help me to achieve a warm retro sound with a high-end roll-off. I'd also opt for a more coloured console or tube mic pre-amps, which I could drive hard to produce a more saturated end result. Alternatively, I might also decide to track it all to tape. However, your decisions aren't just about engineering – they should be about production too. For instance, I could:

- hire a retro Ludwig drum kit and tune it appropriately for the era;
- dampen the drums more to get a less ringy tone;
- ask the drummer to hit more quietly as to be more in keeping with the style; and
- augment or replace the hi-hat with a tambourine.

Can you see how your vision is starting to affect your choices? And how the more pre-production and research you do, the better the end result will be?

EXERCISES

1. Pick a particular instrument and try to make the capture of the sound match as close as possible to the sound you hear in the room.

2. Now try to make the same source sound as good as possible (e.g. 'larger than life') only using mic choice and placement.

3. Set yourself a challenge of reproducing a classic song and make it sound as close to the original as you can with the equipment you have available. To do this, you will need to listen critically to the original and cross-reference it against what you are hearing when reproducing the track. Any research you can do into the recording processes used at the time or even specific information of the recording of that particular track would be beneficial.

10.4 PHASE AND POLARITY

When recording a source with multiple devices, there are two different phenomena that can cause interference and result in a sound that is not representative of the source: polarity and phase. The issues of polarity and phase are two distinct problems, but in audio engineering circles they are often confused, or worse, even used interchangeably. Because of this, let's first outline the science behind each phenomenon and then discuss how to correct these issues in practice.

10.4.1 Phase

In Chapter 9: The Critical Listening Environment, we discussed constructive and destructive phase relationships in relation to room acoustics. Now we will discuss how phase differences affect multi-mic'd sources.

Let's reacquaint ourselves with the meaning of *phase*. In practical terms, phase is defined as how far along its cycle a given waveform is. We should already know that waveforms are cyclical and that they progress through regular cycles of peaks and troughs. The measurement of phase is given in degrees, where 360 degrees comprise a complete cycle. 180 degrees is defined as halfway through a cycle. Any time multiple waveforms of the same frequency are unaligned in terms of phase, they are said to be 'out of phase' with each other. As explained below, this usually occurs due to physical distances and the time differences this introduces, but it can also be introduced through equipment that causes a phase

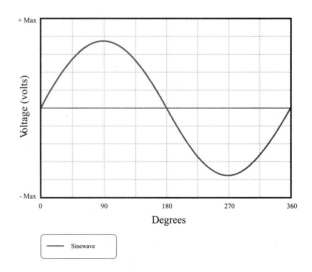

Figure 10.1 Sine wave life cycle

shift, and other processes that introduce short delays. These discrepancies in phase are expressed by their difference in degrees.

As we learnt in Chapter 9: The Critical Listening Environment, the way that identical sine waves sum is easy to comprehend. However, when we are dealing with the summing of broadband signals (i.e. signals that consist of multiple frequencies), phase difference varies according to frequency. This inconsistent phase difference results in comb filtering. Comb filtering causes constructive and destructive interference across many frequencies, leaving a sound that is not a true representation of the source. Comb filtering can amplify or attenuate any frequency range but its worst effects are often heard in the bass and low-mid frequencies.

In practice, phase differences can occur because of many variables, including distance, reflections, speaker/microphone properties, latency, etc.

Time Delay

There are two main scenarios in which phase inconsistency can arise during recording: multi-mic'ing an individual source, and mic'ing multiple sources that are in close proximity. These phase inconsistencies are caused by different physical distances between a source and any microphones that are capturing said source.

Let's look at an example of a multi-mic'd source. Imagine you are recording a guitar amp with two mics, one placed farther away from the amp than the other. Because it takes time for sound to travel any distance, the two mics will be simultaneously capturing the sound at different points on the sound wave.

In this situation, the closest mic's waveform would be ahead (in time) of the farthest mic's waveform. This means that each frequency within the closest mic's waveform will be ahead in terms of its phase rotation with respect to the far mic. Some frequencies will be ahead by a number of complete cycles and thus will be at the same point in phase at both mic positions; other frequencies will be ahead by a different degree of phase, resulting in some amount of cancellation. Frequencies that are ahead by a number of complete cycles plus a half-cycle will experience up to complete cancellation. The result? Yes, you guessed it: comb filtering. Even minute differences in distance can make a very perceivable difference to the overall tonality. In any circumstance where there is a difference in physical distances between a source and its multi-mics (no matter how small), some degree of phase inconsistency will occur. The frequency areas where this happens depend mainly on the distances involved. On a single source, the greater the differences in distance between the source and the mics, the further the phase problems will spread down the frequency range, so larger differences can make the problem worse.

Figure 10.2 Time delay between two mics on a guitar cabinet

Now let's look at an example of mic'ing multiple sources that are in close proximity. Imagine mic'ing two guitar amps with a single microphone each – let's label them amp A and mic A, and amp B and mic B. If they are in close proximity, each mic will pick up some bleed from the other mic's source. Our intention may not have been to multi-mic each source but, due to bleed, *each source will arrive at two different mics*. Amp A will arrive at mic A as a direct signal, and at mic B as bleed. Amp B will arrive at mic B as a direct signal, and mic A as bleed. The distance from each amp to its 'direct' mic and the distance to the other 'bleed' mic will be different, even if it's by only a small amount. As in the first example, this difference in distances causes the sound to be captured at different points in its wave, so there will be phase inconsistencies between the 'direct' waveform in one mic and the 'bleed' waveform in the other mic. That is to say, along with its own direct signal, each mic contains a phase-inconsistent version of the other mic's direct signal. When the two mics are combined into the mix, the bleed signals will potentially cause comb filtering to the direct signals.

If you are mic'ing up multiple sources in the same environment, getting the sources farther apart, and thus the mics too, will reduce the phase problems because the bleed into each mic will be reduced. Obviously, this is not possible in some cases such as on a drum kit but any extra space you can create will help (but be sure to make sure the performer still feels comfortable). Alternatively, gobos or other types of screens can be used to acoustically isolate sources.

When we multi-mic a single source, we usually pan the mics to the same position in the mix (there are exceptions, for example piano, where we might pan the mics across the stereo field). Having phase-inconsistent sources 'sitting on top of each other' often makes the identification of the resultant comb filtering apparent; there will be some degree of hollowing of the sound, which will extend through the sound from the high frequencies downwards, depending upon 'time difference'.

When we mic multiple sources simultaneously, we usually end up panning them to different points in the stereo field to create a sense of separation. This often has the effect of allowing potential comb filtering to go unnoticed. Parts that sound fine in the stereo mix can suffer from significant comb filtering when the mix is played over a mono system. Because of this, in any situations where bleed may have occurred during tracking, it is highly advisable to check your mix in mono.

While there will always be phase inconsistencies, not all phase shift is bad. For instance, slight inconsistencies between close mics and room mics on a drum kit often aids the sense of depth in a mix. Working out if phase shift is causing significant damage to your audio is more difficult as you increase the number of microphones you have on a source. When tracking drums, there are often at least 10 microphones, all picking up the same performance from different positions. While phase problems are more obvious between close mics, they can also occur within more seemingly innocuous partnerships such as between a kick drum mic and an overhead mic. Comb filtering occurring between many channels can be very difficult to identify to the untrained ear and they can vary heavily in strength. A telltale sign for me is the loss of punch or clarity on a drum kit. Hearing the effects of comb filtering, especially within the context of a whole mix, can be difficult. Even now, I am sometimes surprised by certain phase interactions. There have been occasions when I have tweaked EQs and compressors for hours, dissatisfied with a sound, only to discover that phase shift was the cause of the problem.

Phase in Practice

In practice, the fact that it's impossible to physically position mics in exactly the same place, and also the technical differences in equipment, mean that some degree of phase shift is unavoidable. Thus, some degree of comb filtering is unavoidable, and therefore, in practice, it is impossible for two distinct broadband signals to completely cancel each other out. Inversely, this also means that two distinct broadband signals cannot completely reinforce each other either. When an engineer says a multi-mic'd source is 'in phase', he or she is effectively telling you that the phase interference is minimal to the point that it is no longer a sonic concern. As phase cannot be manipulated for each frequency individually, it is a case of finding the best compromise that controls the amount of phase cancellation and also manipulating it so that it doesn't occur in key frequency areas.

Learning to recognize phase interference in your audio in the tracking phase is absolutely crucial to your progress, whatever specialism you are working in. Even a producer who might not be 'hands-on' with any of the engineering, mixing or mastering still needs to be able to identify this because it can severely alter the tonal decisions they are making at the source. For instance, a set of guitar mics out of time alignment might cause comb filtering that knocks out a substantial amount of low mid tone. An instinctive reaction to this is wanting to add more on the amp's settings. This is a very inefficient way of trying to counteract comb filtering because, despite the fact that you are sending more signal into the room, there is still phase-shift, and such brash use of equipment can also cause damage. If this audio is then time-aligned in the mix stage, you will end up with an overly muddy tone that will take more time and processing for the mix engineer to get it to sit right. If the sound is significantly different in the live room to the tone you hear in the control room, this is a red flag to alert you to a potential comb filtering and/or polarity issue.

When considering multi-mic'ing (or DI plus mic), the first question that you should be asking is whether you actually need to use multiple devices to get the tone you require. Don't multi-mic for the sake of it or because other people do it; just like any other decision, there should be a reason for doing it. I don't want to put you off here – I often like to multi-mic sources but I am always listening and analysing if there is a need for it. For instance, I love to blend a ribbon mic with a dynamic mic for guitar recording to get a 'fat' bottom end from the ribbon and an 'edginess' from the dynamic – a result I can't usually achieve with a single mic. On some occasions, though, the bass guitar might be taking up enough space to not need the extra 'fatness', and therefore the ribbon is redundant, and muting it could prevent potential comb filtering in the higher frequencies.

Once you have decided that you want to try to use multiple mics, the key is to make sure that any comb filtering isn't affecting a key frequency area adversely. If comb filtering is immediately apparent, I will physically move the offending microphone(s)' positioning to get them as coherent as possible; this usually means moving them closer together. Dealing with comb filtering is effectively a damage-limitation exercise because moving the mic(s) to different positions is simply going to cause comb filtering in different frequency areas.

On rare occasions, doing this might affect the quality of the tone captured by the microphone too much. In this case, you might want to rethink your approach and handle the phase correction in the DAW. These days, there are many post-production tools at our disposal to help further hone the phase relationships that were not available in the early years of multitrack recording. This greater control of phase is one of the many reasons why modern music can sound 'larger than life'. There are three ways that I have combated phase shift in the DAW:

1. Manual movement of waveforms or using a delay plug-in to time-align.
2. Phase-rotation tool to 'sweep the phase'.
3. Auto-alignment tools.

Manual movement or using a delay to correct time delay should be obvious, but the phase rotation and auto-alignment tools are worth quickly defining.

Phase-rotation tools such as the Little Labs IBP are found as both hardware and plug-ins and allow the user to manipulate the phase in a manner other than by time alignment. IBP works by rotating the phase of a set centre frequency (which is often switchable between low-frequency and high-frequency modes). As the phase is rotated, it will affect the phase relationships of all other frequencies, with the aim of finding a suitable balance by forcing phase interference into less audible areas. As we are dealing with a broadband signal, you cannot rotate the phase of all frequencies simultaneously, so it is important to understand that even a phase rotation tool is still less accurate than time-aligning audio. However, its main benefit is the fact that it will maintain the sense of space in a recording. IBP also has controls to time-delay and correct polarity problems, which we will be discussing shortly.

Auto-alignment is available in plug-in form, and is a tool that analyses the multi-mic recording and automatically compensates for the delay between the microphones. This is great for aligning close mics but some more care and attention should be given to ambient/room/overhead mics so that the sense of depth isn't lost. Examples of auto-alignment plug-ins are Waves InPhase and Sound Radix Auto-Align.

I am so vigilant to the negative effects of comb filtering that I often choose to check all multi-mic'd sources before pressing record. I generally do this by sweeping the phase and/or delaying a signal to hear if it starts to sound 'better'. As previously mentioned, where there are problems, I will first try to correct them by manipulating mic positioning. Remember, though, your ears are the ultimate judge, and on occasion I have reverted back to leaving each of the mics in their optimal positions (tone-wise) and using the phase rotation tool because it achieved more pleasant results.

These days, I tend to avoid manually moving waveforms and prefer to use less destructive means to manipulate phase relationships. Therefore, I tend to use phase rotators or auto-alignment tools. Where there are a number of complex phase interactions and sense of depth like on a drum kit, I will often reach for the phase-rotation tool first. On sources that are mic'd more closely, I find the auto-alignment tools extremely handy.

Even if you are anally retentive when tracking, sometimes comb filtering still slips through the net. Comb filtering can even start to occur during the mixing process, so always keep an ear out for it. One way to do this while tracking or mixing is to regularly switch your mix from stereo to mono, as destructive phase relationships are often more audible in mono. It also helps to check that your mix will not be destroyed on mono playback or compromised stereo systems (such as those found in bars or shopping centres).

Once you are more experienced, you can sometimes use phase relationships to your advantage. For instance, on occasion, I've purposely knocked the bottom end of guitars 'out of phase' so that the mix is less cluttered without resorting to corrective EQ.

Other Phase Shifts

As well as time differences between multi-mics, other factors can cause a degree of phase shift. The most obvious and damaging is comb filtering caused by room reflections. As well as damaging your critical listening environment, uncontrolled room reflections will damage the quality of your audio capture. Room acoustical issues are also caused by time delay except this time delay is going on in the environment itself and is captured by the mic(s) that way and therefore cannot be aligned later. This can be visualized simply by imagining a guitar amplifier, its direct sound and a reflection from the floor arriving later.

Significant thought should be given to mic placement and treatment to minimize uncontrolled reflections, which will be covered in various points through the rest of the course.

Figure 10.3 Guitar amp reflection and direct sound as captured by a mic[1]

As well as room reflections, a lot of the equipment and environment will introduce subtle phase shifts that can affect the end result, particularly when blended with another version of the same source going through a different chain of equipment. For instance, let's think about a bass guitar track that has a DI input and also a microphone on the cabinet that is blended together to taste. Phase interference will be most obvious from the time difference between the more direct sound from the DI and the sound captured by the mic, but other factors will also play a part, such as:

- the phase response of the speaker cabinet;
- the phase response of the microphone; and
- phase shift caused by EQ (subtle phase shift is caused by any analogue EQ and any non-linear-phase digital EQs).

In general, this means that you should try to keep equipment chains as short as you need them and also add processing once the devices that have captured the source have already been summed together.

Phase shift through equipment/plug-ins is particularly apparent with EQ. If you have EQ'd individual channels differently and then summed them, you will probably experience significant comb filtering. This is because phase shift is actually a by-product of the way they work (apart from linear-phase digital EQs). These days, I only EQ sources once they have already been summed to avoid such problematic artefacts.

ITB Comb Filtering

Also note that if you are mixing ITB (in the box) and using different plug-ins on each track of a multi-mic'd source, you may end up with comb filtering caused by the differences in latency between channels. *Latency* is another word for time delay caused by 'a journey through a system'. In this specific instance, latency is caused by the time it takes for the audio to be processed by the plug-ins or the round trip out of the computer and back in again in the case of hardware inserts.

So if you have a different processing time between multi-mic'd channels, the result is phase interference. Pretty much every up-to-date DAW software has a feature called automatic delay compensation (ADC) or similar, which will sort out plug-in latency. Sometimes you may need to turn it on; in other cases, you might have exceeded the level of delay compensation on heavily processed channels, so you will need to increase it in your DAW's settings or reduce processing. Hardware insert latency has to be calculated manually. It is really crucial to be aware of the problems related to this type of phase shift as it can ruin all of your good tracking work and actually leave the final mix sounding worse than the raw tracks.

Note: if you are using an outdated version of your DAW or one that doesn't feature ADC, don't worry as you should be able to find a simple delay plug-in that allows you to delay a signal by individual samples. By delaying the channel with the least latency to match the one with the most, your signals should lock in phase again.

10.4.2 Polarity

Although most of the interference you will encounter will be caused by comb filtering due to phase shift, sometimes another phenomenon called polarity will rear its ugly head. Polarity refers to whether voltage is positive or negative with respect to a reference.

An alternating electrical signal varies through positive and negative voltages. When we wish to describe a varying phenomenon, such as sound pressure, we assign polarities to represent each opposite state. When working with microphones, positive voltage is usually assigned to positive pressure, and negative voltage is usually assigned to negative pressure. Thus, the sound wave can be represented by an analogous electrical waveform.

When working with multiple mics, it is important that the resulting electrical waveforms obey the same convention in terms of how sound pressure is represented (i.e., one polarity (usually +ve) representing positive pressure; the other polarity (usually –ve) representing negative pressure).

If the convention doesn't match in every microphone on that source (including bleed), the offending waveform will not sum properly with the other waveforms.

A break in the convention could occur due to a fault in a microphone, whereby polarities are assigned 'incorrectly'. It could occur at some point after the microphone, whereby a fault in a piece of equipment or wiring inverts the polarity of the waveform (+ve becomes –ve), and vice versa. Whatever the cause, the resulting waveform will not represent the sound wave 'correctly', and in fact will be a mirror image of what it should be.

Summing this inverted waveform with the other waveforms will result in cancellation rather than reinforcement.

Let's look at a simple example involving sine waves. Imagine two sine waves of the same frequency starting simultaneously. When the polarities are matched, summing these signals creates amplification (which is the desired result). However, if the polarities are unmatched it instead creates complete cancellation (extremely undesirable).

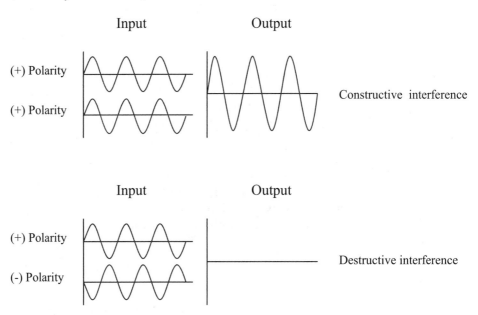

Figure 10.4 Sine wave summing and polarity

Now imagine that there is an element of time delay too. Figure 10.5 shows how this might affect the resultant signal.

The aim, therefore, should be to make sure that all devices that capture or express audio (microphones, cables, speakers, etc.) have matched polarity. However, in practice, wiring mistakes can happen and conventions can differ. Therefore, audio devices such as mixing consoles have buttons to counteract this, which flips the polarity so +ve becomes –ve, and vice versa. This is often called 'polarity inversion' or 'flipping the polarity'. A common misnomer is to call it 'flipping the phase', which is incorrect and is probably one of the causes of the confusion between these phenomena. Using an invert function in your DAW will do the same thing.

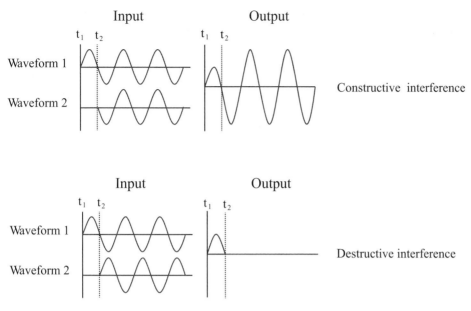

Figure 10.5 Time-delayed sine wave summing and polarity

Because polarity problems are caused by the way that pressure is represented as an electrical voltage, polarities can only be matched or unmatched. They also have no time element so all frequencies are affected equally. There is no such thing as degrees of polarity. Unlike phase, polarity issues are only caused when signals are represented electrically (i.e. it isn't a naturally occurring phenomenon).

However, there are situations where an audio engineer might need to utilize the phase inversion button when multi-mic'ing, even if the wiring of equipment is correct. This occurs when two microphones are capturing the same source but are positioned in such a way that each mic is receiving opposite pressure and therefore creating summing issues. The most common occurrence of this is when you are mic'ing both sides of a snare drum; in this case, the signals from the top and bottom mics are almost always receiving different pressure. Let's examine why this is the case. With every snare hit, the initial impact makes the skin move away from the top mic and towards the bottom mic. This means the top mic sees negative pressure while the bottom sees positive pressure, and vice versa. The mics will be effectively creating signals that are the mirror image of each other. When these signals are summed, it causes massive reduction in level. Simply reversing the polarity of one of the mics via the mic pre-amp or console stops this cancellation from happening. If this was a theoretical exercise, these two signals could completely cancel. In practice, however, they won't. The reason for this is that the actual waveforms captured by the two mics will be subtly different. This is due to the tonal attributes of their positioning and also a certain amount of phase shift because of distance. Because the two mics contain phase shift, the reversed pressures will never completely cancel when summed. The amount of phase shift and variance in tonality will affect how drastic the sound degradation is when the audio is summed. What we have in practice is a problem with both polarity and phase shift.

Let's delve deeper into what happens when polarity issues and phase shift occur simultaneously. In the two examples that follow, I'm assuming each mic is producing the same amplitude. For differing amplitudes, read 'reinforce' as 'sum', and 'cancel' as 'difference'.

If two mics have the same polarity and there is (theoretically) no time delay (i.e. no comb filtering), they will reinforce each other across all frequencies, resulting in double the overall amplitude. Moving them apart introduces comb filtering; the further you move them apart, the further down the spectrum comb filtering extends. As you bring the two mics back together again, the comb filtering retreats up the spectrum, so the nulls begin to disappear until all comb filtering disappears and we once again have complete reinforcement.

If two mics have opposite polarities (for whatever reason, e.g. a mic/cable/channel problem) and there is (theoretically) no time delay (i.e. no comb filtering) they will cancel each other across all frequencies, resulting in zero overall amplitude. Moving them apart introduces comb filtering; the further you move them apart, the further down the spectrum comb filtering extends. But it's the opposite of the in-polarity example. Frequencies that are out of phase now reinforce each other because of the opposite polarity. So where we previously had peaks, those frequencies remain cancelled. Where we previously had nulls, those frequencies become peaks. So we end up with something audible, but the mirror image of the first example. If the comb filtering extends very low, we start to recover closer to half the sound.

As you bring the two mics back together again, the comb filtering retreats up the spectrum, so the peaks begin to disappear until all comb filtering disappears and we once again have complete cancellation.

In the example of the mics having the same polarity, eradicating the phase shift removes the comb filtering and thus restores the signal; the closer they get together, the more reinforcement there is. In the example of the mics having opposite polarity, moving them closer in order to eradicate the comb filtering results in more cancellation until they completely cancel one another. If moving mics closer together and thus into phase alignment is not helping, and is possibly

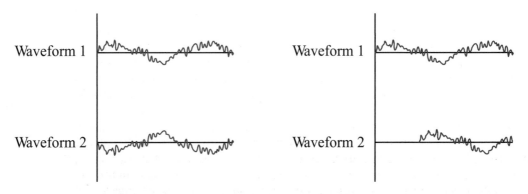

Figure 10.6 Unmatched polarity, but in time,

Figure 10.7 Matched polarity, but out of time

making things worse (your entire signal is beginning to disappear), the chances are there is a polarity disagreement between the two mics.

Hopefully, these examples clearly demonstrate that phase and polarity are two distinct issues and it is clear that in theory signals can have 'unmatched polarity' but be in 'in time', or vice versa.

Sometimes you may even encounter waveforms that are 'unmatched polarity' *and* also significantly 'out of time'.

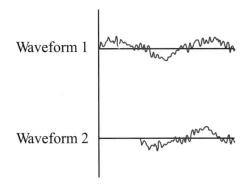

Figure 10.8 Unmatched polarity and out of time

EXERCISES

1. Multi-mic a guitar cabinet. Monitor it in the control room and ask another person to move one of the mics around while the guitarist is playing, and listen to the way that comb filtering is affecting the sound.

2. Now try to make the result as 'in phase' as possible by only moving the mics around.

3. Now fine-tune this further using a phase rotator and/or delay.

4. Assuming the polarities are already matched, flip the polarity of one of the channels and move the mics around again, listening for the position where the sound cancels the most.

10.5 RECORDING WITH COMPRESSION/EQ

I get asked a lot whether you should use EQ on the way into the box or save it for mixdown. Unfortunately, the answer is not a simple yes or no.

There are situations where there is a technical advantage to certain EQ and compression techniques before the signal gets converted to digital. Removing 'rumble' and restricting any excessive 'spikes' leaves you more headroom for the important stuff. This was very important in the days of 16-bit recording as you had to record a little hotter anyway, but with 24-bit recording now the norm we have enough clean headroom to be able to record everything without worrying so much about the noise floor and lost resolution. However, in my opinion, there is no reason why you can't utilize high-pass filters on instruments that don't contain any low-end information. Also, use of compression on the way in to deal with instruments that have too large a dynamic range is not necessarily a bad thing either (but needs care as it cannot be 'removed' once recorded). I've often used a fast-attack, fast-release compressor to shave a dB or so off of excessive peaks on vocals, clean guitars, acoustic guitars and high-frequency percussive sounds such as shakers. Remember, though, this is only compressing the excessive peaks. All of this helps to make the signal a little more stable on the way in. The distinction is that I am doing stuff on the way in for stability and headroom, not for tone-shaping.

I am a big believer in trying everything I can to get the sound as close to my vision as possible without using EQ or compression. Sometimes using processing on the way in can make you lazy with the front end of the engineering process (source, room, mic choice/placement, pre-amp selection). I am not saying that you should never EQ or compression to tone-shape on the way in; I am just saying that there are few practical advantages to doing it at source. Despite this, there are some situations where I would consider using processing on the way in:

- When I am tracking in a studio that has high-end gear that I cannot replicate in the box, I will sometimes choose to process key parts on the way in for the superior tone.
- Sometimes a piece of hardware has a 'sound' even without it processing anything, and that can get it closer to my vision.
- When I am super-confident on my vision *and* confident with my monitoring and critical listening environment, I will sometimes do some on the way in to make my life easier further down the line. This again depends on the quality of analogue gear available to me.

Even if I am not processing prior to the digital conversion, I will still approximate the processing I am looking for by using plug-ins, but only once I've got as close as I can get with my front-end engineering. When you are dealing with vision and how the tracks are going to stack up to create a pleasing arrangement, it is difficult to picture the end result and make decisions unless you are trying to replicate roughly what the mixer is going to do. Therefore, I do some quick panning, EQ and compression to give myself a better image to visualize from. This is particularly true with drums when there is often a lot of processing going on.

When it comes to the typical project/home studio, there is nothing wrong with just recording everything straight and processing in-the-box, and to do this you only need to have an interface that has nice, clean pre-amps and satisfactory 24-bit conversion. These days, the quality and usability of plug-ins versus the cost of analogue hardware that does a better job means that it is simply not practical for a hobbyist to process out-of-the-box. Scrimping on other more crucial areas such as monitoring, room acoustics, interface and basic microphones just to get a basic hardware EQ or compressor is going to be a waste of money and actually be counterproductive for the quality of your output. Of course the other matter is that, generally, people that can't afford analogue hardware are likely to be less experienced so the non-destructive element of working with plug-ins in real time is another potential advantage!

10.6 ANALOGUE IS BETTER THAN DIGITAL (OR IS IT?)

Back in Chapter 7: Demystifying Recording Levels, we discussed some considerations when working with a hybrid system so you can probably guess my views on the analogue-versus-digital debate. However, I want to finally put this argument to bed. On websites and Internet forums, you will find many disputes between engineers about a number of topics, including whether analogue sounds better than digital or if recording at 44.1 kHz or 96 kHz is better. Of course, these variables have some impact on sound, and it is therefore a relevant debate. However, a by-product of the popularity of this argument means that novices might conclude that you need lots of analogue equipment or a higher sample rate to get a commercial-grade result, and that simply is not true.

I have been lucky enough to be around some very high-end studios, some great-sounding acoustic spaces, as well as some not-so-high-end studios, and I can honestly tell you that the advantages gained from monitoring and room acoustics far outweigh the advantages of top-level processing equipment.

As the analogue-versus-digital debate rages on, I want to dispel some myths. If you asked me if processing in-the-box can sound more sterile than working purely in the analogue domain, then at the time of writing this course (2014), I would have to say yes. However, there is a huge caveat here: digital doesn't sound 'worse' and there isn't anything inherently wrong with digital processing. In fact, quite the contrary is true – digital processing is actually more accurate than its analogue counterpart. If anything, it can be described as 'too perfect'. The problem is that the non-linearity and distortion characteristics of analogue equipment are often pleasing to our ears. Whether this is due to our cultural exposure to older recordings that used analogue technology or simply because the rich harmonic by-products are more interesting to our ears, it doesn't matter – we just need to be aware of it. To get the best out of working in-the-box, we just need to bear in mind the 'cleanness' and 'perfectness' of digital audio and inject some analogue colour/warmth. This can be done with many stock compression/saturation/distortion plug-ins in your DAW or, even better, via plug-ins that emulate many of the most revered analogue products. These include:

- Models of classic EQs/compressors such as the Neve 1073 EQ, SSL Channel EQ, UREI 1176, Teletronix LA2A and SSL Bus Compressor.
- Tape machine emulation such as Slate Digital UAD, Waves Kramer Tape, UAD Studer A800 Tape and Massey Tapehead.
- Emulators of mixing consoles such as Slate Digital VCC or Waves NLS.
- Plug-ins that model the ways that analogue distorts such as SoundToys Decapitator or Softube Saturation Knob.

Sometimes a simple solution such as rolling off some top end using EQ on a channel can produce a result that sounds less 'digital' and more like what would have naturally happened when it was tracked to tape, and therefore closer to its analogue counterpart. Of course, this processing is more important for the mix engineer. However, knowing its use and existence can help you make decisions in the tracking stages. Sometimes I'll even add such processing to hear the difference it makes to the source sound. If a recording has been tracked using a transparent and less coloured interface, it can be coloured quite effectively after the fact using these tools. While I still admit that the select top-end analogue equipment is currently unbeatable from a sonic perspective, plug-ins are already getting very close (and are catching up remarkably fast). On the flipside, the set-up and recall of this equipment is much more laborious – so much so that many renowned mix engineers are starting to work purely ITB.

In my opinion, a well-tracked, all-digitally processed recording is not going to prevent your record from being a hit. The song, performance, arrangement, source sound and the mix engineer's talents are all far more important to the finished product than whether it was done in or out the box.

10.7 HEADPHONE MIXES AND LATENCY

One of the most important elements of recording that often gets overlooked is the headphone mix or, as it is sometimes known, the monitor or cue mix. With timing, tuning and dynamics being so crucial to the success of a performance, the performer needs to hear the right blend of instruments at a reasonable level without significant distortion to be able to truly give it everything. Telltale signs of a poor headphone mix include:

- Low signal levels where the artist struggles to hear the headphone mix over the sound in the live room, or headphone mixes that start to distort before reaching that point.
- A lack of definition on important pitch or rhythmic elements.
- An insufficient level of the instrumentalist's own parts.

10.7.1 Headphone and Amplifier Selection

First, a good headphone mix starts with a good set of headphones and headphone amplifier.

Headphone-wise, you will be looking for a decent closed-back design that reduces spill and can handle high signal levels. Something that is comfortable on the ears over long periods is also a bonus.

With the headphone amplifier, you need to make sure that it has enough outputs for a set per performer and that it provides sufficient output level for the performer to hear everything comfortably. There are four main types of headphone amp to choose from:

Headphone output on interface/monitor controller. Many interfaces or monitor controllers might have a dedicated headphone output. Sometimes these might be of low quality, but on the whole they are pretty usable. However, they are mainly suited to one-room set-ups, and often the interface/controller will only have one or two dedicated headphone outputs.

Distribution amp. A simple distribution amp is usually a single stereo mix. Left and right input jacks feed a number of independent amplifiers (often four), with each having an individual headphone jack output with its own volume control.

Figure 10.9 Audio Technica ATH-M50 closed-back headphones

'More me' amp. This is very much like a simple distribution amp except that each output channel will have a second input so that the performer can set the level of a key instrument. This is usually their own instrument, hence the nickname 'more me' amp.

Personalized mixer. This a monitoring system where the performers themselves can have control over their own mix. In these systems, the distribution box will have several inputs to which you will send individual instruments or groups of instruments. Each performer will then get a separate mixer box that will have pots that control the amount of each of these inputs to their headphones to give them their own personalized mix.

I have lost count of the number of project/home studios and even smaller commercial facilities that have scrimped on headphone amps and headphones, and suffered because of it. Remember that, as engineer/producer, a large portion of your job is to get the best out of the performer, and a poor monitor mix can leave you fighting an uphill battle. Some of the negatives that result from a poor headphone mix can be:

- a demotivated and frustrated artist;
- longer hours spent tracking to get suitable results;
- more editing for the engineer; and
- a less natural performance.

It is somewhat ironic, then, that the headphone chain is one of the most undervalued parts of the recording chain for some engineers. Having a headphone system that delivers high output levels is particularly important if you are regularly recording live performances of drums, bass guitar and electric guitar. These situations are likely to need more sound isolation and more level from the headphone amplifier to be able to compete with the level of sound in the live room itself. Even when recording quieter sources, the quality of the headphone chain is still important. For instance, I've lost track of the number of times I've battled with metronome bleed on subtle sources such as a fingerpicked acoustic guitar track. Most of this can be avoided with effective isolating headphones and the rest of it can be reduced by automating the metronome on particularly quiet parts (such as the way that the final notes of the song decay).

10.7.2 Gain-Staging and Pre-/Post-Fader

Whether you are setting up a simple headphone send, a 'more me' system or even a personalized mixer system, you should be aware that there are several ways of setting up a headphone mix. You could do it from within your DAW, through a console or through a software mixer that is developed by the interface manufacturer. The pros and cons of each method will be explained, but let's first get two issues out of the way that might affect your choices regardless of how you implement the headphone mix.

Gain-Staging

Regardless of the way that you set up the headphone mix, you need to make sure that you gain-stage it properly. The golden rule for this is still the same as it would be for any mix: thou shalt not clip. The amount of headroom depends on the situation. For instance, when using a simple headphone distribution system with a single stereo input, I would gain-stage so that my signal into the headphone amp is strong and has less headroom so that the amplifier doesn't have to work as hard (generally –3 dBFS to –6 dBFS peak). When creating 'more me' sends, I would generally aim for my peak levels to be a little less strong (around –6 dBFS on the stereo and the additional channel), and with a personalized mixer I might go lower still (around –10 dBFS on each channel). This is so I am allowing headroom for the summing of each independent input. Obviously, the headroom of the headphone amp and the headphones themselves will, to some degree, dictate the best operating levels, so treat these only as rough guidelines. As long as you aren't clipping, you shouldn't run into too many issues.

Pre- or Post-Fader?

The next thing that we need to consider is whether we want to send the signal to the headphones pre- or post-fader. *Post-fader* means that the level adjustments on the channel's fader affect the signal being sent to the mix, and *pre-fader* means that it gets sent to the mix prior to the channel fader so it doesn't affect the headphone mix. The advantage of post-fader means that you can just set the sends at unity and you have a fairly decent mix in very little time. However, there are significant drawbacks to working post-fader:

1. The performer will often want a significantly different mix to the one in the control room.
2. The production team will often want the freedom to tweak levels while the take is happening without affecting the performer.

Therefore, for practical reasons, the majority of engineers will set their headphone mixes pre-fader.

10.7.3 Methods of Creating a Headphone Mix

Here is a breakdown of the three main methods of creating a headphone mix:

Directly from the DAW

If you are creating a headphone mix directly from your DAW, I recommend bussing it through an auxiliary channel. For instance, you could send all the instruments for the headphone mix to bus 1-2, and then create an auxiliary channel with inputs set to bus 1-2 so that it receives the headphone mix. The final step is to then send the outputs of this auxiliary channel to the output of the interface to which the headphone amp is connected.

This has two major advantages:

1. You can see the output level of the headphone mix.
2. You are able to process the whole send with EQ, compression, etc.

It is possible to send signals directly to the output(s) of your interface/converter, but you will find it hard to get an idea of the resultant signal level being sent out of the system. Doing your headphone mixes inside the DAW is easy to set up and also great for workflow as you don't have to switch to any other software. The big drawback of working this way, though, is latency.

As we discussed in the phase section of this chapter, latency is a time delay caused by a journey through a system. The time it takes for a signal to get all the way through a system is called *roundtrip latency*. All recording systems will have some form of latency; remember, electricity still has to pass through wires in analogue systems, which, while it isn't instantaneous, is almost as close as it gets! This few nanoseconds is way too small a delay to cause any audible problems. In digital systems, on the other hand, the roundtrip latency is much more problematic. Not only does the analogue signal have to be converted from analogue to digital and back again, but all the processing has to happen in between. The demands of interfacing of the hardware and software and doing the digital signal processing means that there is significant pressure on the computer, which it has to keep up with. If the system is unable to keep up with the demand put on it, it can cause audible clicks and pops or even crashes. Obviously, this should be avoided at all costs. To avoid this, a region of memory is allocated to temporarily hold data as it is moved around, which is called a *buffer*. There are a number of these in the digital audio chain, but in the context of this course, knowledge of them is not important. The buffers effectively 'buy the computer some time' so that all this processing can happen without causing such detrimental artefacts. However, the increased stability has a cost – you guessed it, added latency. In your DAW, you will have a setting that controls the hardware buffer size. In Pro Tools, this is called Hardware Buffer Size, and is found in Setup > Playback Engine > H/W Buffer Size.

Figure 10.10 Headphone send via an auxiliary channel using Pro Tools

All of the options in the playback engine window will help to optimize system performance, but the hardware buffer size is the key setting to help keep latency under control. For tracking, the aim is to get the lowest buffer size you can without the system glitching or crashing. When buffer sizes get too big, the artist will start to hear a delay between him or her performing and what he or she hears in the foldback. This could be a distinct 'echo' sound as the roundtrip latency gets nearer to 20–30 ms, which is the threshold where our ears start to hear events as distinct sounds. More often, though, there will just seem to be a small lag between when he or she plays his or her instrument and how he or she hears it. This can be very off-putting for timekeeping and performance. The more plug-ins you use, the harder it will become to use a buffer size that doesn't cause noticeable latency. The buffer size will typically range from 32 samples to 1,024. As we learnt in Chapter 7: Demystifying Recording Levels, when an analogue signal is turned into a digital signal, samples are taken thousands of times a second to reproduce the sound wave accurately. So when the hardware buffer size is set to, say, 256 samples at 44.1 kHz, this means that the latency of the buffer is 5.8 ms:

44,100 (samples per second) / 1,000 = 44.1 samples per millisecond

256 (hardware buffer size) / 44.1 = 5.8 ms buffer latency

When you add this to other system latencies such as A/D and D/A conversion, your roundtrip figure will be even higher. Generally, you will want to keep the total roundtrip latency under 12 ms. To do this, 256 samples hardware buffer size is typically the largest setting you can get away with. However, when recording myself, I have started to feel like my performance was being affected slightly by 256 samples, so I prefer to run at 128 samples or below wherever possible.

Integrated Software Mixer

There are two ways around the issue of digital system latency. The first is to utilize a software mixer that controls the hardware interface. These days, many popular interface manufacturers create software mixers that come with their products that allow you to route the audio around without having to do it in your DAW. Say you are recording a vocal; you can send the incoming vocal signal straight back out of the required output of the interface without having to be routed through the DAW. In some cases, it is even routed back out of the interface without it even being converted to digital, meaning zero latency (well, at least as good as). On top-end interfaces, you can even process the signal with EQ, compression or reverb and delay effects while maintaining a very low latency due to an onboard DSP chip. Despite the fact that you are monitoring the performance before it reaches the DAW, it is still important to keep the buffer size low so that the other parts of the mix are not out of time.

Using a Console

Finally, you can use a mixing console to create headphone mixes. This is much like using a software mixer with zero latency because the signal is sent to the headphone mix from the console before it reaches the DAW. Setting up a cue mix on a console is also very quick and easily tweakable as all the controls are at your fingertips. Just like with a software mixer, you need to remember to keep the buffer size low so that there isn't latency on the other channels of the mix.

10.7.4 Headphone Mix Needs

Finally, we need to consider what the performer actually wants to hear in his or her headphone mix. The blend of instruments is likely to change for each musician. For instance, the bass player is likely to want a lot of drums, and the vocalist is likely to want lots of him or herself. When the drums and bass are being laid down, an almost deafening click track is advisable, but beware of the click track spilling into the microphones. For many of the harmonic/melodic instruments, I will also bias the mix to be very rhythm track-heavy as this will dramatically improve their timing. This means that the kick drum, snare drum, hi-hat, bass guitar and perhaps a rhythm guitar will all be prominent. Your aim with a headphone mix is to get the best performance out of each musician. You need to treat each individual differently, so you shouldn't just assume that the current headphone mix is fine. Always ask the performer what he or she needs and tell him or her that you won't be offended if he or she asks you to change anything.

Dependent on the experience level of the performer, you might also find that he or she doesn't know what he or she wants, or even encounter a musician who struggles on, not realizing that his or her performance could be markedly improved with a tweak to his or her headphone mix. You should be very vigilant to the calibre of performance the artist is giving; if it is not up to the standards you are expecting judging from his or her ability in pre-production, there is a high probability that he or she isn't hearing crucial elements clearly enough. Whenever I am setting up a headphone mix, I like to listen to the send rather than setting levels 'blindly'. If at all possible, it is great to do this with the same model of headphones and a similar amp.

NOTE

1. Mike Senior, Phase Demystified, Sound On Sound, April 2008.

BIBLIOGRAPHY

Chuck McGregor, Phase & Polarity: Causes And Effects, Differences, Consequences, accessed February 2014, www.prosoundweb.com/article/polarity_and_phase_explained.

Figure 10.11 SSL auxiliary channel, called a 'cue send' (picture courtesy of Metropolis)

11

TRACKING DRUMS

So, we are ready to start recording. First off, due to their rhythmic nature, I am assuming that you know what the fundamental parts of a drum kit are and what each part does. Just in case you don't know what the parts are called, Figure 11.1 is a quick explanation.

Figure 11.1 Drum kit

I also assume that at this stage of the production process, you have either recorded a scratch track (a guide track, usually lead vocals and guitar, for the drummer to play along to) or you are intending the band to play together live during the recording. In the latter case, you need to be particularly confident with all the tones you are recording.

Drum recording is a massive part of modern rock production. These days, the drum sound heard on most commercial releases is pretty far removed from what I would consider a 'natural' sound. Techniques such as sample enhancement/replacement, tuning the drums to the key of the piece, recording cymbals separately to the main shells of the kit, use of large rooms for ambience while isolating close-mics using screens, and heavy compression on room mics are all used to help produce a larger-than-life feel to a drum kit. I often use some (or all!) of the above techniques to help hone a drum kit's sound. However, if I want to capture a more 'retro' or 'natural' sounding kit, I'll use just a stereo pair of overhead mics and a kick drum mic, and use minimal processing and isolation.

The trick is to visualize what you want to achieve and adjust your methods to help guide the track in that direction. Put simply, this is the key to music production. Have this in mind at all times while producing. To be able to make the best decisions, you need to have a broad knowledge of as many techniques as possible. In the following chapters, I will discuss a variety of options for recording each instrument in a typical rock track. I will also include videos of the techniques I actually used while making 'Take Her Away'.

Before we delve into the technicalities of choosing a drum room and selecting and placing microphones, we must focus on how the drums themselves sound. This is important because you cannot assume the drummer of the band knows what sound is best, let alone how to go about achieving it. As a producer, you need to understand the tonal qualities that are appropriate to the genre you are recording, and be adept at achieving your envisaged sound. You need to know the different types of instruments available, the peripherals and consumables that help shape the instruments' tone, the rooms you'd ideally record in and the amount of ambience you're likely to require, and also any effects that may be suitable. This is a lot of knowledge to acquire and will build up through experience. The following few chapters should help reduce that learning curve significantly.

11.1 DRUM SELECTION

When recording an unsigned or small indie label band, you're unlikely to have the luxury of budgeting for several hired drum kits. Therefore, it is useful to know how to get the best out of the drum kit that you have to record. I like to have a well-maintained house kit at hand to fall back on if I'm presented with a completely unusable kit. If you have the luxury of choosing a specific drum kit, consider some of the following factors as you make your choice:

1. The genre of the music you will be recording.
2. The keys of the pieces you will be recording.
3. The style of drummer you will be working with.

11.1.1 Shells

A drum shell is comprised of a cylindrical drum with a drumhead attached. In a typical rock set-up, these are your kick drum, snare drum and tom-toms. In terms of the tone of the drum itself (excluding the sound of the drumhead), there are two main factors that need to be considered: the size of the drum and the material it is made out of.

Size

Generally speaking, the bigger the diameter of the drum, the lower its potential pitch and the longer its sustain. Small drums are better at producing higher pitches. A similar rule applies with depth: the greater the depth, the louder the drum. If you are planning to tune the drum kit to the keys of the songs you are recording (this is covered in more depth later), you must be able to physically tune the drums to the required notes; this means that you may need to have several different-sized drum kits to find the right 'sweet spots' for particular songs. If you plan to tune the drums to a more 'neutral' key, one that will stay the same throughout the recording process, you will need to think about the overall 'vibe' of the band. For instance, a jazz band is more likely to suit a kit with smaller drum sizes tuned to a higher pitch than a rock band, which typically requires bigger, lower-tuned drums with a lot of sustain.

Figure 11.2 Rack tom drum shell (picture courtesy of Metropolis)

The most common sizing (diameters) for basic kits are:

Rock kits: 22" kick drum, 14" snare drum, 12" first tom, 13" second tom, 16" floor tom

Jazz kits: 20" kick drum, 14" snare drum, 10" first tom, 12" second tom, 14" floor tom

When looking at drum kit specs, make sure you know which parameter is the depth and which is the diameter, as sometimes manufacturers list them differently. Usually, the depth is listed last, but this is not always the case. For instance, you might see the following drum size:

- 18" × 22" kick drum
- 9" × 12" first tom
- 10" × 13" second tom
- 16" × 16" floor tom
- 5.5" × 14" snare drum

This is a pretty standard rock kit spec with the depth listed first. One easy way to quickly work this out is to look at the snare drum dimensions, as they usually have a much smaller depth than diameter, and sometimes they even use half-inches.

Material

The material the drum is made out of also has a significant impact on the tone the drum creates.[1]

Typically, kick drums and tom-toms are made out of maple, birch or mahogany. Softer woods produce lower tones but less amplitude, and harder woods produce higher tones with more amplitude. Out of the three choices, mahogany is the softest and birch is the hardest; maple is somewhere in the middle and is a good all-rounder.

Snare Material

When discussing drum material, special attention should be paid to the snare drum. Your choice of snare drum can really help the drum sound stay matched to the genre of a track. Snare drums are most frequently made out of wood, but many popular types are made of metal, and some niche snares can even be made of glass, stone or acrylic materials. Wooden snares are often warmer and deeper than metal snares, which are usually brighter and louder (and often also carry more overtones). Here is a breakdown of a few of the most popular metals:

Steel snares are bright, rich in overtones and have a long sustain. They have a unique piercing rim-shot, which almost sounds like a timbale. This means they can be used well in reggae genres, while their brightness also suits rock genres. Due to the low cost of steel, you often find them on entry-level kits.

Aluminium snares are bright, but are 'drier' than many other metal snares, meaning they have fewer overtones and less sustain. This makes them more sensitive to intricate styles, and usually less in need of dampening than most other types of metal snares. They are very versatile and can be used on a range of genres, making them a favourite of recording engineers.

Brass snares are another favourite of recording engineers. They are renowned for their brightness, coupled with a more pronounced bottom end than steel or aluminium snares. They are also more pronounced than other metal drums when tuned to a low pitch.

Copper snares are less bright than other types of metal snares, and are considered very 'warm' sounding. Copper snares are most often found in orchestras.

11.1.2 Drumheads, aka 'Skins'

Although the size and material of the drum itself is important, it is actually the drumhead that makes the biggest and most striking difference to the tone of a drum. The drumhead is the membrane on the top (the batter head: the surface you hit with the stick) and often the bottom (the resonant head) of the drum. When choosing the skins for your drum kit, there are several factors to consider:

Figure 11.3 Uncoated tom head (picture courtesy of Metropolis)

Coated or Uncoated?

An uncoated head is usually transparent and sounds 'brighter' than a coated head, and has more ring and overtones. Tom and kick drum batter and resonant heads are often uncoated.

A coated head is usually a whitish-grey colour and has a warmer sound, with fewer overtones than an uncoated head. Coated heads are favourable when using brushes. This is because the head's coating creates a friction with the brushes. The top skin on a snare drum is usually coated, and many drummers like to use coated skins for all the batter heads of their drums.

Single-Ply or Two-Ply?

As you may have guessed, a single-ply drumhead is made from one layer of a plastic material – usually Mylar – and is found in 7, 7.5, 10 and 12 mils thicknesses. Mils should not to be confused with millimetres – a mil is a thousandth of an inch (0.001'). Single-ply heads usually have a brighter tone than two-ply heads, and have more trebly overtones. Generally, the thinner the single-ply head, the more sustain and less attack. Two-ply drumheads have two layers of material that are usually 7 mils each. Two-ply heads are usually darker sounding, with fewer overtones. They have a less pronounced attack and shorter sustain than single-ply heads. Single-ply heads are often preferred in jazz or other genres where the drummer hits less aggressively. Two-ply heads are usually preferred in heavy metal and rock music where a less ringy 'punch' is important, and also when tempos are high, so a long

Figure 11.4 Coated tom head (picture courtesy of Metropolis)

sustain is not important. Another reason that two-ply heads are used more with heavy hitters is that they are more durable than single-ply heads. This means that they maintain their tuning better and have to be replaced less frequently than their single-ply counterparts.

While on the subject of drumheads, there are also plenty of speciality heads on the market, which include other materials such as Kevlar, which is less tonal, has less sustain but plenty of attack, and is even more durable than Mylar. Other speciality skins include pre-dampening technology (often found on kick skins), vented holes to aid punch and velocity, and also those with reinforced 'hit spots' to improve durability, punch and hit accuracy (often found on snare skins).

11.1.3 Cymbals

On a basic standard kit, you would find hi-hats, a ride, two crashes and possibly a splash too. Used more frequently in subgenres are speciality cymbals such as the china, commonly found in a heavy metal drum kit. Before I mention a few specifics relating to certain types of cymbals, I want to outline the factors that are common to all cymbals.

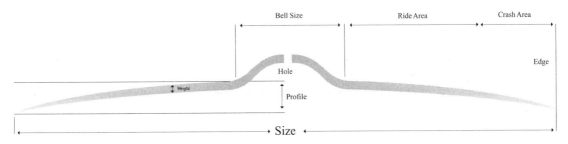

Figure 11.5 Cymbal parts[2]

Size

Small cymbals respond quickly to being struck, have less sustain and are quieter than large cymbals. Large cymbals have a 'bigger' sound but less controlled sound than smaller cymbals.

Thickness/Weight

Thick cymbals have a higher potential volume than thin cymbals, are more durable to hard hits (they last longer without cracking) and require a heavier hit to create their desired sound. Thin cymbals tend to also have a lower pitch. Rock drummers tend to have large, thick cymbals, while jazz drummers tend to have smaller, thinner cymbals.

Material

Cymbals tend to be made from three materials: brass, B8 bronze and B20 bronze (although you can find some variations of bronze between those).[3] Bronze cymbals are a mixture of copper and tin, although you sometimes find higher-priced cymbals with a tiny percentage of precious metals. Brass cymbals are cheap and nasty, and should be avoided at all costs. B8 bronze cymbals are composed of roughly 8 per cent tin and 92 per cent copper; they are cheaper, brighter sounding and more brittle than B20 bronze cymbals, which are composed of roughly 20 per cent tin and 80 per cent copper, and have a warmer, more musical and more controlled sound.

Profile

The higher a cymbal's profile, the higher its pitch will be.

Bell Size

Large bell sizes produce more overtones and a louder sound than small bells.

11.1.4 Cymbal Types

Now that we have discussed the variables common to all cymbals, let's consider the specifics of each cymbal type:

Hi-Hats

When matching cymbals to a kit/player, I usually start with the hi-hats. This is because it is the most used of all the cymbals and is crucial to the rhythm of most drumbeats. Hi-hats typically hold steady rhythmic patterns (rather than more isolated hits) and are usually 13"–15". In my experience, drummers tend to use 14" hi-hats more often than any other size. Hi-hats can be struck in an open or closed position using sticks, or played using a foot-pedal; this allows for more tonal variation than any other cymbal. Having chosen the hi-hats, I find other cymbals that have a complementary timbre that 'gel' well together despite their difference in pitch.

Ride Cymbal

After the hi-hat, the ride is the most used cymbal and, like the hi-hat, is typically used to hold a steady rhythmic pattern. It can be used instead of the hi-hat where variation is required. Drummers often use the bell of the ride cymbal to provide accents in more complex drum patterns. Ride cymbals usually range in size from 20″ to 22″.

Crash Cymbals, aka 'Crashes'

On a standard drum kit, a drummer usually has two crash cymbals – one smaller, darker and less pronounced, and one bigger, brighter and bolder. The main purpose of a crash cymbal is to provide the occasional accent and impact to a beat, rather than a steady rhythmic pattern, but drummers can play crashes as they would a ride cymbal, particularly where a more brash and raw vibe is desired; this is called 'riding a crash'. So when choosing your crash cymbals, it is important to know how your drummer is planning to use them. Crashes vary a little more in size than most other cymbal types, and commonly range from 14″ to 20″. In most rock kits, you usually find one 17″ crash and one 18″ crash. Because crash cymbals are often hit with a lot of force and are also hit at the edge, they are usually the least durable and often the first to crack.

11.1.5 Drumsticks

As well as the types of drums and the materials they are made out of, the sticks used by the drummer have an impact on the tone that is created. Often drummers get used to a certain brand and thickness of stick, and their playing adapts with the stick.

A standard drumstick is made of four parts: the tip, shoulder, shaft and butt.

The majority of the time, the tip is the part of the stick that comes into contact with the drum. This is often made using the same wood as the rest of the stick, but nylon tips are also common, which give a more aggressive tone. The shoulder is the part of the stick that tapers towards the tip; it is often used in a glancing motion to the edge of a crash cymbal and used to play the bell of a ride. The shaft is usually the part of the stick that the drummer holds but can be used to play side stick (rim clicks). The butt is the opposite side to the tip, and on rare occasions it can be used as a tip to create a less aggressive timbre.

Standard drumsticks are labelled with a letter followed by a number (e.g. 5A). The number signifies the thickness/diameter of the wood (although exact diameters aren't totally consistent between brands); smaller numbers will usually indicate thicker sticks. The letter represents the type of taper, shoulder and tip that gives the stick its overall 'feel'. The suffixes 'S', 'B' and 'A' stand for street, band and orchestral, respectively, although these classifications are somewhat out of touch with the modern drummer. A breakdown of the attributes and sonic characteristics that they give when striking the drums is shown in Table 11.1.

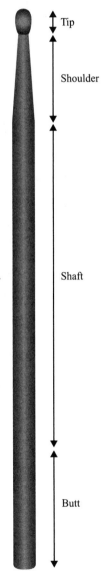

Figure 11.6
Drumstick anatomy

Table 11.1

Style	Orchestra – 'A'	Band – 'B'	Street – 'S'
Attributes	Articulate, lower volume	Durable, powerful, medium volume	Extreme strength, high volume
Body size	Small body	Medium body	Large body
Neck size	Narrow neck	Medium neck	Wide neck
Taper size	Long taper	Medium taper	Short taper

The most common types of drumstick are:

- 7As are thin and light, and are good for jazz drummers or marching bands.
- 5As are thicker than 7As, and are a very versatile and probably the most common drumstick.
- 5Bs are thicker sticks used in rock music.
- 2Bs are very thick and used when the drums are being hit hardest such as in heavy metal.

As well as conventional wooden drumsticks, there are many types of speciality products such as mallets, brushes and rods.

Mallets are handheld beaters with large tips made out of yarn, rubber and a number of other materials. Mallets are often used on keyed percussive instruments such as xylophone and glockenspiel. Yarn mallets are often used on cymbals to create cymbal 'swells'.

Brushes are made from a series of bristles attached to a handle. These bristles can be made of plastic or metal, and are sometimes coated with rubber. Brushes are generally used lightly in either a dragging motion or with a more conventional strike. However, on some occasions, both methods are used simultaneously using both hands. They achieve a more laid back and warm 'swooshing' tone often heard used in jazz, swing and big band music.

Rods or rutes consist of many thin strips of material (usually wood or plastic) held together to form a single stick. Their sound can be considered halfway between a stick and a brush.

11.1.6 Hardware

Often overlooked is the importance of good-quality hardware. Drum hardware consists of the stands or pedals used with a drum kit. There are many types of frames and mounts on the market these days, and the specifics of these are not important. What is important is that the hardware is sturdy and well maintained. Cheap, unsteady or unmaintained hardware will often break, squeak or loosen, and may not provide a comfortable angle for the drummer to play properly (there will be more on hardware set-up later). It is always best to get hold of good-quality hardware if the drummer doesn't own any.

Kick Drum Pedal

Probably the most important decision when choosing drum hardware is the kick pedal because it can have a significant impact on the sound of your kick drum. What makes the biggest difference is the type of beater used on the top of the bass drum pedal. The beater is the part of the pedal that makes contact with the skin. They are usually made out of either rubber, fibreglass, felt, plastic or wood. Heavier beaters generally produce more volume; harder materials generally produce more attack and 'click'. These days, there are multiple-sided beaters available that provide more flexibility in material and tone.

11.1.7 My Kit Selection

I'd like to tie this all together with a few thoughts about the drum tones that I favour. I tend to like a big, over-the-top sound, which suits shells with large dimensions. However, I also like shells with a very focused decay, so I prefer two-ply coated heads on all drums (sometimes even on snares that are typically single-ply coated). I like drummers to use two floor toms instead of a second rack tom. Although this is a huge generalization, as I like to make decisions based on the songs and drummer at hand, my perfect kit for the genres I typically produce would be:

- 22″ kick drum, 14″ snare drum, 13″ rack tom, 16″ floor tom, 18″ second floor tom.
- All coated two-ply batter heads, with single-ply resonant heads.
- Cymbals (all B20): 14″ hi-hats, 20″ ride, 17″ and 18″ crash.
- Tri-tone (which has three materials on the beater that can be switched to find a desired sound) kick beater.

Before we move on, it's worth noting that even if you have your perfect kit at your disposal, you should still listen to the drummer play in the room and decide whether the selection works. Sometimes it may be a case of changing parts of the kit (for instance, large cymbals can be too bright in small lively rooms), or it could simply be a case of moving the kit to a different location in the room (there will be more on this later).

EXERCISES

1. Go to a drum shop with a drummer and get him or her to play on as many different sizes, types and brands of kits as possible to listen to their tonalities.

2. Listen to different types of cymbals and drum skins to hear the effect that they have on tone.

3. Now that you have an idea of lots of the variables, write down a list of preferences and also any ideas you have regarding the different genres you would think different types of kits would be suitable for.

11.2 DRUM TUNING AND MAINTENANCE

Once the kit has been selected, there is still a long way to go before it is ready to record. This could involve replacing drumheads, tuning drums, eradicating rattles and squeaks, and dampening drumheads and cymbals. I've often had to make sure that the drummer has adequate felts and wing nuts on cymbal stands.

11.2.1 Replacing Drumheads

So, you've found that the drumheads are worn beyond belief or are just not right for the style. Usually, you will only need to replace the batter head, but the resonant head can also dull over time, so a good rule of thumb is that every third batter head change, you should also replace the resonant head. Replacing a drumhead is simple, and all you need is a drum key, the replacement skins and a flat surface (some lubricant and a cloth may be required if the drummer is a slob).

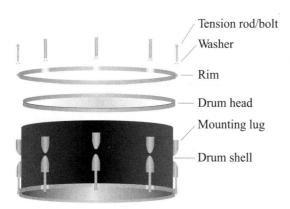

Figure 11.7 Drum shell parts

- Unscrew the tension bolts from the lugs around the shell with a drum key by turning it anticlockwise; there are usually six or eight lugs for each head, spaced evenly around the shell. Each tension bolt should be loosened half a turn at a time before moving to the lug directly opposite; this is to help evenly reduce the tension on the rim and prevent stress on the drum. In Figure 11.8, you would loosen the lugs in this order: 1, 5, 2, 6, 3, 7, 4, 8.
- Once the bolts are all removed from the lugs, remove the rim and the old skin from the shell.
- If the bearing edge (where the skin is placed) looks dirty, wipe it down with a cloth to ensure a flat surface on to which the new skin will be placed.
- Place the new drumhead on the bearing edge and then reinstall the rim.
- If the rods are dry, lubricate them lightly with a light oil, paraffin wax or Vaseline. Then finger-tighten the rods to the lugs in the same order that you loosened them and make sure that all the lugs are the same tightness.

Figure 11.8 Drum lugs

- Place the palm of your hand in the centre of the drum skin and apply pressure. Don't be shy here; really push it hard – it won't break! If you hear some stretching, popping and crackling sounds, don't be alarmed – this is normal. In fact, when doing this to kick drums, I often stand on the skin! This process is called 'seating the drumhead', and it helps settle the head in, allowing it to sound better and also hold tuning better straight away.
- Tune the skin to the desired pitch.

11.2.2 Tuning the Shells

One of the big issues surrounding drum sound is that it is ultimately a very subjective process and one that will dramatically change with the style of drummer, as well as the kit and room you use to record. As well as this, most drummers have their own opinion of what sounds good. This is a particularly hot topic when it comes to tuning the drum kit. I've heard many drummers say that you should tune the shell to its natural resonance and fine-tune from there; some kits made by DW even have a note of the natural resonance printed inside the shell. Other drummers tune purely by ear, worrying solely about intervals between shells and not about tuning to specific notes. Whether the tuned pitch falls in between notes doesn't matter to them as long as the drums sound good. Some drummers, when recording, tune the drum kit to the key of the track. Where do I stand? Anyone who says that drums do not produce a definite pitch is just plain wrong. Yes, they produce a lot of overtones, but you can tune a drum to a specific fundamental note. I can, however, understand why some drummers may not want to tune their drums like that: there is definitely an increase in attack, definition and decay when you find a sweet spot for a certain drum. In reality, I use all of these methods at times when working with kits. What helps make that decision? Usually, time and money. If I have to record 10 tracks of drums in a couple of days, I won't have time to mess around with several kits and tunings, so I find relative intervals that the drum sounds good in, and run with that. On the other hand, if I have the budget and the time to experiment, more often than not I will tune drums to be sympathetic to the key of each song. Sometimes a drum will not suit being tuned to a desired note because of its physical dimensions: it will either be too tight or too loose at that tuning. When this happens, it's best to tune to a note that's more appropriate for that drum, while still being within the key of the song. I do this kind of tuning because the decay of the drums (their notes) are then sympathetic with the music, which means I can apply less dampening to the skins and/or use less gating in post-production, as the resonance doesn't stick out unpleasantly.

So I am going to show you the method I use to tune the drums to a specific key. This method can then be adapted should you want to tune to intervals where your drums naturally sound most solid. When tuning drums for a particular track, all you are really aiming for is to make sure that each drum's fundamental is in key. To do this, you will need a drumstick, a drum key and some sort of instrument to take a reference pitch from. I find that bass guitars or electric pianos are good for this because of their accuracy in the lower octaves. I also use a drum dial to further hone in my tuning by ear (there will be more on the drum dial later).

So you have changed the drum skins, seated them and left the tension bolts all finger-tight as I instructed in the section on replacing drumheads. At this point, you need to start thinking about the pitches you want to set each drum to and also the relationship in tone between the batter head and resonant head. You have three choices when thinking about how your drums decay. The first is to tune the resonant head to the same pitch as the batter head to get a consistent decay that will stay at the same pitch. The second is to make your resonant head looser than your batter head to create a pitch bend that makes the drum's tone get lower in pitch as the sound decays. The third is to make the resonant head tighter than the batter head, resulting in a pitch bend that gets higher in pitch.

In terms of tunings in relation to scales, I like to use the tonic triad of the key as much as possible; this is, again, down to the harmonic series and what we find pleasing to the ear. This means that, wherever I can, I use 1st, 3rd and 5th degrees of the tonic. If this is not possible, 2nd and 4th degrees are usually fine.

Just to avoid confusion, I am using scale degrees here, which relate to the distance between notes in a scale in relation to the root of the key. This is not to be confused with the distance between any two notes (intervals).

Let's reacquaint ourselves with the scale degrees:

Degree	Name	Meaning	Note in C
1st	Tonic	Tonal centre. Note of final resolution	C
2nd	Supertonic	One whole step above the tonic	D
3rd	Mediant	Midway between the tonic and dominant	D
4th	Subdominant	Lower dominant. Same interval below the tonic as the dominant is above the tonic	F
5th	Dominant	Second in importance to the tonic	G
6th	Submediant	Lower median. Midway between the tonic and subdominant	A
7th	Leading tone	Leading tone. Melodically strong affinity for and leads to the tonic. One half-step below the tonic	B
8th	Tonic	Tonal centre. Note of final resolution	C

On a basic four-piece kit, the batter heads would be as follows:

- Snare drum: 1st (root)
- Rack tom: 3rd
- Floor tom: 5th
- Kick drum: 1st (root at two octaves below the snare drum)

Sometimes tuning the snare to a 5th can give a nice relationship between the kick and snare:

- Snare drum: 5th
- Rack tom: 3rd
- Floor tom: 1st
- Kick drum: 1st (an octave below the floor tom)

On a five-piece kit, I could stick to the triad by using this tuning:

- Snare drum: 1st (root)
- Rack tom: 5th
- Second rack tom: 3rd
- Floor tom: 1st (octave below the snare)
- Kick drum: 1st (root at two octaves below the snare drum)

Or I could throw in a 4th:

- Snare drum: 1st (root)
- Rack tom: 5th
- Second rack tom: 4th
- Floor tom: 3rd
- Kick drum: 1st (root at a lower octave than the snare drum)

As you can see from the examples above, you have a lot of options at your disposal. Follow what suits the music, the drums themselves and the time you have available.

While recording 'Take Her Away', James, the drummer of Luna Kiss, was using a five-piece kit and I decided to stick to the tonic triad (1st, 3rd, 5th). The key of the piece is A minor, so it worked out as follows:

- Snare drum: A (root)
- Rack tom: E (the 5th pitched below the snare)
- Second rack tom: C (the minor 3rd, pitched below the snare)
- Floor tom: A (octave below the snare)
- Kick drum: A (two octaves below the snare drum)

For the resonant head, typically you would also stick to triads. For the kick drum and the snare drum, I usually keep the pitch of both heads the same. For the toms, however, I like to have the resonant head at a different pitch. Just as you would choose sympathetic notes when constructing chords, you will be tuning the resonant head to the 3rd or 5th degree of the batter head's pitch and not of the tonic of the piece. For instance, we know that 'Take Her Away' is in A minor. In a minor key, we know that chords 1, 4 and 5 are minor; chords 3, 6 and 7 are major; and chord 2 is diminished. This would mean:

Tom	Scale Degree	Note	Chord Type
Rack	5th	E	Minor
Second rack	3rd	C	Major
Floor	1st	A	Minor

The bigger the difference in pitch between the batter head and resonant head, the greater the pitch bend will be. More often than not, I like to use 5ths because it gives a bigger classic rock sound and it also means that the tuning is more neutral as the 5th of a chord is the same whether the key is major or minor. During the 'Take Her Away' recording session, I decided to use a 5th below the batter head, which worked out at

Drum	Batter Head Tuning	Resonant Head Tuning
Kick	A	A
Snare	A	A
Rack	E	B
Second rack	C	G
Floor	A	E

Now that we know what notes we want to tune the shells to, we have to do it! You may find this trickier than it sounds. I like to use an electric piano or a bass guitar to tune the drum to; usually, I have my assistant playing the instrument near to me or rope in another member of the band to help out. You could also get a piano app for your phone or use a sample library if you are close enough to a computer or other sound source. Most of you will have some musical background and can probably discern when two notes are the same pitch; with a drum, it is a little more difficult. This is because the pitch of the drum is covered by many overtones that mask the pitch. So when tuning the drum, we have to minimize the overtones heard so that we can tune the drum quickly and efficiently. The process set out here applies for each drum:

- Lay the resonant head of the drum on your knee or other surface, dampening any resonance. Make sure that the drum is flat.
- Take a drum stick in your writing hand and pick a lug to tune.
- Every time you hit the drum, use your other hand to dampen the skin in order to stop overtones. The best way to dampen the skin is to place your hand anywhere on the skin apart from next to the lug you are tuning; remember to place that hand very lightly so you don't alter the tension in the head, and therefore its pitch. I tend to end up placing my hand on the adjacent lug anticlockwise to the one I'm tuning.
- Hit the drum a couple of times and listen to its pitch. Remember it.
- Get your assistant/band member to play the reference pitch you have decided to tune that head to. Obviously, if you are on your own, you will have to do this yourself (duh!).

- At this point, you can work out the interval difference between the two notes and use a drum key to tune the drum to that reference pitch.
- Once you have one lug in tune, tune each of the other lugs (again, working in opposites as previously described).
- Once all of the lugs are in tune, listen to each one again quickly and fine-tune any that are slightly out with the others.
- At this point, I often use a device called a 'drum dial' (Figure 11.9) to further aid the accuracy of my tuning. Once I have the drum tuned by ear, I place the drum dial on each lug in turn and average out the readings, tuning all the lugs to the same value. This will give you a great, even drum tone at the required pitch. You will find that the drum can drift slightly sharp or flat of your reference pitch as you adjust each lug. To rectify this, simply increase or decrease the tension on each lug evenly, until you reach your reference pitch again.
- Repeat this process for the resonant head.

Figure 11.9 Drum dial (picture courtesy of Flipside)

A couple of things to remember to make this easier: make sure that you pick the relevant octave of reference note to the drum you are trying to tune. It also helps to remind yourself that hearing an absolute pitch on a drum can be quite difficult sometimes. Sympathetic notes (particularly 5ths) will often sound like the correct ones, so refer back to the drums you have previously tuned to reference the intervals between them.

Further Resources www.audioproductiontips.com/source/resources

 Drum tuning tutorial. (80)

11.2.3 Dampening Shells

Dampening or muffling are methods available to reduce unwanted overtones and ringing on a drum skin. If a drum is evenly tuned, you may not need to dampen it at all. If your drums are sympathetic to the song's key, you stand an even better chance of avoiding dampening. Before I explain how to dampen drums, I want to warn you of the dangers of over-dampening. It is very easy to make the drums sound too 'dead', leaving a drum that's devoid of any sustain or character. What might seem like unwanted ring in isolation becomes unheard in context with the rest of a drumbeat, let alone the rest of the music! Remember when you are dampening that every bit you add affects not only sustain, but also attack, presence and velocity. One way to make sure that you are not overdoing the dampening is to listen to the drums in context of a whole beat, from the opposite side of the room to where they are being played. From this position, it is easier to hear if the decay of the drums is obtrusive to the overall tone of the kit. Before dampening excessively, think about whether you could find a better tuning for the drums, or, if the drum kit has single-ply heads, consider trying two-ply heads. Even if you are inexperienced at judging the amount of dampening needed, I always advise to err on the side of caution. This is because it is easier to use tools such as compression, gating and/or strip-silencing to help reduce ringing than it is to correct an overly dead drum. In fact, the best cure for an overly dead drum is to completely sample-replace it, which certainly should be considered a last resort!

Now that I have warned you of the occupational hazards, here are a few methods to help reduce snare and tom ring (the kick drum will be covered later):

Shop-Bought Products

1. *Drum rings*. These are rings of material, usually about 1 inch wide, that are cut to the size of the drum and placed on the batter head of the drum shells.
2. *Moongel*. This is a self-adhesive gel that sticks to the surface of the drums. It can be moved anywhere around the drum and is very flexible in the amount of dampening it creates. For instance, if you place the gel closer to the edge of the drum, it dampens less, but sounds more controlled with minimal impact on the drum's decay; a few centimetres further towards the centre produces more dampening, but the drum has less decay. You can even cut Moongel into smaller chunks and/or place multiple pieces of Moongel on to a drum to further optimize its performance. I have even used Moongel on cymbals to make them sound less 'thrashy'.

Figure 11.10 Moongel (picture courtesy of Flipside)

Resourceful Ways of Dampening Drums

- *Gaffer tape*. Sticking gaffer tape on the drum skin in similar places to where you would place Moongel can be effective at removing overtones. Like with Moongel, it is easily customizable with amounts used, and it also has the advantage of being able to be used freely on resonant heads without the risk of it falling off. Remember not to use other types of tape – duct/duck tape is hard to remove and leaves a sticky mess, and electrical tape will not stick sufficiently to the drum. Some drum techs like to make loops with air gaps out of the gaffer tape; this further enhances the dampening effects.
- *Socks* (or any cloth) placed on the edge of the skin, again, in a similar place to where you would put Moongel, can add a significant amount of dampening. This is often too much for toms, but is perfect for overly bright snares. I've even asked drummers

Figure 11.11 Gaffer dampening with air loops (picture courtesy of Flipside)

to take socks off for this purpose mid-session (ewww!). The picture below is from a session where I asked precisely that.
- *Old drumhead* turned upside down and placed on top of drum skin. This is perfect for snare drums and not only dampens, but also makes the snare sound unique. Give it a try – it sounds to me like a tightly controlled funk snare with this trick. You can also take old drum skins and make your own drum dampening rings by measuring 1 inch from the edge evenly around the drumhead and cutting the middle out of it!

Figure 11.12 Crusty sock used as snare dampening (picture courtesy of Flipside)

11.2.4 Other Maintenance

Now that you have re-headed, tuned and dampened the shells, it doesn't mean you are 'home and dry' when it comes to maintenance – but with just a few more simple checks, you can have your drums ready to go in no time.

Eliminating Rattle

When your kit is set up and ready to play, you may find that there are suspect squeaks, rattles and vibrations causing a metallic ring to the kit. Most of this will come from the snare drum, and in particular the actual snares (the metal beads on the bottom of the drum). A little snare rattle when different parts of the kit are played is pretty much unavoidable, and in fact can help to 'glue' the kit together when recording. However, the rattle should not be heard while in the context of a full drumbeat. If you cannot isolate the direct cause of the rattle, it is best to start by looking at some of the easiest fixes:

- Check all hardware stands and shells to make sure that all nuts and bolts are correctly tightened, and that no parts of the drums are inadvertently touching each other, which can cause a sharp metallic ring. You should even check toms that have already been tuned because tension bolts can often come loose from the lugs for no apparent reason.
- Sometimes vibrations from the floor cause rattles in the snare drum. Try placing the drums on a mat to isolate the drum vibrations from the floor. If this isolation helps reduce the snare rattles, use the smallest drum mat available as a hard floor can make the ambience of the kit sound better while recording.
- If a particular drum is making the snare rattle, then it is probably due to the harmonic relationship between the tom and snare. In this case, try slightly tightening or loosening the resonant head of the drum in question and see if the problem goes away.

If you are still fighting rattle, it is time to look at the snare drum itself:

1. *Check the clamp is holding the snares properly.* On the side of the snare drum is a clamp that you tighten using a drum key. The clamp holds little plastic

Figure 11.13 Snare tightening knob (picture courtesy of Metropolis)

strips looped around a metal hole to fasten the snares against the resonant head. If these plastic strips are not taut and held in firmly, then it becomes impossible to tighten sufficiently.

2. *Tighten the snare beads.* It sounds simple, but sometimes just tightening the snares is enough to remove unwanted noise. Remember that having some rattle from the snares is desirable, hence why the drum is designed that way! In certain genres such as pop-punk or funk rock, the snares are often over-tightened to give the drum more 'crack' and less rattle. In jazz, folk, acoustic and easy listening, the snares are often quite loose to give a rattle on snare rolls when brushes are used. To tighten the snares, simply turn the knob on the side of the snare drum.

3. *Remove suspect snare beads.* Sometimes snare beads get caught, tangled up and bent out of position. When this happens, it is usually best to just cut them off – removing one or two beads should not alter the tone sufficiently enough to ruin its sound.

4. *Tape snare beads.* If removing beads does not help, or there are too many to remove, you can take a small amount of gaffer tape and stick it to the snares on either side of the drum (not in the middle). This should help to reduce rattle. However, I only use it as the last resort before replacing the snare beads entirely because it is very easy to over-dampen the sound of the snares with tape.

5. *Replace snare beads.*

6. *Check seating.* Rattles on the snare drum can often come from not properly seating the drum – particularly the resonant head, as this is often forgotten.

7. *Try a different snare.* If you have still not sufficiently reduced rattle to a reasonable and 'musical' level, then maybe just admit defeat and try a different snare drum!

Checking Cymbal Stands, Accessories and Drum Angles

As well as making sure that hardware stands are correctly tightened, ensuring the best angle and the correct accessories are used can help avoid weird drum artefacts. Many drummers get stuck in bad habits that should be shaken out of them; this usually starts with them becoming accustomed to an unhelpful drum set-up. While an unconventional drum set-up can on occasions add personality to a band (and in these cases you should embrace it), most of the time it sucks! If you are working on a longer-term project and have enough time in pre-production, you should try to rectify some of these bad habits. Here are a few things to look out for:

- Washers, felts and wing nuts are important on cymbal stands: they stop unwanted rattles and they also make the cymbal more rigid and controllable when hit. Cymbals need felts both sides of the cymbal itself. Each stand (apart from the hi-hat) should have the pieces shown in Figure 11.14.

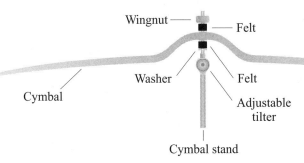

Figure 11.14 Cymbal stand set-up

- Hi-hats should have the following (Figure 11.15):

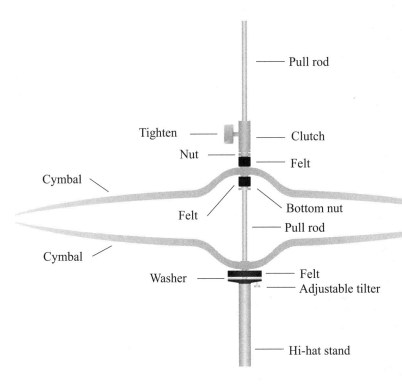

Figure 11.15 Hi-hat stand set-up

- Reduce shell angles. Drummers often set up their toms and snare so that they are heavily angled towards the drummer. This actually makes the shells sound worse and makes them wear faster. The angle of the shells should be virtually flat with a slight tilt towards the drummer. I have found that shorter drummers tend to angle the toms more. In this scenario, I'd recommend raising the drum seat as much as possible (without adversely affecting the drummer's ability to use the kick drum pedal comfortably) before you tilt the toms.
- Cymbal angles, just like the angles of the drum shell, are important in producing the best cymbal tone. Cymbals should be angled slightly towards the drummer so that you can see the cymbal's logo from sitting on the drum stool. Angling the cymbals this way also helps prevent damage to your drumsticks and the edges of the cymbals themselves.
- Get the drummer accustomed to playing cymbals as high above the shells as possible. This is more for the benefit of the recording engineer than the drummer. Placing the cymbals higher has two main advantages. First, it helps provide more room for the placement of microphones and their stands. Second, it provides more isolation for the cardioid close mics, as the cymbals are in the mics' dead spots (the places in a mic's polar pattern where it picks up the least amount of sound).
- Even when you have dampened the kick head, you may feel that the kick drum sounds a little 'flabby', 'boomy' or even 'hollow'. To make the drum sound more 'focused', try placing a blanket or pillow inside the drum. Doing this right will give your kick drum a better 'thump'. However, be careful not to use too much as that can kill its tone almost entirely. These days, there are some products on the market to help with this, such as the Evans EQ pad.

Further Resources www.audioproductiontips.com/source/resources

 If you still feel that you need some help with the basics of drum set-up, here are some sites to help. (81)

EXERCISES

- Reskin a drum kit.
- Tune the drum skins to a pitch that suits the shells, with degrees of 1st, 3rd and 5ths between batter and resonant heads.
- Now try tuning the skins to many specific keys and see what limitations there are and how much ring or buzz you get when you start to take a drum out of its 'comfortable range'.
- Set up the drum kit correctly, taking note of shell and hardware angles. Also make sure that the cymbals have the correct washers, bolts and felt. None of the drums should touch each other, and there should be no squeaks and rattles coming from the kit.
- Dampen the kit, trying out as many (if not all) of the dampening methods described in this chapter. Try under- and over-dampening the kit to hear the effects on its tonality, before settling on a usable amount of dampening.

11.3 PLACING THE DRUM KIT

When you are starting to think about the placement of a drum kit, you can perform the clap test to give you an idea of the 'best spots' in the room for the drum kit. There is a nice little trick to fine-tune this process: ask the drummer to hit his or her floor tom in the 'best spots'; imagine where you would place the room microphones (as described in Chapter 10: Decision-Making and General Recording Techniques) and listen from that position, discerning the best-sounding location; set up the drum kit around the position of that floor tom.

11.3.1 Fine Adjustments

Even after you have replaced the drumheads, and tuned and dampened the drums ahead of the studio session, you may need to spend some time fine-tuning the drums on the day. This is because every room sounds different. Transportation can also have an effect on drums, as can temperature. When using large rooms, you can use gobos, blankets and reflection filters to help prevent too much room sound getting into the close mics. I suggest you try these isolation methods as soon as you've decided where the drums are going. Once ambience gets into a recording, it is very difficult to reduce it – it's much easier to add reverb to a 'dry' recording than it is to remove unwanted reverb after the fact.

11.3.2 Dealing with Small Tracking Rooms

Sooner or later, you will be presented with the challenge of recording drums in a small room. While this should be avoided wherever possible, don't despair. In fact, the room I recorded 'Take Her Away' in was one such room. The main problem with this room was the low ceiling (it is lower than in an average room). This is probably the worst issue you can come across in a drum room, as it creates significant early reflections and comb filtering that leads to a 'boxy' drum sound. To add to this, rumble from the floor was creating rattle, so I placed the drums on a small riser. This left only a few inches from the top of the overhead mics to the ceiling. Unusually, this meant that the best place for the drums was about two-thirds of the way back in the room and almost directly in the centre. Theoretically, it's a bad idea to place the kick drum anywhere near the centre of a room (either length-wise or width-wise), as bass frequency cancellation is rife in this position (as learnt in Chapter 9: The Critical Listening Environment, in the section on monitor placement). In this case, however, it was the lesser of two evils, as I was battling with the severe effects of early reflections, comb filtering and bass response from being near the corner of a room with a low ceiling.

Since the 'Take Her Away' recording session, I have built gobos (portable absorption panels) that are placed around and in particular above the drums, reducing the amount of early reflections from the ceiling and sidewalls that reach the mics. Consider this rule of thumb: wherever an instrument or microphone is within 2.5–3 metres of a wall or ceiling, use gobos to reduce the impact of reflections.

Figure 11.16 'Take Her Away' tracking room

Figure 11.17 Drum reflection frame on a subsequent session in the same studio (picture courtesy of Flipside)

EXERCISES

1. Experiment playing on a drum kit in many different drum rooms. Try to find optimum positions in each room to set up the kit.
2. Use gobos to isolate and reduce room sound in the close mics.
3. Alter drum tuning to work sympathetically with the drum room.
4. Listen to famous drum recordings and consider what kind of acoustic environment they were recorded in, then research to find out what kind of room they were actually recorded in.

11.4 DRUM MICROPHONES AND PLACEMENT

11.4.1 Overheads

When mic'ing up the drum kit, I usually start with overhead mics. This is so that I can get a natural-sounding kit that gels and breathes as one instrument. If I have a good room to work in, I will aim to get an overhead sound that gives an accurate and defined representation of the whole drum kit, and not just a good cymbal sound. These days, a lot of novice engineers rely too much on close mics. In reality, with a great room and great drummer, you could get a top-level drum sound with just three mics: a stereo set of overheads and a kick drum mic. In a room that isn't so good, you may need to rely more on close mics, but you should exhaust all overhead placement first.

For overheads, I usually choose a pair of matched condenser microphones. Large diaphragm condensers tend to capture a more balanced overall tone and have a wider pick-up radius. Small diaphragm mics are more directional and more 'trebly'. In smaller rooms, I almost always use small diaphragms with a cardioid pattern, which helps minimize pick-up of room reflections. In larger, more treated rooms, I tend to go for large diaphragm condensers. Occasionally, I even use ribbon mics as overheads to get a darker drum sound. I would usually use ribbons on something that I wanted to sound more retro such as Motown and '60s British Invasion pop rock. Ribbon mics also tend to work well for drum overheads in acoustic and folk music.

Here are some recommendations for potential overhead microphone choice:

Microphone Type	Budget End	High End
Large diaphragm	SE 2200	Neumann U87
	Audio Technica 4033	AKG 414
Small diaphragm	Rode NT5	Neumann KM184
	Oktava MK012 (Michal Joly mod)	AKG451
Ribbon	Cascade Fat Head	Coles 4038
	SE VR1	Royer 121

In terms of microphone placement, there are also many options available.[4] If you haven't experimented a lot with mic placement, you will be surprised by the amount of tonal, stereo-width and attack and decay difference between different placements.

Listed below are a few of the most common overhead mic positioning methods, many of which are used successfully in other applications where stereo microphone techniques are required (i.e. grand piano, strings):

Spaced Pair

Probably the most common is the *spaced pair* overhead technique, where a set of two mics are spaced apart from one another. Usually, this is set up with one mic over the hi-hat side of the drum kit and one over the ride. Some engineers advocate having the mics angled towards the snare drum; others prefer them to be parallel with the floor.

The spaced overhead technique is great for stereo width but it is notoriously difficult to deal with phase issues. To avoid phasing issues, it is important to experiment with the position of each overhead. When setting up any stereo pair of mics, it is advisable to monitor in mono to easily spot any potentially disastrous phase relationships. I have been known to make a slight compromise to the position of the stereo overhead mics in order to minimize phase cancellation in mono – welcome to the art of recording! Thinking about the placement of each overhead mic in relation to the distance they are away from the kick or snare is useful to get a phase-correlated sound to the most important parts of kit. This will make the snare's phase relationship tighter between overhead mics. Sometimes producers place the mics equidistant from the kick if, for a particular track, they consider phase coherency on the kick to be of greater importance than the snare's phase relationship. This is much rarer, but I would certainly consider it in dance-rock hybrids. One more aspect to note is that the phase relationship of the cymbals themselves is, of course, less crucial, since the majority of their content is high in the frequency range.

Figure 11.18 Spaced pair overhead set-up

XY/Coincidental Pair

The next most common overhead set-up is called *XY*, or *coincidental pair*. This technique involves positioning two mics at a 90-degree angle to one another, with the capsules meeting at one point above the kit. The point directly below where the capsules meet will be the centre of the stereo image. Some engineers choose the snare drum as the centre of the stereo image, while others choose the kick drum. A coincidental pair set-up has very little stereo image but it is great for phase reliability or mono applications. Setting up an XY overhead pattern is easy and accurate with a stereo bar microphone holder.

Figure 11.19 XY pair overhead set-up

ORTF/Near-Coincidental Pair

With the *ORTF* or *near-coincidental pair* technique, as with the XY pair, the microphones are placed alongside each other. This time, however, the microphones are angled at 110 degrees away from each other, and the centre of the capsules are 17 cm apart. This creates a wider stereo field while maintaining a good phase relationship.

M/S (Mid-Side)

The *mid-side* technique takes advantage of the adaptability of microphone polar patterns. This is particularly useful with a matched pair of large diaphragm condenser microphones that have selectable patterns. On the first microphone, select a cardioid pattern, then position it above the kit with the mic's polar pattern pointing downwards. This will give a mono track to act as a central image. To create stereo 'side' images, place a second mic above the first, but on this mic select a figure-of-eight polar pattern, and ensure that the polar pattern is facing outwards to the sides of the kit (not front-to-back!). With most large diaphragm condenser mics, this placement will result in the two mics being at 90 degrees to one another: the centre mic lying horizontally, and the figure-of-eight, 'sides' mic being vertical.

Figure 11.20 ORTF overhead set-up

To extract the stereo 'side' images from the mono figure-of-eight mic, a little trickery is required:

- Duplicate the channel containing the figure-of-eight mic's signal.
- Pan the original hard-left and the duplicate hard-right.
- Flip the polarity of the right-hand channel.

Figure 11.21 M/S overhead set-up

The major advantage of mid-side recording is that you can control the middle content independently from the sides. To put this into perspective with a drum kit, it allows you to EQ, compress and level the middle content that is likely to have more fundamental kick and snare information differently to the sides that are likely to have more cymbal content. A disadvantage of M/S recording is that, when in mono, you lose all side information, meaning that if you have a lot of cymbals in the sides, you lose that on a lot of systems.

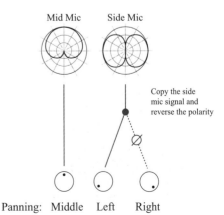

Figure 11.22 M/S stereo image creation

Recorderman Technique

Recorderman is one of the more modern overhead techniques and is fast becoming my favourite overhead microphone configuration. Its peculiar name comes from the pseudonym of an Internet forum member who popularized this technique. It is an adaptation of the Glyn Johns method described below. It is great for reducing room ambience in the overheads, is heavy on both kick and snare information, and places the kick and snare in the centre of the stereo image. It works by placing each overhead mic at an equal distance from the kick and snare. The first mic goes 16 inches directly above, and pointing towards, the middle of the snare drum; the second mic goes over the drummer's right shoulder, again 16 inches from the middle of the snare drum, and at a 45-degree angle. Here are some simple instructions on how to do this with your drumsticks:

Figure 11.23 Recorderman overhead set-up

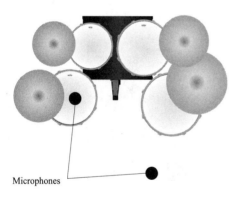

Figure 11.24 Recorderman overhead set-up, bird's-eye view

- Get your drummer or assistant to sit on the drum stool.
- The drummer should now hold the sticks end-to-end vertically (so it stands double the length of a single stick) in the middle of the snare drum facing directly upwards. This is where you place the first mic.
- With the sticks still end-to-end and in the middle of the snare, ask your assistant/drummer to move them to a 45-degree angle pointing towards/over his or her right shoulder. This is where you place the second mic.
- If all is done correctly, the mics will also be the same distance from the kick drum. To fine-tune, take any XLR cable or tape measure and measure the distance between the kick beater and each mic.
- Adjust the mics so that they are equidistant to both the kick and snare.

Glyn Johns

Although the *Glyn Johns* method of drum mic'ing is not strictly an overhead technique (as it uses four mics spread over the full kit), this is probably the best place for it. Glyn Johns is a British engineer who worked with many of the all-time greats, including The Who, The Rolling Stones, Eric Clapton and Led Zeppelin. It was his drum sound with the latter that gained worldwide recognition. Like the Recorderman technique, you start with an overhead microphone pointing directly over the snare. Next, place the other overhead microphone near the floor tom facing over towards the snare and hi-hat. Again, like the Recorderman technique, you need to get a measuring device and make sure that both mics are equidistant from the snare drum. Finally, close-mic the kick and snare to further enhance the key elements of the kit, and voila!

Before moving on to the kick drum, I just want to state that although these overhead techniques are the most common, this doesn't mean that you shouldn't experiment for yourself. They also stand as a good starting point. Remember to trust your ears – I've had some weird and wonderful-looking overhead placements in the past that sounded great. Our eyes always want to make things look more symmetrical. Avoid this urge and trust your ears. Phase coherency is important so always experiment with placement, check in mono, and reverse phases where needed.

Figure 11.25 Glyn Johns set-up

11.4.2 Kick Drum

Once the overheads are placed, you should listen to the sound you have captured and work out, according to genre and your vision for the track, what elements are missing from the overhead channels. Sometimes you will have captured a healthy amount of kick, and some bottom-end punch or some top-end click might be all you need to add. Once you have decided which additional elements you need to capture, go into the live room and listen to the kick and decide how best to achieve this. It could be with a mic just outside of the kick drum itself, inside the shell or even next to the resonant head. Remember, though, if you have enough inputs, it might be best to capture the kick drum in a couple of places just in case.

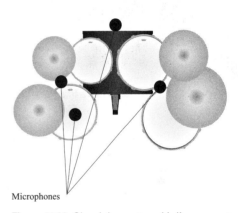

Figure 11.26 Glyn Johns set-up, bird's-eye view

There are many types of microphones that you can use on a kick – in fact, just about any microphone can be used to capture a good kick sound when used correctly. Typically, you want to try to capture two types of sounds from a kick drum. The first is the attack of the beater hitting the skin (approximately 4–5 kHz); second is the more rich, warm and natural bottom end you associate with the kick (60–100 Hz). To do this effectively, you usually need two or three microphones. Let's explore the options:

Dynamic

There are many dynamic microphones on the market these days that are tailor-made for kick drum. These mics tend to have a pre-equalized nature to them, removing a lot of the mid range content. Dynamic mics are great inside the kick shell as they can handle the high SPL (sound pressure level) that most condensers can't. When placed closer to the beater, they sound 'clickier', while when placed further away they sound warmer and less pronounced.

Here are three of my favourite modern dynamic kick drum mics and their characteristics:

Microphone	Characteristics
AKG D112	This is the least pre-EQ'd of the three. It sounds more mid-heavy and is great when the kick drum should not be a main focus of a mix. Its main con is that on its own it often sounds lost in metal or hard rock tracks.
Audix D6	This sounds the most pre-EQ'd: it is deep in the bottom end and clicky in the top. It should be used when attack and punch is essential and where you want it loud and proud in a mix. Usually my favourite kick drum mic.
Shure Beta 52	Despite it having a bigger presence peak than the D6, it sounds somewhere between the two above. This is because it has a lower but less pronounced bottom end, and the presence peak is at 4 kHz on the Shure as compared to 5.5 kHz on the Audix.

You could also try some of the more neutral multipurpose dynamic mics if even the D112 has too much punch. These include: EV RE20, Shure SM7B and Sennheiser 421MD.

Large Diaphragm Condenser

Large diaphragm condensers are great at providing a natural kick drum sound. This is because their frequency response often extends to 30 Hz. The drawback here is that because of the SPL a kick drum shell creates, it would be impossible to place most LDC mic in the same position as a dynamic mic without breaking it. For this reason, most engineers place the mic outside, at least half a metre from the resonant head of the bass drum. This has a couple of unfortunate side effects:

- *Bleed.* As the microphone is placed in a less isolated position, the mic is prone to picking up high amounts of other sources, including the rest of the kit, and often guitars if you are recording live. To get around this, engineers often build a 'bass drum tunnel' to isolate outside sources.
- *Lack of click.* The further away from the beater, the less pronounced the 'click' of the kick, and this can be exaggerated by the bleed from other parts of the drum kit masking the click.

You can make a simple drum tunnel out of many readily available materials, such as studio foam, blankets and old kick drum shells with both heads removed.

Microphone	Characteristics
Neumann FET47	This is the most sought after and also one of the most elusive kick drum mics. Neumann have recently announced they are re-issuing it, but I've not used one of the new models so I can't comment on how faithful it is to the original.
Sontronics DM-1B	This is a modern FET47-style mic. It is pre-EQ'd with kick drum and bass guitar cabs in mind. One of the best features is that it is capable of withstanding 155dB SPL, which makes it one of the only LDCs – if not the only – that could actually go inside the kick drum shell.

Mojave MA201FET	Designed by David Royer, the Mojave MA201FET is often likened to the Neumann. It is more neutral sounding than the Sontronics but responds well to EQ.

Pressure zone microphones (PZMs)

As well as dynamic mics, PZMs (pressure zone microphones) can be used inside a kick drum. Because of their physical constraints, PZM microphones aren't great at capturing a natural-sounding low end when placed inside a kick drum. This means that it is almost always used in conjunction with a low-end heavy mic (such as a D112) for the purpose of adding extra 'click'. They are predominantly used in live sound rather than in studios, but they are great at achieving processed kick drum sounds for genres such as metal.

Microphone	Characteristics
Shure Beta 91	The Shure Beta 91 is very top-heavy and even sounds 'plastic' on its own. However, it is the perfect companion to a dynamic or condenser mic.
Sennheiser e901	The e901 is much more natural and full sounding than the Beta 91. It doesn't quite have the aggression of the Beta 91 but it is usable as a single kick mic, which is something I couldn't say about the Beta 91.

Speaker Transducers

Another 'old-school' trick to get more sub-response from your kick drum is to convert an old nearfield monitor speaker into a microphone by wiring it in reverse. These days, Yamaha make a commercial product based on this idea, called a 'subkick'. To use a subkick, simply place it against the resonant head of the kick drum. Similarly to PZMs, you wouldn't use a subkick on its own because of its complete lack of usable mids and no treble at all. I made my own subkick from an old tom shell, a cheap 8 inch speaker and an XLR socket. If you make your own, remember to either build a limiting resistor into its design, or to only use it with a mic pre-amp with a dedicated –20 dB pad, as the signal level is very hot.

Placement

As is the case when mic'ing any instrument, a small change to the mic's angle or position can make a huge difference to the tone of your recording. For this reason, experimentation is the key. There are, however, a few things I've noticed while mic'ing the kick:

- *Resonant head on or off?* While taking the resonant head off creates easier access for finding the 'sweet spot', it means that more spill will get into the microphone. If the resonant head does not have a hole, try cutting a small circular hole in the skin that is big enough to place a microphone through.
- *Inside mic at the front or back of the shell?* The closer to the beater/batter head, the clickier the sound. The further away from the beater, the warmer and less pronounced the sound.

- *Where to place the outside mic?* I suggest just outside the hole if one is present – this is as close as I recommend if you are using a condenser microphone. If you consider placing the mic further back, be wary of spill. If the mic is over 0.25 metres from the resonant head, a drum tunnel is probably a necessity.
- *Blending multiple mics?* A general rule of thumb is to consider the inside mic as the punchy mic that will capture the click of the beater hitting the batter head. The outside mic should capture the 'beef' of the kick (i.e. the low end). When placing these mics, I position them so that each mic's tonality and frequency response is sufficiently different to complement and reinforce the other. This way, they need less processing when it comes to mix. Also, don't forget to phase-check the mics.
- *On-axis? Off-axis?* As on-axis means a mic is in line with the sound source, for kick drum this is in line with the beater. On-axis generally produces a bassier and 'rounder' response, and off-axis (offset, or at an angle) produces a more 'trebly' response. Take this knowledge and experiment. One thing to note is that when recording a high-SPL source such as a kick drum, you can use off-axis placement to help protect a delicate mic, as it reduces the pressure on the mic's capsule.

11.4.3 Snare Drum

I know what you are thinking: 'My brain can't take in all these options!' Thankfully, snare drum mic choice and placement is more straightforward. The snare drum usually needs two microphones: one on the top (batter head) to pick up the impact of the stick hitting the drum (this is often called 'crack'), and one on the bottom (resonant head) to capture the rattle of the snare beads on the drum. The vast majority of producers stick to the trusty workhorse that is the Shure SM57 on top of the snare. On the bottom, an SM57 is also used for most recordings. However, when the snare needs to be more prominent or crisp sounding, or when the snare is played lightly in easy-listening genres, a pencil condenser mic such as an AKG 451 is often used.

Microphone	Characteristics
Shure SM57	The most popular snare mic by a considerable distance. It is great on the top or bottom, and has depth and attack.
Audix i5	The Audix i5 is probably the main competitor to the SM57's crown. It is more present between 5 kHz and 6 kHz, making it a great mic for metal and harder rock.
Neumann KM184	Great on the bottom of the snare for capturing the crispness of the snares themselves. It can handle higher SPLs than most pencil condensers.
AKG 451	Just like the Neumann, it is great for capturing the beads and it can handle high SPLs.
Rode NT5	Again, bright and crisp like the two above. One of my favourite budget pencil condensers, although a bit edgy and very high input. This can handle such high SPLs it could even be used as a top snare mic when an SM57 or i5 sound a little dark.

A good starting point for the top snare mic is to position it between 10 and 11 o'clock from the drummer's perspective, with the capsule close to the edge of the skin and aimed towards the centre of the drum. This helps avoid it being hit by the drummer's stick and helps reduce bleed from the hi-hat; these are the main issues that you tend to encounter when mic'ing the snare. Another trick for reducing bleed is to position the hi-hats as high and as far away from the snare as possible; this places the hi-hats in a dead spot in the snare mic's polar pattern. A good starting point for the bottom mic is to copy the distance from the snare and angle used for the top mic (again at 10–11 o'clock). The capsule should point towards the middle of the snare beads.

As already discussed, top and bottom snare mics will almost certainly have phase issues, and if they are the same distance from the snare and angle as suggested, the bottom end will be almost completely out of phase. The snare will sound very thin and unnatural, lacking body and fullness. Reverse the polarity of one of the mics by hitting the 'phase-flip' switch on your channel-strip or microphone pre-amp.

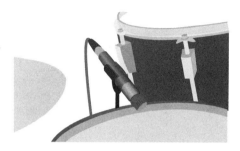

Figure 11.27 Snare top positioning

At this point, it is worth mentioning that once I have nice phase relationships between close mics on the kick and snare, I also regularly experiment with 'phase-flipping' to check for any problematic phase issues between these close mics and the overheads/room mics. I am usually monitoring in mono while doing this as it makes any phase cancellation more obvious. If I find some problematic phase cancellation, I will often move the overheads or room mics' placement slightly to alleviate the problem, rather than leaving the polarity reversed.

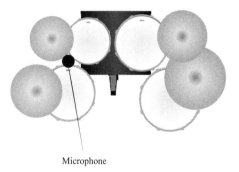

Figure 11.28 Snare top positioning, bird's-eye view

11.4.4 Toms

As with the snare, toms are almost always mic'd with a dynamic microphone. The Sennheiser MD421 is the most popular tom mic because of its superior bass response when compared to other dynamic mics such as the SM57. Similar to the snare mic'ing technique, toms are usually mic'd close to the batter head, with the capsule aimed towards the middle of the drum. Engineers sometimes use large diaphragm condensers such as the AKG 414 or I often prefer the smaller-framed Audio Technica AE3000; this produces a more sonically pleasing response but is prone to a lot of bleed because of its sensitivity. Condensers are also more fragile than dynamic mics so their use on toms can cause breakages if inadvertently hit by drummers. When placing the mics, it is best to position them so that the bleed from the cymbals is kept to a minimum. You can accomplish this by placing the mics close to or directly under the cymbals themselves, which maximizes rejection. Remember, you should have already asked the drummer to play with his or her cymbals as high over the kit as possible, so you should normally be able to squeeze the mic in without the cymbal hitting the mic when played. Remember to again check phase relationships with the overheads and room mics.

Microphone	Characteristics
Sennheiser 421MD	The most popular tom mic because of its pronounced bottom end.
Shure SM57	Again, this versatile microphone works great on toms, but has a tendency to sound thin on certain drum kits.
Audix D2	Very similar in timbre to the 421, but like all Audix drum mics, it has a more pre-equalized top end than its competition.
Sennheiser e604	Sounds a lot like the SM57. However, because of its clip, it can be mounted directly to the drum shell itself. This practicality has made the mic a staple in the live sound world, and can also be hugely advantageous to the project/home studio market.
AKG 414	More sonically pleasing low end and top end than a dynamic mic but is more prone to picking up a lot of bleed from other parts of the kit (particularly cymbals).
Audio Technica AE3000	Very similar to the AKG 414. However, the smaller frame and tough grille makes it a lot more durable and less likely to break if hit by a drumstick.

11.4.5 Room Mics

As with overheads, you're likely to want a good balance of the whole kit (i.e. no individual elements should be too prominent or lacking from the sound). My general belief regarding room mics is: the room mic(s) should sound much like how you imagine your final drum mix sounding. Unlike overhead placement, though, there are very few 'rules' regarding the number, type and placement of room mics. To find potential room mic positions, you'll need to walk around your room, standing and crouching, listening for any positions that give you this balanced drum kit sound. Sometimes you will hear one spot that jumps out; other times you might find two positions, either side of the room, where you could get good results from a stereo pair of room microphones. My drum mixes tend to be pretty shell-heavy relative to the cymbals, so I often position room mics low and close to the floor to avoid overbearing cymbals. Try starting out with the room mic(s) at the same height as the top of the bass drum. Experimentation is the key to getting great sounding results from room mics!

Once you are in the optimal position, remember to check for phase issues with the rest of the kit mics. By this point, there will be a lot of microphones on the kit and the distance between the room mics and the other parts of the kit will be larger, which means the phase relationships between the room mic(s) and the other parts of the kit will be more complex. This means that you are likely to be fine-tuning rather than having to make massive adjustments to help get back bottom end. In fact, one of the major reasons for the use of room mics is that the slight phase inconsistencies contribute to the perception of depth.

In terms of microphone selection for room mic(s), remind yourself of your vision for the track. If your vision is of a warm and retro-sounding kit, then consider classic ribbons. For a modern punchy sound, consider large diaphragm condensers, and for a garage rock sound, consider dynamic mics. Without wanting to sound like a broken record, experimentation is key! I can't predict what your player, room or kit sounds like, so it's up to you. Remember: trust your ears!

11.4.6 Hi-Hat and Ride

Finally, sometimes you may want to close-mic the hi-hat or ride. Both cymbals are usually well captured by the overheads, but you may want to enhance them with a close mic that will sound more crisp and direct. In this case, a pencil condenser microphone placed just on the edge of the hi-hat should suffice. To cover my back and provide options, I tend to close-mic the hi-hat, but 99 per cent of the time I find it unnecessary when it comes to mixing. In regards to the ride, I've only found the need to close-mic when it is particularly prominent in the song or when the bell is used. In this scenario, I place a condenser mic at a similar distance from the bell as I would with any drum shell, at about a 30-degree angle.

Figure 11.29 Hi-hat mic placement

Further Resources www.audioproductiontips.com/source/resources

 Microphone type comparison and examples of each placement, including stereo overhead techniques and room placements. (82)

11.5 'TAKE HER AWAY' DRUM CHOICES

One purpose of this course is to help you get the best out of the gear you have available to you. I could have chosen to hire out Abbey Road to record the track and use the world's best equipment to track the song, and while this would have produced a better result, it would have been an unfair representation of the hurdles you are likely to face with the budgets you have available. For this reason, a modest studio with a modest microphone selection was used to record in. I also chose not to hire any external equipment and dealt purely with what this studio had available. This meant that the amount of I/O (inputs and outputs) I had available, the sound of the room and also the mic selection was more limited and far from perfect.

I also wanted to show that you could get good-to-great results with budget gear. We all want the best gear. However, lack of gear should never be an excuse for poor recordings.

It's also worth noting that, over time, your skills, tastes and preferences change. In this section, I'm going to outline my mic choice for the drums on 'Take Her.

Away'. Some of these choices would be rather different if I were to record the track today. I hope you find this warts-and-all approach to be refreshing as it offers a different perspective to what you usually read in the pages of a music production text.

Before we get started, I recommend that any project studio should invest in a couple of high-quality pre-amps, preferably with built-in compressors. If you are recording only one musician at a time (apart from recording the drums), it is very rare that you need to use more than two inputs. Having compression on input helps to keep transients under control when you are close-mic'ing a dynamic source such as vocals or acoustic guitar (covered more later). In this case, the specialist pre-amps were two Avalon 737 valve pre-amps.

11.5.1 Overheads

A lot of my decisions were based on the knowledge that the room sounded very heavy in the mid range (particularly 500 Hz–1 kHz), mainly due to the room's low ceiling. This led to the drum shells sounding boxy and the cymbals sounding thrashy. To counter this, I decided that I would favour close-mic'ing more than usual while mixing, so I kept the boxy sound of the shells out of the overheads as much as possible. To do this, I decided to use pencil condensers placed in a spaced pair arrangement. I used the studio's Rode NT5 mics instead of its Oktava MK-012s or Audio Technica 4033s because they were the brightest and had a less prominent mid range that counteracted the sonics of the room.

As I now use this studio regularly, I've had a lot of time to perfect my placement of drum overhead mics. I now favour using the Recorderman overhead technique, which keeps room tone to a minimum. Despite this, I still favour the Rode NT5. On a couple of occasions where I wanted more of a retro sound, I have used the warmer Oktava pencils. I also bring in my own pre-amps to use on the overheads and room mics.

11.5.2 Kick Drum

For the kick, I used a single AKG D112. I used a single mic for several reasons:

- To keep the track count low as I was maxing out the studio's input count.
- I had always planned to sample-enhance the kick with a subkick sample (the D112 is quite a thin-sounding mic in my opinion, considering it's tailored for kick drums). As subkicks operate in frequencies you feel as much as you hear, this enhancement does not denaturalize a drum kit – in my experience, it actually gels the drums together.
- The studio lacked other specialized dynamic mics. I tried the multipurpose Shure SM7B on the kick, but I didn't like it – it lacked punch. I also decided that the studio's Violet Globe (an LDC) would be best used as a room mic, despite its potential as an outside kick mic.

To further enhance the bottom end, I decided to use the first of the Avalon tube pre-amps on the kick drum, using a little compression on input for dynamic consistency, and using the EQ to cut some of the mid range (500 Hz) and boost some of the upper-mids for extra snap (2.4–4 kHz).

11.5.3 Snare Drum

I used the hyper-cardioid Electro-Voice RE18 on the snare top as James, the drummer of Luna Kiss, likes his hi-hat very low. This hyper-cardioid polar pattern picks up less from the sides and rear than a standard cardioid mic. While I prefer the sound of an SM57, due to its 'thicker' low end, I felt the extra hi-hat rejection was needed in this case. Due to the lack of thickness, I decided to use the other Avalon on the top of the snare to add tube warmth to the bottom end. For the bottom snare mic, I used the tried-and-tested SM57. Overall, the snare sound was one of the weakest parts of this recording, and during mixing this was the element that I had the most problems with. Not only did I find the mic short of 'bottom end', but I also felt the choice of drum (a drum with a wooden shell and hoops) meant that there wasn't enough punch and ring to give the more aggressive feel I desired.

11.5.4 Toms

The Sennheiser e604s were used on the toms. I find the sonic differences between e604s and SM57s to be negligible, so I opted for the clip-on e604s as they don't require a mic stand and this helps manage clutter around the kit. There were no Sennheiser 421s available.

11.5.5 Room

Violet Globe. The Globe is a lovely crisp and bright LDC, without being fatiguing, which is perfect to counterbalance the boxy and mid-heavy sound of the room. We swept around the room and found that, height-wise, level with the top of the kick was the best place for balancing cymbals and shells. We placed the mic the farthest we could from the kick, but as the room was fairly small, we needed to listen carefully for comb-filtering side effects.

11.5.6 Hi-Hat

Although I was very confident of not needing a hi-hat mic, I placed an Audio Technica 4033 close to the bell of the hi-hat, which produces a sweeter sound than when placed near the edge.

Further Resources www.audioproductiontips.com/source/resources

 Drum microphones and placement used on 'Take Her Away'. (83)

EXERCISES

1. Set up each overhead technique, record a sample, and compare and contrast results.

2. Try to get a great sound using only three mics, a kick drum and overheads. Try using both pencil condensers and large diaphragms and analyse results.

3. Fully mic the whole kit and experiment with gobos to isolate the close mics from the room reflections.

4. Experiment with various kick drum microphones and mic placement options.

5. Try compressing individual channels on input with analogue gear, and then compare the results to doing the same using plug-ins.

6. Record a band live in one room and try to use gobos, positioning, microphone polar patterns and any other methods you can think of to reduce bleed.

11.6 GETTING THE BEST OUT OF THE DRUMMER

By now, you are several hours into a recording session – the drums are tuned, the room is chosen, the kit is set up and the microphones are placed. Not a single note has been recorded! Relax, though – these are probably the most time-consuming and tricky tasks a studio engineer ever has to deal with. It is at this point that a producer really starts making a difference to the session: it is time to get the best out of your drummer.

11.6.1 Full Band or Multitracked?

There is magic in a recording where the band are playing together live, providing the band are talented, tight and enthusiastic. You can create a bond that cannot be captured as beautifully in any other way. However, as recording time is becoming cheaper than ever, home studio technology continues to advance, and digital tools are developed to make bands sound better than they are, producers are having to do more and more to get great results.

A large proportion of bands are not only deluded as to how good they are, but also expect more from the recording process than is reasonable or practicable. Combine this with the fact that studio time is less expensive and yet premises are more expensive to rent and maintain, and you can see why it's difficult to justify recording many bands live – it's rarely cost-effective to rent enough space, and for some bands it's just a bad idea!

In the case of Luna Kiss, they are certainly talented enough to record live, but because of limited space and the bleed that would ensue, I deemed this as inefficient. Due to the complex nature of the song's time signatures and tempos, a more regimented sense of timing (snapping to the grid) could prove to be an advantage.

When recording instruments individually, you must make sure that you still get the performance you require from the drummer. It is very easy for him or her to feel more placid and isolated when tracking on his or her own, and his or her performance suffers because of it.

11.6.2 Hit Harder

The first potential pitfall with drum playing in the studio is how hard the drummer hits. In most rock/metal, the harder you hit the drums, the better. Be warned, though, this might not be the case with folk, jazz, country or any more reserved genres. When you hit harder, drum tone becomes more stable, with a more punchy and musical decay. Most drummers outside of a live performance with their band members have a noticeable drop in aggression and performance. It is the producer's job to make sure that the drummer does not forget that he or she is still performing. I regularly say to drummers before the first take: 'If I don't see you sweating by the end of this first take, I'll keep you drumming all night long.' On other occasions, I've even got the drummer shouting and screaming before a take and given them a superfluous item from the studio to break into smithereens just to bring out the aggression!

11.6.3 Hit the Centre of the Drum

Another problem is that a lot of drummers don't actually think about exactly where they are hitting each drum. Making the drummer aware that you are watching his or her technique, and that he or she should be hitting same spot every time, is usually enough to keep him or her mindful of where his or her hits land. Hitting the dead centre of the drum produces the fewest number of overtones. In modern pop and rock, this can be optimal, but in some other genres, for example funk, where you want a bit more ring, a position slightly off centre might be better. If you think the drummer has a serious technique problem, you should approach this gently in pre-production. On occasion, I've even drawn a target on the drum to help the drummer hit the correct spot.

Sometimes skilled drummers use other areas of a snare drum to add extra dynamics and excitement to their playing. It is important to distinguish between what is poor technique and what is a conscious playing style. For instance, many drummers alternate the position they hit during a snare roll to give a more 'ringy' quality.

11.6.4 Groove and Dynamics

As alluded to above, hitting as hard as you can and hitting the centre of the drum is not always the right approach for the entire song. The very hardest hitting should be reserved for choruses and solos, and the lightest for intros, breakdowns and verses. A general rule of thumb is that as a song progresses, hits should become slightly more aggressive. How much more depends on the arrangement and the genre.

Finding the groove and dynamics that the band naturally possess onstage is part of the art of recording, and is again the producer's responsibility to bring out when necessary. This is much harder when tracking parts individually.

In the case of Luna Kiss, much of James's playing on 'Take Her Away' is groove-based rather than aggression-based. He uses his hi-hat and, in particular, his toms in a way that is almost melodic. Adding extra drive and urgency to a track as it progresses is not just down to velocity, either. In 'Take Her Away', I added very slight tempo increases in the latter half of the song. This way, James could maintain groove and style without having to make his hits harder to make the track feel like it was gaining momentum.

On the subject of groove and dynamics, be aware of the danger of over-repetition in the drumbeats. I often find two types of drummers:

1. those with no flair; and
2. those with only flair.

All good musicians know when to step forward and when to step back in a performance. A mixing engineer has a certain degree of control over track dynamics and instrument balances, but the arrangement should already do most of this work. Some drummers play very repetitively from one verse and chorus to another. I say it's good to change it up *slightly*, but not a whole different beat entirely for each repeated section. Subtle changes in dynamics, hi-hat flicks and differing fills can make all the difference to a performance. Drums shouldn't always be the most noticeable and flamboyant part of a recording. Yes, if that's what the song calls for, do it! And yes, if your band is about virtuosity, go for it! Just remember, however, that the *song* should determine the performance, *not* the ego of the player. If you find yourself recording an egotistical musician, brace yourself for one of the toughest challenges! If you haven't built a sufficient rapport with this kind of person, approach discussing his or her performance with extreme caution.

11.6.5 Hand and Foot Coordination

This final hurdle should ideally be corrected in pre-production. Weaker drummers not only have issues with the intensity and consistency of their dynamics, where they hit the drums, and playing in the groove, but they also struggle to coordinate their hands and feet. Typically, it is the feet that fail to act in cahoots with their hands. This leads to slight unintentional flams between parts of the kit (creating a grace note because elements weren't played together). If the drums don't seem to groove, even though the drummer is smacking the living daylights out of his or her kit, you can almost guarantee that his or her kick is inconsistent and is creating a flam with cymbals, toms and, at worst, snares, when they should be well-synchronized. This is probably the trickiest challenge a drum editor faces, and should preferably be prevented through improving the drummer's technique during pre-production, rather than cured through editing. Just like dealing with a narcissistic drummer, dealing with a bad one is pretty tricky, so, again, proceed with caution! The main reason this kind of drum editing is so tough is because it affects the phase relationship of the whole kit. It usually results in you having to trigger artificial samples a lot more than you may want to. This leads us perfectly into a discussion on drum editing.

11.7 DRUM EDITING

A lot of the editing procedures mentioned here are best learnt through video instruction. Luckily for you, I've prepared such tutorials. But let's first discuss the options. I happen to use Pro Tools, so I will be using the terminology of that DAW. The information here is not a 'how-to' – I will save that for the videos. It is more of a discussion of the options available and their pros and cons and is not solely aimed at Pro Tools users. It is possible to do these techniques in whichever DAW you use.

11.7.1 Methods to Tighten Timing

The obvious editing tools are the ones relating to timing adjustments. In Pro Tools, the three most popular methods are manual editing, Beat Detective and Elastic Audio. Each method has its advantages and you may choose different methods across your instruments, dependent on the particular attributes of the instrument's waveform, or on the amount of tightening you require.

Manual Edits

The first and most obvious way of editing material together is by manually selecting areas in the timeline to cut, and then moving the audio into the correct place. You should then crossfade either side of the edit points, to ensure that there are no clicking or popping artefacts, or any interferences that suggest an edit has been made to the artist's performance.

A major advantage of the manual editing procedure is that it can be used regardless of whether the track was recorded to click. In other methods, audio is quantized or 'snapped' to a grid – the grid being the beats and sub-beats of the track's tempo. These are shown in the edit window of Pro Tools as a semi-transparent grid underneath the audio itself.

Figure 11.30 Pro Tools manual edit controls

Although working with automatic algorithms on the grid is very quick, effective and accurate, many argue that it sounds unmusical. Even if you have recorded to click, you can still edit manually and use the grid as a reference. This means that you can easily tighten to taste, allowing you to correct only what you really deem to be out of time. This will sometimes leave whole chunks of music untouched, while in other methods all of the audio might be tightened, on occasions to 100 per cent accuracy. When combined with other editing methods, however, this can make the end result very mechanical sounding. Manual editing is also probably the best method when you have recorded a band live. This is because every edit you make will potentially throw the other instruments out of time. Having fine control over the relative placement of each instrument, rather than a grid, allows for a more natural-sounding end result.

The major drawback with working manually, however, is the time it takes. Typically, manual editing takes at least four times the amount of time and brainpower compared to other methods. It also takes a lot of practice and patience to determine with your ears and eyes what is out of time when you don't have a grid to guide you.

Beat Detective

The next two methods of timing correction are based on automatic algorithms to detect transients (hits) and move them to the correct position on the grid based on parameters set by the user. These parameters include note values that the program material contains, groove and exclusion settings, and also the strength of the quantization or 'snapping'. This method is called *Beat Detective*, and it works via the following steps:

- detect transients;
- split audio files;
- conform them (snap them to the grid); and
- smoothing (filling the gaps and crossfading).

Beat Detective does not alter the audio files – it simply splits them up into little chunks, moves them around and then crossfades to smooth any gaps. Because this method doesn't stretch or compress the audio files, it is clean, smooth and isn't prone to artefacts. However, it's not particularly intuitive to use, is fairly time-consuming and requires a certain degree of precision to work effectively. Particularly poor drum tracks will require editing prior to using Beat Detective. In this situation, look to loop good sections or ones you have manually edited in order to reduce this pre-editing time.

Elastic Audio

Elastic Audio works in a very similar way to Beat Detective, but, having detected transients in the waveform, it quantizes them to the nearest beat or sub-beat, and time-stretches the audio to reduce or fill any gaps. It provides a workable means of correcting performances from less capable drummers who are regularly sloppy. The main downside of this method is that it is prone to producing 'phasey' or 'stretchy' sounding artefacts if you are not careful. Having said this, I have found that, with a good knowledge of what Elastic Audio is capable of, you can edit drums with no noticeable side effects.

As well as for quantization purposes, Elastic Audio is also extremely flexible for manual editing duties as it uses transient indicators (called warp markers in Pro Tools) that can be moved around in real time, which automatically adjusts the amount of time stretching applied.

11.7.2 Considerations for Drum Tightening

First of all, we have to decide what editing is required. You might guess by now what I am going to ask: What is your vision for the track?

If you record live, with a vision for the track to sound natural, it would be foolish to snap everything to grid, undoing the authenticity that you set out to achieve. However, if the performances are loose and sound out of time, you should adapt to the situation. I can recall several times when I've had to change my tactics due to unforeseen factors such as imperfect performances.

When I record each member of the band separately, I often do any required drum editing before other instruments are recorded. This helps reduce the

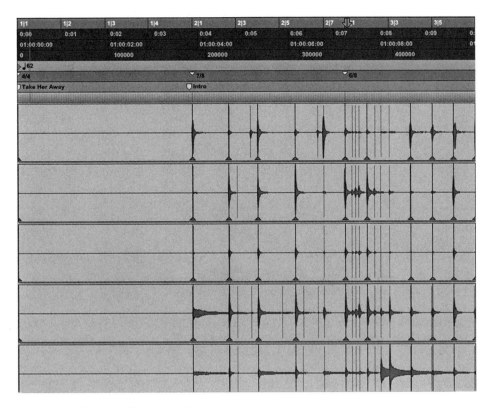

Figure 11.31 Elastic Audio warp markers

number of tricky edits you may have to do later on in the recording process. I use Elastic Audio quite heavily on drums and bass guitar, snapping most of the main beats to the grid (possibly leaving some wiggle room by adjusting the *exclude* and *strength* parameters), and allowing for 'groove' in the tom fills and snare rolls, which can sound robotic when over-tightened. When a drummer has inconsistent dynamics, Elastic Audio can sound particularly unnatural when tightened 100 per cent to the grid, as one expects a drummer with good timing to be dynamically consistent, so consider leaving these performances slightly looser.

There are a lot of sceptics when it comes to using Elastic Audio and other time-stretching-based solutions, because it loses authenticity and also reduces natural groove. I personally believe that you should judge it on a case-by-case basis. While I agree that for a tight, raw indie band it may be the wrong decision to heavily tighten, for a modern metal band it is almost compulsory.

You will also find many bands (particularly when you are just starting out and working with smaller local bands) that simply aren't good enough to get a good result without using these heavy editing methods. These bands are the sort of bands that you should identify in the pre-production stage and recommend recording each instrument separately.

Whichever method you decide to use for drum editing, remember to group all of your drum channels so that they are edited together. This maintains the natural phase relationships between mics. This is particularly important when using elastic audio, where stretching different mics by differing amounts will result in phasing.

As a final note on Elastic Audio, there are several algorithms designed for different types of transient material. I edit all of the drums using *Rhythmic* mode, which is designed for strong and fast transients. Once this has been completed, I switch the overhead, room and hi-hat mics to *X-Form* mode. Cymbals show up 'stretchy' artefacts more than anything else, and X-Form limits these artefacts. X-Form is a highly complex algorithm and takes a while to process, so switch to this mode when you are about to go for a decent break. If your computer's processor is particularly slow, consider letting the X-Form render overnight.

As well as grouping the drum tracks together, I also break up the song into smaller, more manageable chunks, based on when drum grooves change. This is advantageous for two reasons:

1. You don't have to try to quantize different drum grooves with the same parameters. This reduces the likelihood of incorrect quantizations, and allows you to easily see where you are in the song.
2. When you quantize a long section, one or two misplaced beats can lead the quantization to get progressively less accurate, and can even change the entire groove.

Finally, while it is good to use the grid as a reference and experiment with automatic methods of tightening, timing is really about how the band 'lock together' and move as a unit. Don't lose sight of this! As long as there aren't massive fluctuations in tempo, simply moving instruments around so that they are in time with each other is usually sufficient to create a great groove. These moments in a track are particularly important:

1. rhythmic stabs and pauses;
2. the first beat of new sections;
3. lead riffs and fills;
4. the initial introduction of an instrument in the song; and
5. the ending.

When listening for things to edit, use your ears more than your eyes. To quote an old Joe Meek phrase: 'If it sounds right, it is right.'

11.7.3 Drum Augmentation/Replacement

As well as timing issues, you may feel that the drums just don't 'come alive' even after heavy EQ, gating and compression. This could be due to deficiencies in the drum kit itself, the acoustic environment you recorded in, the placement or choice of microphone, or the drummer's competence or style. In reality, a lot of top-level engineers use drum samples even on the best drum recordings as a matter of course. There is certainly no shame in deciding to enhance the sound you already have by supplementing the performance with samples. It is one of the key 'hacks' used by the top guys in the business. You should never use sample enhancement as an excuse for being lackadaisical in your drum recording and failing to apply the processes this chapter has discussed. However, it does mean that you might have a 'get-out-of-jail-free card' if your drum recording has weaknesses.

You have some options when considering drum enhancement: you can augment the drum performance with samples (i.e. place triggered samples into your mix along with the natural drum sound). You could sample-replace, which involves completely replacing the close mics with samples while leaving the original overheads (possibly applying a steep high-pass filter to preserve cymbal information while clearing up most of the fundamental drum sound). You also have the option, while recording drums, to take samples of the kit itself, which can be a huge benefit as it allows you to sample-replace close-mic'd drums without bleed from other drums and cymbals. I regularly sample the kit I am using – in fact, I now have a library of drum kit samples that I have recorded – and I suggest you do the same. One of the big advantages of sampling the kit you are using is that not only are you sampling the drums themselves, which will always sound more natural with the other mics than pre-existing samples, but you are also sampling the drummer's hitting style, which will further mesh with your drum performance.

Being open-minded to drum triggering can give you a lot of advantages:

1. These days, drum-triggering software is very sophisticated, and each particular drum sound may be multi-sampled to such an extent that it sounds as dynamic as a natural performance. You can also blend samples of a room in with the close-mic samples for an extra sense of depth.
2. Conversely, if your drummer was very inconsistent, you can reign in the dynamics to get a more consistent result. This is better than overly compressing your drums and losing dynamic range.
3. You can drastically EQ samples without affecting other parts of the kit because they contain no bleed. For instance, you might want to get more top-end crack into your snare drum, but when you try to do that it brings out the hi-hat too much. One solution could be to use a sample of that particular snare itself, level the natural snare and trigger, and then EQ the sample to add the punch you require.
4. If you don't have a great set of mics, a good room or you have even made mistakes while tracking, using samples can help improve the mix.
5. You can abuse samples in a creative way to make songs more contemporary sounding.

A lot of the time, drum augmentation/replacement is actually performed by the mixer of the project. However, to give the mixer a good idea of what the producer wants, you may want to consider providing him or her with not only some samples of the kit itself, but also some pre-existing samples that you like. I feel this forethought requires mentioning in this section, rather than consigning it wholly to literature dedicated entirely to mixing.

Another reason why drum triggering should be brought up here is that in several genres, triggering is a key component to the overall sound. For instance,

Figure 11.32 Steven Slate Trigger 2

it is common in metal to tighten 100 per cent to the grid, and then completely sample-replace the shells of the kit. To do this, producers often use trigger devices that attach to the kit and connect to a drum module that converts the trigger information to MIDI notes. This means that it is easier to trigger samples without the bleed that you get from close mics. However, the trigger module will take a bit of time to set up for the style of each drummer. Triggers are also useful on pop productions where a performance might make use of electronic drum samples and claps as well as a natural kit.

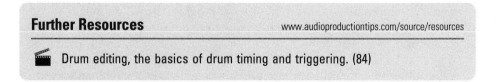

Further Resources www.audioproductiontips.com/source/resources

Drum editing, the basics of drum timing and triggering. (84)

11.8 SUGGESTIONS

Just to summarize, the drum sound is so crucial to most recordings that it takes up huge amounts of time and patience to set up. It is often the making or breaking of a great recording. Most people with a good musical ear and creativity can record guitars, keyboards and vocals fairly quickly. The drums, however, take a lot of experimentation. My advice is to spend a few days, maybe even weeks if you can afford it, just exploring, making mistakes and even trying to break the 'rules'. If you can memorize the sonic parameters, you can exploit them for the benefit of the recording, and then you can record anything. Start by learning to record drums and everything else will seem easy.

One of my biggest problems with my own formal education experience was that I spent too much time learning about EQ, compression and microphone types, and not enough time actually listening and making my mind up about how I wanted things to sound. Too much emphasis was on the tools the producer has to fix issues, rather than preventing them in the first place. The earlier in the chain you can make something sound great, the easier it will be to mix, and the better the end result will be (in the right hands, of course).

To be able to listen and make informed decisions, you need to know the variables you are dealing with. While this chapter has explained many of the variables you will encounter while recording drums, you will only really know what you are doing when you have tried these things out for yourself. While the insight I have given you in this chapter is a great starting point, you will only truly understand what I mean when you have experienced them for yourself.

I can tell you honestly that I spent my first couple of years in the industry working in a small commercial studio, sticking to one type of microphone set-up, one room and one vision. While this is great for churning out large quantities of material, it is catastrophic to quality and creativity. What was worse was that I chose these particular methods *without* having tried out the alternatives. You are never too experienced or too good to learn – in fact, I learnt the Recorderman technique *after* I recorded 'Take Her Away', so there really are no excuses to stop experimenting and striving for better.

You will by now have become very aware of my general feelings towards music production, and I am going to repeat many of the same principles throughout this course. Sorry (okay, I don't mean that), this may bore you, but it *needs* to sink in. These principles are what pop into my mind before recording each instrument, so here they are again:

1. Decide what you want to achieve. Your vision and the genre of music dictates pretty much every decision you will make.
2. Use your ears, rather than doing something because you are told to or because that is the way that you usually do it. Think of your vision and use your ears to make it a reality.
3. Get as close to this vision as early in the chain as possible. For example, if you want a less 'boxy' sound from your kick drum, don't just jump for the EQ on your plug-in or desk first. Try getting as close to your vision by experimenting with different mics, placements, shells, beaters and rooms (if you have the luxuries). By having knowledge of the tones available to you, you will begin to make correct decisions very quickly.
4. Know the limitations. I take huge pride in my work. I find it very difficult to switch off, and I constantly compare my work to that of the greats. While this is a great ambition, and great for advancing your skill set and ultimately achieving your goals, you will often work past the point of no return where you will be making things worse, getting yourself frustrated and losing

confidence. Being realistic about your skill level, the players, the songs, the technology available to you and also your environment will save you a whole heap of stress (although this is easier said than done). When recording anything, be aware of the limitations.

Bear all of these points in mind in the proceeding chapters, where we will discuss the other instruments you are likely to encounter recording popular music genres.

EXERCISES

1. Use the drum-tightening session files provided to tighten up the drums to 'Take Her Away'. Experiment with doing edits manually, as well as using Beat Detective and Elastic Audio.

2. Take a multi-mic'd drum recording and start using Elastic Audio destructively. Purposely push it past its limits to hear the stretching artefacts occurring and learn the signs.

3. Start editing the drums without them being grouped, just to hear the phase inconsistencies.

4. Use a drum replacement tool to completely replace the close mics of the kick, snare and toms in a recording, and try to make them sound as natural as possible.

5. Now try using the drum replacement tool to augment the natural drum sounds by bringing in the samples underneath the original hits.

6. Finally, make some custom presets out of the samples of James's kit I've provided with the drum-tightening session files. Use these to augment and/or replace the drum sounds from 'Take Her Away' and compare the results to using stock samples bundled with the drum replacement software.

NOTES

1. Shell materials, accessed February 2013, www.thomann.de/gb/onlineexpert_page_snare_drums_shell_materials.html.
2. Anatomy of a cymbal, accessed February 2013, www.sabian.com/en/pages/anatomy-of-a-cymbal.
3. Cymbal materials, accessed February 2013, www.thomann.de/gb/onlineexpert_page_cymbals_materials.html.
4. Matthew Mcglynn, Drum overhead microphone technique comparison, accessed February 2013, http://recordinghacks.com/2010/04/03/drum-overhead-microphone-technique-comparison/.

12

TRACKING ELECTRIC AND BASS GUITAR

After recording the drums, the microphone set-up procedure for recording guitars may seem like a walk in the park. However, you are not out of the woods yet! When recording guitars, we have a completely different evil to deal with: the search for the perfect signal chain. As you are about to learn, the tonal variations that can be achieved through the combination and configuration of guitar, amp and effects chain are astonishing!

For instance, if you are going for a lightly driven late '60s-sounding tone in the verse of a song, I'd consider tracking with a Fender Telecaster guitar through a Fender amp with the majority of drive coming from a classic overdrive pedal. In contrast to this, for a double-tracked classic rock tone in a chorus, I'd probably start with a Gibson Les Paul through a Marshall JCM series head, and use a cabinet with Celestion Greenback speakers. Sounds pretty complicated? Well these are probably the simpler of signal chains!

Throughout this chapter, there will be links to demos of the various guitars, pedals and amps described. It is important to listen to these and internalize the options so that, when the time comes, you can sculpt the desired tone in the quickest way. Trust me – it'll save hours of trial and error and a lot of head-scratching.

As with drum recording, we cannot assume that the player will know what is best. You should, however, quickly work out if he or she is of strong and useful opinion. A great producer will have a technical knowledge of each of the core instruments in his or her genre and should be able to make informed decisions on instruments, tones and parts, in cahoots with the band.

12.1 THE ELECTRIC GUITAR

Let's start from the top and look at how the choice of gear can help you achieve your desired tone, from the choice of guitar, amplifier and effects pedals, through to your microphone selection and placement.

Although the bass guitar is usually the first guitar to be recorded, I will focus first on the electric guitar, as many of the principles can be directly transferred to the bass. I will discuss the specific considerations for bass guitar after we have covered the electric guitar.

While I would like to go into a huge amount of detail here, I will just cover the most important considerations and the most popular types of gear.

> **Further Resources** www.audioproductiontips.com/source/resources
>
> 📖 *The Ultimate Guitar Tone Handbook: A Comprehensive Guide You May Want to Check Out* by Bobby Oswinski and Rich Tozzoli. (85)

12.1.1 The Instrument

The guitarist(s) you are recording is likely to have a personal favourite model of guitar. They will probably be keen to use that particular model for the majority of the recording and will feel more comfortable playing it. The player's guitar (and amplifier) will probably be appropriate for the band's style. However, when recording less experienced or eclectic bands, you may have problems. I can usually tell, even before I hear the band play, whether or not I will want to use my own gear, or need to hire specialist equipment. Here are some warning signs:

1. Age (14–20). While some young bands are well equipped, a lot of them are stuck with their starter equipment or their tastes have changed while growing up, leaving them with stylistically unsuitable equipment.
2. Cheap equipment. Enough said, really.
3. Bands where each member has a different style of dress. While I am sometimes surprised, I usually find this means their musical influences – and therefore their equipment – will be equally different in style.

Before we begin, familiarize yourself with a typical electric guitar (Figure 12.1).

Most guitars are based around classic designs from Fender and Gibson, the guitar industry's two heaviest hitters. For this reason, I will be talking mainly about the features of the following five guitars:

- Fender Stratocaster
- Fender Telecaster
- Gibson Les Paul
- Gibson SG
- Gibson ES-335

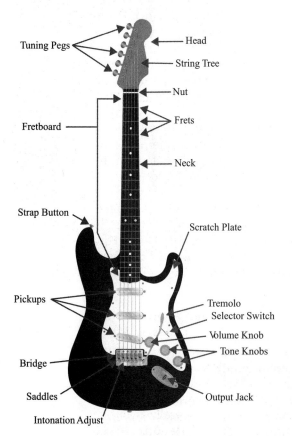

Figure 12.1 Parts of an electric guitar

I estimate that 95 per cent of all guitars closely resemble one of these models. Fender guitars tend to be light, small, 'bright' and have less sustain and bass response compared to Gibsons, which are heavier, less noisy, with more sustain and 'warmth'. I recommend that when you are starting to buy guitars for studio use, you cover your bases by getting a Fender-style guitar (Stratocaster or Telecaster), a Gibson-style guitar (Les Paul, SG) and also a semi-acoustic-style guitar (Gibson ES-335). Don't worry if you don't own any of these; try to find friends or local rental companies where you can hire them when needed.

Fender Stratocaster

The Fender Stratocaster (or Strat) is probably the most recognized guitar of all time. A standard Stratocaster features three single-coil pickups (bridge, middle and neck), 21 frets and a tremolo arm (more correctly called a vibrato arm, since it is used to adjust pitch, not volume). The Stratocaster features a five-way pickup selector, which allows you to select the following combinations of pickups:

Position 1: Bridge (selector switch closest to bridge)

Position 2: Bridge + Middle

Position 3: Middle (selector switch in the middle)

Position 4: Middle + Neck

Position 5: Neck (selector switch closest to neck)

In simple terms, the closer to the neck, the more 'thick' and 'warm' the sound, and the further towards the bridge the 'brighter' and 'twangier' the tone. If you choose one of the positions where two pickups are active, you get a weaker but distinct sound that some guitarists like.

The Stratocaster also has two tone controls, one for each of the middle and neck pickups. The tone knob works by filtering out high frequencies: when a tone knob is on full (10), the signal is unfiltered; at minimum (0), it has the most filtering (i.e. attenuation of the top end), and is therefore 'bassier', 'warmer' or 'muddier'.

Stratocasters are solid-bodied guitars and are usually made from ash or alder, which is lightweight (especially compared to Les Pauls). The neck is usually maple and is bolted on to the body; this typically transfers less vibration between body and neck, which gives you a little more note definition and is more 'cutting'.

Figure 12.2 Fender Stratocaster (picture courtesy of Metropolis)

It is a very versatile guitar and is used in many genres. You can find very silky-smooth tones easily and even make it go slightly edgy. It is typically favoured by country or blues players. Users of the Strat include Jimi Hendrix, David Gilmour (Pink Floyd), Eric Clapton and Mark Knopfler (Dire Straits).

Heavy metal guitarists often use guitars styled after Stratocasters, but with a humbucking pickup in the bridge position to get extra warmth and sustain out of the guitar. In fact, Fender make a range of Stratocasters for this purpose called Superstrats. As well as humbucking pickups, other features such as Floyd Rose tremolo systems and roller nuts appear on guitars by specialized heavy metal guitar manufacturers such as Ibanez to maintain stable tuning when tremolo bars are used excessively.

Fender Telecaster

Fender's second most popular design is called the Telecaster (Tele). It is a 21-fret design with a bolted neck and is made of ash or alder, just like the Stratocaster. However, the Telecaster is bigger in body than the Strat (so it's a little heavier) and it features two single-coil pickups (bridge and neck). The bridge pickup is fixed in a metal casing that forms part of the bridge, which is one of the reasons for the classic 'edgy' sound of the Telecaster. Other differences to the Stratocaster include a single master tone knob that affects both pickups at the same time, and a three-way tone selector:

Position 1: Bridge

Position 2: Bridge + Neck

Position 3: Neck

Figure 12.3
Ibanez Strat-styled metal guitar (picture courtesy of Metropolis)

The Telecaster is famous for its 'edgy', 'gritty', 'twangier' sound. Because of this, it is mainly favoured by country musicians, but it is also a favourite of British alternative and indie bands. Users of the Telecaster include Keith Richards (The Rolling Stones), Jeff Beck, Andy Summers (The Police), Graham Coxon (Blur) and Jonny Greenwood (Radiohead).

Gibson Les Paul

The Gibson Les Paul is often viewed as the 'holy grail' by those who want a 'thick', 'warm' and 'fat' sound. This is mainly due to its two humbucking pickups, but its heavy body and glued neck also contribute to its particular sound. The guitar features 22 frets, a heavy body traditionally made from mahogany, and a glued neck (again, mahogany) rather than the bolted necks of the Fender-style guitars. The advantage of the glued neck is that vibrations transfer more readily between the neck and the body, allowing the guitar to have a longer and stronger sustain. The guitar has a three-way selector switch positioned just above the end of the fretboard near to the neck pickup, and you can select either rhythm (bridge), treble (neck) or a blend of both.

Figure 12.4 Fender Telecaster (picture courtesy of Metropolis)

Figure 12.5
Gibson Les Paul (picture courtesy of Metropolis)

Figure 12.6
Gibson SG (picture courtesy of Metropolis)

The Les Paul is a standard for heavy rock but is also used in other genres. Notable users of the Les Paul include Jimmy Page (Led Zeppelin), Slash (Guns N' Roses), Zakk Wylde (Ozzy Osbourne), Bob Marley and Noel Gallagher (Oasis).

Gibson SG

The Gibson SG ('Solid Guitar') shares many design features with the Les Paul. These include a glued neck, 22 frets, two humbucking pickups and the same controls and pickup selector switch (although this switch is placed next to the tone and volume pots). The main differences are:

- much thinner and lighter-weight body; and
- the neck joint has a double cutaway to allow easier access to the top frets (and elongating the neck length).

Because of the similarities to the Les Paul, it retains some of its 'thickness' and sustain but also benefits from being lighter, 'edgier' sounding and easier to solo high up the fretboard. However, many players still prefer the Les Paul due to its superior sustain and bottom end. Compared to other guitars, the SG's neck-to-body weight ratio is high, which can take some getting used to.

Figure 12.7
Epiphone ES-335 (picture courtesy of Metropolis)

The Gibson SG is a very popular model for rock guitarists. Those who favour the Gibson tone but dislike the weight of a Les Paul often choose it. Notable users of the SG include Angus Young (AC/DC), Tony Iommi (Black Sabbath), Pete Townshend (The Who), Frank Zappa and Dave Grohl (Foo Fighters).

Gibson ES-335

The Gibson ES-335, unlike the other four guitars, is not a solid-bodied guitar, but it shares the same electronics, woods and necks with the other Gibson guitars. This guitar typifies a semi-acoustic, or hollow-bodied, design. As you can see in Figure 12.7, there are two f-shaped sound holes as you would see on orchestral string instruments. These make the sound much louder acoustically (i.e. when not plugged in), but the volume is not substantial enough for the ES-335 to be considered usable as an acoustic guitar. The hollow body design gives a warm tone with responsive dynamics, but lacks the 'edge' of solid body guitars.

Because of its warm tone and dynamics, the ES-335 tends to be favoured by old-school blues, jazz and fusion guitarists. Recently, hollow-bodied electric guitars have been used by some British indie performers, for example Oasis, The Libertines and The Kooks, presumably for its looks rather than its sound. Notable users of the Gibson ES-335 include B.B. King, Roy Orbison and Chuck Berry.

> **Further Resources** www.audioproductiontips.com/source/resources
>
> Guitar type examples and comparison. (86)

12.1.2 Pickups

As I've described the tone of the guitars, I should mention the most common types of magnetic pickup found on electric guitars: *single coil* and *humbuckers*. You will find at least one of these two types of pickups on pretty much every electric guitar. Electric guitar pickups are built by wrapping many coils of copper wire around a magnet. The proximity of the metal strings to the magnetized coil causes the strings to magnetize. When a guitar string is played, it disturbs the magnetic field and causes an electrical current to be generated in the wire.

Passive or Active?

Both single-coil and humbucking pickups can be found in passive or active varieties. *Passive* pickups are the original guitar pickup design and are still by far the most popular. They require no power to work, but use more coiling to produce the required output level. *Active* pickups use less coil and would therefore output a much lower signal level. To combat this, a pre-amp stage is incorporated into the design, which boosts the guitar output before being sent to the amplifier. This pre-amp is commonly powered by a 9-volt battery and is usually located on the rear of the body of the guitar. The use of the pre-amp stage in active pickups means that the signal level sent to the amp is usually higher than a passive design.

Active pickups have several advantages over passive varieties: they are less prone to hum and background interference and less prone to feedback. To add to this, passive pickups can occasionally cause intonation issues due to their larger magnetic pull on the strings. Active pickups have a lower dynamic range than passive pickups, giving a more consistent volume and extra detail that helps to distinguish notes in complex and quiet passages. Think of this as a natural compression effect. The increased output before feedback means that active pickups are often used in alt rock, hard rock and metal.

Despite these advantages, a lot of guitarists don't think active pickups sound as good as passive pickups, with players describing them as 'lacking warmth' or sounding 'sterile'. This is due to the fact that passive pickups actually roll off some high-frequency detail, making them subjectively warmer. Passive pickups also have a greater dynamic range, and are therefore great for guitarists that like to play subtly as well as loudly.

Single Coil

The original and simplest type of guitar pickup is the single coil. Often found in Fender styles of guitar, they are brighter in sound than humbuckers, with greater note definition between strings. A word of warning: they are very prone to noise, particularly from lights and other electrical appliances.

As well as the traditional single-coil shape shown in Figure 12.16, Gibson also produced a passive single-coil pickup called a P90 that was the standard pickup in the Gibson range until the humbucker was introduced in 1957.

Humbucker

Humbucker pickups are basically two single-coil pickups with opposite polarity and windings, which are then connected in series. This design massively reduces noise and interference, and produces a higher output than a single-coil pickup. The tone they produce has less high end, and is darker and fatter than that of a single-coil pickup; they also overdrive amps more easily due to their higher output. They are typically found in Gibson-style guitars but have gradually migrated to a lot of Strat-style guitars that are built for heavier rock. Humbuckers can sometimes be found with or without a metal cover. Humbuckers without covers are slightly less prone to feedback than those with covers.

Figure 12.8 Single-coil pickup (picture courtesy of Metropolis)

Figure 12.9 Humbucker without cover (picture courtesy of Metropolis)

Figure 12.10 Humbucker with cover (picture courtesy of Metropolis)

12.1.3 Strings

When recording electric guitar, I insist that players restring their instruments just before the recording session. I also advise having a spare set on hand (particularly a high E) as breakages can and do happen. As a producer, it is good to know what different strings sound like so you can help guitarists choose a set that will suit them. While everyone has their favourite manufacturers, it is more important that you determine what gauge and material is best for the musician you are recording.

Gauge

The gauge of electric strings is measured in 1,000th of an inch. For instance, the thinnest string on a light set of strings is gauge 9, and this is .009 of an inch thick; the higher the gauge, the thicker and heavier the string. When asking for a set of strings in a guitar shop, you use the gauge of the highest-pitched string as a reference. For example, if I wanted a light set, I would ask for a set of 'nines'. While gauges are similar from one manufacturer to the next, the names of the sets often vary. The table (*left*) shows an example of what the gauges might be called.

String	1	2	3	4	5	6
Super Light	8	11	14	22	30	38
Light	9	11	16	24	32	42
Regular	10	13	17	26	36	46
Medium	11	13	18	30	42	52
Heavy	13	16	26	36	46	56

There are advantages and disadvantages to both light and heavy strings:

- The lighter the string, the more easily they bend and vibrato, which makes them preferable for some lead guitarists.
- Lighter strings are easier on the musician's hands.
- Lighter strings are more prone to breakages.
- Heavier strings reduce fretboard buzzing when actions are set low.
- Heavier strings have more volume, sustain and a thicker sound.
- Heavier strings perform better when detuned.

Due to the demand of modern metal players, you often see hybrid sets of strings, including thicker low strings that are good while detuned, and standard thinner strings that are good for lead work.

I tend to like gauge 9 for most applications. However, when a guitarist plays almost completely rhythm, I often recommend sets of 10s or 11s.

Material

Electric guitar strings come in a few different metals. However, the effect of material on tone is less dramatic than the gauge of strings and the guitar itself. There are three common metals for electric guitar strings: stainless steel, nickel-plated and pure nickel:

- Stainless steel strings sound brighter and have more volume than nickel strings. However, they are heavier to play and more prone to finger and fret noise.
- Nickel-plated strings are effectively stainless steel with a thin nickel plating to offer a 'faster' and 'smoother' texture and 'softer' and 'warmer' tone.
- Pure nickel strings consist of nickel wrapped around a small steel core. They are an older way of making guitar strings and are suitable if you desire a vintage sound. These are typically the dullest of the three types.

12.1.4 Guitar Maintenance

In Chapter 11: Tracking Drums, I discussed in detail how a studio engineer should be involved with the maintenance of a drum kit. For guitar maintenance, however, it's best to let the player restring and clean the instrument him or herself. If you have a reason to believe that a guitar is in need of attention (beyond changing strings), then urge him or her to get it professionally set up before the recording session.

You could write an entire book on guitar maintenance, as there are a lot of different mechanisms and styles of guitars. If you want to learn more about guitar maintenance or how to set up a guitar, please check out the following:

Further Resources www.audioproductiontips.com/source/resources

 Guides to setting up Gibson- and Fender-style guitars. (87)

However, while you should probably get a professional to perform maintenance, it is imperative that you are able to identify problems with guitars so that they do not plague a recording session. Here are some of the potential issues that you should listen out for in a recording session:

Fretbuzz

This sounds like a metallic rattle and happens when the strings vibrate against another part of the guitar, most commonly on one or more of the metal frets. Fretbuzz can be caused by a number of issues:

- A player's technique.
- Low action (action is the distance between the string and the fret). Low action is probably the number-one cause of fretbuzz. Here is a step-by-step guide to adjusting the guitar's action:
    - First, you will need a flat work surface, a small screwdriver and a metal ruler. If you have a Fender guitar, you will also need a small Allen key.
    - Lay the guitar flat on the work surface, making sure that you have no pressure on the neck of the guitar.
    - Find the guitar's bridge.
    - Slacken off each string so that you are not adjusting the action while the strings are under tension.
    - Use the screws on either side of the bridge (on some Fenders, there might be six screws across the bridge) to raise or lower the height between the string and the frets. Make sure the action is uniform on both sides. When I adjust the action of my Les Paul, I lower it until it starts to buzz and then raise it a little. Rhythm or slide players may opt for slightly higher than this. You could set the action with the help of a metal ruler; I've also heard people talk about levels such as 4/64s above fret 17, or metric units of 2.5–5 mm from fret 12. I generally gauge the action by the feel of the guitar and the technique of the player.
    - If you have a Fender-style guitar, you can fine-tune each string's action by inserting an Allen key in the top of the string's saddle.
- An incorrectly set truss rod. The truss rod is a metal bar inside the neck of the guitar that is slightly curved to allow a low action across all frets. Adjusting the truss rod is one of the more risky things to do yourself, and for this reason I am not going describe the process. If you decide to attempt it, make adjustments in small increments to avoid permanent damage to the guitar.

Figure 12.11 Truss rod in guitar[1]

- The top nut of the guitar is cut too low or worn. If you are encountering fretbuzz only when the string in question is played open, then it is likely to be a problem with the top nut being too deeply cut.

- A warped neck.
- A badly set Floyd Rose system, the tremolo system I mentioned earlier that helps maintain guitar tuning.

Dead Notes

If a certain position on a fretboard produces a weak or buzzing sound when fretted, it is called a 'dead note'. While fretbuzz is quite common, dead notes are not so. Strangely, if the musician is playing an electric guitar, sometimes you can hear a dead note acoustically, but it sounds fine when amplified. Dead notes can occur for a number of the same reasons as fretbuzz. They can also occur when you have uneven frets on the guitar; with cheap guitars, this is usually due to a manufacturing problem, but it can also occur on heavily used instruments. If you encounter a specific problem with a solo or arpeggio before the recording session, send the guitar to a specialist. If you are mid-session, use a different guitar for that particular part.

Intonation

This is a particularly common issue and unfortunately it can go unnoticed while recording, especially during a long and demanding session. The signs of intonation issues are a sharpening or flattening of the guitar's tuning when the strings are fretted. It can be worse the higher up the fretboard you move. This can be adjusted easily on a string-by-string basis by lengthening or shortening the string with a screwdriver on the bridge saddle. If the guitar's tuning drifts sharp as you go higher up the fretboard, you need to adjust the string to be longer. You do this by moving the saddle closer to the bridge. If it drifts flat, you need to shorten it by moving the saddle closer to the pickups. Remember, though, if you have not set up the guitar for a long time, you may want to get a professional to do this as there could be problems with the truss rod or spring tension on the bridge. Here is a step-by-step guide on setting intonation:

1. First, you will need an electronic guitar tuner, a flat work surface and a small screwdriver.
2. Lay the guitar flat on the work surface, making sure that you have no pressure on the neck of the guitar, as pressure can slightly affect the strings' tuning.
3. Find the guitar's bridge saddle screws.
4. Tune the guitar so the strings are precisely in tune when played open.
5. Play each string while fretting on the 12th fret and check tuning.
6. If the tuning is now incorrect, move that string's saddle using the screwdriver to compensate. The more out of tune, the bigger the adjustment needed.
7. Repeat steps 4–6 until every string is correct when played open and on the 12th fret. This may require a little trial and error, but if you really struggle to get it right, it suggests a more major issue is at hand, and I suggest you send the guitar to a technician.

Circuit Issues

As well as issues regarding the playability of the guitar, there can also be some common issues that cause 'crackles', 'pops' and electrical noise:

1. Loose input socket – where the jack input is plugged into the guitar is notorious for causing dropouts or 'cracks' and 'pops'. Always make sure that the input nut is tight and there are no dropouts when you wiggle the input jack.
2. Crackly or scratchy volume or tone pots, or switchers – quite often just a quick wiggle can remove it, but if not try some contact fluid on the joint. If all this fails, get a specialist to replace the pot or switch in question.
3. Pickup noise – sometimes there is excessive pickup noise from a guitar, particularly cheaper guitars. This is usually due to an internal ground loop. A specialist should easily be able to shield the cables more effectively.

If you decide to attempt maintenance, you should do it in the following order: electronics, truss rod, action, intonation.

12.1.5 The Plectrum

Figure 12.12 Plectrum

Typically, most rock musicians will play with a plectrum, which creates a brighter sound than using the fingers. Plectra are usually made out of types of plastic, although you can also find some metal ones.

The thickness of the plectrum has a greater affect on tonality than most beginner guitarists think. Thinner plectra tend to create a 'washier' sound with less presence and attack. Thinner plectra are more versatile because, by holding the plectrum closer to the tip, you can make it behave a little more like a thicker plectrum. Thicker plectra, however, are generally more suitable for heavy rock genres, as they easily provide a more defined tone, and also pinched harmonics and squeals can be produced more easily.

Plectrum thickness is measured in millimetres. Most commonly, they are between .5 mm and 1 mm for electric guitar use, with values between .6 mm and .8 mm being the most commonly used.

12.2 ELECTRIC GUITAR AMP SELECTION

Amp selection can be a very daunting and complicated process. Matching the right guitar and amp to the style of guitarist is an art, but it doesn't have to be such a daunting task. As with everything in music production, your ears should be the judge. However, knowing the basics of amplification can help reduce the learning curve and the process of trial and error.

12.2.1 Amp Type

There are two types of configuration for amps: combo amps (combined amps) and 'head plus cab'. A combo comprises an amp and speaker in a single enclosure; in a head plus cab set-up, the head (amp) and cab (speaker) are housed in separate enclosures.

There are also two common types of amplifier: solid-state and valve. Please note that in the US, valves are called tubes – they are the same thing. Although solid-state technology is constantly improving, most guitarists tend to prefer valve amps. Below are the pros and cons of each type:

Tube/Valve Amp

Tube amplifiers use vacuum tubes to amplify the sound. They were the original breed of guitar amplifier that we have all come to love and cherish. Tube amps are chosen for their tone, which is often described as 'warm', 'thick' and 'fat'. However, they are heavy, prone to breakage, more expensive, and tubes periodically need replacing. The tone that tube technology creates can also be widely different depending on the age and wear of the tubes.

Solid-State Amp

Solid-state amps use transistors to amplify the sound. By and large, they have replaced tube amplifiers, other than in the professional arena. This is because solid-state equipment is very light in comparison to the weight of vacuum tubes and the extra circuitry that goes with them. Solid-state amps are much quieter than a tube amp of equivalent wattage. In the early solid-state days, they were known for sounding 'sterile' and 'uninteresting'. Solid-state technology has improved massively in recent years. You can get very close to a tube sound with a modelling amp that simulates the character of tubes. However, for me, I don't think they have quite perfected the technology. While solid-state amplifiers are more common in the cheaper price range, a well-designed solid-state amplifier is still a very good piece of equipment and I would probably choose a top-end solid-state over a budget tube amp. It is worth testing all sorts of different amps when auditioning. A couple of notable examples of musicians that have used solid-state amplifiers include B.B. King (Gibson Lab Series L5) and Dimebag Darrell (Pantera) (Randall Warhead solid-state amp).

The Tubes

Contrary to what you might think, different types of tube are usually used in each stage of a tube amp. The choice of tube plays an important role in shaping the sound of the amp and its distortion characteristic. Fender and Marshall are the two most famous amplifier manufacturers. Their original circuitry and choice of tubes shaped the way later amplifiers were built. I will now discuss the different tube stages found within a tube amplifier, and give a brief overview of different tube types.

Pre-Amp Tubes

This is the first tube stage of the amp. The gain knob on an amp controls the amount of signal sent to the pre-amp tubes. This determines how much the tubes are driven, and thus how distorted the tone will be. The 12AX7 is a common choice for pre-amp tubes.

Effect Tubes

Some amps have tubes built into their effects loop. A 12AT7 is usually used here.

Power Amp Tubes

This tube stage handles the most amplification. The tubes are worked harder as you increase the volume of the amp. Common power amp tubes include the EL-34, EL-84 (Marshall) and 6L6, 6V6 (Fender). The power tube plays a key role in the tone of a tube amplifier, and for this reason you will see the greatest variety of tube types in this stage.

Figure 12.13 Electro-Harmonix EL-34 power tube, inside a Marshall JCM800 head (picture courtesy of Flipside)

Rectifier Tubes

In every amp, there is a rectifier stage. The rectifier converts the AC current from the wall outlet into DC power that the other valves need to operate. To reduce costs and physical weight, many amps have solid-state rectifiers, but some amps use tubes. Due to an electrical phenomenon called sag, solid-state rectifier stages are more efficient than valves. In practical terms, this means that valve amps have less of an immediate 'attack', resulting in the sound possessing some compression characteristics. Due to sag, less voltage gets to the power tubes, which creates an 'edgier' driven sound with less headroom. This makes tube rectifiers perfect for classic rock or bluesy applications. Solid-state rectification, on the other hand, is more transparent and has more attack, and is therefore great for high-gain applications or any time when you want to maintain transient information.

Common rectifier tubes are the 5Y3 and 5U4.

Picking suitable tubes for the amp is definitely a great way to alter the guitar tone. However, it takes a lot of experience or research to achieve the results you want. Certain players may state that they find a certain tube stage more sonically important to the overall tone than the rest. In my personal experience, because there are so many variables involved, it is not quite as simple as that. Plus, you also have to make sure the amp is biased correctly for the tube; this is what regulates voltage in the tube and stops the tubes breaking prematurely. Different brands and sizes of amps, and even the level at which the amps are played, means that replacing one tube with another may not produce the results you'd expect. Therefore, if you don't have a lot of time to experiment or don't have a lot of experience with electronics, I would leave this to a specialist.

12.2.2 Classic Tube Amps

Fender

Fender amps are known for their bright and sparkly tone and for having more clean headroom before breaking up. The sparkly and present tone of the Fenders is often referred to as the American sound.

Marshall

Marshall amps are known for their 'crunchy' distortion tone. They break up more easily than Fender amps do, and are 'grittier' sounding in the mid range. The gritty mid-range tone of the Marshall is often referred to as the British sound.

As well as these two behemoths of the amp world, the 1970s and 1980s set the precedence for guitarists who wanted to achieve more and more heavily distorted tones. This led to amplifiers with more gain stages. Mesa/Boogie amplifiers became prominent after developing the first three-gain-stage guitar amp.

Figure 12.14 Fender Hot Rod Deluxe combo amplifier (picture courtesy of Metropolis)

Mesa/Boogie

While Mesa/Boogie started in 1969 as a company that modified Fender combo amps to give them more input gain, they quickly started to design their own amplifiers. In 1972, they released the Mark I. It featured an extra tube gain stage to the pre-amp and three gain controls at different points of the circuit. This was the very first high-gain amplifier. Mesa/Boogie went from strength to strength, and their unique gain-staging started to be copied in many other manufacturers' amps. Although Mesa/Boogie are known for their heavy sound, the list of players that are using or have used Mesa/Boogie amps is surprisingly varied, perhaps due to the fact that their Fender roots combined with more gain stages led to a

Figure 12.15 Marshall JCM800 amplifier head (picture courtesy of Flipside)

Figure 12.16 Mesa/Boogie Triple Rectifier amplifier head (picture courtesy of Flipside)

very versatile amplifier. With the gain dialled back, its clean tone sounded more like a Fender than a Marshall, due in part to its scooped sound (less mid and more bass and treble). As a high-gain amplifier, the Mesa/Boogie came to define the tone of American heavy metal.

Before moving on, I should say that the variation between amp models is massive. Use these stereotypes to help you anticipate what amp may be best for your project, but remember to use your ears to judge what is best, and not to merely trust the reputations of the manufacturers. There can even be significant differences in tone between two identical models made in the same era – such is the impact of the manufacturing process of the tubes themselves.

As an extremely rough rule of thumb, I tend to like Fender or Vox amps for cleaner tones and bluesy stuff, Marshalls or Orange for classic rock and very British-sounding music, and Mesa/Boogies or ENGL for metal.

> **Further Resources**   www.audioproductiontips.com/source/resources
>
> Guitar amp examples and comparison. (88)

12.2.3 The Speaker (Driver)

The loudspeaker used in guitar cabinets is also a critical part of an electric guitar's tone, and it is as important as the choice of amplifier itself. Part of the difference between the sound of a Fender and a Marshall is due to the speakers themselves.

The cones of these speakers start from about 6.5 inches and go up to 15 inches. However, 10- and 12-inch cones are by far the most popular.

As with amps themselves, speakers are chosen for their particular 'voicing'. Jenson are a famous manufacturer of speakers that produce the 'American' voicing associated with Fender amps. Their flagship 12-inch design was the P12N.

Celestion speakers are famous for having a 'British' voicing, and were used in classic Marshalls. Their flagship 12-inch design was the G12M, nicknamed the 'Greenback'.

While Jensen speakers are no longer being manufactured, Celestion speakers are still in production, with modern adaptations of their classic 'Greenback' design still widely used in Marshalls and Mesa/Boogies, among many others. Today, a lot of the classic designs that are no longer available can be replaced with copies made by a company called Weber.

The frequency range of a speaker cone roughly matches that of a typical electric guitar: 80 Hz–6 kHz. A lot of speakers are designed with a presence boost at 2–3 kHz (British voicing) and some have a low frequency boost for extra 'warmth' (typically American-voiced speakers).

I would like to briefly discuss a couple of the biggest factors (other than impedance and wattage, which we will discuss later) that you should observe when selecting a cabinet, or choosing a speaker:

Sensitivity

Speaker sensitivity contributes significantly to the potential volume of the guitar rig. Typically, sensitivity specs for guitar speakers range from 92 to 102 dB. This spec is how loud the speaker will be at a distance of one metre with one watt of power. Every 3 dB higher is like doubling the power supplied by the amp. A speaker sensitivity of more than 96 dB is considered to be good. As well as being louder, higher-sensitivity speakers tend to sound 'brighter' and have a 'tighter' bottom end.

Resonant Frequency

Typically, the resonant frequency of a guitar speaker is between 50 Hz and 150 Hz. The lower the resonant frequency, the lower the bass extension. However, there is a trade-off between lower resonant frequencies and speaker efficiency. Speakers with a higher resonant frequency produce a more 'aggressive' sound, with an increased risk of 'boxiness' or 'muddiness'. Because of this, the ranges between 50 Hz and 100 Hz are generally preferred. For detuned guitars, consider using speakers with a lower resonant frequency. Some bands, such as Queens of the Stone Age, are known to use bass guitar cabs as guitar cabs to exploit this.

Magnet Type

There are three common types of metals used in speaker magnets:

Alnico. An alloy consisting of aluminium, nickel and cobalt. This was used in all of the vintage speaker magnets. It produced a 'warm' bottom end and was good at lower volumes. Guitar players also liked them for their fast response due to their strong magnetic field. The downside of alnico magnets is that they are expensive, and many amp manufacturers began to opt for less costly options.

Ceramic. These magnets were designed as a direct alternative to alnico at a lower cost. They produce a wider range of tones than alnico and are better at higher volumes. However, they are less warm and weigh more.

Neodymium. This is the newest of the magnet types, and its cost is more than ceramic designs but less than alnico. It weighs less than either of the other two designs but has a stronger magnetic field. They have a well-balanced frequency response at various volumes and have a fast response much like the original alnicos.

12.2.4 Matching Amp and Speaker Sizes

When it comes to amps, bigger doesn't always mean better. In a studio setting, an amp's volume isn't as important as its tone. In fact, Eric Clapton famously recorded Layla with a tiny 5-watt Fender Champ – and that sounds pretty good, right?

Although, generally speaking, the more watts, the louder the sound, this is not always true. Speaker variables (among other factors) make a big difference.

As a rule of thumb, it takes 10 times the wattage to double its perceived loudness. This means that a 10-watt micro-amp is half the volume of a 100-watt head. Shocking, eh? This means that the volume difference between a 50-watt and 100-watt head is not as drastic as manufacturers want you to believe. You should also factor in that the human ear has a natural compression mechanism at louder volumes (Fletcher-Munson curves), meaning that the perceived volume difference is even less.

When recording a guitar amp, what matters is that you can achieve a good signal level with no noise and a great tone. When auditioning amps in a studio, I prefer to audition them in the control room at a matched volume level, and using the same brand and model of mic, to make the test consistent. This helps me make a more objective call on what amp has a better tone for the job.

'The bigger the amp, the better the sound' is a popular misconception created by marketing departments to sell big amps. However, it is true that they look cool on stage in big venues. Additionally, manufacturers of amplifier heads often neglect to mention that an amp's output wattage is *not* the only contributing factor to tone and volume. Cab design, speaker size and efficiency, as well as several other factors, have at least as much of an impact on perceived volume as amplifier power does. For instance, the higher the speaker sensitivity, the less wattage it needs to move, which is more desirable. *This means that the cab is as important as the amp.*

In the studio, you are likely to try many different combinations of equipment when searching for that 'new sound'. But, when using a head/cab solution, remember that you need to match the speakers to the amp. I've blown a couple of speakers and amps in the past from improper re-amping techniques, such as mismatching impedances or distorting the signal to the point of a square wave through a low-wattage speaker. Trust me, you don't want to make a habit of this unless you have a healthy bank balance. In Appendix A, I discuss voltage, current and impedance. We are now going to expand on those concepts so that you don't leave a trail of blown amps and speakers.

There is a lot of misinformation on the Internet about matching guitar amplifiers to cabinets. This is partly down to confusion between the different operation of tube amps and solid-state amps.

As well as this, you have to consider that matching amplifiers to cabinets is different for guitars than for sound engineers matching amps and cabinets in a studio or live sound environment.

In a guitar amp, you want to make the system foolproof. Tube amps sound better cranked, because they add those 'musical' even-order harmonics that we learnt about in Chapter 7: Demystifying Recording Levels. Therefore, if a guitar amp goes to 11, guitarists will play it at 11. So always make sure that when selecting a guitar amp and speaker combination, it can be abused without damaging either the amp or the speakers.

When solid-state technology was developed, manufacturers were left with a problem. Solid-state distortion sucks: it sounds horrible due to its harmonic content, and it starts and stops very abruptly and contains odd-order harmonics – meaning chords, in particular, would rapidly dip in and out of extreme and unpleasant distortion. Because of this, solid-state amps are prevented, through design, from passing the point of distortion. This presents a problem to manufacturers: distortion is an integral part of electric guitar tone. To overcome this, manufacturers implement additional circuitry to mimic the distortion of tube amps. Some sound good; some sound bad. It's rarely as satisfying as true tube distortion.

Let's look at the interval values of the harmonic series again, to see why the even-order harmonics of tube distortion sound better than the odd-order harmonics of solid-state distortion:

Even-Order Harmonic Intervals	Odd-Order Harmonic Intervals
2 – Octave	3 – Fifth
4 – Octave	5 – Third
6 – Fifth	7 – Minor Seventh
8 – Octave	9 – Second
10 – Major Third	11 – Tritone
12 – Fifth	13 – Minor Sixth
14 – Minor Seventh	15 – Major Seventh

As you can see, even-order harmonics contain more closely related pitches of the fundamental note, especially lower down the harmonic order; therefore, our ears perceive this as less dissonant than odd-order harmonics. This is one of the reasons that far more distortion can be introduced on tube amps before it becomes problematic than can be introduced on solid-state amps.

The difference between solid-state distortion and tube distortion is also at the root of what is often a puzzling observation: a tube amp will sound considerably louder than a solid-state amp of the same power rating. The answer to this is very simple: whether the two amps are set to 5, 8 or 10, the tube amp will be delivering more power to the speaker. The key to understanding this is the fact that power is rated at the point at which the circuit starts to distort. A 100 W solid-state amp will deliver 100 watts of power before distorting, and so, because we don't want solid-state amps to enter distortion, this is the maximum it will deliver; dial it up to 10 and you're getting 100 watts. Of course, it will usually sound distorted; but remember, this is distortion created by the additional internal circuitry. On the other hand, tube distortion sounds very pleasing, so manufacturers and guitarists want to push a tube amp into this range. A 100 W tube amp will deliver 100 watts before distorting; that's the definition of the power rating. Distortion on a tube amp may start to kick in at around 5, but when you crank a 100 W tube amp up to 10 it will be pushing considerably more than 100 watts. Tube amps are designed to be pushed past their power rating. This difference in maximum output is reflected across the whole volume range. At 5, a 100 W solid-state amp will output 50 watts; at 5, a 100 W tube amp will deliver half its maximum output, which will be more than 50 watts.

Typically, a ratio of 4:1 (solid-state to tube) is a useful guide for determining how much power is needed to volume-match. For example, a 100 W solid-state amp will produce a similar perceived volume to a 25 W tube amp.

In a PA or studio monitoring system, your aim is completely different. Any distortion in the chain should be there as a creative effect and should be controllable by the engineer. Distortion or colouring from driving power amplifiers is detrimental to making correct mix decisions; therefore, solid-state amplifiers are almost exclusively used for these purposes (their smaller frames and lighter weight helps too). This means that in these systems, amplifiers are over-spec'd and run well away from their clipping point.

So let's get into discussing how to match guitar amplifiers with cabinets. When dealing with matching amps with cabinets, we need to be aware of both the impedance (ohms) and power (wattage) of the equipment we are using. With the examples that follow, I have assumed that wattages and impedances are the same for each speaker being used. This is the case with the vast majority of guitar speaker cabinets. If you want to build your own cabinet with different speaker ratings, then please research further before trying anything out.

Matching Wattage

The first thing we need to discuss is the fact that the wattage rating on an amplifier is calculated by its clean power. When distortion is introduced, through pedals, pre-amp or power amp stages, it results in a higher power output. A distorted guitar tone has the potential to go well over the wattage rating of the amplifier. When matching amplifiers and cabinets, we need to take this boost in wattage into account. This means that we need to make sure that the cabinet can handle at least 1.5 times the power rating of the amplifier, or 2 times to be absolutely safe. If we had a 100 W Marshall head, we would want a speaker cabinet that could handle at least 150 W, although 200 W would be optimal.

If the system has more than one speaker, the power will be spread out evenly between the speakers, so you need to make sure that *total* power of all the speakers is higher than the power output of the amp. A pre-bought guitar cabinet will specify the total power rating for the speakers.

When you connect an amplifier to a cabinet, you need to be sure that you have the correct figure for the amp's power output. This figure should be located near the quarter-inch jack outputs of the amplifier. If you find a wattage figure near the input power lead, this is most likely the AC power consumption of the amplifier and is not the amp's output wattage.

Even if you are looking in the right place, sometimes your amp or speaker will have several output wattages. These could include peak, RMS, continuous and/or program. Based on what you have already learnt from this course, you can probably make an educated guess to what some of these mean; but, just in case, here is a quick breakdown:

Continuous/RMS power. This is the amount of power the amp can deliver, without distortion, using a continuous test tone; just as with other types of calibration, this is usually a sine wave at 1 kHz. This is the figure you need to take into account when you want to match an amp with your speakers.

Peak power. This is the maximum transient power. In a speaker, this is effectively the amount of power needed to give the speaker cone its full excursion (its most forward position). Peak power looks impressive on paper, but it is not really relevant when matching cabs to amps.

Program power. This is the amount of power delivered when the input signal varies in both frequency and voltage (amplitude and time). In practical terms, this is its power rating under more normal music operation.

What you really need to remember is when you are matching amps to speakers, deal with RMS watts, not peak or program.

Matching Impedance

So far, so good, but when we start considering impedance it gets a lot more complicated. Let's quickly explain what is happening between the amplifier and the cab. It can be summed up in one sentence: the amplifier pushes electrical current through a speaker or set of speakers. This system disperses the electrical energy coming from the amplifier, converting it into physical motion by moving the speaker. Just like any component, a speaker or set of speakers will have a certain amount of resistance to the flow of electricity. When matching impedances between an amplifier and cabinet, we want to meet two conditions:

1. Your amp is driving into an optimal resistance (ohmage). We need to make sure that you don't use too low an ohmage on the amp compared to the resistance rating of the cabinet. This would mean the circuit is not absorbing enough electrical current and the amp would blow because too much current is now flowing through it.
2. Your speakers have optimal power. When resistance changes, so does the amount of power in a circuit. If you use too high of an ohmage, you are restricting the flow of electricity too much and your speaker produces less sound. However, better this way than breaking the amp!

Basically, we are looking for an optimal power transfer between the amp and speakers; or, put simply, we are searching for a 'sweet spot'.

Remember that as ohmage changes, so too does the amount of power in the circuit; so the total speaker wattage needs to be able to handle the power being given to it. *This means that your safe wattage level of your speakers is directly related to your choice of ohmage.*

Due to circuitry requirements, tube amps have an impedance switch on the back of the amp, which you need to match to the total speaker impedance of the cabinet.

Solid-state amps don't need such a switch; therefore, you are likely to see either a minimum load warning such as 'Minimum load 4 ohms' or a power output listed at its minimum load such as '200 watts at 4 ohms (min)'.

The bottom line is this:

- *If you are using a tube amp, you must set the amp's impedance switch to match the total speaker impedance. This is because a mismatching impedance either side of its optimal value can cause damage to the amp.*
- *If you are using a solid-state amp, make sure that the total speaker impedance is no less than the minimum rating stated on the amp.*

Let's take this theory and use a real-world example of a cabinet and match an amplifier to it.

I have a 4 × 12 guitar cabinet with two inputs, and it can run at 200 W/16 Ω (mono) or 2 × 100 W/8 Ω (stereo). Mono mode would give all four speakers the same signal; in stereo, each input would go to a different pair of speakers.

With this cab, we would need an amplifier that can operate at 16 Ω if connected in mono, or 8 Ω in stereo.

If I chose a tube amp, I would look for one that is switchable between 8 and 16 ohms to make sure that I didn't overload the amp, and I'd make sure that it was around 100 W to give the headroom that will allow the amp to be cranked without breaking the speakers.

If I chose a solid-state amp, and was planning to use this in mono mode, I would be looking for one that could deliver no more than 100 watts at 16 ohms.

In live scenarios, guitarists often power two speaker cabinets from the same head. Provided the impedance is the same on each speaker, the combined speaker impedance value is half that of each individual speaker. This isn't really as applicable in the studio as we are only really searching for tone, not volume.

The calculation of impedance between speakers is more difficult than wattage, and unlike wattage, it changes depending on how the speakers are wired. Luckily, the total impedance is usually listed on the cabinet. If you are wiring your own, in order to calculate impedance, you need to take into account the way that they are wired.

Speaker Wiring

If you have one speaker, the calculation is easy: the output impedance of the amplifier should be equal to the impedance of the speaker.

So with an 8 Ω speaker, you would need an amp with an 8 Ω output.

If you have more than one speaker wired up, there are a few different ways to wire them. Please note that for the sake of understanding the calculations the 50% safety buffer has been omitted from the following figures.

Series. This is where each speaker is daisy-chained into the next one. By summing the speaker impedances, you get the value at which you should set the amp's output impedance.

In the example above, the amp is set to run at 8 Ω – the sum of the two 4 Ω speakers.

Parallel. Parallel wiring is found in two-speaker set-ups. Each of the two speakers receives signal directly from the amplifier. Providing that they are of equal impedance, dividing impedance by 2 (the number of speakers) calculates the correct amp impedance setting.

In the example above, the speaker's impedance (16 Ω) divided by the number of speakers (2) gives you an overall impedance setting of 8 Ω.

Series/Parallel. Multiple speakers (typically four or eight) can be wired in both series and parallel. For instance, in a standard 4 × 12 guitar cabinet, the speakers

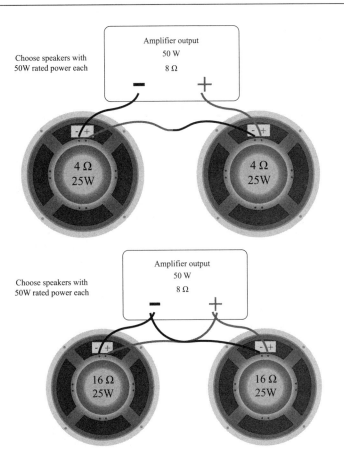

Figure 12.17
Series speaker wiring[2]

Figure 12.18
Parallel speaker wiring[3]

are split into two sets of two, wired in series, and then both sets are connected to the amplifier in parallel. With this wiring, the overall impedance remains the same as one speaker's impedance (see Figure 12.19).

In the example above, you need to do the calculation in two stages:

1. Add the speakers in series together – two speakers at 8 Ω equals 16 Ω.
2. Now divide this number by the number of sets of speakers (2).

When working this out, I always imagine each set of speakers to be one speaker, so in this case the two 8 Ω speakers with a 12.5 W power rating is like having one 16 Ω with a 25 W power rating.

This means that you need to run the amp at 8 Ω.

So, to summarize, when dealing with impedance with more than one speaker you need to remember that:

1. If you have more than one speaker in series, the impedance in the circuit will *increase*.
2. If you have more than one speaker in parallel, it will *decrease*.
3. In series/parallel, the impedance will *remain the same*.

Figure 12.19
Series/parallel wiring[4]

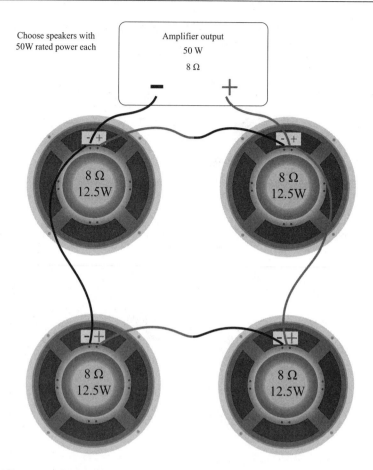

Tonal Differences between Wirings
While the tonal differences are definitely more negligible than other factors, the way that the speakers are wired does impart slight tonal differences. This is something that ends up being quite subjective and will depend heavily on the type of amplifier driving the speaker combination. Here is what I think based on some rather limited experimentation:

1. Series: Sounds grittier with gain; slightly less low end.
2. Parallel: A little more low end; more present.
3. Series/parallel: Seems to me like a middle ground between the two.

12.2.5 The Cabinet (Enclosure)

As much as the speaker is a big factor in guitar tone, it needs to be matched with the right enclosure. An unsuitable enclosure can easily ruin the sound of a good speaker and amp combination. While I don't expect you to be making your own amplifiers or cabinets, some of these specifications can impart changes in tonality that an engineer should be aware of.

There are many factors relating to the enclosure that affect the tone produced by the interaction of the speaker with the cabinet. These include the material, size, thickness of the enclosure, how the speakers are mounted, and whether the cabinet has an open or closed back.

Cabinet Size

Probably the most fundamental factor about the cabinet's size is that it determines the lowest frequency that can be reproduced. If the cabinet is too small, the tone sounds 'thin'; too large and it sounds 'boomy'. This is why bass cabinets are usually larger than guitar cabinets.

Open or Closed Back?

Guitar enclosures can be either closed-backed or open-backed. An open-back cabinet has the rear of the speaker(s) exposed to air, whereas in a closed-back design the whole rear of the speaker(s) is enclosed.

Each type of design has its merits. The open-back sound is more complex (especially in the very top end of the speaker's response), is less directional and is 'looser' in the bass end. Closed-back cabinets are more directional; they have a more controlled bass end and are typically punchier but less 'airy'.

Open-back cabinets lend themselves well to a lot of country, pop, light rock and less rock-infused blues. Closed-back cabinets lend themselves well to rock and metal.

Figure 12.20 Fender amp with open-back design (picture courtesy of Metropolis)

The type of enclosure often affects the choice of speaker. A fine illustration would be Celestion's two most revered products: the Greenback and the Alnico Blue. While the Greenback is one of the most desirable guitar speakers to be used in closed-back cabinets, it often sounds 'trashy' in open-back cabinets. The opposite can be said of the Alnico Blue: the top end sounds more complex and 'airy' when used with open-back cabinets, but its sound is 'constricted' in closed-back cabinets.

Cabinet Material

As with guitars, amplifiers and speakers, cabinet manufacturing has evolved from early designs to maximize results from lower-cost materials. Despite these advances and the bang-for-your-buck they can deliver, there is no substitute for high-quality materials. Because of this, older models, for which mass-market cost was not a design criterion, are widely revered and sought after.

First, it is common for cabinets to be made out of some sort of plywood. Plywood is not a type of wood: it is a manufacturing process whereby thin sheets of wood are layered with each new layer's grain being glued to the last at a 90-degree angle. This means that the wood is strong and less prone to cracking and shrinking, as well as being inexpensive and flexible. The layers, called veneers, can be found made out of many woods but in guitar cabs, pine and birch are popular.

With smaller open-back cabinets and small speakers, the choice of wood is less important, as there is less surface area to vibrate (and therefore less tonal difference). This is why lightweight, cheap and more sonically neutral woods (pine, MDF (medium-density fibreboard)) are chosen in these applications. In bigger, higher-powered and closed-back designs, the wood becomes more important, and 11- and 13-ply birch is often found.

Vintage Fender cabinets were made out of finger-jointed pine, and Marshall used 11-ply Baltic birch. These days, most budget cabinets are made from MDF, and high-end cabinets are types of Birch composite.

Speaker Mounting

The mounting of the speaker to the enclosure is called a *baffle*. The material it is made out of and the way it is attached to the cabinet have a profound effect on the tone. Thinner baffles vibrate sympathetically with the speaker, producing a louder and more coloured tone, usually with a greater bass response. Thicker baffles have less of an effect on the speaker's tone. To further colour the tone, thinner baffles tend to be attached to two sides of the enclosure: top-to-bottom, or side-to-side. This allows the baffle and the enclosure to vibrate more and thus alter the tone more drastically; this type of baffle is called a floating baffle. To achieve the required strength without adding excessive weight, speaker baffles are almost always made out of plywood. The baffles in vintage designs from Fender and Marshall used pine-ply and birch-ply, respectively, just like the enclosures themselves. In closed-back cabinets, a centralized piece of wood attaches the baffle to the back of the enclosure, which ensures that the baffle and enclosure resonate in phase with each other.

Figure 12.21 TC Electronic Spark Boost

12.3 ELECTRIC GUITAR EFFECTS

While there are many types of guitar effects pedals, I am going to concentrate on some of the effect types you may be familiar with. With some experience in production, the operation of EQ, compression, reverb and delay pedals should be obvious. Therefore, I want to focus quickly on distortion before moving onto less well-understood modulation devices such as wah-wah, flange, chorus and phasers, as well as pitch-shifting and harmonizing effects.

12.3.1 Clean Boost

The purpose of a clean boost pedal is to take the signal going into it and boost it, but not enough to clip the signal. A clean boost, therefore, imparts no tonal changes to the signal other than amplitude.

Clean boosts are used to improve the intelligibility of subtle playing, or used on more extreme settings as a way of overdriving an amplifier's input stage (usually a tube amp) so that it starts to distort. You should not confuse a clean boost with an overdrive pedal.

12.3.2 Distortion Pedals

I am sure we all know what distortion sounds like. There are a few different types of distortion pedal, but they all share the same type of operation. In simplistic terms, they amplify a signal to the point of internal clipping. The amount of amplification the circuit adds affects the amount of clipping, which can shape the sound in rather different ways. This is why there are a few different categories of distortion:

Overdrive. An overdrive pedal is the least extreme of the three main kinds of distortion pedal and is usually a soft-clipping circuit that 'warms up' the tone. The effect is fairly similar to the way in which a tube amp starts to distort when overdriven, hence the name overdrive. It can also be used with a tube amp to drive it even further than a clean boost can. An example of a classic overdrive pedal is an Ibanez Tube Screamer.

Distortion. A distortion pedal is more of a hard-clipping circuit and is more extreme than an overdrive pedal. You will notice that a distortion pedal will sound a lot more edgy and compressed than an overdrive pedal, and will impart much more distorted tones on subtler settings than an overdrive. An example of a distortion pedal is a Pro Co RAT.

Due to their similar nature, extreme settings on overdrive pedals and subtle settings on distortion pedals can sometimes sound very similar. A good rule of thumb to remember is:

- *If you want to drive the amp a little harder to produce some distortion, try a clean boost pedal.*
- *If you want to drive the amp to produce distortion with a little extra colour, try an overdrive pedal.*
- *If you want to significantly distort the signal before even hitting the amp, try a distortion pedal.*

Figure 12.22 Electro-Harmonix Big Muff with tone wicker

Fuzz. A fuzz pedal is a special kind of distortion. It is an extreme hard-clipping circuit, to the point where the signal is pretty much 'brick-walled' and becomes nearly a square wave. This means it has a very distinct sound which, well, sounds 'fuzzy'. Used in the right way, it is extremely effective, but can quickly become 'mushy'. An example of a fuzz pedal is the Electro-Harmonix Big Muff.

12.3.3 Wah-Wah

Wah-wah, or simply wah, is a staple electric guitar effect in the genres of blues rock, hard rock and funk. The wah pedal uses a band pass filter to mimic the 'aah' and 'oooew' sound of the human voice, and the frequency at which it boosts can be manipulated with the foot pedal. With the pedal fully up (heel to the floor) the frequency boost is applied in the bottom end and sounds like an 'oooew', and with the pedal fully down (toe to the floor) it applies the filter to the top end, which sounds like the 'aah'. By rocking the pedal, you can create unique and expressive effects that cannot be created with the guitar alone.

In blues rock, the pedal is used with a crunchy tone. This can be heard on 'Voodoo Child (Slight Return)' by Jimi Hendrix and 'The White Room' by Cream.

With hard rock, the wah-wah pedal is used with a high-gain distorted tone like in Metallica's 'Enter Sandman' solo and Guns N' Roses' 'Sweet Child O' Mine'. It has been used in combination with detuning and the tremolo bar and Floyd Rose system. Virtuoso guitarists Steve Vai, Joe Satriani and Eddie Van Halen epitomize this use of the wah pedal, and it can be heard on the track 'Bad Horsie' by Steve Vai.

In funk/disco, the wah-wah is often used with a clean or slightly driven tone. Examples of this are 'Night Fever' by The Bee Gees, 'If You Have to Ask' by Red Hot Chili Peppers and 'Get Down on It' by Kool and the Gang.

As well as these uses, leaving the wah on in one position (usually halfway) can act as a type of guitar tone control, and this effect was used a lot by Jimmy Page (Led Zeppelin) in songs such as 'A Whole Lotta Love' and 'Black Dog'. It can also be heard in 'Ziggy Stardust' by David Bowie.

12.3.4 Modulation Effects

There are many different types of modulation effects. So what is the definition of a modulation effect? A modulation effect is a processor that splits the original signal and duplicates it or mimics it with a particular alteration to the duplicated signal. In fact, to modulate means to *adjust* or *adapt*. Some of these modulation effects can sound similar to one other on subtle settings or very different on extreme settings.

Technically, some delays and echo effects with tonal shaping or modulation features could be classed by the definition above as a modulation. However, the important distinction between delays with shaping and true modulation effects is that a modulation effect does not sound to the ear like distinct repetitions of the original source; it is simply a blend of the original and duplicated signals to make one altered sound.

Flanger

Flanging is an effect created by taking the original sound source, duplicating it and then delaying the duplicated sound by a gradually changing amount. This creates phase cancellation between the two signals, which changes frequencies based on its current delay time. Typically, these delays are all less than 20 milliseconds. The threshold for human hearing to distinguish a delayed signal as two distinct sounds is approximately 30 milliseconds.

On a flange pedal, the delay variation can be controlled and adjusted. The delay time applied to the duplicate signal cycles repeatedly between zero delay and a maximum delay time, set by the depth control. The higher the depth control is set, the further down the frequency spectrum the comb filtering extends during each cycle. The amount of time taken to complete each cycle is set by the rate control, where higher rates produce more cycles per second.

The result of a flange effect is a 'swooshing' underwater sound, as the phase cancellation sweeps across the frequency spectrum. Faster rates create a stronger-sounding effect but slower rates sound more like a sweep.

On extreme settings, flange is a very noticeable and extreme type of modulation effect, and is often used on clean guitar and also when an effect needs to be prominent. A good example of flange effects can be found on '90s rock/alt/grunge records such as 'Are You Gonna Go My Way' by Lenny Kravitz, where, in the breakdown, it is used on guitar and drums, and 'Breakout' by Foo Fighters, on the intro guitar. An example of a flanger on a cleaner sound is 'Purple Rain' by Prince, in combination with a chorus effect.

Phaser

A phaser or phase shifter works in a very similar way to the flanger but instead of the the phase cancellation coming from the modulation of the duplicate signals delay, in a phase pedal the delay stays static and the signal is modulated by rotating the duplicate signal's phase. This creates a much more subtle comb-filtering effect than flanging and although it still sounds like a 'swoosh', it is not so dramatic. Much like the flanger's rate knob, phaser pedals often have speed knobs to create a stronger effect or lower sweep. The phaser is a particular favourite of many heavy metal and virtuoso guitarists.

Notable examples of phasing used in tracks include 'Little Wing' by Jimi Hendrix, 'Soma' by Smashing Pumpkins and 'Eruption' by Van Halen.

Figure 12.23 MXR Phase 90

Chorus

While flangers and phasers are modulated to introduce obvious comb-filtering effects, chorus pedals use modulation to create thickness and mimic the natural timing and pitch variations of an ensemble. The slight discrepancies in pitch and timing are what make ensembles such as choirs and string ensembles sound so 'rich'. A chorus pedal may duplicate the original signal several times with slight variations in pitch and time. Typically, chorus pedals use a slightly longer delay than flangers to avoid overly dramatic comb-filtering artefacts. Chorused tones are 'thick' and 'rich', but tend to lack attack and punch compared to the dry signal. Pop and rock productions in the 1970s and 1980s used chorus extensively. Notable artists include The Police, Prince, The Smiths and Bryan Adams.

Notable examples of chorus effects are 'Message in a Bottle' by The Police, 'Smells Like Teen Spirit' by Nirvana (verse) and 'This Charming Man' by The Smiths.

12.3.5 Tremolo

Tremolo was one of the first electric guitar effects along with vibrato (which it is often wrongly confused with or used interchangeably). It works by quickly modulating between full volume and nearly off; this makes the guitar sound like it is 'pulsing'.

As well as tremolo in pedals, some vintage amps have built-in tremolo.

In fact, the misnomer of confusing tremolo with vibrato most likely occurred because Fender amps labelled their amp's tremolo effect as vibrato. Modern tremolo effects can be achieved easily through tempo-synced gates in DAWs.

Some notable uses of tremolo include 'The Bends' by Radiohead, 'How Soon Is Now' by The Smiths and 'Boulevard of Broken Dreams' by Green Day.

12.3.6 Vibrato

Comparing vibrato back to back with tremolo, it is easy to see how the two are often confused. While tremolo is the oscillation of a signal's amplitude, vibrato oscillates quickly between pitches slightly sharp and flat from the original source. This creates a sort of 'warbling' effect. Because these states are often programmed to happen very quickly, the differences between tremolo and vibrato are more subtle.

Many modern players tend to choose tremolo over vibrato, and therefore notable uses of a vibrato pedal are more limited: 'Crimson and Clover' by Prince and 'Did I Let You Know' by Red Hot Chilli Peppers.

12.3.7 Pitch Shifter, Harmonizer and Octaver

A pitch shifter creates a second signal by raising or lowering the pitch of the input signal by an interval in tones or semitones as set by the user to create a two-tone harmony. However, there is no algorithm to determine the scale of the harmony created. Therefore, if the user sets the interval to a major 3rd, playing a note of A will result in a second signal playing a note of C♯ (the major 3rd in the A major scale). However, playing a B will create a second note of D♯, which is out of key for the scale of A major, and is therefore not a 'true' two-voice harmony for a major 3rd interval. This type of effect will therefore sound very artificial but can be used to create 'spacey' special effects.

A harmonizer, however, allows the key and scale of the song to be set, and the duplicate signals can therefore be automatically tuned to stay within the notes of that scale. In this case, if the scale was set to the key of A major, playing a B note would instead create a second note of D, keeping the harmony in key and replicating a 'true' two-voice harmony in the key of A major.

Most modern pitch shifter pedals offer both some pitch shifting and harmonizer functions. On many, two or more voices can be added to create multi-voice harmonies. Harmonizer pedals are more commonly used in live performance than in the studio because the algorithms are prone to errors, plus the fact that a recording of a real harmony part will always sound more real.

However, there is one particular type of pitch shifter that is often used in recordings: the octaver pedal. An octaver pedal duplicates the source signal and transposes it up or down an octave. This can be done with a normal pitch shifter, but guitarists favour specialized octavers for their highly accurate tracking. An octave pedal used in conjunction with fuzz or heavy distortion can produce really heavy riffs. The signature guitar tone of The White Stripes often incorporated an octaver, but the effect has found favour with many heavier rock bands.

Some notable songs with octaver pedal usage include 'Seven Nation Army' by The White Stripes, 'Little Sister' by Queens of the Stone Age, 'Sledgehammer' by Peter Gabriel (bass part) and 'Know Your Enemy' by Rage Against the Machine.

Further Resources www.audioproductiontips.com/source/resources

 Guitar effect pedal examples and comparison. (89)

12.4 SOME GOOD TONAL STARTING POINTS

We've learnt that great guitar tone is very subjective and genre-specific. While I advocate trying out as many different combinations as possible, it is often difficult to know where to start – so here is a selection of gear chosen by some of the most famous guitarists in their genre.[5]

12.4.1 Jimi Hendrix (Vintage/Blues/Rock)

Jimi Hendrix is famed for his use of a Fender Stratocaster. However, what many people don't know is that his preferred amplifier was a Marshall. His tone was a hybrid between the edginess of the single-coiled Strat with the mid-heavy Marshall amp and the Celestion Greenbacks inside the Marshall 4×12. Hendrix was also a heavy user of fuzz pedals, which created a very squashed tone. He was also heavy-handed in his use of effects such as the psychedelic Uni-Vibe, and the Vox wah pedal. The Uni-Vibe is a foot pedal-operated effects unit that adds phasing, chorus and vibrato to the signal in an attempt to emulate the sound of a Leslie speaker in an organ. Despite its failure to truly sound like a Leslie, it became a popular effect in its own right. This Fender guitar and Marshall amp selection is also used by John Frusciante in the 'Blood Sugar Sex Magik' era Red Hot Chili Peppers, proving a great combination for blues-infused funk rock.

Key Guitars	Amps/Speakers	Key Pedals
Fender Stratocaster	Marshall Super Lead	Fuzz Face
	Marshall 4×12	Uni-Vibe
		Vox (wah)

To get this tone out of modern equipment, I'd use a USA-made Strat, with a Marshall Super Lead head and a Marshall 1960A 4×12 cab. For effects, I'd suggest going with a Dunlop Fuzz Face, Vox wah and Roger Mayer Voodoo Vibe.

12.4.2 John Mayer (Modern Blues/Rock)

John Mayer is known for his smooth yet bluesy tone. The edginess to his tone is provided from the single-coil pickups of his primary guitar, which is a Fender Stratocaster just like Jimi Hendrix. However, he prefers to go with less gritty and less mid-heavy amplifiers such as classic Fenders. I might have expected him to go for a Jensen type of speaker. However, his guitar tech states he favours a Celestion Alnico Blue. Much of the distortion is added through vintage/classic pedals rather than multi-gain-staged amplifiers. He also relies quite heavily on classic reverbs and delays.

Key Guitars	Amps	Key Pedals
Fender Stratocaster	Fender Bandmaster	Marshall Blues Breaker (overdrive)
	Two Rock – John Mayer Signature	Uni-Vibe
	Dumble – Steel String Singer	Vox (wah)
		Way Huge Aqua-Puss (analogue delay)

If I was trying to achieve John Mayer's tone out of readily available modern equipment, I would plug a USA-made Fender Stratocaster into a Fender open-back valve amp, preferably of the Blackface variety, and use a vintage-sounding overdrive pedal before the amp. As the original black-cased Marshall Blues Breaker is no longer made and is very sought after, an alternative you might want to try is the Fulltone FullDrive-2 Mosfet. Then to make the tone a little more interesting, I'd add an analogue delay such as a Strymon El Capistan, a Vox wah and a Uni-Vibe.

12.4.3 David Gilmour – Pink Floyd (Blues/Progressive/Pop)

David Gilmour's tone is very unique considering he plays through a lot of the same sort of equipment as Hendrix and John Mayer, but with his choice of pedals and settings he manages to get a lot of sustain to single-note licks and also a lot of depth. As mentioned earlier, a lot of this is down to his style and consistency, but also a compression pedal set to really aid sustain. To add to this, he uses a lot of rhythmic delays (particularly dotted eight notes) from a delay pedal, with the feedback control turned up to add further depth and sustain. He is another guitarist that likes to use a lot effects such as a flanger and Uni-Vibe. You can hear these effects on tracks such as 'Breathe'. His prominent use of delay became one of his signatures, and The Edge (of U2) took this to a whole new level.

Key Guitars	Amps/Speakers	Key Pedals
Fender Stratocaster	Fender '56 Tweed	MXR Dyna-Comp
	Hi-Watt DR 103 (head)	Uni-Vibe
	4×12 with Fane Crescendo speakers	Vox (wah)
		Electro-Harmonix Electric Mistress
		Chandler Tube Driver (distortion)

Out of readily available modern equipment, I'd go for a USA Strat, with a Fender Tweed amp, MXR Dyna Comp and a delay pedal with a tap tempo/programmable tempo such as the TC Electronic Nova Delay. Then add modulation effects to suit the type of song.

12.4.4 Jonny Marr – The Smiths (Pop Rock/Shoe Gaze)

One of the defining images of Johnny Marr of The Smiths is the Rickenbacker 330 black-and-white guitar. However, Johnny Marr is known for his diverse range of 'jangly' guitars, and a lot of The Smiths songs were actually recorded with a Fender Telecaster, including 'This Charming Man', in which the 330 is shown in the video. In later years, Johnny Marr is known to have preferred the Fender Jaguar. In terms of amplifiers, he has always preferred Fender amps and has used many models. Like Andy Summers of The Police, he is also known for his strong use of chorus and his almost exclusive use of Boss effects.

Key Guitars	Amps/Speakers	Key Pedals
Rickenbacker 330	Fender Twin Reverb	Boss CE2 (chorus)
Fender Jaguar	Fender Deluxe Reverb	Boss OD2 (distortion)
Fender Telecaster	Fender Bassman	Boss TW1 (touch wah)

Recreating Johnny Marr's tone is fairly easy, and if I was to recreate the classic Smiths tone I'd use a USA Fender Telecaster with a Boss CE-2 Chorus into a Fender Twin Reverb.

12.4.5 Jonny Greenwood – Radiohead (Alternative Rock/Indie/Experimental)

Before moving away from Fender players, let's discuss the tone of another Telecaster player. Jonny Greenwood from Radiohead is famed for his jagged and edgy tone with many creative effects. One of the most important details of his set-up is that his clean tone and distortion are actually from two different amps, the clean being a Vox AC30 and the distortion being a Fender Deluxe 85, which, unusually, is a solid-state amp.

Key Guitars	Amps/Speakers	Key Pedals
Fender Telecaster	Vox AC30	Marshall Shred Master Boss Super Overdrive (distortions)
	Fender Deluxe '85	Digitech Whammy
		Roland Space Echo
		Electro-Harmonix Small Stone
		DOD Envelope Filter

To try to recreate a Jonny Greenwood tone, it is important to play like Jonny Greenwood. He plays very aggressively and creates a lot of attack on the strings, which makes an already 'edgy' Telecaster even edgier. He has a thin, high-gain overdrive compared to most guitarists due to his choice of pedals and solid-state amp. Another big feature of his playing is the use of delay to loop phrases while tweaking modulation effects to create synth-like sounds without having to play his guitar. This is probably because Radiohead have two other guitarists in the band, and in passages he often uses the guitar to create 'soundscapes' rather than using a guitar in a more traditional rhythmic way.

12.4.6 Zakk Wylde – Ozzy Osbourne/Black Label Society (Hard Rock/Classic Rock)

Moving on from the bluesy and edgy Fender guitarists, we come to the pretty much universal sound of classic rock as typified by Zakk Wylde of Ozzy Osborne fame. Zakk's set-up is quite simple: a supercharged Les Paul with higher-gain EMG 81 pickups paired with Marshall JCM800 and Marshall 4×12. As well as this, he also uses drop tunings such as E♭ and drop-D. The effects chain is also very simple: a Dunlop Cry Baby wah, Boss CH1 chorus and a Boss SD1 super overdrive; it creates a lot of punch and warm sustain.

Key Guitars	Amps/Speakers	Key Pedals
Gibson Les Paul	Marshall JCM 800	Boss CH1 (chorus)
	Marshall 4×12 w Greenbacks	Boss SD1 (overdrive)

12.4.7 Jack White – The White Stripes and The Raconteurs (Alternative Rock/Lo-Fi/Garage Rock)

As mentioned earlier, a lot of Jack White's tone is controlled by the combination of an octaver pedal such as the Electro-Harmonix POG or Digitech Whammy, and a fuzz pedal like an Electro Harmonix Big Muff. Solos are usually played with the octaver set at two octaves above. Heavy rhythms, like 'Blue Orchid' and 'Seven Nation Army', use a one-down sub-octave. These octave tricks can be further enhanced by drop tunings. This makes a valve amp break up and creates a heavily distorted bottom end and mid range. When dealing with such high gain it is also often advisable to gate the signal so that it doesn't constantly feedback. Amps-wise, Jack White uses a rare Silvertone amp (which is almost impossible to find) and a Fender Twin Reverb. He runs these amps simultaneously, with the Silvertone providing the crunch and the Twin providing some presence and depth with its reverb.

Key Guitars	Amps/Speakers	Key Pedals
Gretsch Triple Jet	Silvertone	MXR Micro Amp (gain boost)
	Fender Twin Reverb	Electro-Harmonix POG (octaver)
		Digitech Whammy (various pitch-shifting effects)
		Electro-Harmonix Big Muff (fuzz)

To recreate this tone, place the micro-amp first in the chain followed by the octaver/pitch shifter, then the Big Muff. In place of the Silvertone (assuming you aren't very rich or very lucky), use something crunchy such as an Orange Crush 30R and use a Fender Twin simultaneously. I'd recommend using a guitar with humbucking pickups, especially if it is hollow-bodied like a Gretsch or Gibson ES-3312.

12.4.8 Dimebag Darrell – Pantera (Metal)

A suitable guitar for playing heavier styles will have an enhanced bass response, while the whole neck needs to be easily accessible with a shape that aids fast playing. This is why metal guitars often combine Gibson-style pickups with a Strat-style neck. Unlike pop and blues, the tone of metal guitar usually comes from the amplifier rather than pedals. Metal guitarists typically go for an amp model with three or four gain stages. What was unusual about Dimebag Darrell's rig was his solid-state amp head.

Key Guitars	Amps/Speakers	Key Pedals
Dean ML	Randall Century 200	MXR Flanger/Doubler
	Randall Jaguar Cab Celestion with speakers	Rocktron Guitar Silencer (noise suppressor)
		Furman PQ4 (parametric EQ)

12.4.9 Wes Borland – Limp Bizkit (Nu Metal/Modern High-Gain)

Just like metal guitarists, nu metal gear was tailored towards fast fretting and a fat humbucking sound through high-gain amps. Wes Borland is known for using Ibanez guitars through Mesa/Boogie Dual Rectifiers in tandem with Orange amps and cabs. More recently, he has had signature models and endorsements with Jackson and Yamaha. The high gain and scooped sound of nu metal relies heavily on these Superstrat-style guitars and three- or four-gain-stage amps. Like many nu metal guitarists, to further deepen the sound, he often uses C# tunings. In terms of pedals, Borland prefers Boss effects and uses a noise suppressor to reduce unwanted noise at high gain, as well as a graphic EQ to further control tone. He also uses a phaser, plus a reverb and delay pedal.

Key Guitars	Amps/Speakers	Key Pedals
Ibanez AXR 7-string	Mesa/Boogie Dual Rectifier	Boss GE7 (graphic EQ)
	Orange Thunderverb	Boss NS2 (noise suppressor)
	Orange 4×12s	Boss PH3 (phaser)
		Boss RV5 (reverb and delay)

EXERCISES

1. Go to a guitar shop (with a guitarist if you don't play) and try out as many guitar and amp combinations as possible to learn the differences.
2. Experiment and memorize tonal differences between pickups on guitars.
3. Try a Strat, Tele, Les Paul and SG in as many amplifiers as possible.
4. Audition each different type of effects pedal.
5. Listen to some of your favourite records and try to work out what effects, amplifiers and guitars were used to make it.
6. Record some cover songs and get the guitar tones as close to the original tracks as possible.

12.5 INSIDER INFORMATION

Before we move on to how we consider microphone selection and placement, let's discuss a few trade secrets that will help with tone selection.

12.5.1 Tune Regularly/Every Take?

A lot of the most well-known engineer-producers I've watched ask the guitar player to tune at every given opportunity. Tuning every single take might be a bit over the top, but I've certainly seen it happen. While most listeners won't hear a guitar string that is out of tune by a couple of cents, they will subconsciously sense a cohesion between instruments when the tuning is bang-on.

12.5.2 Re-Amping/Amp Modellers

With the popularization of the DAW, many new techniques related to guitar recording have emerged. Potentially the most revolutionary of these is the ability to record a clean DI guitar signal directly into your DAW, and then either send it to an amplifier (which is termed *re-amping*) or process this clean signal through an amp simulator plug-in within your DAW. Over the past few years, many amp simulators have come on to the market, ranging from basic amp styles to processor-draining plug-ins that emulate many types of amp, effects and microphone set-ups.

Figure 12.24 IK Multimedia AmpliTube guitar amp and effects modelling plug-in

There are many advantages to taking a clean DI signal:

- Using clean DI signals (and amp simulators when recording to give the guitarist a suitable tone in his or her headphones), and then re-amping later, is particularly great in small studios when you want to retain a more live feel without the obvious drawbacks of using amplifiers near to drum kits.
- You can capture the perfect performance without having to worry about perfecting the tone prior to recording.
- A clean signal is a safety net: you always have the ability, tone-wise, to go back to the drawing board.
- It also affords you the option of using your DAW's effects plug-ins in place of guitar pedals; many modern plug-ins are superior to guitar pedals in terms of 'tweakability' and tempo locking.

- If you have a band who are technologically sound, with a good knowledge of production, you can have them record their guitar parts to a click track prior to entering the studio; this can help cut costs and speed up sessions. This, however, does require that pre-production is done to work out suitable tones and tempos, and also some vetting before you get to the studio.

There are a couple of drawbacks, though:

- It is not possible to create feedback with re-amping techniques, so when feedback is needed, or you desire the sustain that comes from the guitar being next to the amp, it is better to record the guitars in the traditional way.
- Levels, output types and grounding issues: When you are recording the clean guitar parts into your DAW, you will want to record at a normal gain level for studio equipment, usually line level (the optimal gain level with the best signal-to-noise ratio, as discussed in Chapter 7: Demystifying Recording Levels). The problem is that once you have recorded it at this level, it will be far hotter than the typical input signal level that a guitar amplifier is designed to receive. If that were the only problem, then we could just attenuate the output signal to match the level that a guitar amp would be expecting. However, with the output of most pro-audio equipment being balanced, and the input of most guitar amplifiers being unbalanced, another conversion will also need to take place. To add to this, you will probably encounter a lot of noise if both the output from the DAW and the amp are grounded, or if you are running long lengths of unbalanced cable. The solution is a re-amp box. These devices automatically attenuate the level of the output from your DAW and also convert the output from balanced to unbalanced so that your DAW and amp are now compatible with each other. To minimize noise, place the re-amp box next to the guitar amp and use a long, balanced lead from the output of your DAW. Re-amp boxes also have a handy ground lift switch to remove ground loops.

Figure 12.25 Palmer DACCAPO re-amp box (picture courtesy of Palmer)

You can think about implementing a clean DI signal into your recording workflow in three ways:

1. You can record a clean DI signal instead of using an amplifier, and then use an amplifier simulator plug-in to get a similar tone to using an amp. This is perfect for home studios (where space and noise is an issue) and for creating scratch tracks (Figure 12.26).

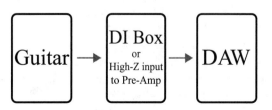

Figure 12.26 Direct recording to DAW

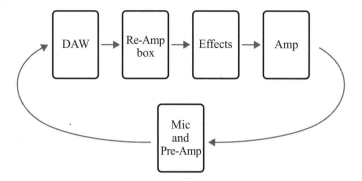

Figure 12.27 Re-amping output stage

2. As above, you can use a clean signal initially and use an emulator during the take to get in the ballpark of the guitarist's tone, with the intention of removing the simulator and sending it back out of your DAW to an amplifier at a later date (Figure 12.27).

3. You can split off the guitar's signal even if you are using an amplifier live in your session, so that you have a 'safety net' in case you are unhappy with the decisions that you made initially. You can do this with either a DI box or an amp switcher pedal (Figure 12.28).

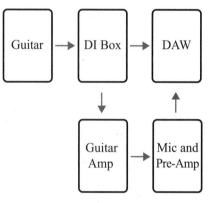

Figure 12.28 Splitting guitar signal

A word of warning before we move on: while it is theoretically possible to use a passive DI box in reverse to re-amp, you run a very high risk of overloading the amplifier due to level and impedance issues (leaving you with a lot of blown speakers or worse) and also encountering horrendous noise. Do yourself a favour and if you are planning to re-amp guitars: buy a re-amp box.

Re-Amping Bass Guitar

Roughly 75 per cent of the time when recording bass guitar, I feel no need to record more than a DI signal. This is because low-frequency information can quite commonly clog up the bottom end of a recording, and often that extra low-mid warmth gained from integrating some bass cab mic'ing is overkill. If I want to give the bass guitar more edge, I will often duplicate the bass channel and use a SansAmp (plug-in or hardware unit) or another type of distortion plug-in to create extra punch; this tends to thicken up the 1–4 kHz region where it is hard to get the bass to cut through the guitars. I then mix together the dry and distorted tones to taste.

There are, however, occasions where bass cab mic'ing is useful. If I am trying to achieve a full yet raw bass tone (the kind you associate with garage rock, where it's common to only have one electric guitarist), the extra low-mids from the cabinet can provide an advantageous 'gelling' effect. In folk and acoustic music, the extra low-mids can be advantageous too. I also quite often mic bass cabs when recording thrash metal and other high-gain guitar music. This might seem counter-intuitive considering how scooped heavy rock genres can be, but I do it to give myself (or the mixer) flexibility: you often end up with very scooped guitar and bass tones, which can often leave a hole in the mid range if you aren't careful, and the bass mic can often fill that gap nicely.

Re-Amping Electric Guitar

I prefer to use real amps rather than amp simulations for electric guitar. Although plug-ins are becoming increasingly more realistic, they aren't the real deal yet. This is particularly noticeable on distortion tones. I find it similar to the difference between a solid-state amp and a valve amplifier. You *can*, however, get a more than usable clean tone, particularly where reverb and delay effects are used. When recording a clean DI signal, it is advisable to compress on input as a guitar amplifier always has a compression type of effect from its circuitry and speaker (this is particularly noticeable on distorted/overdriven settings).

12.5.3 Compression

Whether re-amping or not, one of the big secrets of guitar tone is compression. You may well be thinking, 'I always compress the guitars in my mixes!' Well, I'm talking about the use of compression before the signal hits the amplifier. David Gilmour of Pink Floyd is a good example of a guitarist who used compression (an MXR Dyna Comp) in his pedal chain; it ironed out kinks, brought the guitar closer to the forefront and added sustain to guitar leads. Be mindful that when you overdrive an amplifier, the amplifier itself introduces a degree of compression to the signal; you therefore may not need much additional compression. In any context, this technique can be a very useful tool in achieving the best possible tone.

When I re-amp a guitar signal, I generally pass it through a compressor plug-in within Pro Tools prior to sending the signal out to the guitar amp. This gives me a lot of flexibility as I can use fast attack times to help control inconsistent transients, and use medium/slow attack and medium release times to add sustain. I'd say that 95 per cent of the guitar tracks I re-amp end up with at least one compressor in the chain, and often with erratic players I use two compressors in a chain, each with different attack and release times.

12.5.4 Effects

When dealing with electric guitars, effects pedals will often be a major part of a guitarist's sound, yet quite often they are chained together in an uninformed and random order. While I believe that you should always follow your ears and vision when producing, and know when to break the rules, I've found the following configuration to provide the best results.

As you can see from Figure 12.29, I prefer to place modulation effects, pitch shifting, delay and reverb *after* distortion to help prevent the tone getting 'mushy'. To have these effects occur post-distortion, you can add them via an effects loop in the amp (which places them after the amplifier's distortion section), or via your DAW. In the latter case, you would re-amp and record with distortion only, and then apply the effects to the recorded signal. As I mentioned earlier, there are benefits to implementing these types of effects within your DAW: they can

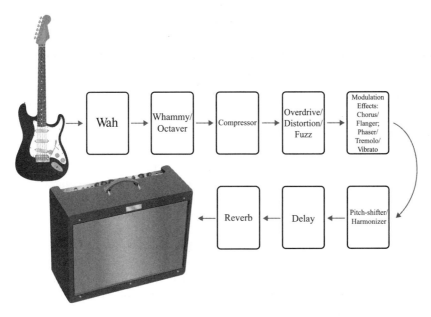

Figure 12.29 Ideal guitar pedal chain

be automated/tweaked/synced to the tempo of the song, and they often also have more parameters than typical guitar pedals. As previously mentioned, you can add these effects before leaving the DAW for maximum 'tweakability'. My advice here is that unless you have a lot of experience producing, or have a very clear vision, try to err on the side of caution and refrain from adding too many effects on input.

Although this is my go-to chain of effects, I sometimes place modulation effects such as chorus before overdrive. I usually do this for two reasons:

1. When I want to add a small amount of modulation to the signal, just enough to make the tone a little more interesting, but without drawing the listener's attention to it too much. I find this order helps to blend the effect with the rest of the signal, as opposed to hearing the modulated signal cleanly on top of the dry signal. However, you can easily lose definition, particularly when using high-gain settings.
2. When I want to go all-out dirty and make crazy sounds. A lot of Eddie Van Halen's famous mechanical and animal-like sounds came from combining heavy distortion settings with very aggressive use of the tremolo bar, flanger, phaser and pitch-shifting effects. So, when trying to get virtuoso sounds or early 1980s sounds, try chaining modulation before any overdrive.

12.5.5 Dial Back That Gain – and Do You Really Need an Amp That Big?

A common faux pas by inexperienced (and sometimes even experienced) guitarists is that they do not control their gain staging very well in their amplifier.

Just like in the studio world, you can suffer horrendous side effects when the amplifier's gain staging is set badly. These include feedback, thinness to the tone, mushiness and lack of definition. Many of today's guitar amps have multiple gain stages. This means that, at a minimum, you will have a control for gain (also called drive) and the master volume.

The first gain stage determines the signal level that is sent to the pre-amp stage; this has a significant effect on the *tone* of the amplifier. You can create distortion with higher gain levels, which overdrive the pre-amp stage. The gain level does affect the overall output volume, and you usually need at least a tiny amount of gain to get a decent clean signal out of the amp. However, the majority of the volume should come out of the second gain stage – the master volume. The master volume is the amount of signal sent to the power amplifier, which is what comes out of the speaker. This is a cleaner type of gain and does not distort easily.

So what does this mean in terms of setting up the amp? To set an amp correctly, I always start by trying to find a suitable clean tone that I like. I use very little gain and a large amount of volume to get a clean and defined sound with just a touch of warmth from the gain. If the amp only has one channel, as opposed to separate channels for 'clean', 'crunch' and 'lead', remember to make a note of the clean setting before you begin work on the overdrive tone.

Now that I have an idea of the overall volume level I want to work with, I find an overdrive sound I like. I do this by initially setting the controls in the same way as for the clean, then I gradually increase the gain whilst decreasing the master volume level until I have the tone I want.

A great way of working out where to set gain levels is to listen to some reference music; I think that you will be surprised at how little distortion is used on many of your favourite tracks (particularly when referencing some of the earliest heavier music such as Led Zeppelin and Black Sabbath).

If the amp is particularly vintage, you might have only one gain stage. You then have to gauge the best balance between volume and distortion for your required purpose. Remember, you could always add more overdrive with a guitar pedal if need be. If the amp is a modern multi-channelled amp, then you will probably have three gain stages: a gain and volume stage for each channel and then an overall master volume.

Another common mistake is to pick the biggest and most powerful amp possible. As mentioned earlier, bigger is not always better. This is particularly true if noise levels are a concern. Smaller amps harmonically distort at a lower volume, so you can play in the 'sweet spot' without irritating the neighbours. You may also battle with comb-filtering artefacts from room reflections when recording a loud amp in a small room.

But, most importantly, many guitarists actually *prefer* the sound of smaller amps and single-speaker cabinets.

12.5.6 Double Track with Different Equipment

Double tracking guitars (recording and layering the same part twice) is a simple and powerful way to achieve a larger-than-life and in-your-face rhythm guitar sound. It is one of the most popular techniques used to enlarge and thicken the sound of a rock track. Inexperienced players and engineers tend to confuse this tone with distortion and try to recreate the sound by adding excessive gain. Indeed, double tracking gives the illusion that the amp's gain is higher than it is. This is another reason why you should exercise caution when setting your gain levels.

Some producers like to quadruple track guitar parts. This is great for some types of music such as grunge or punk, but it can inhibit the natural nuances of the part and make transients less pronounced. To a degree, double tracking can help to make a guitar sound more pad-like or compressed. If you are double tracking (or more), you may want to experiment using a different combination of equipment for each take – a different pickup, guitar, amplifier, pedal or microphone combination. Not only does this add more interest, but I've found it helps to preserve transient detail in the mix.

Double tracking isn't a 'fix-all' technique, and it can be easily overused. I rarely double track guitars in verses and breakdown, saving the technique for choruses and the loudest sections. If your arrangement is dense already, you may not need to double track at all. Remember that you should do everything for a specific reason and not out of habit. If your arrangement is boring, it is unlikely to be fixed by double tracking; new parts should be written that have different tonal qualities in different pitch registers. If your arrangement lacks stereo interest, or the guitars need to sound bigger, heavier and thicker, then double tracking might well do the job.

12.5.7 Multing

It is quite likely that at least one guitarist's part will require a different chain of effects for different sections of the song. By splitting these sections on to individual channels, known as multing, you can make mixing more manageable, handling the tonal changes without having to do substantial level rides while mixing. As I always do some form of pre-production before recording, I am able to plan out and set up the channels in advance, rather than splitting them up after recording. The advantage of setting out a player's parts before tracking is that you can easily see which parts need to be recorded with different types of settings; this will help speed up your workflow, as you focus on one part at a time, minimizing the overall number of tonal adjustments. The downside of this is

EXERCISES

1. Experiment with using amp modellers and comparing the tone to recording a real amp.
2. Try placing guitar effects in different orders and memorizing tonal attributes.
3. Experiment in a studio with re-amping and using DAW plug-ins in place of effects pedals.
4. Compare directly the pros and cons of DAW effects against pedals.
5. Try comparing a single take of distorted guitar against a double-tracked guitar part with a little less gain.

that it makes the musician's performance more fractured. If you are dealing with an impatient musician or one who has a short attention span, you may want to record a clean DI'd performance into your DAW, and then later re-amp and tone-tweak to your heart's content without the musician present.

12.6 MICROPHONE SELECTION AND PLACEMENT

We are now ready to discuss microphone selection and placement in relation to electric guitar recording. Thankfully, as I have previously mentioned, this is a much simpler task than when recording drums. You will use fewer microphones to record an electric guitar than a drum kit – usually between one and three mics per guitarist.

Considering that a speaker cone is a very directional source (at high frequencies) it might shock you that there are several schools of thought regarding the best way to mic a guitar cabinet. What I think every producer would agree on, though, is that the sound coming out of the amp is the most important aspect to get right. In this next section, I will talk about the options available and my own personal preferences.

12.6.1 Placement of Microphone in Relation to the Part of the Cone(s)

When mic'ing a guitar cab, the difference of just a few centimetres across the speaker cone can make a huge difference to the microphone's sound. I like to think of mic placement as an organic form of EQ. The closer the mic is to the dust cap (the centre of the cone), the cleaner, brighter and crisper the tone. This area generally gives you the most accurate reproduction of the speaker's tone. As you move more towards the edge of the cone, it gets progressively duller in the top end of the speaker's response and the transient attack gets softer, making the overall tone seem a bit more distorted and diffuse in a lo-fi/garage rock kind of way. When recording an amp with a number of different cones, I usually listen closely to each cone (with the volume low) to pick the best tone and then mic only in front of that cone. I do this because even if the speakers are all matched, they might sound slightly different due to age, wear, position in the room and other factors.

12.6.2 Distance from Cone

As well as the placement across the cone, the mic distance from the cone has a huge impact. The general rule is that closer results in less room sound and a greater bass response due to proximity effect. Now, in some genres, more bottom end in the guitars is very advantageous to gelling the drums, bass and guitar together. In other genres, it may be muddy and distracting. Some producers love the sound of the proximity effect; others hate it. I tend to mic close to the cabinet when recording in very small and acoustically imperfect environments: I would far rather roll off some bass with a high pass filter (HPF) than battle room tone and the comb filtering that will invariably occur with a distant mic.

For hard rock and metal, I generally like to make use of the proximity effect. If I was recording country, easy listening or jazz, I would most likely place the mic(s) further back.

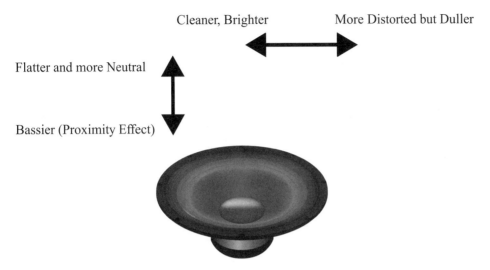

Figure 12.30 Guitar mic placement tonal attributes

12.6.3 Microphone Choice

When dealing with electric guitars, you can use combinations of different microphones and mix and match them to achieve a blend of different tonal qualities. Dynamic mics are often first choice – and again, the Shure SM57 shines here. A dynamic mic can deliver gritty upper mids/treble and punchy attack (2–5 kHz); this gives the recording an 'edgy' and 'raucous'. They can, however, in isolation, sound 'thin' and 'brittle'. Large diaphragm condensers (LDCs) create a more natural overall tone and make a guitar 'shine' in the top end (particularly for a clean tone). However, they can often lack the 'character' and 'grit' that a dynamic mic gives. Lastly, engineers turn to ribbon mics to produce a rich bottom end that the dynamic lacks. Ribbon mics also have a high-frequency roll off so they can sound very smooth – almost compressed. The Royer 121 is a popular choice because of its robustness, its ability to handle high SPLs, and its rich bottom end and less drastic high-frequency roll-off. (Table 12.1.)

Let's remind ourselves of the fact that if one mic gives you the perfect tone, there is no need to put any more mics on a guitar cab; this will save you time and potential issues. However, it can prove wise to record a few mics and give the mix engineer some choices. I've found that most producers will use a dynamic microphone plus at least one other mic, and blend them together. Remember, though, when using more than one mic, you need to be rather anal with your phase coherence.

Table 12.1

Microphone(s)	Type	Characteristics
Shure SM57	Dynamic	Punchy, edgy but lacks bottom end.
Audix i5	Dynamic	Very similar to the SM57. However, it is more pre-EQ'd and has a more pronounced treble.
Sennheiser 421MD	Dynamic	It is known for its mid range and surprisingly good low-end extension for a dynamic mic.
AKG 414	Large diaphragm condenser	Very natural sounding but lacks a bit of personality compared to the dynamics.
Royer 121	Ribbon	Great warmth and punch but lacks the 'edgy grit' of a dynamic or the 'sheen' of a condenser. Is most often used in conjunction with another mic.
Shure SM57 + Sennheiser 421MD	Both dynamic	The SM57 is used for the top end and the 421 for the mid range and bass. Lacks a little warmth compared to the condenser or ribbon but has more 'raw aggression'.
Shure SM57 + AKG 414	Dynamic and condenser	The SM57 provides the upper mids and lower top-end character. The 414 is used to give some more bass and extreme top. The resulting tone is natural but punchy.
Shure SM57 + Royer 121	Dynamic and ribbon	The SM57 provides the top-end 'grit' and punch with the Royer adding the bottom end.
Shure SM57 + Sennheiser 421 + Royer 121	2 dynamic and 1 ribbon	The SM57 is used for the top end, the 421 for the mids and Royer for the bottom. It results in a punchy, full sound but it is a little less natural and can need EQ to sound less 'muddy'. Also, the phase consistency between three mics is harder to align.

12.6.4 On-Axis/Off-Axis?

We learnt in Chapter 8 that if a microphone is not directly pointing at the source, it is called off-axis. On an electric guitar amp, this is typically ±45 degrees and produces a darker, more resonant sound. One well-known trick to get the thickest guitar tone possible is to point the microphone off-axis and then add some more treble to get a thicker yet more 'airy' tone than is possible on-axis. It is also popular, when using more than one mic, to turn one of the mics off-axis and have the rest on-axis to get the maximum contrast between mics.

12.6.5 The Room

While the acoustic treatment of the live room when recording guitars is less critical than when recording drums, it is definitely an overlooked variable in guitar tone. While I tend to record a lot of rock and metal, I often prefer an extremely close-mic'd sound (with the mics almost touching the grille sometimes), there is still value in spending a few minutes getting to know the environment and finding some sweet spots for room mics. This is even more critical, of course, in genres where a more natural sound is required. If you want your recorded signal to be less roomy, you could:

- move the mic(s) closer to the cab;
- use absorption materials and gobos to reduce space/amount of reflections;
- move the amp to a different spot in the room; and
- change the angle of the guitar cab.

To clarify the last two points: guitar amps can get really loud. You can reduce room sound by placing the amp away from immediate boundaries and angling the cabinet so the sound does not reflect so easily back into the microphone, thus reducing the possibility of comb filtering.

12.6.6 Phase

After you have settled on the best amp settings, placed the cab in the best spot in the room and mic'd up the cab with a decent selection of mics in a tried-and-tested configuration, you may find the tone in the control room is dramatically different to what you heard in front of the amp. This will more than likely be due to the phase relationships between the mics. As we've already learnt in Chapter 10: Decision-Making and General Recording Techniques, the effects of phasing can range from extremely mild to an almost unrecognizable tone. Therefore, I *always* use phase rotation or alignment plug-ins to check the phase relationships between multi-mic'd sources; this gives me a good indication of the amount of phase correlation I'm dealing with. From there, I tweak the positions to reduce the amount of phase cancellation as much as possible. The easiest way to get a close correlation is to have the mics as close to each other as possible. If the mics are still causing phase issues, you may want to try moving one and also consider taking one of them off-axis. This will rotate the phase of the mic and it may correct the phase relationship. It may also leave you wanting to slightly change the tone of the amp.

All placement issues are a compromise between optimal placement for each mic and their relationship to each other. Although you should be able to place multiple microphones in sympathetic positions, you may find that some further fine-tuning with a phase rotator plug-in may improve results.

Bear in mind that you shouldn't always be afraid of phase cancellation: sometimes you can use it to your advantage, either by creating special effects or by using a phase rotator as a type of EQ when mixing. To explain the latter, I sometimes use a phase rotator to find a spot where the top end is very coherent and has a lot of presence while the bottom end is less 'muddy'; this can *sometimes* produce more effective and more natural results than using EQ. I would say, however, it is always best to track instruments in phase and reserve any creative phase abuse for the mixing stage!

Further Resources www.audioproductiontips.com/source/resources

 Guitar cab mic'ing. (90)

EXERCISES

1. Record a guitar part using a single SM57, placing the amp in many different areas of the room. Then listen and analyse the tonal differences between the takes, and choose your preferred position.

2. Now, with the amp in the best position in the room, experiment with using gobos/blankets/absorption materials to further isolate the sound and analyse the results.

3. Then experiment with the mic placement of the SM57, moving it first across the speaker driver, then on- and off-axis, and finally closer and further away from the cabinet. Analyse the results.

4. Use a ribbon mic and a condenser mic on the guitar cab and listen to the tonal differences between the dynamic, ribbon and condenser.

5. Finally, try to combine these mics to get the thickest and most exciting overall sound, making sure that the mics are correctly in phase with each other. This might mean removing a mic (or even two) that isn't adding anything worthwhile to the tone. Remember, though, different genres may also require a different combination.

Figure 12.31 Hofner V3

12.7 'TAKE HER AWAY' GUITAR CHOICES

During the recording of 'Take Her Away', we were lucky to have two guitarists who had a very good idea of the tone they wanted and who had the majority of the equipment to realize this. We used re-amping techniques for both lead and rhythm guitarists, with most of the effects being added as plug-ins ('in the box') before being re-amped. This included compression (on all but the POG octaver lines), slight EQ, distortions, reverbs, delays and a quirky Leslie effect that I will talk about later. Rather than talk in detail about all of the plug-ins on the guitars, these settings can be seen on the clean guitar DI channels that are inactive within the Pro Tools session.

Chris, the rhythm guitarist, used a rather vintage Hofner V3 for his main part, which gave the track that slightly 'muddier mid range' retro tone I was looking for.

An Electro-Harmonix POG 2 pedal was used to create the organ-type tone throughout the choruses of the track. This was recorded before hitting the computer, as it was so integral to the sound of the track. Usually when re-amping, guitar effects are chained post-recording on its way back to the amp. Special effects such as wah, which are integral to the player's performance, are an exception: these must be captured at the outset.

Wil, the lead guitarist, used a Fender Jaguar HH, with two humbuckers rather than single-coil pickups. Its tone, in my opinion, falls somewhere between a Telecaster and a Gibson SG.

Even though we were multi-layering guitars, we decided to only use one guitar for each player to preserve some tonal consistency, and to help give the impression it had been played live. If I were to go for a more produced sound, I might have asked Wil to use a Stratocaster on the neck pickup for the first, cleaner 'bluesy' solo.

All the guitar effects were created in the box, except for those in the breakdown section, where Wil used an Electro-Harmonix Freeze pedal to sustain a single guitar note indefinitely. He also used a Digitech Whammy to manipulate the pitch of the note, in a similar way that Jonny Greenwood might.

The amplifier used was a Marshall Mode Four on the clean and crunch channels. This was by no means the perfect amp for the job (a Fender Deluxe would have been a better choice), but it was the best amp available without hiring. Because the two overdrive channels on the Mode Four were completely inappropriate for the track, any additional gain/distortions were provided in the box and were a mixture of AmpliTube 3 stomp box effects and the stock SansAmp PSA-1 distortion plug-in. This type of gritty distortion can be heard on the second guitar solo and was mostly due to the SansAmp distortion you would usually hear on bass guitar lines.

Another cool effect used was the swept chord at the end of the first guitar solo. For this, we used the Pro Tools BS33 organ plug'in's Leslie cabinet emulation in conjunction with some reverb and delay to produce a fluttering type of effect.

We used two mics on each guitar line: the Violet Globe in the centre of the cone for a cleaner and more hyped top and bottom end, and a Shure SM57 slightly off-axis and on the edge of the cone for a grittier/edgier tone. Both of these mics were placed the same distance from the cone, almost touching the amp's fabric covering. The equidistance of the mics reduced phase issues, and the proximity to the amp kept the sound tight, as we intended to add ambience artificially.

Figure 12.32 Fender Jaguar HH

12.8 GUITAR EDITING

Just as with editing drums, in your DAW you will find the same basic options: manual edits, Beat Detective and Elastic Audio.

I prefer to edit guitars manually. This is because of the following issues:

- Using Elastic Audio, phase problems between multi-mic'd sources can create strange 'sucking' effects even while grouped.
- Beat Detective is often unable to correctly find the correct transients on overdriven guitars.

There are some exceptions where I might use Elastic Audio or Beat Detective:

1. When the timing is really sloppy and I have a clean DI signal available to then re-amp.
2. When recording metal or mathcore music where an almost robotic sense of timing is not only perfectly acceptable, but even preferred. In these cases, you would normally insist on taking a clean DI signal for re-amping, even if it is just for a safety net.
3. On sections of music that have a lot of brief stabs that can easily be lined up using Beat Detective.
4. If you encounter tuning problems, it is possible to tune a clean DI guitar part using Melodyne (more on tuning in Chapter 14: Tracking Vocals) while simultaneously editing their timing within the plug-in.

Taking a clean DI of the guitar recording not only acts as a safety net whenever you wish to re-amp for a change of guitar tone, you can also edit guitars more forensically when working with a clean DI signal: not only can you see the transients more easily, but the Elastic Audio algorithm is less 'confused' by the clean signal and is less prone to stretching artefacts and glitches. Typically, I mult (multing) guitar parts in sections, whereby the monophonic algorithm is used to edit lead parts; polyphonic for single strummed chords, stabs and palm-muted parts (or Beat Detective, in the event of polyphonic-causing artefacts); and then manual editing techniques for strummed chord parts. While editing guitars this way, you may want to monitor how the guitar sounds through an amp or through an amp simulator, to make sure that the edits you are making are inconspicuous. You may find certain moments of edited guitar playing that sound artificial with every algorithm or stretch you try; in this event, you can try copying and looping previous sections or doing some manual editing before using Elastic Audio. Particularly in the case of heavy metal guitars, editing with Elastic Audio is advantageous: you have a lot of control over timing and the way that notes decay. This means that for genres such as nu metal or post-'90s thrash, short stabs and quick muted passages can be perfected.

It may seem like there are a lot of positives to using Elastic Audio, and you may be wondering why I don't use it more regularly. Well, this is for three simple reasons:

1. It kills the 'feel' of a track: sometimes the odd imperfection is musical, and snapping every instrument in the track to the grid is over the top (apart from in the genres just mentioned).
2. Not only is it over the top, it is often unnecessary. Distorted rhythm guitars are easy to move manually and crossfade inaudibly. The timing of rhythm parts is often more accurate than lead parts and may only need the odd tweak.
3. When only a couple of minor tweaks are needed, I find it more efficient to re-amp, move on and make any minor edits before mixing.

When using Elastic Audio on guitars, you need to make sure that you have a clean DI signal to edit. It is almost impossible to edit guitars that have already been mic'd, as the stretching of the distorted signal is usually audible and unavoidable. To further compound the issue, when using Elastic Audio with multi-mic'd guitars, it causes clearly audible phase problems between the mics, due to the complex nature of the harmonic content of electric guitars. If you have to time-stretch an already recorded guitar part, bounce the multi-mics down to a single channel first; this avoids the phase issues explained above.

Most of the time, I edit drums and bass quite heavily in time with each other (and/or the grid), but I will usually leave guitars a little 'looser' and only correct what I need to manually (apart from in the situations above). When doing this, you will need to know how to crossfade effectively and master a few tricks for hiding these fades. First, let's discuss the different places you could put a crossfade, then we will look at the types of fades available.

12.8.1 Places You Can Put a Crossfade

- At a chord or note change: This is probably the most popular and the place you should try first. At the point that a chord changes, your ear is not only used to a bit more variation in tone, but it is also the point where the sound has decayed most (sometimes there is only ambient noise remaining). Short fades that start at the end of a chord and are completed just as the transient of the next chord begins are the most transparent.
- At the transient hit of a lead instrument or drum: I use this when I just can't find a point where the takes gel nicely. I often hide the crossfade at a point where there is a big transient; this could either be the attack of a lead instrument or a transient of a kick or snare drum. Even if these edits are audible in isolation, other instruments can mask the effects very effectively.
- During a big chord where the timbre is similar: If neither of the two above methods work (and 99 per cent of the time they will), you might have to resort to creating a long fade during a chord or note. This is particularly tricky because the overtones of each take will be slightly different. To perform an edit such as this requires patience, skill and experimentation.

As well as these, you should look out for the following:

- When doing edits on multiple different instruments at the same point in the timeline, try to use slightly different locations and types of fade to avoid the edits becoming too audible.
- When dealing with multi-mic'd sources, it is best to group the tracks together, so that any fade you make is also applied to the other microphones at the same spot.
- Sometimes when there is a gap between two notes and each take was played sloppily, it might be easier to simply fade out the first note and then fade in the second in the correct place.
- Clean and acoustic guitars are often more difficult for placing inaudible crossfades – especially when they are exposed in the mix – so be ready for a lot of trial and error.
- Remember, the fade only has to be inaudible in the context of the whole mix. However, you should bear in mind that nearfield monitors often do not expose

bad fades as much as headphones. So, if you are unsure of a particular fade, try auditioning it on a set of mixing headphones. Or, getting into the habit of making periodic checks on headphones is even better (this is great for spotting metronome bleed too).

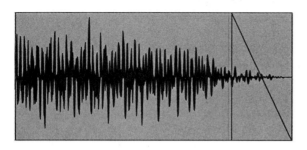

Figure 12.33 Linear fade

Figure 12.34 Logarithmic fade

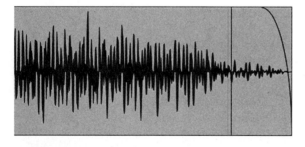

Figure 12.35 Exponential fade

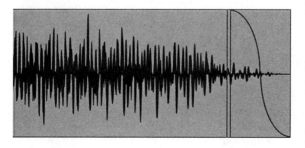

Figure 12.36 S-curve fade

12.8.2 Types of Fades

Before discussing different types of fades, I want to mention that you should audition several fade types when dealing with complex situations as, in my experience, it takes some trial and error in deciding which fade type works best with the program material. Below is a list of the fade types that you'll find in your DAW. We will then cover how they apply to crossfades.

Linear

The most basic and common form of fade is a linear fade. A linear fade's rate of attenuation stays constant over the length of the fade (Figure 12.33).

Logarithmic

With a logarithmic fade, the rate of attenuation is high at the start, then slowly tapers towards the end of the fade (Figure 12.34).

Exponential

An exponential fade is the opposite of a logarithmic fade: the rate of attenuation is low at the start, and then increases rapidly towards the end (Figure 12.35).

S-Curve

An S-curve fade is named after the shape it resembles. It eases in and out of the fade, with the middle at 0 dB (Figure 12.36).

When dealing with quick fade-ins and fade-outs in the middle of a busy mix, you are unlikely to hear much of a difference between the fade types. For longer fades (often used at the start and end of a track) or for fades that happen across many channels simultaneously, some experimentation is usually needed. I tend to start with an

S-curve fade when attempting to replicate the natural decay of a sound; if that doesn't work, I'll experiment. Remember to audition fades before moving on to the next one.

When I am attempting a 'classic long fade-in or fade-out' of a track (with all the instruments fading at once), I will usually use a linear fade.

Now let's discuss the Pro Tools Crossfade window.

When dealing with crossfades, you can set the fade-in and fade-out to be whatever type you want (in Pro Tools, you can even set different kinds of fades on each side if you wish). I usually tend to work with S-curves or linear 99 per cent of the time. One other interesting feature of a crossfade is that you can set it to be one of two modes:

Equal Gain

Equal gain mode should be used when you are dealing with phase-coherent material; in music production, this is the mode that you should use 90 per cent of the time. If you are fading between two takes of the same instrument, it will usually provide the smoothest transition between instruments of a similar timbre, particularly when the phase is consistent between takes.

Equal Power

Equal power mode is used more in other types of post-production where you are fading between two different timbres or even crash edits of different tracks.

When dealing with fades, if you notice a swell in volume at the centre of the crossfade, it is a good indication that you should be using either a shorter fade and/or equal power crossfade mode.

So here are some good rules of thumb when dealing with crossfades:

- Start small; try using a very small crossfade.
- Try to find a transition between two chords in which to place the crossfade; if that's not possible, try to hide it around a major instrument's transient.
- Use equal gain mode first.
- S-curve and linear fades are the most common and are easier to make sound inaudible.

EXERCISES

1. Take a guitar track that you have already fully recorded (i.e. it has been recorded through the amp and mic'd) and edit it manually, experimenting with different types, lengths and placements of crossfades.

2. Use Elastic Audio on a multi-mic'd and fully recorded guitar part (remember to group the channels), then listen out for the problematic phase relationships that will most likely occur even while moving grouped warp markers.

3. Edit a clean DI track using manual editing and Elastic Audio where necessary, then re-amp it. Compare and contrast the results with the other types of editing.

12.9 THE BASS GUITAR

Now that we have covered the electric guitar in such depth, it is time to run through the options regarding the bass guitar. As I've said, many of the same factors apply when recording bass guitar, and therefore I will mainly focus on outlining the differences.

12.9.1 Playing Approach

Figure 12.37 Fender Precision Bass (picture courtesy of Metropolis)

Unlike electric guitarists, who predominantly play with plectra, bassists use either their fingers or a plectrum; for 'slap' bass, they use their thumb. A bassist's choice of playing approach has a huge effect on gear choices. For instance, if he or she plays using fingers, you may consider using a slightly brighter amp and/or more trebly settings to pronounce the top end: finger-style playing results in a mellower tone with less attack, which is often preferable for a bass instrument. Playing with a plectrum gives a lot of attack and punch, and works particularly well with a little bit of distortion, but it can easily lead to a loss of bottom end and a more brittle tone; to counteract this, rock bassists are known to use less bright strings and favour tube amplification. Working with slap bass is tonally much like using a plectrum, except that you may want to over-pronounce the top end. Slapping is also prone to pretty extreme fluctuations in level and attack. You can use a generous amount of compression with a quick attack and release to tighten the dynamics, and also use a graphic EQ to tame any overbearing 2–5 kHz 'pop'.

12.9.2 The Instrument

Most bass guitars have four strings, but five- or even six-string basses can be favoured, particularly in progressive rock and metal genres. The bass guitar world is dominated by two fretted designs by Fender: the Precision Bass and Jazz Bass. A lot of the world's top bassists own and use both varieties depending on the tonality of the song.

Fender Precision Bass

Figure 12.38 Split-coil pickup (picture courtesy of Metropolis)

The Precision Bass (or P-Bass) was the earlier of the two designs and it was first introduced in 1951. It features a single-coil pickup and ash body, although nowadays the P-Bass uses a split-coil pickup (to act like a humbucker) (Figures 12.37 and 12.38).

This is the bass guitar of choice for classic rock with its big bottom end and generous sustain. Users of the P-Bass include Sting (The Police), Roger Waters (Pink Floyd) and Pino Palladino (a famous session musician who has worked with the likes of The Who, John Mayer and Eric Clapton).

Fender Jazz Bass

The Jazz Bass (or J-Bass) was first introduced in 1960 and features two bipole pickups (meaning two poles of a magnet per string) connected in parallel. It is usually made with an alder body (Figures 12.39 and 12.40).

This is the favoured bass guitar for jazz (who would have thought?) and funk. It sounds more trebly, thinner and has a nice 'slap' to it. Because of its trebly nature, it is also often used by rock bassists who like to be more prominent in the mix, particularly in three-piece bands where there isn't a wall of guitars to compete with. Users of the Fender Jazz Bass include John Paul Jones (Led Zeppelin), Adam Clayton (U2), John Entwistle (The Who) and Noel Redding (The Jimi Hendrix Experience).

While the majority of considerations for choosing a bass guitar are similar to electric guitars, there are a couple of additional factors that make a big difference to the tonal quality of a bass guitar. These are:

(a) Active pickups (which are more popular in basses) versus passive pickups?
(b) Fretted versus fretless?

Active or Passive Pickups?

As previously stated, active pickups give a brighter and clearer sound than passive pickups: this suits slap players, pop players or those who want to make the bass more prominent. A passive bass has a warmer and more 'gritty' tone, which suits most types of rock and metal.

Figure 12.39 Fender Jazz Bass (picture courtesy of Flipside)

Fretted or Fretless?

Just like traditional stringed instruments such as double bass or violin, a fretless bass is an instrument that doesn't have the steel bars at intervals across the fingerboard. This makes the positioning of fingers more difficult to a beginner and also imparts a tonal difference. While most rock and other contemporary styles rely mainly on fretted basses, genres such as jazz, funk and RnB often favour fretless basses. This is partly down to look, but also down to the fact that a fretless bass sounds 'warmer' and more 'smooth' than fretted ones. This means that the fretless bass sounds even more similar to a double bass.

Strings

Typically, bass guitar strings are made from the same types of materials as guitar strings – nickel, which is softer and warmer, and stainless steel, which is harder and more trebly. They are also measured in the same way (thousandths of an inch) but they are thicker and are processed in a few different ways. In my opinion,

Figure 12.40 Bipole pickup (picture courtesy of Flipside)

the contrasts between types of strings on guitar are more exaggerated in the case of bass guitars. For bass guitars, the choice of string type is often of greater importance than with a regular guitar. This is mainly due to the hugely differing sonic attributes of different playing styles. For instance, bassists will often choose a different type of string dependent on whether the player uses a plectrum, his or her fingers or performs slap bass. Here is a list of some of the options:

Figure 12.41 Round-wound string dissection[6]

- Round-wound consists of a round wire wrapped around a core. This type of string is great for a brighter tone with more top-end sustain. Funk bassists may prefer the excessive brightness of steel strings with a round-wound process, whereas rock players may prefer nickel round-wounds.

Figure 12.42 Flat-wound string dissection[7]

- Flat-wound consists of a flat-edged wire wrapped around a core. This type of string is less bright and has a more mellow sound, which is perfect for finger-style bassists. Flat-wound strings are preferred among fretless players so suit folk, jazz and swing music.

Figure 12.43 Half-wound string dissection[8]

- Half-wound is a middle ground between the two above. It consists of a round wire that is filed down before being wrapped around the core. This type of string is often used in pop and acoustic music.

Here is a table of the different types of string gauges commonly found on bass guitars:

String	1	2	3	4
Extra/Ultra Light	30	50	70	90
Light	40	60	80	100
Medium	45	65	85	105
Heavy	50	70	90	110
Extra Heavy	55	75	95	115

12.9.3 Bass Guitar Amp Selection

In the electric guitar world, there are many different manufacturers producing products that are lauded in their own genre. But in the bass guitar world, there are none that are as universally lauded as Ampeg for their rich and complex sound and bright but non-fatiguing top end. Their SVT series of heads and 8×10 cabinets really are the flagship amp for bass amplification.

Other high-end choices for bass amplification include Eden, which produces an 'American' sound, and Trace Elliot and Ashdown for a more mid-heavy 'British'-sounding amplifier.

The tube versus solid-state debate rages on in the bass guitar world too. While it is generally considered that valve amps are again superior, due to their size and weight it is often seen as impractical for unsigned and modest touring bands to have a full valve rig. In fact, in the world of bass, you will often see hybrid amps that feature a tube pre-amp section (normally using 12AX7 tubes) and a solid-state power amp section that helps keep size and weight to a minimum while maintaining some of the tube warmth.

Bass Guitar Cab Selection

Bass cabinets share many traits with guitar cabinets: their enclosures are often made from similar materials, and their speakers are often produced by the same manufacturers, such as Celestion (though Eminence are also popular). There are, however, some fundamental design differences you need to be aware of:

Figure 12.44 Ampeg SVT stack (picture courtesy of Metropolis)

- The full frequency range that a bass cabinet is required to reproduce cannot adequately be delivered by a single speaker. Because of this, you might find multiple different sizes of speaker within an enclosure. This collection could comprise, for example, a subwoofer of 15' or 18', a 10' or 12' mid-range speaker, and a horn to produce the top end. Remember that when you are dealing with multiple different styles of speaker, you will need to have a crossover system built in to be able to drive each part in the correct way.
- A lot of bass cabinets have only 10' speakers. This might seem strange, especially since a lot of guitar speakers are bigger than this, but the primary intention in these cases is that the cabinet efficiently reproduces the mid range and top end of the bass guitar. In the studio, this is usually then supplemented and blended with a DI signal from the bass guitar, which provides the missing low-end content. The result is a 'best of both worlds' combination of the round, smooth low end you get from a DI signal, and the punch and definition typical of a mid-sized speaker.

- Some bass amps are ported, which helps to improve the system's bass response in terms of both low-end extension and efficiency. The trade-off, however, is that they often have problems with their transient accuracy, and they also smear the bottom end. Personally, as with nearfield monitors, I prefer sealed enclosures.

12.9.4 Bass Guitar Effects

While all of the types of effects are the same for the bass as they are for guitar, many of the circuits and/or algorithms are optimized for bass guitar use. Modulation effects are sometimes used on bass guitar, but they are not as widely employed as on electric guitars. Things you typically find in a bass player's rig include:

- *Compression*: This is a bass player's best friend, and helps to maximize sustain or reduce problematic plectrum attack. Typically, higher ratios (3:1 or 4:1 roughly) are used with a medium attack time and medium release to de-emphasise the 'muddy' decay and give a thick sustain and release. Alternatively, you sometimes see a compressor with a fast attack and release to combat excessive pick attack or slap. Compressors can come in the form of a rack-mounted unit, a pedal placed before the amp, or a processor on the bass amp itself. The DBX 160X half-rack unit is widely lauded for bass compression.

Figure 12.45 DBX 160 compressor (picture courtesy of Metropolis)

- *Graphic EQ*: This is great for tonal shaping. You often see a smile-type curve on the graphic EQ to reduce mid range and pronounce the top and bottom; sometimes you also see a dip between 2 and 5 kHz to reduce some slapping/popping sounds. Again, these are sometimes pedals or a feature on the amp. The Boss GEB-7 is popular.
- *Octaver*: An important factor with the bass guitar is being able to get a thick enough bottom end. An octaver focused on generating an octave lower helps give a strong sub-bass response. The Aguilar Octamizer is a well-respected bass octaver.
- *Distortion*: A staple part of a bass player's rig. It is usually used either as a subtle drive to give the bass presence and help it cut through a wall of guitars,

or a full-on drive serving as a noticeable effect. A popular bass drive is the SansAmp (which also features a useful DI output).
- *Fuzz* is also widely used on bass, and the combination of the octave pedal and fuzz is a classic. The Electro-Harmonix Bass Big Muff is a popular bass fuzz.

12.9.5 Microphone Selection and Placement

In terms of microphone placement, the parameters are the same as with guitar. However, you may wish to place the mic a little further back from the speaker as the proximity effect might result in an overly bassy tone.

You will quite often also leave the DI signal in the mix. The DI has a very accurate transient response, which is great for an articulate sound with more presence in the upper-mid range. Due to the mixture of the speaker and the microphone, the transient response will be slightly more smeared, giving a warmer and thicker sound. Mixing the two together can give a larger-than-life sound and result in you having the best of both worlds. Therefore, picking the right microphone that works in sympathy with the DI channel is important. When selecting mics, many of those that are typically chosen for bass guitar also fall comfortably within the realm of kick drum microphones.

Microphone	Type	Characteristics
AKG D112	Dynamic	Has the punch of a dynamic. More pre-EQ'd than the RE20 but less than the Audix.
Audix D6	Dynamic	Very punchy and pre-EQ'd dynamic suitable for metal and hard rock.
Neumann FET 47	Large diaphragm FET	Just as when used on the kick – back it off from the source slightly and listen for its balanced and smooth tone.
EV RE20	Dynamic	A remarkably flat dynamic but with enough punch for most genres.

12.9.6 Bass Editing

Unlike guitars, I will often use Elastic Audio when editing bass, and then re-amp after. This is because not only does the bass guitar rarely play chords, which means that it has fewer artefacts when using Elastic Audio, but also, in a lot of genres, I like to heavily tighten the drums and bass to create a rock-solid beat. Folk, indie, garage rock and grunge are genres for which I seldom consider this approach suitable. I will typically use monophonic mode, but if there are any parts that are playing chords or have some harmonic overtones that create artefacts, I will mult them to a separate channel and use polyphonic mode on them. Often, bass players in heavier genres like to play aggressively, which affects tuning slightly; when this is evident and distracting to the song, I will use Melodyne to make the bass closer to the intended pitch.

EXERCISES

1. Audition both a Jazz and Precision Bass and compare and contrast their tones (make sure that both models are fretted and are standard models and not customs).

2. Try out different types of strings and note their tonal attributes.

3. Build your experience in using compression and graphic EQ on bass guitars either in your DAW before re-amping, as pedals before it hits the amp, or on the amp itself.

4. Experiment with microphone placement as suggested in the guitar section, using a pre-EQ'd dynamic mic (such as the Audix D6 or Shure Beta 52) and then again with a large diaphragm condenser mic (FET if possible, and also make sure that it can handle high SPLs). Compare the results.

5. Edit a clean DI bass line with Elastic Audio before re-amping.

6. Edit the clean DI bass line manually and compare and contrast results and also time taken. Edit the bass manually once it has already been recorded and compare with the previous two results.

NOTES

1. Truss rod, accessed June 2013, http://en.wikipedia.org/wiki/Truss_rod.
2. Speaker Impedance, Power Handling and Wiring, accessed June 2013, www.amplifiedparts.com/tech_corner/speaker_impedance_power_handling_and_wiring.
3. Speaker Impedance, Power Handling and Wiring, accessed June 2013, www.amplifiedparts.com/tech_corner/speaker_impedance_power_handling_and_wiring.
4. Speaker Impedance, Power Handling and Wiring, accessed June 2013, www.amplifiedparts.com/tech_corner/speaker_impedance_power_handling_and_wiring.
5. Accessed June 2013, www.guitargeek.com/.
6. Choosing bass guitar strings, accessed June 2013, www.hqslapbass.com/choosing-bass-guitar-strings/.
7. Choosing bass guitar strings, accessed June 2013, www.hqslapbass.com/choosing-bass-guitar-strings/.
8. Choosing bass guitar strings, accessed June 2013, www.hqslapbass.com/choosing-bass-guitar-strings/.

BIBLIOGRAPHY

Bobbi Owsinski and Rich Tozzoli, *The Ultimate Guitar Tone Handbook: A Definitive Guide to Creating and Recording Great Guitar Sounds* (Alfred Music, January 2011).

13

TRACKING OTHER INSTRUMENTATION

We have covered some of the fundamental instruments that you will find in many recordings, particularly for my specialism of rock genres, but there are also many other instruments that could feature in popular music. Although you will tailor your approach to capture the nuances of each instrument, you will, once again, see that the tracking mindset is the same: *envisage how you want the instrument to sound, and achieve it as early in the recording chain as possible.*

Many of the instruments discussed here are acoustic instruments, so don't forget: *optimal placement of the instrument in the acoustic environment and the position of the microphone are more important than the choice of microphone or pre-amp.*

To keep this course to a manageable size, this chapter is deliberately brief. Yes, these instruments are just as important to capture correctly as any other. However, I have already given you the tools to make accurate judgements. Try adapting some of the placement ideas from the previous chapters to help guide you. In essence, the information here will be guidelines on how these instruments might typically be recorded from a tracking engineer's perspective and will not delve quite as deeply into considerations such as instrument choice, maintenance and performance advice.

13.1 PIANO

Most of you will be well aware of the sound of a piano and roughly how it works, but recording a piano is a significant challenge for any engineer. On a standard full-size piano, the lowest note's fundamental will be well inside sub-bass territory (around 30 Hz), and even though the highest note's fundamental frequency is only just over 4 kHz, the rich harmonic content extends past any human's hearing range. Then you have to consider the physical constraints of the piano; not only the size of the room needed to house the piano, but also due to the frequency range just described, the room should be as flat as possible across all frequencies to produce optimal results. We haven't even mentioned the high cost and maintenance of a decent calibre of piano – yikes! In reality, this leaves two viable options for the average home, project or even small commercial studio owner:

Chapter 13 TRACKING OTHER INSTRUMENTATION

1. Hire a top-level studio that own and maintain a quality piano. This won't come cheap and you may have to pay an additional fee for the piano to be tuned before the session.
2. Rely on sample-based solutions and use a quality MIDI controller with software samples. Alternatively, sample modelling technology as found in the best stage pianos and software instruments (this technology is catching up with the best sample libraries). To do this right, get a controller/keyboard with 88 keys, high-quality weighted keys and a sustain pedal such as the Roland a-88 MIDI controller.

We will be discussing sample libraries later in this chapter, but a good guideline is: if the piano track is likely to stay in the background of a piece, budgets are tight or I'm looking for an ultra-perfect-sounding end result, I am not going to lose any sleep by choosing the software route. With rock bands, I've often found that the musician playing a piano part is self-taught and it is not his or her primary instrument, so making it a point of focus may be the wrong decision.

Figure 13.1 Native Instruments Grandeur piano sample library

However, if I were recording Elton John, Ben Folds or Billy Joel, I'd be doing the artist a disservice by not fully presenting the subtleties of their performance, including the often enriching irregularities of recording a real piano, such as slight tuning variation and pedal noise.

13.1.1 Grand versus Vertical Pianos

If you do opt to record a real piano, first you need to be aware that there are two main types of piano: grand and vertical.

As you can see, a grand piano has the frame and strings arranged horizontally, with the strings extending away from the keyboard and player. The mechanism of the hammer action hits the strings from underneath and utilizes gravity to return to rest. Grand pianos can come in a number of sizes, most notably the larger concert grand and the much smaller baby grand.

Vertical pianos are smaller and easier to manoeuvre than grand pianos. Their mechanism is fundamentally similar to a grand piano's, but as the frame and strings are arranged vertically, the hammers have to move horizontally. Springs return each of the hammers to their positions of rest. The most common variety of vertical piano is called an upright piano, which is any vertical piano over 1 metre. However, the term upright piano is sometimes used interchangeably to mean any vertical piano. Other sizes of vertical piano include studio and console.

Figure 13.2 Yamaha C7 baby grand piano (picture courtesy of Metropolis)

As grand pianos have longer keys and an action that operates vertically, their hammers return to rest more quickly after each strike, and they afford more control over each strike. This means that grand pianos are much more responsive. Sonically, I also find grand pianos superior (assuming the same quality of build and components). This is down to a couple of main factors:

1 *Soundboard placement.* In a grand piano, the soundboard is horizontal and you get the sensation that the sound is raising up into the room and filling the environment. On vertical pianos, the soundboard is vertically placed, so sound travels out of the rear of the piano. Vertical pianos are often placed against a wall, which can cause a lot of comb filtering due to early reflections.
2 *String length.* Grands will typically have longer and thicker strings, which produce a richer sound. This is similar to guitar strings where thicker gauges sound comparatively fuller. The reason for increased thickness on longer strings is because a longer string requires a higher tension to achieve a certain pitch. Another advantage of a longer string length is they suffer from less inharmonicity. Inharmonicity is the phenomenon where overtones will start to sharpen from the exact multiples of the fundamental. The greater the inharmonicity, the more harsh the sound will likely be perceived.

Despite my favour of grand pianos, a quality vertical piano is still very usable; the combination of a good vertical piano in a well-treated acoustic environment can produce stunning results. The advantages of vertical pianos are their cheaper price range, smaller frame (so less studio space is needed) and portability, making them much more suitable for homes, project studios or environments where they are likely to be moved more often.

Before moving on to mic'ing examples, it is also worth noting that the pedal configuration is often different between grand and vertical pianos despite the fact that they may both have the same number of pedals.[1] On a grand piano, you are very likely to find three pedals, and on an upright piano you are likely to see two or three. Table 13.1 shows a breakdown of the operation of the pedals.

13.1.2 Piano Microphone Choice and Placement

In terms of mic'ing, you have a lot of options available. Typically, engineers stay away from dynamic microphones as the heavier diaphragms struggle to capture the top-end richness of the piano; they also seem to blur the transients a little. When you also factor in the large frequency range the piano covers, this means that condensers or high-end ribbon mics are the most popular for piano applications. Large diaphragms are probably the most popular, but sometimes pencil condensers are used too. In tighter situations, boundary microphones are a good and convenient choice because they can be attached to the lid of the piano and reduce bleed and comb-filtering effects.

As my preference is to record grand pianos, my advice will be focused on mic'ing them. However, most of these concepts can be copied or slightly adapted for use with vertical pianos.

Now let's factor the physical size of the grand piano when mic'ing. Due to the large distance between the bass strings and treble strings, engineers will

Table 13.1

Pedal	Grand Piano	Upright Piano
Una corda (left)	Also known as the soft pedal, it shifts the whole keyboard to the right so that the hammers can strike fewer of the strings associated with each note. Because not all of the strings for each note are hit, the effect is softer tonality.	In an upright piano, the una corda pedal doesn't reduce the number of strings hit. Instead, it moves the hammer closer to the strings. This results in a sound that is best described as quieter rather than a softer one.
Sostenuto (centre)	Sustains only those notes that are being held down when the pedal is pressed, which means that other notes played after this are not affected. The sostenuto pedal is the least used of the three pedals.	Sometimes you will find pianos with only two pedals. In these cases, it is the sostenuto pedal that is missing. This is particularly common in upright pianos. It is a rare occurrence for an upright to have a true sostenuto pedal. On rare occasions, you might find uprights and even grands that have a middle pedal that doesn't operate as a sostenuto. Therefore, it is always best to research the model or listen carefully to the sound.
Damper (right)	Otherwise known as the sustain pedal, it raises all of the dampers from the strings, allowing them to sound even after the note has been released.	The sustain pedal works in the same way on an upright as it does on a grand.

usually use a stereo pair of microphones each capturing half of the piano. When looking at the ways of capturing each half of the piano, you can consider some of the techniques we used on drum overhead mics, but before we do this I want to outline two factors regarding why you will probably want to use your large diaphragms in omnidirectional mode:

1. Because of the size of the piano, sound will need to be picked up as evenly as possible from across the front and sides of the microphone; therefore, off-axis performance is critically important.
2. Cardioid modes exhibit proximity effects that omnidirectional modes don't When you factor that sound is arriving at the microphone from varying distances and angles, the proximity effect would not be advantageous.

Now let's talk about placement techniques. Out of all of the stereo mic'ing techniques we discussed in the drum chapter, coincidental pair techniques such as X-Y or spaced pair are by far the most popular for use on piano. Additionally, a Blumlein Pair is also a common occurrence on piano, and we will cover this technique later in this chapter. Just to recap, the X-Y technique is great for when you want to maintain a tight phase relationship. However, it lacks some of the stereo spread and overall excitement that the spaced pair creates. If your track is densely arranged, mono compatibility is important and you might want the piano to have a narrow stereo width. In this scenario, X-Y is probably the way to go. Inversely, if your track is heavily piano-driven, a spaced pair technique will occupy more space in the mix and you will hear a greater depth and excitement when you pan them widely. The trade-off, however, is a compromise in their mono compatibility. In modern rock and pop music, pianos are usually mic'd much closer than they would be in the classic world. They are usually placed inside the piano with the lid open, kept between one or two feet from the strings to allow for a natural and full sound.

X-Y Placement Advice

If you are using the X-Y technique, a good starting point is to place the mics between 0.3 and 0.6 metres from the strings, placed at the central point near to the music stand (rather than at the foot of the piano). For ease of placement and phase-locking, sometimes producers prefer to use pencil condensers for this rather than large diaphragms.

Figure 13.3 Piano close-mic'd with X-Y technique

Spaced Pair Placement Advice

If you are using a spaced pair, try placing each mic a third in from either side. Sometimes engineers choose to place the mic that is on the lower strings further down towards the foot of the piano, which captures a more accurate bass response because of the longer length of the bass strings. When setting up a spaced pair, remember to watch out for phase issues.

Figure 13.4 Piano close-mic'd with spaced pair technique

Although these simple techniques have been proven to sound great, you may want to go further by adding additional microphones. Just like the drums, you may want to use ambient room mics to help capture a thicker tone. Your options here are almost unlimited, so experiment with finding sweet spots before committing. In a recent project that was piano-heavy, I used a combination of close, mid and more distant mics and automated them in the mix dependent on the type of feel we were going for during different sections of the song.

13.2 ACOUSTIC GUITAR

While many of the same variables exist between electric and acoustic guitars, recording the acoustic guitar is a very different art. Recording the acoustic guitar closely resembles the techniques that are used to record other acoustic instruments, and we will be covering a lot of these later in this chapter.

First, when recording an acoustic guitar, as with all acoustic instruments, the quality of the instrument itself is arguably more critical than it is when recording an amplified instrument or synthesizer. This is due to the fact that the instrument alone is creating the entirety of the sound. With unamplified instruments, it is difficult to escape the fact that you need a quality instrument that suits your vision.

Many engineers/producers buy a decent acoustic guitar for their studio, as it is often a low priority for musicians outside of the folk or acoustic scene. At the very least, you should have contact with someone who has a decent acoustic available for hire/loan for such occasions when the band does not own a suitable instrument. Acoustic guitars can vary hugely in tone; the variables are so complex that even two guitars of the same model can often sound very different. This means that while price is a good rule of thumb with regards to quality, you should audition all acoustic guitars prior to purchase in order to avoid a shock (or to find a diamond in the rough).

Figure 13.5 Simon and Patrick SP6 cedar dreadnought-style acoustic guitar

13.2.1 Acoustic Guitar Types and Accessories

Although there are many types of acoustic guitar bodies, two types are far more common than the others. These are the *dreadnought* and the *grand auditorium*. A dreadnought has a larger body, and therefore the low end and low mids are emphasized. Dreadnoughts are great if you need a warmer and louder acoustic guitar tone, particularly if you are playing strummed rhythms. Dreadnought guitars usually sound best with medium- to heavy-gauge strings.

If you favour a more balanced overall tone centre, or require the ease of playing fingerstyle when sitting, then a grand auditorium may be the best option. In my experience, grand auditoriums sound more sonically exciting in the top end. Grand auditoriums usually suit lighter-gauge strings.

Just like electric guitars, acoustic guitar string gauges work with the scale of 1/1000th of an inch (assuming that you are not using a classical guitar with nylon strings). You will also want to replace acoustic strings before the session and make sure they are settled before starting to track. Remember to tune up before takes; this is even more critical because of the complexity of an acoustic guitar's tone. The thinnest acoustic strings start at around 11 gauge for the top E and the thickest strings that I've ever used were 15 gauge. Thinner strings are easier to play but have a thinner sound; thicker strings will have a louder and warmer sound, but are hard to bend and can also lack the exciting overtones you expect from the acoustic guitar. Just like electric guitar strings, acoustic guitar strings can come in several varieties:

Phosphor Bronze – This is the most common type of acoustic guitar string. They are known for their well-balanced tone.

80/20 Bronze – They get their name from being composed of 80 per cent copper and 20 per cent tin. The 80/20s tend to be initially brighter than phosphor bronze, but tend to dull very quickly with age.

Silk and Steel – These strings have a copper wrap combined with silk, which gives a mellower tone than the two types above. They are also softer and make for easier fretting with the left hand. The main drawbacks with silk and steel strings are the fact that they aren't as loud as most fully metal strings.

Nickel – Even though nickel strings are often cited as an electric guitar string, you can also find nickel strings made specifically for acoustic guitar (and even dual-purpose ones). They have a less harmonically complex sound, so are less bright and could be considered 'warmer'. Therefore, they would work very well in some applications where the body of the sound is important but not where the top-end ring is crucial to the sound. According to many acoustic guitarists, another disadvantage of nickel acoustic strings is that they lose their life very quickly and need to be restrung more often.

Figure 13.6 Washburn J28SCE/BK grand auditorium-style acoustic guitar with cutaway

As well as these different string compositions, you can also buy coated acoustic guitar strings, which last longer. The coating creates a smoother texture to the strings that some players prefer. The string coating helps the strings to resist rust. These benefits come at a cost, and coated strings are usually more expensive than regular strings.

Although you probably won't encounter classical-style guitars too often, it might be of interest to you that nylon strings are not classified by gauge, but by tension, and are usually found in the following sets: soft, medium and hard tension string.

Finally, plectrum thickness or the use of the fingers makes a huge difference to acoustic guitar tone. If you are a fingerstyle player, you are likely to create a much softer and quieter tone with much less attack. The softness and weaker attack has the effect of making the tone sound warmer overall. When recording a fingerstyle player, you will also have to be much more careful with your recording technique, as you will likely have to use a lot of gain on your pre-amp to get optimal levels. This means that unwanted noise is more likely to be picked up. This could be acoustic noise such as bleed or prominent finger sounds or buzz, or secondary unwanted noise in the recording chain. If you are using a plectrum, you are likely to create a much more aggressive and louder tone, with improved note clarity. The optimal thickness of plectrum may vary depending on how much you want the acoustic to stand out among the other layers in the arrangement or due to the style of musician. Generally, the thicker the plectrum, the more pronounced the attack and the louder the overall tone. When the guitar needs to be subtler, or where there are several layers of acoustic guitar, I would advise choosing a thinner plectrum.

You can see from all the variables above that it is not just the instrument itself that makes a substantial difference to its tone. Therefore, there are many ways to achieve the sound that you require. A little forethought can go a long way, so envisage your desired guitar tone and consider all of the variables above to get the best results. For instance, if I were to record an indie band in the style of Oasis, I would probably get the best results out of a dreadnought-style guitar, using medium to thick strings and a medium plectrum. If I were recording a fingerstyle guitarist, on the other hand, I'd likely choose a grand auditorium style with light/medium strings. An acoustic guitar style that is gaining popularity involves playing fingerstyle but adding percussive 'drum like' effects by hitting the body of the guitar. In these instances, the warm bottom end of the dreadnought-style bodies may be best in combination with some light to medium strings. Even when you have envisaged what you want, you still might need to tweak the sound so still use your ears even when you think you have the perfect combination.

13.2.2 Acoustic Guitar Environment

As well as the instrument itself, the acoustic environment is also critical. The acoustic guitar is an instrument rich in sustain, harmonic content and brilliance in its top end, which is best captured in a controlled but bright space with more ambience than might typically be desired when recording drier, close-mic'd instruments. These requirements can make the acoustic guitar a difficult instrument to record in a small home studio. However, there is usually a sweet spot in most rooms that can achieve good results – it may just take a while to find it. If your room overall is too dead sounding, then you might want to try placing some wooden panels on the floor or on walls, making sure they are not positioned at early reflection points (in relation to the instrument). Doing this will help make the area a little bit more reflective and make the guitar sound more exciting. The biggest problems I've encountered in project studio live rooms are where there are more reflections in certain frequency areas than others.

Before you blame the recording chain and upgrade mics and pre-amps, think about some of the DIY room treatments suggested in Chapter 9. Even though these treatments are suggested for critical listening or control room environments, they can be a great improvement in the live room too. Just remember that you probably won't need so much absorption, and use a little more diffusion too. If you don't have much of a budget, then hanging a couple of duvets in choice places can make a huge difference. You can use the clap test to help outline problem areas.

13.2.3 Acoustic Guitar Mic Choice and Placement

So we have a good player and a decent and well-maintained instrument in the sweet spot of a good room. Let's consider mic choice and placement. First, in terms of microphone choice, I almost always favour condenser mics. Just like with the piano, dynamic mics aren't as functional because the acoustic guitar produces a large frequency range, with a lot of harmonic complexity in the top end. To add to that, the acoustic guitar can also create extremely 'spiky' high-frequency transients that the microphone needs to respond to.

Theoretically, then, small diaphragm condensers would be the best choice for the acoustic guitar as the smaller diaphragm has a more accurate and detailed transient response. However, as we all know, what sounds good to our ears is not always the most obvious scientific choice; sometimes I prefer the sound of a less responsive diaphragm even on sources with sharper transients. Doing this can help reduce the amount of compression needed because it smoothes out the transients, as well as bringing out other harmonics that would usually be smeared by these high-frequency spikes. A great example of this is using a ribbon mic instead of a condenser: a ribbon's roll-off softens the harshness that is often undesirable in acoustic guitars, cymbals, etc. Because of this, a large diaphragm condenser will often provide a happy medium and get the results I require without artefacts produced from heavily compressing the signal.

When mic'ing the acoustic guitar, you need to make a decision on how many microphones you use to capture it. Just like any mic selection/placement decision, this needs to be based on a few factors:

1. Can I capture the sound fully with the number of mics proposed? This could be determined by the physical distances concerned, like with the piano. Alternatively, it could be affected by the tonal character of different areas of the instrument, like the distinct difference in sound between the 'crack' of the top of a snare and the 'rattle' of the beads on the bottom.
2. How much space is there for the sound in the mix?
3. How wide does the sound need to appear in the mix?

In the case of the acoustic guitar, I've achieved more than satisfactory results from a single condenser mic placed well; in tracks that are dense already, this is often all that is needed. However, in folk music, country and often ballads, the acoustic guitar needs to not only sound 'massive', but also to hold a significant excitement in the stereo field. In this case, I've often used two or three mics to allow the mixer to give it more of a stereo feel.

Single Mic Placement Advice

When placing mics for the acoustic guitar, there are two rules of thumb I like to remember:

1. The magnetic pickup, piezoelectric pickup or microphone built into electro-acoustics usually sounds terrible on all but the most expensive guitars. Therefore, I typically only use it in live scenarios out of convenience as isolation and portability are important. Some producers like to take the feed from it and use it in conjunction with other mics but it is almost never used on its own.
2. It is usually a bad idea to face the microphone on-axis directly in front of the sound hole. Now this might sound counterintuitive (yes, it is where the sound comes out of), but due to the body resonance of the guitar itself, the result will almost always be a 'boomy' mess. As acoustic guitars seldom need any more bass or low mid, it is usually better to place the mic at least 0.3 metres away from the instrument to avoid proximity effect.

Figure 13.7 Single-mic'd acoustic guitar, 12th fret

In terms of placement, many engineers start by placing the mic at the 12th fret, and then gradually tilting the microphone from its position on-axis until they find a sweet spot. I find that this spot is usually with the microphone facing somewhere between the sound hole and where the neck joins the body of the guitar. Generally speaking, the closer the mic is pointing to the sound hole, the fuller and warmer the sound will become, and the closer to the neck, the brighter it will become; for an effective acoustic guitar sound, you need the right balance of both.

Other engineers favour placing the mic directly above or below the guitar with the mic off-axis facing towards the sound hole.

Multi-Mic Placement Advice

As I mentioned previously, the acoustic guitar sound needs the right balance of brightness and warmth, and using more than one mic to capture this balance can help make a track more exciting, especially with creative use of panning and effects. You can use many of the stereo mic'ing techniques we've previously discussed. However, acoustic guitars are normally overly exaggerated to provide excitement rather than being a true-to-life stereo source. This can be done easily by finding one mic focused on the sweet

Figure 13.8 Single-mic'd acoustic, above guitar

Figure 13.9 Horizontally exaggerated spaced pair

Figure 13.10 Vertically exaggerated spaced pair

Figure 13.11 Wide spaced pair

spot for the warmth of the body and another for the brightness, and then hard-panning these either side.

A convenient way to do this accurately is by using a spaced pair on the horizontal plane. One mic is placed somewhere near the sound hole of the guitar to capture the body, and another mic is placed near the 12th fret to capture brightness. I then pan these hard to create an exciting pseudo-stereo effect. Remember to tilt off-axis as required (Figure 13.9).

Another pseudo-stereo effect can be achieved by using two mics positioned vertically, one above the other, with the top mic capturing the low strings, and the bottom mic capturing the high strings. The two mics can then be hard-panned left and right (Figure 13.10).

If you want to use a true stereo placement, you may want to consider a widely spaced pair, one at the bridge position and another at the neck, tilted off-axis and towards each other. This creates a similar effect to the horizontally exaggerated pair and, while it is not as direct or exaggerated, it is a more true-to-life stereo sound (Figure 13.11).

You could also use an X-Y pair near the sound hole. As the microphones are 90 degrees off-axis, there isn't as much of an issue with boominess, but if you do encounter some you can offset it slightly towards the neck. Standard Blumlein or M/S configurations are also options.

Placing acoustic guitar mics can be a lot of fun, and some unusual positions can work out to be the best in your acoustic environment. So take some time to experiment not only with the examples above, but with less conventional placements. Also, do not forget about our good friend phase when placing acoustic guitar mics, especially if you are planning to include the electro-acoustic's internal mic.

13.3 CLASSICAL STRING INSTRUMENTS

First, I want to begin by taking a little time to discuss the instruments themselves, as many of you will not have had any experience working with these instruments. When I talk about classical string instruments, I am talking about those that we hear all over our favourite Western classical tradition pieces, such as the violin, viola, cello and double bass – better known as the *violin family*. Each of these instruments has the same fundamental mechanics.

We should already know that lower frequencies have longer wavelengths, and to create such wavelengths takes a longer string, and to sufficiently amplify this takes a larger body. Inversely, a shorter string is better at creating higher-pitched sounds and a smaller body resonates better at higher frequencies. So, despite their similar aesthetics, each member of the violin family is optimized to work in specific frequency bands.

Figure 13.12 Violin family

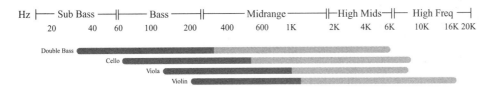

Figure 13.13 Violin family frequency range[2]

With each member of the violin family, you can stimulate the string in two ways. The first and most common way is to drag a bow across the strings. Alternatively, you can pluck the strings with your fingers, called *pizzicato*, much like you might on a bass guitar.

Bows are made out of horsehair and players use a resin-like material called *rosin* to make it slightly 'sticky', which helps the bow stick to the strings and creates the friction necessary to make the string vibrate properly. Using a bow, you can create many different timbres (tonal colour) and articulations (attack).

Let's explore the bowing variables that affect the volume of the instrument, each of which will also alter the tonality of the sound:

- *Pressure of bowing.* This is the most common way to increase the loudness of a stringed instrument. The more pressure you place on the strings, the louder the sound. This can come at a price, as increasing the pressure on the string tends to produce a harsher sound.
- *Point of contact.* Bowing closer to the bridge of the instrument has a louder, more intense sound that emphasizes higher harmonics; closer to the fingerboard creates a softer, more delicate sound, which emphasizes the fundamental frequency.
- *Speed of bowing.* The faster you bow, the louder the sound. However, in combination with the other techniques above, you can still increase bow speed and play softly.

Using these three variables in conjunction with the amount of bow being used and the way that the bow is 'bounced' or stopped can create all sorts of different emotions. These different techniques are called *articulations*. Because the vast majority of string players are classically trained, and the fact that there are many ways to achieve different timbres, it is a big advantage to speak to string players in terms found on traditional stave notation. Obviously, if you seldom record musicians that have this kind of classical training, it is rather redundant. However, the basic knowledge I am about to impart has been advantageous to myself on several occasions, despite the fact that I am usually recording rock bands. I am not going into tremendous detail, so if you are a string player you may want to skip ahead a little. Unfortunately, for those of you still reading, most of these dynamics or articulation marks are written in Italian (although occasionally French or English). First of all, let's examine some dynamics terminology. These will be useful not only for string players, but for any classically trained musicians:

ppp or *piano pianissimo / pianissimo possibile* meaning 'softest possible'

pp or *pianissimo* meaning 'very soft'

p or *piano* meaning 'soft'

mp or *mezzo-piano* meaning 'moderately soft'

mf or *mezzo-forte* meaning 'moderately loud'

fp or *fortepiano* meaning to attack the note and then quickly reduce the intensity of the note after the initial attack

f or *forte* meaning 'loud'

ff or *fortissimo* meaning 'very loud'

fff or *forte fortissimo / fortissimo possibile* meaning 'loudest possible'

As well as dramatic changes in dynamics, you can also have gradual changes:

crescendo (often abbreviated cresc.) meaning 'gradually becoming louder'

decrescendo (often abbreviated decresc.) meaning 'gradually becoming softer'

Next, let's look at some articulations that relate to the way that these instruments are bowed. Some of these are only applicable to string players:

Détaché is the most basic stringed instrument articulation and indicates separate bow strokes for each note. This makes the sound very smooth and consistent. If no articulation is marked in the notation, then détaché is assumed. The definition of détaché in French is 'detached'. This can often lead to the meaning being misconstrued or confused with staccato. However, détaché simply means that the notes are *not* slurred into one another. Détaché passages are usually played with the middle section of the bow. Two variations of détaché are commonly found: a punta d'arco and al tallone:

A punta d'arco uses the tip of the bow and gives the passage a lighter feel.
Al tallone uses the heel of the bow and gives thicker more aggressive feel.

Legato, meaning 'tied together', indicates the notes should be smoothly connected, played either in one or several bows. In musical notation, slurs are often used to show the legato bow stroke. A legato passage sounds very 'fluid' and has no pause.

Collé, meaning 'stuck' or 'glued', indicates a very short stroke, beginning with the bow lightly contacting the string with a short, sharp pinch. The bow is then lifted between notes.

Martelé, meaning 'hammered', indicates each note is percussive, starting with a sharp accent known as a 'pinch' followed by a quick release. A stroke with an even stronger pinch is referred to as *martellato* or *marcato*.

Staccato, meaning 'detached', indicates the notes are shorter and stopped more abruptly. In violin playing, the staccato bow stroke leaves the bow on the string, deadening the sound completely. There are several variations of staccato, including staccatissimo and mezzo staccato:

Staccatissimo indicates a very short, separated staccato.
Mezzo staccato is a slurred staccato, meaning that a series of individually articulated notes are connected with a single bow stroke.

Spiccato is like a staccato stroke but the bow is lifted off the string, which creates a little more 'ring' or decay. Spiccato is characterized by a 'light' bouncing sound.

Sautillé (or *saltando*) is a bow stroke played rapidly in the middle of the bow with one stroke per note so that is bounces slightly off of the string. Spiccato and sautillé strokes are very similar, especially at higher tempos, and are sometimes used interchangeably.

Portato (or *louré*) is the combination of legato bowing with individually articulated notes to create a 'pulsing' and 'magical' effect; played loudly, it is much more 'dramatic' and 'emphatic'.

Jeté (or *richochet*) means to 'throw' the bow on the strings so that it bounces; this produces a series of rapid notes. Jeté is a technique most often used in solo passages or by virtuoso players.

Glissando (or *gliss*) means to distinctly slide from one pitch to another. This creates a gliding effect that can be very emotive but tiresome if overused. An alternative to gliss is *portamento* (or *port*), which is a less exaggerated slide often used out of technical necessity. Portamento is usually much quicker than glissando and provides a way to smoothen the transition, being two large intervals on a single string.

Pizzacato (or *pizz*) means to pluck a note with your fingers rather than use the bow. Because of the decay of the instruments, pizzicato is a very short sound on violins and violas. However, the cello and double bass are capable of a longer decay when played pizz. With larger string instruments, further rhythmic effects can be created by using the fingers to mute the string after plucking. The double bass is often used pizzicato in contemporary style.

As well as the bowing technique, the other hand that fingers the string can have an effect on the tonality and expressiveness of the instrument, much like on a guitar. One such signature texture is vibrato, a hallmark of orchestral music. Like how a guitarist creates vibrato with his or her fretting hand, vibrato is created by the string player rocking his or her finger on the string while playing the note. This gives the instrument a much more emotive quality. Passages where expression is to be exaggerated (including vibrato) are often marked with *espressivo*. If only vibrato is required, then the music is marked with *molto vibrato*. Another common technique is to rapidly fluctuate between adjacent positions creating a *trill*, which is again similar to techniques used by lead guitarists. A trill is notated with a *tr* above the note and sometimes followed by a wavy line. If you use greater intervals, the trill becomes a *tremolo*, which is used to create tension in pieces. Tremolo is notated with extra strokes through the notation.

Now that we know more about how the instruments themselves are likely to be played, you should be able to communicate more effectively with the player. This will free you up to concentrate on other aspects of the production. However, the dynamics and articulations described here only scratch the surface of what you might find in notation. Because of this, I usually make a point of asking for the string players' opinions and also ask them to mark out any extra expression in scores given to them. After the session, I will often take home their notes so that I can review any changes they make and learn from them. Before we move on to some mic'ing techniques for string instruments, let's examine the differences between the violin, viola, cello and double bass.

13.3.1 Violin

The violin is the highest-pitched instrument in a classical string section and is therefore the smallest in size. The violin is played from a position resting on the shoulder of the musician and is tuned in fifths to the notes G-D-A-E (from lowest string to the highest). The body length of a 4/4 (full size) violin is approximately 14″ (356 mm). The range of a violin is concert G3 to roughly E7 and is written in treble clef. Although the violin is capable of a producing a fundamental frequency just under 200 Hz, the sound is far more controlled and cleaner sounding in the top end of its register. This is because the body is not sufficiently sized to amplify the fundamental frequency of the lower notes; the result of this in the low end of the violin's range is that the fundamental frequency is significantly quieter than some of its harmonics. This particular problem is one that occurs in all the instruments of the violin family, and is one of the main reasons why there are four different instruments in a traditional string section. In the upper end of their register, string instruments have loud harmonics that go way above the highest fundamental. For instance, the violin can create prominent harmonics up to around 10 kHz, which is much higher than its highest fundamental of approximately 3.5 kHz.

13.3.2 Viola

The next highest in pitch is the viola. Like the violin, it is played on the shoulder and it is tuned a fifth below the violin, again at intervals of fifths, the open strings being C-G-D-A. The viola is usually notated in alto clef and has a practical range of concert C3 to roughly E6. Unlike the other members of the violin family, a viola does not have a standard full size and is instead categorized by its length in inches, which is usually between 15″ and 18″ (this makes the viola between 1″ and 4″ bigger than a 4/4 violin). Viola sizes are less standardized than those of the violin because creating an instrument with the same acoustic properties as a violin for the pitch of the viola would make it too large to play on the shoulder. This means that a viola will often be chosen to suit the size of the player. Adult violas are most popular with a total body length of between 15″ and 16.5″ (394–420 mm). They can produce a lowest fundamental frequency of around 130 Hz and a highest of approximately 1.3 kHz. As you have probably already hypothesized, the fact that violas are undersized in comparison to a violin makes the frequencies of the lower fundamentals even less pronounced. This disparity between body resonance and frequency range of the strings gives the viola its unique 'nasal' quality.

13.3.3 Cello

Next is the cello. It is played from an upright position and is the highest pitched of the family to be played this way. The player is usually seated with the body of the cello positioned between the player's legs, but on rare occasions soloists might play from a standing position. A full-size adult cello is approximately 48″ (121 cm); it is tuned in fifths at an octave lower than the viola (C–G–D–A). The cello is usually notated in bass clef but higher passages can be tenor clef or even treble clef for extended higher passages. The cello has a larger range than smaller violin family instruments from concert C2 to roughly G5. With the lowest fundamental of the cello being at 65 Hz and the body resonating just above 100 Hz, there is again an issue with amplifying sounds at its lowest extremes. The upper range of the cello is around 1 kHz and can produce harmonics up to around 8 kHz.

13.3.4 Double Bass

The double bass is the lowest-pitched member of the standard violin family and is again played from an upright position. Although the double bass from a quick glance is similar to the rest of the violin family, it is perhaps more similar in construction and tuning to the older viol family of instruments. This includes its tuning, which is fourths rather than fifths; its tuning of E-A-D-G will be instantly familiar to bass guitarists. The double bass is notated in bass clef and the pitch played is an octave lower than it is written (we will be discussing transposition instruments in more depth shortly). The double bass has a practical range of concert E1 to roughly D5, although some instruments use an extra string to get to either B1 or C1. Some double bass players can even utilize a device called a c extension to lower a standard four string double bass to C1. With a length of approximately 75" (199 cm), the 4/4 double bass is far less manageable for most players, so the 3/4 size is more popular, having a length of approximately 72" (182 cm). While the other members of the violin family are predominantly played with a bow (*arco*), certain styles of double bass are played by plucking the strings (*pizzicato*). In orchestral music, the double bass is most often bowed and played from a sitting position (often alternating between arco and pizzicato passages). In more contemporary genres, the double bass can be amplified and plucked from a standing position. Genres that utilize this are 1950s-style blues and rock and roll, rockabilly, bluegrass and some variants of folk music. The range of the double bass is from 41 Hz to around 250 Hz, and it produces harmonics up to about 7 kHz.

13.3.5 Single Player, Ensemble or Samples?

Now that we have dissected the attributes of the modern violin family, we can discuss capturing the sound that they make. First, your budget is going to have a huge impact on which direction you choose to take. Hiring a full orchestra of session musicians is only going to be suitable for major label projects and, to be honest, you can achieve similar results by multitracking smaller sections. Many achieve this by hiring a quartet or quintet of session musicians and double or triple tracking them; this will be in reach of those with a moderate budget. When recording a string ensemble, you will usually use a number of room mics to capture the whole string section and also mics positioned to capture each instrument (or instrument type in the case of larger ensembles). These mics that are used to help augment certain instruments (or sets of) are called spot mics. Just like other acoustic instruments, the size and sonic quality of the acoustic environment is critically important when recording strings. This is due to several factors:

- There is a greater sense of ambience expected from strings, so the room needs to be of sufficient size to give some sense of space (although more can be added with artificial reverb).
- There also needs to be sufficient space between the musicians to minimize bleed between spot mics.
- Because there is more ambience required, room mics are often ridden higher in the mix than spot mics. Therefore, comb filtering and uneven RT60 times across the frequency spectrum caused by room inaccuracies have a bigger impact on the overall tone than when an instrument is close-mic'd.

Even if you are not recording an ensemble together or are only recording a soloist, there are still disadvantages to using close-mic'ing techniques. In close proximity to the strings, a mic will capture a scratchier and thinner sound that doesn't possess the full, complex sound of a string instrument; this is because a certain amount of distance is needed for the sound to have the chance to develop. Another downside to close mics is the fact that they also increase the audibility of the mechanical noise from the bow.

If you only have a small or untreated environment, then I'm afraid it will be very difficult to achieve a satisfactory sound recording string instruments. If you have this conundrum, then you may want to consider finding/hiring a better space. If that is unfeasible, then you might also want to consider sample libraries. In my experience, a well-programmed string section using a great sample library can sound great. Within a densely arranged pop or rock track, if it is well programmed, it can sound almost indistinguishable from a real string section, but on purely orchestral arrangements it will, of course, be more apparent. I have also found that layering a small number of real string parts with a larger ensemble of sampled parts can deliver even more polished results. However, there are many reasons why using a sample library might not be as attractive as it first seems; these reasons will be discussed in the sampling section of this chapter.

Before moving on to mic choices and placement for string sections and soloists, it is also worth noting that some producers even use samples and real strings in tandem, often recording one or two lines of real strings and using samples for consistency or to beef them up. This can help to make a more realistic-sounding string ensemble possible on a budget rather than relying on samples alone.

13.3.6 Microphones, Placement and Positioning

Strings are both very dynamic and rich-sounding instruments, which sound very forced with anything more than slight post-production tweaking. In fact, a lot of the time there is no EQ used (other than maybe high-pass filtering). Compression is also used infrequently but sometimes the style/dynamics of the player might result in a spot mic needing to be compressed slightly. Room mics are rarely compressed as the amount of space between the players and the mic(s) helps to even out any level discrepancies.

In a more classically orientated work or a piece that hinges around a full orchestra or large ensemble (such as a Bond theme), the room mics are likely to be very high in the mix and spot mics used simply to fine-tune aspects. In the majority of more contemporary styles, though, you are likely to have a little bit more flexibility in your approach.

Spot-Mic'ing

Before we get on to room mics, let's quickly discuss spot mics. A spot mic's main purpose is to help supplement any room mics; for instance, if the room mics are not capturing enough of violin 2, then the mixer is likely to use a little more of violin 2's spot mic. Spot mics can also be used to bring focus to a certain part; if there is a solo violin passage, for example, you might want to ride the violin's spot mic up a bit for that section. The use of spot mics can therefore really help to add what the room mic(s) might be lacking.

In a smaller string section such as a quartet or quintet, you would have a spot mic on each instrument; for instance, a quintet would most likely comprise first violin, second violin, viola, cello and double bass. You've probably noticed my reluctance to call these mics 'close mics'; this is because although they are focused on a particular instrument (or type of instrument), they are not necessarily that close to them. In fact, in my experience, it is best to place each spot mic around 50–80 cm (20–30 inches) from the source.

Specific placement of these spot mics and the mic selection itself also changes drastically depending on whether the instrument is played on the chin (violin, viola) or on the floor (cello, double bass); it can also change depending on the frequency areas that you want to emphasize.

Violin/Viola Spot Mic Selection and Placement

Because the violin and viola are orientated horizontally, held at the chin and shoulder, the sound primarily radiates upwards, above the instrument. This means that they should be mic'd from an overhead position. However, this is not the only consideration. A violin or viola's frequency pattern tends to change drastically even within a small change in position around its axis. For instance, the bottom end of a violin/viola is radiated more evenly around the player, but the top end is much more directional, and is best picked up on-axis. But don't get ahead of yourself: this does not mean that it is always best to place spot mics directly on-axis. In fact, the extreme top end of a violin/viola is often full of undesirable mechanical noise that makes the sound buzzy or scratchy. Therefore, when spot-mic'ing string instruments, special care and attention should be paid in finding a sweet spot where you are getting a nice rich top end without an obscene amount of mechanical noise. Figure 13.14 is a good approximation of how the sound radiates from a violin or viola.

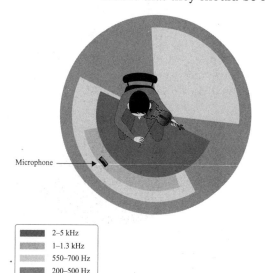

Figure 13.14 Violin/viola frequency-based sound radiation[3]

- 2–5 kHz
- 1–1.3 kHz
- 550–700 Hz
- 200–500 Hz

In terms of microphone choice for violin/viola, the most common candidates are ribbon mics and small diaphragm condensers as they both have a fast transient response. Ribbons are great for a rich warm character, and their roll-off means they are great at reducing top end noise when it is a problem. However, their figure-of-eight polar patterns might also cause some ambient noise to be picked up. Small diaphragm condensers will generally have the fastest transient response of any mic and are good for getting the strings to 'poke out' of a busy rock mix. When you are picking a microphone for the job, it is wise to avoid microphones with a hyped top end as this can result in a scratchy tone very quickly, making placement much more difficult.

Microphone	Characteristics
AEA R84	Warm yet punchy ribbon mic. More modern sounding than a Coles due to its quicker transient response but less top end than a Royer 121. A good all-rounder.
Royer 121	More present than an R84 or Coles but less extreme top end than condenser mics. Great for strings that aren't overly bright.
Coles 4038	Warm and retro ribbon mic with a lot of top end roll-off and slower transient response. Great for vintage styles.
Neumann KM84	A well-balanced small diaphragm condenser with a transformer for warmth and a fast transient response. Great all-rounder. However, it is noisier than more modern mics, so quieter passages may be more of a problem.
Neumann KM184	A newer Neumann SDC transformerless with a top end lift. Another good all-rounder. It is quieter than a KM84 but it is also significantly brighter.

Budget mics that are good on violins/violas include the SE VR1 (ribbon), Beyer M180 (ribbon) and AKG 451 (SDC).

Cello/Double Bass Spot Mic Selection and Placement

Cellos and double basses radiate sound horizontally rather than vertically so microphones are often placed at the level of the f-hole (i.e. the f-shaped sound holes). Because lower frequencies have a longer wavelength, it is usually best to give them a little more distance so that the sound develops, therefore I tend to err towards the latter end of the 20–30″ guideline. As with the violin/viola, cellos and double basses don't evenly radiate frequencies, and again the higher frequencies are more directional. While ribbon and small diaphragm condenser mics are most commonly best suited to capturing violin/viola, producers usually choose either ribbons, tubes or large diaphragm condensers to best capture cello and double bass. Tubes or LDCs have a slower transient response than SDCs, which can be useful as they aren't

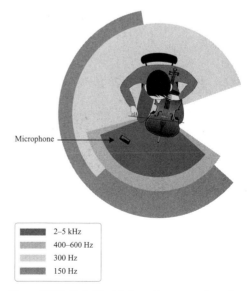

Figure 13.15 Cello/double bass frequency-based sound radiation[4]

prone to sounding as edgy or scratchy in the high frequencies. Using SDCs on violins/violas and LDCs on cellos and double basses can also help to aid the sense of separation between the instruments as each microphone is tailored for the type of sound each instrument creates. Despite this being an excellent rule of thumb, you should still make sure that you remember to test this for yourself and experiment with finding each instrument's sweet spots.

Microphone	Characteristics
Neumann U47	A classic tube microphone, it has a top-end boost but somehow still manages to sound warm and not overly brittle in the top end. However, because of its top-end boost, it might not be the best for budget or thin-sounding instruments.
Soundelux E250	A warm tube mic that doesn't have so much of an airy boost. However, it may sound a little dull on some instruments.

Room Mics and Ensemble Positioning

Now we have got spot mics out of the way, let's get down to the real meat and potatoes of string recording. As we have already discussed, there is typically less post-production trickery used on string instruments. This means that you need to be even more confident with the tone you are capturing at source, and this includes the room itself. The room should offer a large amount of space, and an even RT60 time is important. A lot of wooden surfaces (walls and floor) in the room can be very beneficial because, despite the fact that the wooden surfaces will invariably increase the top end's RT60 time a little, it will also add excitement in the top end, thus reducing the need to add any via EQ, helping to minimize the risk of scratchy-sounding strings.

Ensemble Positioning

Before we consider room mic placement, we should look into how string musicians are placed in both a full orchestra and also small ensembles. This may seem a little irrelevant in a pop/rock production where there are far fewer rules, but subconsciously your brain is excellent at recognizing patterns and is uncomfortable with irregularity. After countless classical and contemporary works have used a traditional orchestral seating, not doing so *might* cause the listener to perceive your mix differently; this is especially true when you have many parts of an orchestra in your piece rather than just strings. The positioning of a large orchestra is very rigid, and Figure 13.16 shows an orchestral seating plan.

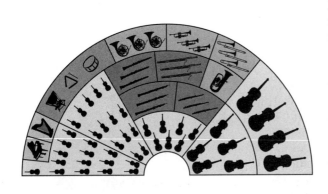

■ Percussion
■ Keyboards/Harp
■ Strings
■ Brass
■ Woodwinds

Figure 13.16 Orchestral seating plan[5]

An orchestra is arranged in this way in order to take into account the volume differences of the various instruments, thus allowing the fullest, most balanced sound presentation for an audience. As well as this, it is optimized for an interesting, balanced and pronounced stereo width. On a contemporary track, you are unlikely to be working with this many musicians; most of the time, producers will be recording quartets and quintets. In these cases, there are a couple of variations in seating plan. Remember, though, whichever option you choose, you need to remember that the spot mics need to be panned to the same position as they appear in the room mics. A more classically minded seating plan for a quartet would have the violins positioned on the left and the lower-pitched strings on the right.

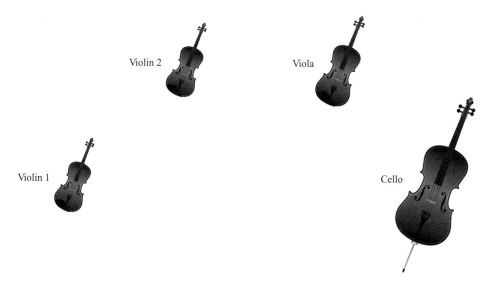

Figure 13.17 Classic string quartet seating plan

Pop and rock producers might choose to 'split' the violins and have the lower-pitched strings in the centre to more closely match the way that you would pan other bass/treble instruments in a contemporary genre. The drawback of this is that many string sections are not used to performing in those positions, and violin players in particular feel that it can affect their intonation when they are not sitting together (Figure 13.18).

Some producers may even go a stage further by placing both the viola and cello dead centre with the cello farthest to the rear (Figure 13.19).

If you happen to be recording an ensemble between the size of an orchestra and a quartet/quintet, you can always use the seating ideas discussed above by putting each type of instrument into rows like you see on the orchestral seating plan.

Room Mics

Now that we are more aware of the way that string players can seat themselves, we can think about room mics. With a full orchestra, you will have a conductor

438 Chapter 13 TRACKING OTHER INSTRUMENTATION

Figure 13.18 Contemporary variation quartet seating plan

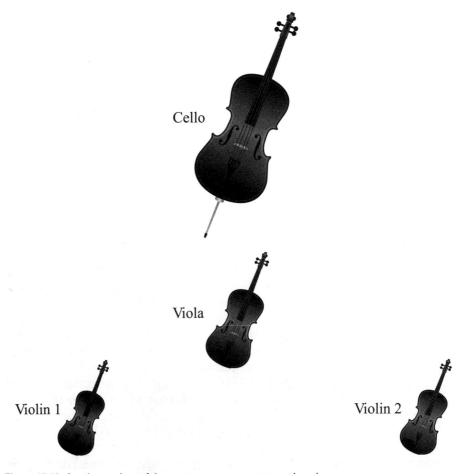

Figure 13.19 Another variant of the contemporary quartet seating plan

who has the job of organizing the players, setting the tempo and listening critically to the overall sound of the orchestra. Throughout the performance, you will see him or her gesture useful instructions to the musicians to improve the overall sonic quality of the orchestra. To be able to do his or her job correctly, the conductor needs to be in a position where he or she is:

- in the line of sight of every member of the orchestra; and
- in an optimum position to hear each section clearly.

If you look at the orchestral seating chart again, you will see that the majority of the loudest and often lowest-pitched instruments are at the back (remember, bass projects further). As well as the benefit to an audience, keeping this sort of separation helps the overall sound arriving at the conductor's position to be more evenly represented. Due to this and the conductor's proximity to the musicians, it is an ideal position to hear each section of an orchestra clearly so it is also a great place to set up room mics. Much smaller string sections such as a quartet won't necessarily need a conductor, but when I'm setting up room mics, in order to help with the positioning, I like to imagine there is one present, occupying a similar position. Even though the conductor's position is the most tried-and-tested location for room mics, the vision for your track might take you in a different direction, so don't think that is your only option. In fact, many producers will have a few stereo room mic set-ups placed at different distances so that they can blend or automate to get a different feel in post-production. Once you have considered appropriate location(s) for the room mics, you will have to think about how the string players are represented in the stereo field. To do this, you could use some of the stereo-mic'ing techniques that you would employ for drum overheads. However, two types of mic'ing techniques are far more popular with orchestral mic'ing, and these are the Decca Tree and the Blumlein Pair.[6]

Decca Tree
Let's start with my personal favourite, the Decca Tree. This microphone placement technique was specifically developed for orchestral recording in the 1950s by the engineers at Decca Records. The Decca Tree uses spaced microphones to give a wide stereo image, although it is sometimes criticized for lacking definition in the centre of the image. It uses three omnidirectional mics in a left, centre and right configuration (placed in a 'T' shape), usually above the conductor. The left and right mics are spaced approximately 2 metres apart and the centre is in between those, approximately 1.5 metres in front. Specialist stands similar to a stereo bar are often used to achieve this but you can also use three separate stands. When mixing, the left and right channels are hard-panned and the centre channel is sent to both the left and right channels at a level such that it fills in the centre without sounding too mono. The Neumann M50 is the classic microphone choice for a Decca Tree. However, a large array of condenser microphones could be workable as long as they are omnidirectional.

As well as orchestral situations, the Decca Tree is widely used in choir recordings and is also a useful room-mic'ing technique for many other purposes.

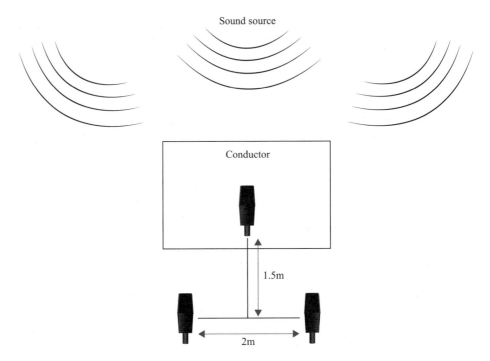

Figure 13.20 Decca Tree microphone set-up

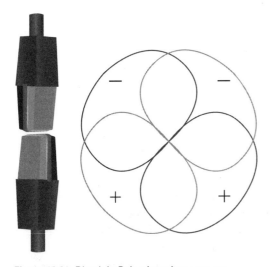

Figure 13.21 Blumlein Pair microphone set-up

Blumlein Pair

The main competitor to a Decca Tree set-up is the Blumlein Pair. Instead of using a spaced technique, it utilizes a coincident pair of microphones to create a stereo image. It provides a rich, stable and highly realistic stereo image but is often maligned for its lack of larger-than-life width compared to a spaced technique. A Blumlein Pair consists of two figure-of-eight-patterned microphones positioned at 90 degrees to each other, in order to create a full stereo representation of the source. When arranging this, the capsules should be kept as close as possible to one another. Ribbon mics and switchable-pattern condenser mics lend themselves well to Blumlein configurations.

13.4 BRASS AND WOODWIND

You might have noticed some other types of instrument in the orchestral seating plan that we haven't discussed yet. As this course has been focused primarily on contemporary styles of music, a huge section on recording harpists would be a bit of a departure from the needs of the reader. However, within many genres of contemporary music such as soul, funk, Motown, ska and ska punk, you will likely encounter a number of brass instruments – or, as they are more commonly called in contemporary music, *a horn section*. On occasion, usually in folk/acoustic music, you might also find woodwind instruments such as flutes, clarinets or recorders.

Due to the physical constraints of brass and woodwind instruments, there are usually a number of different sizes of each type of instrument, each one designed to deliver a specific range and tonality. These different sizes are given names relating to their pitch range such as 'bass', 'alto' and 'tenor', or are referred to by their key/tuning. One of the reasons for referring to instruments by key is so that musicians are aware of the fact that, when music is notated for their instrument, the note that they read is not necessarily the pitch they should sound; these instruments are called *transposition instruments*. Music for brass and woodwind is usually notated with the key of the instrument in mind rather than the actual key of the piece. So, for instance, a piece in the key of E will be written for a B♭ trumpet in the key of F♯. Having to adapt for notation transcribed for other instruments is also a fairly common occurrence for brass or woodwind players. For instance, take a player of a C trumpet who is reading music notated for a B♭ trumpet (the most common variety). The musician would have to transpose each note down a whole tone for his or her instrument – so, when the music reads D, the musician would play a C. Accomplished brass and woodwind players will sometimes be able to do this on the fly, but it is not as easy as it may appear as they will have to consider both the key and any accidentals (sharps and flats outside the key). Transposition on the fly is made a lot easier if the player knows his or her scales!

We will delve into to the common variations of brass and woodwind instruments shortly, as well as discuss the way that both brass and woodwind instruments work. For now, however, I want to quickly discuss the ways in which recording these instruments might be similar to recording other acoustic instruments:

1. The tonality of the room and the player's placement within the room, again, plays a huge part. Just like when recording other acoustic instruments, having a suitably treated room will make your life a lot easier. As with strings, I tend to like recording brass and woodwind in wooden rooms, particularly if they have a slight top end roll-off; this helps to stop some of the harshness that often occurs with brass, particularly trumpets. In the case of brass, larger rooms are also useful as the instruments are extremely loud. I like brass recordings with slightly less room tone than I do for strings. As live rooms suitable for ensembles usually have too long a reverb tail for my tastes, I will often employ gobos, placed a metre or so from the performers, to help 'tighten' the sound.
2. If you record brass or woodwind sections live (particularly in a classical context), you are likely to utilize a similar mixture of spot mic'ing and room mic'ing as you would with strings.

Despite these similarities, recording a contemporary horn section or woodwind instrument is likely to provide a little more freedom than recording a string section. Factors such as the genre of music, the number of parts, the acoustic spaces available, the calibre of musicians and even the time available make a difference to approach. This has meant that I've recorded brass and woodwind instruments in several different configurations, ranging from recording each musician individually, to recording a whole section live in a large live room. For instance, on a Motown-inspired project, I might decide to record a horn section live because songs of that ilk tend to sound more 'live' and 'rougher around the edges', as well as feeling like they were performed in a real space. By contrast, horns on a skater punk track might warrant a more modern set of production values and a 'tighter' sound. Therefore, the best course of action may be to aim for the separation and larger-than-life width that is achieved by recording each musician individually, and to consider incorporating the use of gobos to reduce room ambience.

13.4.1 Brass

The classification of an instrument as being 'brass' is defined by the way the sound is created, rather than whether or not the instrument is actually made from brass. The most common classical brass instruments are the trumpet, trombone, tuba and French horn.

Figure 13.22 Brass instrument family

Early brass instruments were limited to playing one overtone series at a time. If you wanted to play in different keys, it required instruments of different lengths, or one instrument with replaceable sections of tubing. Modern brass instruments are mechanically complex, and various systems are employed to produce different pitches. This includes the use of slides, valves, crooks and keys. These mechanisms change the length of tubing (which vibrates) and thus alters the available harmonic series. Because of these systems, modern brass instruments are now fully chromatic (i.e. not limited to one harmonic series at a time). With brass instruments, sound is made by the musician 'buzzing' his or her lips against a small cup-shaped mouthpiece. Players can control the shape of their lips against the mouthpiece to select a specific harmonic, thereby producing a pitch. The way in which a player controls a brass or woodwind instrument with his or her lips and facial muscles is called his or her *embouchure*. Modern brass instruments are fully sealed except at the mouthpiece and the bell (which is at the opposite end and is where the sound leaves). Because of this, brass instruments are very directional and can produce high volume levels.

A horn section in contemporary genres is likely to consist of a trumpet, trombone and saxophone player. As tubas and French horns are unlikely to be used in a contemporary setting, we will not discuss them here. Unlike the trumpet or trombone, the saxophone isn't technically a brass instrument; this is because it operates with a reed and is therefore classified as a woodwind instrument (although the instrument happens to be made from brass). Therefore, we will consider the saxophone in the woodwind section that follows later in this chapter.

Trumpet

The trumpet is perhaps the most well known of the brass family. Modern trumpets usually have three (or, on rare occasions, four) piston valves. When a valve is pressed, it increases the length of tubing, which lowers its pitch. The fingering between varieties of trumpet is usually of equal interval, meaning that the pitch drop from the instrument's starting pitch is the same on each type of trumpet. This cannot be said for all brass instruments. The first valve lowers the instrument's pitch by a tone, the second valve by a semitone, and the third valve by a tone and a half (three semitones).

Trumpets are notated in treble clef and, as previously mentioned, the most common type is the B♭ trumpet. It is used in contemporary music and orchestral and jazz ensembles. It is known for its warm tone, and beginner trumpeters almost always start on a B♭ trumpet. A B♭ trumpet has a tubing length of approximately 140 cm. The playing range of a B♭ trumpet is concert E2 to B♭5 (notated as F♯2 to C5).

Other common types of trumpets are as follows:

- *C trumpet.* It is tuned to the key of C and is shorter than the B♭ trumpet. It has a brighter timbre (which could also be considered 'shrill') than the B♭ trumpet and is often used alongside it in in orchestral pieces.
- *Piccolo trumpet.* This is smaller than a B♭ trumpet and is pitched an octave higher than a regular trumpet in the same key (it has half the length of pipe). Sometimes piccolo trumpets have a fourth valve that normally lowers the pitch a perfect fourth (five semitones).

Other types of trumpet (which are rarely used today) include D, E♭, E, F and G trumpets.

Trombone

After the trumpet, the trombone is the most widely used brass instrument. Virtually all trombones employ a slide mechanism that changes the length of the tube and therefore changes the pitch of the instrument. Trombones that employ valves are available but seldom used. Like many classical instruments, trombones come in a variety of sizes, from contrabass (lowest) to piccolo (highest). However, tenor and bass trombones are by far the most popular. In contemporary music, it is almost always the tenor or bass found in horn sections. The approximate range of a tenor trombone is concert E2 to F5, and a rough range for the bass trombone is concert B♭0 to B♭4.

Because a slide system can create any length of tube between its minimum and maximum positions, trombones have always been fully chromatic. Due to this fact, no transposition system was ever necessary and trombone parts are notated at concert pitch. Trombone parts are typically notated in bass clef, although sometimes tenor, alto or even treble clef can be used. In British brass bands, the trombone is often notated in B♭ to make it more consistent for players of both the trumpet and trombone.

Mic'ing the Trumpet or Trombone

When you think about microphone choice and placement with trumpet and trombone you need to consider three main factors:

- High sound pressure level (SPL). Or, to put it another way, brass is often tremendously loud! Both the trumpet and trombone are capable of SPLs in excess of 130 dB at close proximity to the bell and typically the trombone is 5 dB or so louder than a trumpet.
- High frequencies have less dynamic range. At higher frequencies, it takes much more effort from the player to create the force required to create the sympathetic vibrations that cause a sound to be produced. This means that high parts of the brass instrument's register cannot be played as quietly as the lower parts of its range. This means that as you ascend, the instrument's register the dynamic range available at each successive pitch decreases.
- Brass instruments tend to have prominent upper harmonics, right up in the areas that can easily become fatiguing (8–10 kHz and sometimes beyond). With brass, the balance of upper harmonics compared to the fundamental frequencies and lower-order harmonics can change depending on volume too. For instance, the upper harmonics found in a trumpet note are significantly less prominent when it is played quietly. This means that a trumpet played quietly will sound much 'softer', 'warmer' or 'muddier' compared to a trumpet played loudly, which will sound 'brilliant', 'airy' or 'harsh'. A trombone, on the other hand, is much more stable, harmonically speaking, and even though there is some fluctuation, the upper harmonics are more present even at comparatively lower volumes.

This all means that you should:

- Avoid 'super-close' mic'ing situations. If you get too close, you can break sensitive mics or end up with a distorted and/or unrepresentative tone. You might see brass mic'd like this in live applications, but this is due to reasons of practicality and spill reduction. For trumpet, I will leave at least a metre between the bell and the mic; typically, this distance is much greater than

that, at around 2 metres. As trombone is a few decibels louder than a trumpet, I will usually place the mic a little further away, somewhere between 2 and 3 metres.
- Avoid the mic being directly on-axis with the bell. As brass is loud and hugely directional (especially at high frequencies), SPL levels can easily reach 130 dB and above. Placing the mic on-axis can seriously damage or break fragile microphones such as ribbons or condenser mics. It is often the physical mechanism that can be damaged, as well as the internal electronics, so even turning on the pad feature on microphones that have it is not guaranteed to protect them. A better way to avoid potential breakage is to either tilt the mic so that it is not directly on-axis with the bell, or place the microphone above, below or slightly to the left or right of the bell. I've even had good results doing a combination of both.

Figure 13.23 Brass mic positioning

In terms of microphone choice, I prefer ribbons, despite their potentially fragile nature. Classic designs such as the Coles 4038 are great because they have a good transient response but roll off the potentially fatiguing top end of the brass in a musical way. I've also achieved good results from the more modern Royer 121, which achieves a little more brightness and is far less fragile than vintage ribbons. If ribbons are not delivering the results you require, large diaphragm condensers are known to work well too; the classic Neumann U87 might be a good starting point in this situation.

13.4.2 Woodwind

Just as brass instruments are not necessarily made of brass, woodwind instruments are not necessarily made of wood. They also share some other similarities;

for example, both instrument families share a basic operating mechanism: making air in a tube vibrate to create sound. Again, like brass instruments, different varieties of instrument are named by the register they reside in. For instance, flute sizes could be piccolo, alto, bass, contrabass and double contrabass. One of the key differences between brass and woodwind instruments is the fact that a brass instrument only allows air to escape at the mouthpiece and bell, whereas woodwind instruments can let air escape out through a number of fingered holes. These holes alter the pitch of the instrument because they prevent the full length of tube vibrating; when holes are open closer to the mouthpiece, they shorten the length of vibration and thus raise the pitch.

In orchestral works, you are likely to encounter arrangements that include flutes, oboes, clarinets and bassoons.

However, with the exception of the saxophone and harmonica, you will rarely hear woodwind instruments in contemporary music. Some notable exceptions are:

1. 'Stairway to Heaven' by Led Zeppelin, which features recorders in the intro.
2. 'California Dreamin'' by The Mamas & The Papas, which features a flute solo.
3. 'I Talk to the Wind' by King Crimson, which features a flute solo and two clarinet parts.

Figure 13.24 Orchestral woodwind family

Woodwind instruments can be subcategorized into two groups: flutes and reeds.

- Flutes create sound by the player blowing across or directly into a hole on the instrument's mouthpiece.
- Reeds' sound is created by the player blowing into or across a reed mouthpiece that then causes the reed(s) and the body of the instrument to vibrate in sympathy.

In the following sections, we will dissect these woodwind families more closely.

Flutes

As previously mentioned, many flutes, including the popular concert flute, are played by blowing air across a hole; this is often difficult to visualize. Figure 13.25 shows the way that air is passed into the instrument.

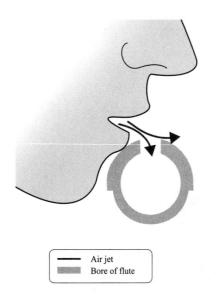

Figure 13.25 Flute air jet

Flutes can be subcategorized into three main groups: *open flutes*, *closed flutes* and *internal duct flutes*. Open flutes have a hole in the middle of each key to allow air to escape when the key is not pressed. A closed flute, on the other hand, has no holes in the keys, and therefore air does not escape this way. Typically, open flutes are more popular as they are able to produce greater tonal variation, a sound that is often described as 'warmer' or 'fuller', and are able to produce more harmonics than are available with a closed flute. Closed flutes are often played by flutists requiring a brighter tone, or by beginners as clearer tones are easier to achieve. Internal duct flutes (otherwise known as *fipple flutes*) operate with the use of a specialized mouthpiece called a *fipple* or *block*. We will discuss the basic mechanics behind fipple flutes in the recorder section.

Concert Flute

The concert flute is the most common type of flute and can also be described as a *transverse* or *side-blown* flute. With transverse flutes, sound is created by blowing across a mouthpiece on the side of the tube (as opposed to blowing into a mouthpiece at the end of a tube). Because of this, transverse flutes are played with the tube lying horizontally. Most flutists prefer open varieties, but closed varieties are also found fairly frequently. A concert flute is usually made from metal (often silver), although wood versions can also be found. They typically have 18 keys, are pitched in C (i.e. they are not a transposition instrument) and have a range of approximately three to three and a half octaves. Most flutes are able to cleanly pitch notes between concert C3 (middle C) and C6. On some flutes, the top extreme can sometimes reach F6, and with the use of an extra key (19th) and a longer foot joint (called a *B-foot*), flutes are able to extend to B3 on the bass extreme. The concert flute is known for its 'airy' and 'bright' tone.

Recorder

The recorder is a fipple flute. Fipple flutes are known for their 'whistle' like timbre. As air is blown directly into the end of the instrument the fipple flutes tube faces vertically. Blowing directly into the slot at the top of the fipple

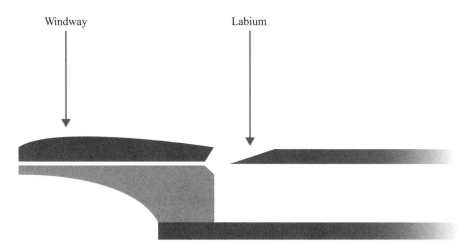

Figure 13.26 Fipple mouthpiece

mouthpiece forces the air down a 'windway' that is directed so that it hits a hard edge called the *labium*; this mechanism creates the recorder's signature whistle-like sound.

Professional-grade recorders are usually made from varieties of hardwood but there are many plastic varieties too, which are cheap to produce, durable and good for beginners. The sound of the recorder is pure and clear due to its lack of harmonic complexity. A typical recorder will have seven finger holes on the front of the instrument and one on the rear. There are many sizes of recorder and their pitches cover a vast range of the audible spectrum. Members of the recorder family are usually non-transposing. However, some sizes of recorder do sound at a different octave than is notated.

Mic'ing Flutes

First, members of the flute family are significantly quieter than members of the brass family, meaning you are much more likely to place the mic closer to the instrument. While exact figures vary on a case-by-case and instrument-by-instrument basis, I often find somewhere between half a metre and one metre usually gets optimal results. One other factor that you are likely to face is the fact that, unlike brass, the sound from a flute radiates from multiple areas of the instrument. In fact, the most common 'mistake' novices make when recording members of the flute family is to only concentrate on the sound that is produced out of the foot of the instrument.

Mic'ing Concert Flutes

With concert flutes, sound radiates from the mouthpiece and the keys, as well as from the foot end. The required blend of each of these elements is likely to change depending on the application and thus dictate the way that you go about capturing the source.

In folk or orchestral music, there is usually an even balance of the sound from the mouthpiece, keys and foot. A good approach here might be to place the mic directly above the instrument, as a concert flute is played with the instrument parallel to the ground. Placement somewhere along the keys can get a nice balance between the mouthpiece, foot and keys (although you may need to move it a little closer to the mouthpiece if the amount of mechanical noise from the keys is excessive; this is particularly problematic on cheaper instruments). Folk music often has a 'closer' sound, so placing the microphone nearer to the instrument, and the use of gobos, might help to make the source sound more forward. With orchestral music the sound is often more coloured with room tone and general ambience, so placement above the instrument, and just a little farther from the source, might be best. A jazz-style flute also has a 'close' sound but is likely to require a much 'breathier' sound than other genres. Therefore, mic placement above or in closer proximity to the mouthpiece can be very effective.

In terms of microphone choice, the best thing to do is to avoid microphones that sound too 'scooped' or microphones that have a significant presence boost. The flute naturally has a lot of top end and easily becomes shrill and 'grating'. Sometimes some of this top end might need taming slightly, which makes a large diaphragm condenser a good candidate as its transient response can slightly 'smoothen' the sound without removing too much of the flute's character. I've found more neutral LDCs, such as the AKG 414, to be a good choice because of this, although some engineers swear by a Neumann U87 for flutes. Alternatively, you can also look at ribbon mics; a vintage Coles 4038 might be noisy, but it can sound nice when a flute is overly 'present' for its intended application in the track, and a Royer 121 will balance the top end of the flute and create some warmth.

Mic'ing Recorder

Just like concert flutes, sound radiates from a recorder in three places. However, instead of sound radiating directly from where you blow into the mouthpiece, it comes from the fipple hole. The sound from the fipple hole is a sizeable component of the overall tone of the recorder and, because of this, I often choose to focus the microphone position towards it. Mic'ing directly on-axis to the fipple is less desirable though, as it dramatically increases 'breath noise', which, unlike that of a concert flute, is rarely wanted. As the recorder is played vertically, I like to place the microphone in front of the player, at roughly the height of the player's nose, and then angle the mic so that it is pointing somewhere between both the fipple hole and the finger holes.

The recorder is a very forgiving instrument in terms of mic choice. I have used dynamics, condensers (both large and small) and ribbons, all with success. However, when it comes to picking a favourite, again, a large diaphragm condenser with a more neutral tonal quality is my personal preference (Figure 13.28).

Reeds

A *reed* is a thin strip of material that vibrates to produce a sound on a musical instrument. Reed instruments come in three main types: single reed, double reed and free reed.

A *single reed* is made from cane (or a synthetic equivalent) that is fixed over an opening in the flat surface of a mouthpiece. When blown, the reed vibrates against the mouthpiece to create sound. The clarinet and saxophone are examples of single-reed instruments (Figure 13.29).

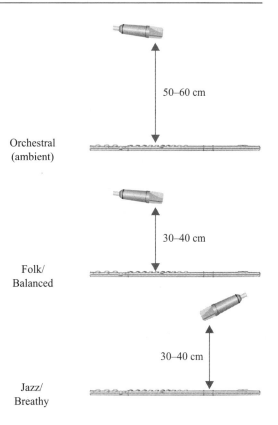

Figure 13.27 Flute mic'ing examples

Figure 13.28 Recorder mic'ing

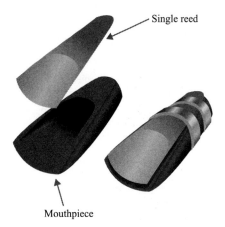

Figure 13.29 Single reed

Figure 13.30 Double reed

A *double reed* is also made from cane (or synthetic equivalent) but consists of two pieces of reed that are attached together. When a double reed is blown, the two reeds vibrate against each other, creating sound. The oboe and bassoon are examples of double-reed instruments (Figure 13.30).

Free reeds consist of a small, thin strip of material (usually metal, plastic or bamboo), which is placed in or over a slot, and fixed at one end. When air is forced to flow over the reed, by blowing or sucking, it flexes back and forth though the slot, producing sound. Many free-reed instruments have multiple reeds of different lengths incorporated into a frame. By focusing air flow over individual reeds, different notes can be produced; longer reeds produce lower pitched notes, shorter reeds produce higher pitched notes. These instruments are categorized as *free-reed aerophones*. Pressure or suction can be applied with the mouth like on a harmonica, or using bellows like on an accordion.

Saxophone

The saxophone is perhaps the most commonly used example of a single-reed instrument in contemporary music. Despite the fact that the saxophone is popular in contemporary music genres, you are unlikely to see it employed in classical works. By acoustic instrument standards, the saxophone is a young instrument. Developed in the 1840s, the saxophone's inception occurred long after many of our favourite classical works were composed. In fact, the more rigid standards of the orchestral world mean that, even to this day, classical tradition composers rarely incorporate it in their music. However, in the early twentieth century, the saxophone did begin to gain a greater popularity, and, during the big band era of the 1940s, it became a key instrument for jazz compositions. From then on, it has maintained its popularity.

The saxophone is a woodwind instrument by operation, but its brass body makes it sound like a hybrid between a woodwind and brass instrument. The distinctive 'rasp' and sound radiating from the keys like other woodwinds is there, but so are the brightness, upper harmonics and projection of a brass instrument. Like other brass and woodwind instruments, saxophones come in a number of sizes to accommodate many ranges.

Saxophones have 23 keys: 13 for the left hand and 10 for the right hand. Saxophones are standardized between types, meaning that the interval gaps between keys are identical, regardless of the size of saxophone. Because of this, many saxophonists play and own a variety of types. This standardization also means that transposition in notation is, again, important. The most common types of saxophone are the alto and tenor saxophones, which are notated in E♭

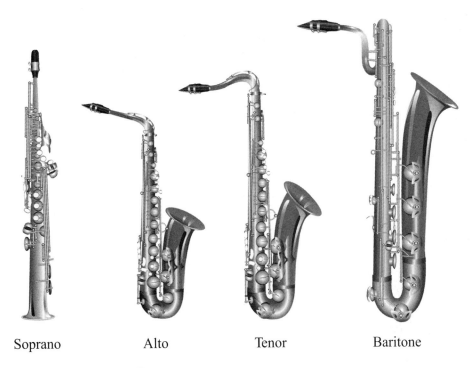

Soprano   Alto   Tenor   Baritone

Figure 13.31 Saxophone family

and B♭, respectively. The alto sax has a range of roughly two and a half octaves from concert D♭3 to A♭5. The tenor sax has a range of roughly two and a half octaves from concert A♭2 to E♭5.

Harmonica
The harmonica (also known as a *mouth organ* or *blues harp*) is a free reed aerophone used in many types of rock, blues, jazz and country. Harmonicas feature a number of different holes where the reeds are tuned to create a specific pitch. There are many ways in which a reed can be tuned to a pitch:

- change the reed's length;
- change the weight near the 'free' end; and
- change the stiffness near the 'fixed' end.

Longer, heavier and springier reeds produce lower pitched sounds; shorter, lighter and stiffer reeds make sounds of a higher pitch. Different pitches can be created by breathing out into the mouthpiece to create pressure (called a *blow*) or breathing in to create suction (called a *draw*).

There are many different styles and types of harmonica but there are two broad types of harmonica that are used the majority of the time; these are called the *chromatic harmonica* and the *diatonic harmonica*, for which, if you read the music theory chapters, you should be able to hazard a guess as to how they operate.

The diatonic harmonica is the most popular type of harmonica. It is usually the smaller of the two types and typically has 10 holes. Diatonic harmonicas are known for their raw and 'bluesy' sound with more prominent harmonics than a chromatic harmonica. A signature aspect of the diatonic harmonica's sound is its 'musical' pitch bend, which the player creates via a particular embouchure technique. As its name suggests, a diatonic harmonica is tuned to a single key (usually a major key). To play all 12 major keys, you need – you guessed it – 12 different harmonicas. Due to their note arrangement, diatonic harmonicas are able to play chords easily.

A chromatic harmonica usually consists of 12 holes (although 14- and 16-holed versions are available). They are typically larger than diatonic harmonicas and are known for their purer, almost 'French' sound. A chromatic harmonica is capable of playing every note in the chromatic scale. This is done by using a button-activated slider located to the side of the harmonica, which changes the shape of the hole and therefore sharpens the sound. For instance, if you were playing a C note, by pressing the button you would change the note to a C♯. Chromatic harmonicas are great for playing melody lines but playing two or three notes at a time is more difficult. To do so relies on notes in key being on adjacent holes (or advanced tonguing techniques that block off specific notes out of key); because of this, even chromatic harmonicas can be bought in different keys, which makes playing multiple related notes easier. However, despite this, diatonic harmonicas are almost always the preferred option when chords are integral to the arrangement.

As well as utilizing different embouchures, harmonicists can also change the tonal attributes of the harmonica's sound by altering their hand positioning. A lot of harmonicists will cup the mic to 'focus' the sound, and also move their hands to create a tremolo effect.

Mic'ing Reeds

This section will be primarily focused upon recording saxophone and harmonica, which are more popular in contemporary music (sorry, clarinettists!). However, similar principles can be adapted for other reed instruments.

Mic'ing Saxophones

Just like the flute and recorder, the sound from a saxophone radiates from more than one place. On the saxophone, the sound radiates from the keys as well as the bell. You also need to be aware that saxophones have similar tonal qualities to brass instruments, in that they are loud and can have 'cutting' upper harmonics. Because of this, a hybrid approach between those you employ for brass and woodwinds is required to get the best from a saxophone.

First, you need to make sure that the microphone is placed so as to pick up plenty of the sound from the bell, but not so close or on-axis as to potentially damage the mic or pick up too many of those 'grating' upper harmonics. Second, to get a balanced saxophone sound, you need to place the mic so that it picks up some of the sound radiating from the keys. A tried-and-tested mic technique for recording the saxophone is to place a mic between 1 and 2 metres away from the player, facing the keys; this should pick up a nice amount of both elements.

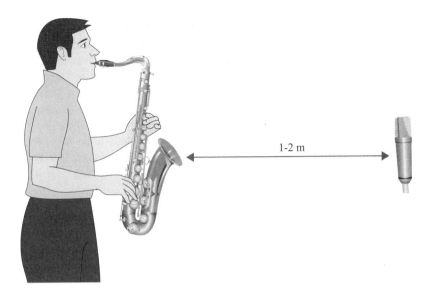

Figure 13.32 Neutral saxophone mic placement

Alternatively, if you want to emphasize the sound from the bell a little more than the keys, you can place the mic between 1 and 2 metres away at the top of the keys, angled so that the mic's diaphragm is directly facing the very top edge of the bell.

In terms of microphone choice, you should consider the same styles of microphone as with brass instruments. I like ribbons as they typically roll off some of the undesirable top end in a musical way. Alternatively, more neutral-sounding large diaphragm condensers are also very good choices.

Mic'ing Harmonicas

There are two basic approaches to recording the harmonica that depend on the intended application:

Acoustic mic'ing techniques. To capture the sound more naturally, a lot of engineers use a mic on a stand placed a short distance from the player. This way of capturing the harmonica is favoured when you want to capture a truer representation of the acoustic properties of the harmonica. This is the common approach in most genres outside of the blues.

Mic'ing harmonicas in this way is rather hassle-free as long as you remember two main factors:

1. Avoid directly placing the mic in front of the sound holes. Just like other brass and woodwind instruments, mic'ing directly on the sound holes is likely to cause undesirable breath noise.
2. Position the mic far enough away so that so the mic isn't in the way of the player's hands. Many expressive techniques are performed with the player's hands and space needs to be given for the sound to develop.

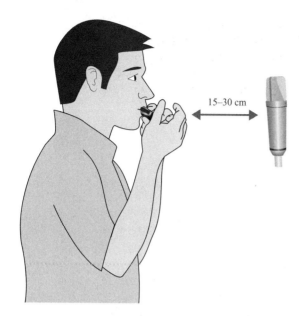

Figure 13.33 Harmonica mic placement

For 'on-stand' applications, condenser mics are often preferred. Most of the time, I opt for large diaphragm condensers for the job, as they seem to have the right balance between warmth and transient response that can help the sound to sit nicely in the song. In terms of distance, I usually find that I position the mic somewhere between 15 and 30 centimetres from the player. I also place the mic a little higher or lower than the harmonica to avoid wind blasts. Typically I prefer to mic lower as this avoids a lot of extraneous nasal noise. However, in some applications, the rawness of this extra nasal noise can add to a track's realism.

Handheld and amplified mic'ing techniques. The player might use a handheld mic to help capture a gritty, edgy harmonica tone. As the microphone is cupped tightly in the hands very close to the instrument, it causes a change in pressure on the capsule and introduces proximity effect, which results in both a top-end roll-off and boost to the lows and low-mids. This means that the same microphone can sound significantly different depending on whether it is used in the player's hands or placed on a stand at a distance. Handheld techniques became popular with electric blues players and were subsequently also used with some of their derivatives, including rock music. Because harmonica players in electric blues often had to compete with drums and amplified bass and guitar, they also started to use guitar amps to further amplify and distort their sound, and to give the tone more 'edge' so that it could compete with the other instruments. This subsequently gave the harmonica a signature gritty and larger-than-life tone that suited heavily amplified music.

Historically, amplified harmonica players used piezoelectric crystal transducer mics. This rudimentary technology produced a very lo-fi tone that became synonymous with the amplified harmonica. The problem with these mics was that their elements were very fragile, as well as being sensitive to moisture. The Shure 520 'Green Bullet' is a classic crystal transducer model. These days, specialist harmonica mics are usually made from dynamic transducers, but are designed to exhibit the coloured and lo-fi tone of their predecessors. The Shure 520DX 'Green Bullet' is such a remake of the classic model, which conveniently has high impedance, as well as a quarter-inch jack output, for direct connection with a guitar amp.

Other standard dynamic mics, such as the Shure SM57 and SM58s, can work well with the harmonica, but are low-impedance devices; this means you need an impedance-matching transformer to maximize efficiency when plugging into a high-impedance guitar amp. If you try to use a standard XLR-to-jack lead, the impedance mismatch will always result in a less than optimal tone.

Any amplifier can be used to amplify a harmonica. However, like guitarists, many harmonica players prefer tube amps, and vintage Fender amps in particular have a good reputation in the blues harp community. Once you have your desired tone from the harp/mic/amp combination, you can use any microphone techniques that you would usually employ when mic'ing an electric guitar amp in order to further hone the tone.

13.5 HAND AND MALLET PERCUSSION

Before moving on to more synthesized instruments, let's consider the use of other percussive instruments, which can be roughly categorized as hand percussion or mallet percussion.

13.5.1 Hand Percussion

Hand percussion is technically any percussion instrument from any culture that is small enough to fit into a hand. In contemporary music, you are most likely to encounter instruments such as tambourines and shakers.

The cowbell could also be categorized as a hand percussion instrument, but in reality you are more likely to see them used while incorporated into a drummer's beat.

In the case of a cowbell, it should be captured well in overheads and room mics, but you may want to place a dynamic mic such as an SM57 closer to it as a safety measure, and/or use EQ on the close mic as a way to emphasize certain frequencies.

Hand percussion might at first seem like a trivial part of an arrangement that is easily recorded, but in reality it may be a little trickier than it first seems. The use of hand percussion can, in fact, become an important part of an arrangement as they are often used to fill space in a sparse arrangement or increase energy/pay-off in choruses. On many occasions, I have added tambourines or shakers to a track where I felt that there was a lack of high-frequency energy or lack of movement/momentum. Before discussing engineering techniques, it is important to think about the type of hand percussion used, as tonalities vary greatly between them; even two types of shaker can have very different timbres. Because of this, I will try out different options until I find one that fits the vibe of the track. If you are running a commercial studio or are producing music regularly, it might be an idea to get a box full of different types of hand percussion.

In terms of recording techniques, hand percussion will often require a fast transient response because of the nuances and intricacy of their sound. Some types of hand percussion can also be very quiet, making selecting a low-noise signal chain crucially important. These two traits make condenser mics, and smaller diaphragm varieties in particular, the most obvious choice. More recent developments in ribbon designs (such as a Royer 121), however, mean that some models will have a low enough noise floor to be useful. On occasions where a slightly 'rounder' tone is required, a ribbon mic could be useful. When positioning the mic, I like to give the hand percussion a little space rather than restricting the player's movement, so I will usually give it at least half a metre of space. The total distance between mic and instrument tends to vary dependent on the

Figure 13.34 Microphone positioning to bring out accents

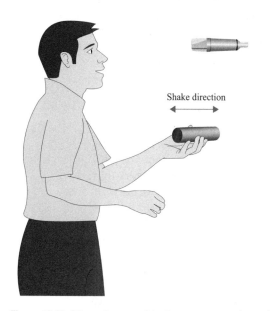

Figure 13.35 Microphone positioning to even out dynamics

volume of the instrument. For instance, the distance between mic and instrument is likely to be smaller with a shaker than with a tambourine. You can also use microphone placement to help make a hand percussion part more prominent or to 'sit it back'. If you place the mic so that the player is shaking the instrument towards the mic, it can bring out the accents (Figure 13.34).

Alternatively, placing the mic so that the diaphragm isn't in line with the direction of movement should help to even out dynamics (Figure 13.35).

13.5.2 Mallet Percussion

Mallet percussion or keyboard percussion are melodic percussion instruments with tuned keys or bars which are struck with mallets to create sound. Popular instruments of this type include the wooden-barred xylophone and marimba, and the metal-barred glockenspiel and vibraphone. The term xylophone is often used incorrectly to describe all mallet percussion instruments; this can in part be blamed on the toy industry where glockenspiels were often mistakenly labelled and sold as xylophones. The keys/bars on mallet percussion are typically arranged as you would find on a piano, in ascending pitch order from left to right. This is easily visible, as higher-pitched notes will have smaller bars. Accidentals are also placed on a row above the pitches that would be 'white notes' on a piano. On larger mallet percussion instruments, or those that need to compete with other instruments, the bars are supported by resonator tubes to help amplify the sound.

Although mallet percussion is not widely used in contemporary music, it can be a nice touch. Contemporary songs that feature mallet percussion include:

- 'Somebody That I Used to Know' by Gotye (feat. Kimbra), which uses a xylophone.
- 'Mamma Mia' by ABBA, which features a marimba.
- 'Little Wing' by Jimi Hendrix, which features a glockenspiel.

Types of Mallet Percussion

I don't want to go into detail, as they will rarely be used in contemporary music, but below is a brief breakdown of the most notable members of the mallet percussion family:

Xylophone

The xylophone is the most common instrument of the mallet percussion family. Its wooden bars are usually made from rosewood and it is played with two mallets (or, on rare occasions, four). The xylophone is known for its bright, short and sharp sound and is played with rubber mallets. It is higher in pitch than the marimba, with a typical range between concert F4 and C8, although this varies dependent on instrument. Incidentally, in order to make them more closely fit the treble clef (and thus be more easily read), notes on the xylophone are written an octave lower than their actual pitch.

Marimba

The marimba is lower in pitch than the xylophone and is more popular as a solo instrument. This is due to its larger range and warm tone. It is most commonly played with four mallets, but is sometimes played with just two. The range of the marimba is usually between four and five octaves. The top note is almost always C7 but the bottom note varies depending on how large the instrument is. The marimba uses resonator tubes to amplify the sound. It is most often played with yarn-covered mallets, but rubber mallets can also be used for effect. Unlike the xylophone, the marimba is notated at its natural pitch.

Glockenspiel

The glockenspiel has metal bars and is much smaller than the other four types of mallet percussion mentioned here. The glockenspiel produces a high-pitched sound, which cuts through other instruments well. It also has a lot of sustain and is often extremely loud, and therefore does not require resonators. Its range is often concert G5–C8 but this again varies between instruments. The glockenspiel is notated two octaves below its actual pitch.

Vibraphone

Like the glockenspiel, the vibraphone is made with metal bars. However, it is more similar in size to the xylophone or marimba. The vibraphone's use of aluminium bars gives it a much longer sustain, necessitating the use of a mechanism, by way of a pedal that engages a damper, to allow the player to control note length. Like the xylophone and marimba, the vibraphone uses resonators. Like the marimba, the vibraphone's range is notated as it sounds and almost always has a range of concert F3–F6.

The vibraphone also has an in-built 'vibrato effect', which can be turned on or off. This effect is achieved with a small metal disc called a fan in the top of each resonator under the note. This metal disc is connected to a motor that allows it to rotate, opening and closing the tube. The speed of the effect is often variable. Technically, this effect is not vibrato as the pitch isn't changing; it is in fact tremolo, as it is amplitude that is altered by the fan. This misnomer is similar to the mislabelling of tremolo as vibrato on Fender amplifiers.

Mic'ing Mallet Percussion

Mallet percussion will generally require a stereo set of mics positioned either above or below the instrument, or sometimes even both. You have the usual stereo techniques such as spaced pair, XY and ORTF. The placement of these mics on larger mallet percussion is obviously more crucial. This is because you need to pay specific attention to making sure that each bar is picked up evenly by the microphones, thus eliminating 'dead spots'. With spaced pairs, it is usually the central pitches that are the most likely to cause issues. With XY and ORTF positioning, it is often the extremes of the instrument's register that are the most likely to suffer.

My personal preference is usually a spaced pair; this is especially true for occasions where the instrument is large or a wide stereo field is required. If the instrument is likely to warrant less space in the mix, I might reconsider this. When placing the mics, I often like to get the instrumentalists playing the central bars to make sure that the middle notes are clearly represented and in even magnitude to the rest of its register. Placing mics above the instrument is likely to achieve more attack (from the mallet(s) hitting the bars). Microphones placed under the instrument are likely to capture a less pronounced attack but more sustain, especially when used on instruments with resonator tubes.

I tend to record mallet instruments with resonator tubes such as the xylophone, marimba and vibraphone from underneath the instrument. If I feel that this lacks attack, I will then add another pair of microphones above the instrument. It is usually best to place these opposite the mics below (mirrored about the instrument), and to make them the same distance from the bars as the mics underneath are. Doing this might result in you needing to reverse the polarity

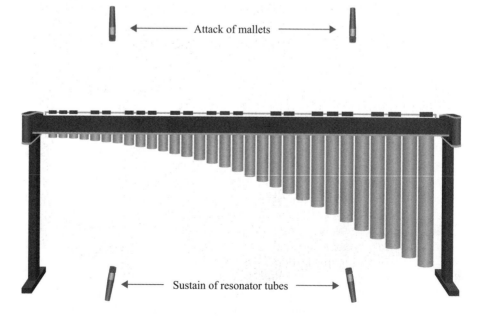

Figure 13.36 Mallet percussion mic placement

of one of the sets to get the best results. With the glockenspiel and other instruments without resonator tubes, I will usually start by placing mics above the instrument. Typically, I like small diaphragm condensers above the instrument; if the top end is a bit much, then I might turn to ribbons. Dynamic mics work well underneath glockenspiels, especially larger dynamic mics such as the Electro-Voice RE20 and Sennheiser 421.

13.6 ORGAN

Organs come in many different types, shapes, sizes and emulations. These include pipe organs, which you are likely to find in churches, reed organs, which you might have found in smaller churches or homes, and a variety of electronic organs. Organs are capable of a large range of timbres and will serve as a good introduction before we get into synthesis and sampling.

13.6.1 The Hammond B3 Organ

When you associate an organ with contemporary music, you will almost certainly be thinking of the sound of the Hammond B3, which was produced between 1954 and 1974.

Figure 13.37 Hammond B3 organ

Hammond organs that were produced pre-1975 create sound via an electrical current generated by rotating metal tone wheels located near an electromagnetic pickup; this current is then amplified via a vacuum tube pre-amplifier stage. This makes their method of operation more similar to an electric guitar pickup than later types of electronic keyboards, which use integrated circuits, op-amps and pots to generate and adjust tone.

The B3 is a dual-manual organ, which means you have two keyboards, each consisting of 61 keys. Unlike on a piano, pressing a key results in a continuous sound being played until the key is released. The keys on a Hammond B3 are not touch-sensitive, so there is no velocity change regardless of how heavily the key is pressed. The overall volume is controlled via a foot pedal (otherwise known as an 'expression' or 'swell' pedal).

Bass notes on the B3 are played with the feet on a 25-note wooden pedalboard with the top note being middle C.

As well as the keys and pedalboard, the Hammond organ has a series of metal sliders located above the manuals (the keyboards played with the hands, as opposed to the pedalboard), called *drawbars*. These drawbars allow the player to create a variety of different timbres. Each of these sliders allow the player to add pure tones (sine waves) to the sound, which combine to create a more complex tone. The B3 contains nine drawbars for each manual, which are coloured and labelled. Each of these drawbars controls the volume level between 0 and 8 of a specific harmonic or subharmonic (which are related tones lower in pitch than the fundamental).

The drawbars are labelled '16', '$5^{1/3}$', '8', '4', '$2^{2/3}$', '2', '$1^{3/5}$', '$1^{1/3}$' and '1'. This follows traditional pipe organ terminology and relates to the height of the pipe needed to create the corresponding pitch for the lowest note on the keyboard. The colouring of drawbars is also slightly esoteric. The white drawbars (8, 4, 2 and 1) represent extremely consonant intervals or, more specifically, octaves. The first white drawbar is the fundamental and the proceeding white drawbars from left-to-right ascend by an octave each time. The black drawbars represent intervals that fall between octaves ($2^{2/3}$, $1^{3/5}$, $1^{1/3}$) and, finally, the opening brown drawbars represent an octave below the fundamental (16) and a fifth above the fundamental ($5^{1/3}$). Table 13.2 is a full breakdown of which harmonic each drawbar creates.

Organists will often write down their drawbar settings as shorthand code that contains a series of nine numbers. Each of these numbers corresponds to a drawbar (as you look at them, left to right) and their volume setting (0–8). So that these are easily readable, they are split into groups of two, four and three digits. For instance, 88 8888 888 would signify all the drawbars set to maximum volume.

Hammond Organ Tone Families

By configuring a Hammond organ's drawbars to emphasize certain harmonics, the player can create sounds that vaguely resemble timbres created by other instruments.[7] Organ enthusiasts often divide the types of sounds created into four families: flute, reed, foundation and string.

Table 13.2

Drawbar	Harmonic	Note (Based on Fundamental of Middle C)
16	An octave below the fundamental	C2
$5^{1/3}$	A fifth above the fundamental	G3
8	The fundamental	C3
4	One octave above the fundamental	C4
$2^{2/3}$	An octave above the fifth ($5^{1/3}$ drawbar)	G4
2	Two octaves above the fundamental	C5
$1^{3/5}$	A major third three octaves above fundamental	E5
$1^{1/3}$	Two octaves above the fifth ($5^{1/3}$ drawbar)	G5
1	Three octaves above the fundamental	C6

Flute

Flute tones are among the most pure of all instruments. This means that, compared to other instrument types, they contain few harmonics, and the ones they do contain tend to be highly consonant. This means that flute tones will usually contain the fundamental drawbar and one of the other white drawbars. By emphasizing the fundamental, the instrument will sound more forward and aggressive, whereas emphasizing one of the other octaves will create a lighter, airier sound. Here are two example organ tones:

- Heavy Flute: 00 6200 000
- Light Flute: 00 3700 000

Reed

The family of organ tones described as reeds also covers tonal qualities that would be considered brass- or woodwind-like. These tones are created by emphasizing not only the fundamental and first harmonic such as the flute family, but also the next few drawbars, which creates a bright and brilliant tone.

- Oboe: 00 4632 100
- Trumpet: 00 6876 540

Foundation

The foundation family of organ tones is less like other instruments and is a distinctive tone that people associate with the organ itself. Their heavy and warm traits make them great for playing chords and for use as padding. Below is a full organ tone, which has all drawbars active to some degree to create a rich and full sound. As well as this, we have a theatre organ that also has each drawbar engaged except it emphasizes the brown drawbars to give it a much bassier and more dramatic tone.

- Full Organ: 54 7878 766
- Theatre Organ: 87 8766 553

String

The string family of tones, like the foundation tones, also have a vast amount of different harmonics. However, they are very flexible. Employing few of the brown drawbars, and plenty of the fundamental and first octave, as well as a small amount of other higher-pitched harmonics, will give a more violin-esque sound. Using more subharmonics and lower frequencies, and a lower but more even amount of upper harmonics, results in a more cello-esque sound.

The Hammond B3 is able to save settings as presets. A preset can be recalled using controller keys, which are to the left of the standard keyboard. These are easily distinguishable because of their reversed colour scheme. The far-left controller key cancels all presets and returns the organ to its default state. If no drawbars are active, this results in no sound coming from that manual. The two controller keys that are furthest to the right (B and B♭) activate the corresponding set of drawbars for that manual. All of the other controller keys produce preselected drawbar settings from the colour-coded wiring found on the internal rear of the organ. Rerouting these colour-coded wires results in different drawbar settings.

Hammond Organ Effects

As you can see, the Hammond B3 is an extremely versatile instrument, but it doesn't stop there. The B3 has built-in chorus and vibrato effects consisting of three vibrato and three chorus settings (V1, V2, V3, C1, C2 and C3). These are selected using a rotary switch, and can be selected independently for each manual. The vibrato settings provide variations in pitch while a note is being played. The chorus effect combines each note's sound with another sound at a slightly different and varying pitch.

13.6.2 The Leslie Speaker

The rotating Leslie speaker used with the B3 is named after its inventor, Donald Leslie, and is almost synonymous with the Hammond organ.

Leslie found that a rotating sound achieved a closer emulation of a pipe organ. The tremolo that results from this became revered in its own right, and the Leslie speaker has since found use with other instruments (particularly guitar). A Leslie speaker will, more often than not, contain both a treble and bass speaker, as well as an amplifier. Each of these speakers features a rotation mechanism, both of which are calibrated so as to achieve a similar and constant speed. The horn (treble speaker) in the Leslie is simply rotated by a motor. The bass speaker, which faces downwards, fires the sound into a rotating wooden drum that has a

Figure 13.38 Leslie 122 speaker

'scooped' shape at the bottom of it; this redirects the sound horizontally out of the cabinet. A Leslie speaker has slatted holes in the wooden enclosure, which are aligned horizontally and positioned close to each speaker to allow sound to escape.

While there have been many varieties of Leslie speaker, the most popular is the Leslie 122. This Leslie speaker has dual speeds that give rise to two very different tones, named 'tremolo' (fast) and 'chorale' (slow). The switch for selecting the speed is found on a half-moon-shaped panel near to the preset controller keys.

Mic'ing a Leslie Speaker

First, I want to state that this section will focus on capturing the sound of a Leslie that has a bass and treble speaker; some models exist with a single speaker, but they are much less common. As we see in Figure 13.39, the horn (treble speaker) is located at the top of the enclosure, and the woofer (bass speaker) is at the bottom. The most common approach is to use a pair of mics on the treble speaker, and a single mic on the bass speaker. Using a pair of microphones on the bass speaker is also an option, and can create an even more extreme effect. However, the reduced stability in the bottom end and technical concerns with vinyl cutting (the needle jumping out of the groove) make this approach much less common. As with guitars, dynamic mics are generally preferred. Mics such as the Shure SM57, Sennheiser 421 and Audix i5 will certainly be up to the task on the treble speaker. Sometimes, however, I will feel this approach lacks a little 'sparkle', and I may reach for a set of small diaphragm condensers. Although the dynamic mic options will also provide good results on the bass speaker, I usually plump for a dynamic microphone that is tailored to kick drum; the Shure Beta 52 and AKG D112 are good choices as they have a less pronounced top end than some mics designed for kick drum. A dynamic mic designed for studio vocal performance, such as the Shure SM7B or EV RE20, is also a viable consideration on the bass speaker. To reduce wind noise, the mics are often placed a little farther back than you would on a guitar cabinet. However, in smaller or more reflective rooms, this might cause excessive room tone. If this is the case, try reducing the reverb time with gobos or blankets. Failing that, if you have to mic extremely closely, sponge windshields can be of some help.

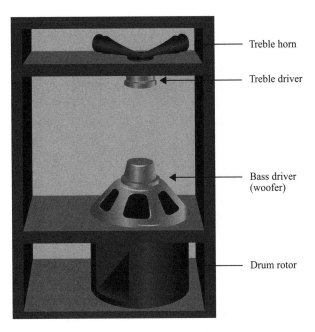

Figure 13.39 Anatomy of a Leslie speaker

In terms of placement, the positioning of the two 'treble mics' can achieve varying degrees of width and excitement. I suggest three main options for this: opposite sides, spaced pair and XY.

The most exciting and widest stereo image comes from mic'ing the middle of a treble speaker sound hole on opposite sides of the Leslie cabinet. Where

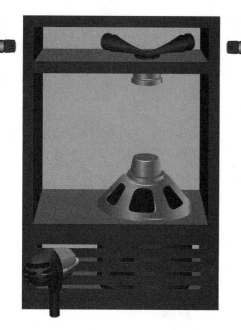

this works, it sounds incredible. However, this can easily become nauseating or annoying, especially when wearing headphones.

A spaced pair on the same side of the Leslie cabinet, to the left and right extremes of one of the treble sound holes, creates a wide stereo image without some of the nauseating downsides of mic'ing opposite sides (Figure 13.41). Admittedly, this option doesn't sound as exciting as the opposite sides approach and is not as mono-compatible as the XY approach.

Figure 13.40 Leslie mic'ing opposite sides

Finally, there is the XY approach. As with drum overheads, an XY configuration (positioned about the middle of one of the treble speaker sound holes) creates a tight and cohesive stereo image that folds to mono extremely well, but it can lack width, energy and excitement (Figure 13.42).

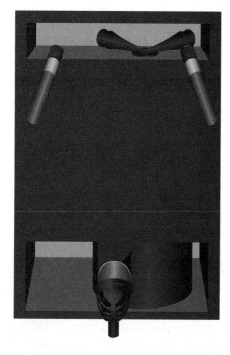

Figure 13.41 Leslie mic'ing spaced pair

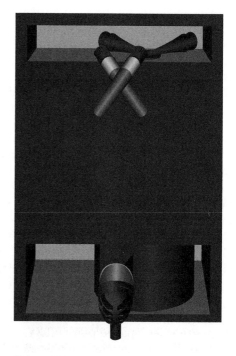

Figure 13.42 Leslie mic'ing XY

With each of these techniques, I've found it best to place the bass mic the same distance from the Leslie as the stereo set, but in the middle of one of the bass sound holes.

13.7 SAMPLING AND SYNTHESIS

The focus of this course has been very much on recording instruments that require mic'ing. However, many genres of contemporary music such as hip-hop, EDM and pop rely heavily on the creation of sounds using electronics and digital methods, which is often categorized as *computer music*. Computer music can be subcategorized into two main groups: sampling and synthesis. These disciplines constitute yet another area of music production that warrants a whole book's worth of content. The following section is extremely brief and covers only the rudiments; despite this, you should still come away with enough information to be able to experiment with a modern sampler or synthesizer.

13.7.1 Sampling

Traditionally speaking, sampling in music is the process of taking a portion of a sound recording (or *sample*) and reusing it in a different context within another composition. We mentioned this in Chapter 6: Arrangement and Orchestration when we discussed the hip-hop genre, which, in the 1970s, popularized the use of looped portions of a sound recording via DJ decks. However, the earliest forms of sampling actually appeared within experimental music of the 1940s, where musicians would manipulate music found on tape and vinyl. Such musicians would often capture everyday sounds that weren't traditionally musical, such as the sound of a train approaching a station. In the late 1960s, manipulating recorded sounds with effects and tape looping became popular with experimental rock bands too. The Beatles, in their experimental phase, were particularly influential in making these techniques more mainstream.

Nowadays, sampling is so widespread that you won't be able to listen to commercial radio for too long without hearing songs that have incorporated a sampled loop. As well as creating and manipulating loops with tape and vinyl, you can also use your DAW to manually move and copy loops in the DAW's timeline.

However, because of the prevalence of sampling, hardware and software manufacturers created *samplers*, devices dedicated to the task, which make it easier for users to trigger and manipulate samples. Samplers in hardware form, such as the AKAI MPC series, are often in the style of control surfaces with multiple trigger pads. In software form, samplers such as Native Instruments Kontakt often use MIDI keyboards as a means of triggering the sounds loaded into them.

Sampled Instruments

As well as a more traditional approach to sampling (i.e. taking a loop of existing music), technological advancements have meant that there are now teams of developers creating libraries of samples that replicate pretty much any instrument you can imagine. These developers go to incredible lengths to capture multiple samples of each note of the instrument, played at many different velocities and/or with different techniques. For instance, a string instrument can be bowed or plucked in many ways, so sample libraries often include different patches for each playing style. The depth of sampling, as well as the 'tweakability' and flexibility of modern software, make it possible to accurately reproduce instruments, and get commercial-grade results, without even having the instruments present.

Despite the obvious advantages of sampling, many producers, engineers and musicians still prefer to play and record instruments acoustically. There are many reasons for this:

- Sometimes a sense of realism comes from controlled imperfections, such as variation in tuning and timing, and imperfect room ambience.
- High-quality sample libraries cost big bucks.
- Samplers are hugely CPU-intensive.
- Creating great-sounding instruments from samples is an art form and will take a serious investment in time, along with knowledge of the instruments themselves (such as their ranges) and the techniques required to use the library's controls, MIDI and keyboards together to create a realistic result.
- The amount of time it takes to properly program, arrange and print the sampled parts might also be inefficient for some projects.

Personally, I generally prefer the sound of live instruments, and choose to record acoustic performances whenever possible. However, I do use sample libraries in certain situations:

1. When a part is more of a pad or background instrument and the time, effort and budget required to locate and organize an instrumentalist would not be economical.
2. If I want to create a hyper-perfect end result with less human element.
3. To use in conjunction with a live performance to fill it out or help solidify the tuning.
4. When I have a talented musical composer/arranger present in the session who specializes in the creation of realistic and dynamic performances using samplers. These people are often talented pianists too, and will play many of the parts, incorporating the use of pedals, control sliders, wheels and buttons on their MIDI controllers to program more human-like expression.

13.7.2 Synthesis

Whereas sampling involves the use of electronics and computing to trigger and manipulate pre-existing sound, synthesis is the artificial creation of sound using these technologies alone. An instrument that utilizes synthesis is called a synthesizer. You can think of a synthesizer much like an organ, as sounds created on it can be designed to roughly replicate the timbres of other instruments. It can, however, also be used to create tones and timbres previously unheard of. There are two types of synthesizer available: analogue and digital.

Analogue synthesizers directly manipulate electrical signals in order to create an analogue signal that can be outputted.

Digital synthesizers use digital signal processing algorithms to create and manipulate signals; these signals then need to be converted using a digital-to-analogue converter before they can be outputted as an analogue signal.

The sound of analogue and digital synthesizers, and players' preferences for one or the other, draw parallels with how audio engineers feel about the analogue or digital processing debate. Analogue synths are often renowned for their rich sound, caused by the slight anomalies in analogue circuitry creating subtle but musical distortion. Digital synths, on the other hand, are more reliable, flexible and often cheaper than their analogue counterparts. While many analogue synths are still revered, many digital synths can compete with their tonality. In recent years, some manufacturers have even started to use a mixture of hardware and software technology to get the best of both worlds, attempting to combine the flexibility and reliability of digital synthesis with the sound of analogue synthesis.

Types of Synthesis

There are many types of synthesis such as additive, subtractive, frequency modulation, phase distortion and granular synthesis, among others. It can sometimes be difficult to understand some methods of synthesis, so let's start with two types that are the easiest to understand: additive and subtractive. *Additive synthesis* works by manipulating sine waves. A sine wave is the simplest tone possible as it only contains one frequency, but by adding together sine waves of different frequencies, as well as manipulating their amplitudes and phase relationships to each other, you can, theoretically, create any possible sound. In practice, though, additive synthesis isn't very efficient and is rarely used. Starting with more complex waves and removing elements that you don't want is much more practical; this approach is called *subtractive synthesis*.

When discussing common forms of synthesis, special mention must go to frequency modulation synthesis. *Frequency modulation* or *FM synthesis* works by modulating waveforms against each other. An audio carrying wave is called a *carrier wave* (which is the wave you hear) and the waves used to modulate them are called *modulators* or *modulating waves*. Frequency modulation creates complex waves that are rich in harmonics. The main problem for inexperienced users of FM-based synthesizers is that they can easily sound terrible. By using modulators with clean ratio relationships to the carrier (multiples), you create a new waveform with musical harmonics, but if you use less-clean ratios you end up with inharmonics, otherwise known as more dissonant sounds.

Despite the variety of synthesis methods, whether analogue, digital or virtual, the majority are based on subtractive synthesis. From this point on, we will be focusing on the programming synths developed with the use of analogue synthesizers that utilized subtractive synthesis; you will also find this terminology crossing over to the digital/virtual world, even though their algorithms may work slightly differently to the physical electronics in an analogue model.

Before we get into how to program subtractive synthesizers, it is important to state that, while synthesizers can theoretically recreate any waveform, in practice they are not ideal for trying to accurately replicate the sound of real instruments. For instance, a synth violin can sound similar in timbre to the real instrument, but it won't trick anyone into thinking it is the real thing. If you are trying to replicate an instrument without having an instrumentalist available, sample libraries are a far more effective solution. Synthesizers, however, have their own merits: they are great at filling the same sonic space as traditional instruments, while at the same time having a unique sound of their own. An example of this is the classic '80s synth bassline that fills the same space as a bass guitar, but can give the track a different flavour.

Synthesizer Programming

Altering a synthesizer's settings to create a new sound is called *synthesizer programming*. Synthesizer programming is another area where a basic knowledge of the science behind audio is important. Concepts such as amplitude, frequency and harmonics are all important. It is also important to hear the effects of the different parameters explained here, so it might be wise to have a synthesizer handy to experiment with while reading this section. Pro Tools comes bundled with Vacuum, which is a neat subtractive synthesizer plug-in that is graphically designed to resemble a classic analogue synthesizer.

In order to successfully synthesize sounds using subtractive synthesis, an understanding of the basic building blocks of a synthesizer is crucial; this can best be visualized with a simple flow chart (Figure 13.43).

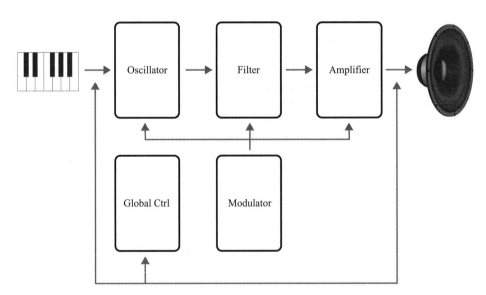

Figure 13.43 Subtractive synth flow chart[8]

Broken down into its simplest form, we are dealing with a three-stage process:

1. *Oscillator.* You start with a sound source generated by an oscillator or noise generator. The fundamental frequency of the oscillator is often controlled by pressing a key on the keyboard of the synthesizer. Sometimes synths will also have multiple oscillators for 'stacking' sounds.

2. *Filter.* The signal is then filtered to remove frequencies that you do not require.
3. *Amplify.* In the amplifier stage you have control of the signal level over time, which is called the sound's envelope.

Modulation can be added to any of these three stages to make a sound more interesting, or to make it more like a naturally occurring sound. Modulation can either be automatically implemented by the synthesizer, or controlled manually with tone wheels or other user controls.

It is easy to get daunted by all the keys, knobs, buttons, sliders and wheels on a synthesizer, but if you think logically about the order in which you are adjusting parameters you will get to a useable setting much sooner. So, before we go into detail on commonly found parameters of subtractive synths, I want to quickly assert the order in which I like to set up the sound:

- Set the basic noise source (oscillator or noise generator).
- Filter the signal to get closer to the sound you desire.
- Set the sound's envelope so that its expression is suitable.
- Modulate the signal to create interest if you require.
- Set any additional effects, such as chorus, distortion, reverb and delay (which are often built into synthesizers).

Oscillator/Noise Generator

The audio signal of a synthesizer is generated by the oscillator or a noise generator. With oscillators, you can normally choose from a selection of waveforms, each with different harmonics, and different amounts of harmonics. As we learnt in Chapter 3: Basic Music Theory, the relationships between the fundamental frequency and its harmonics are responsible for an instrument's unique timbre. Each of these waveforms is named after their visual appearance on an oscilloscope. Let's look at the waveforms commonly found on synthesizers, from the purest to the most harmonically dense.

Sine wave is the simplest wave and consists of only the fundamental frequency; it is great for 'pure' sounds (Figure 13.44).

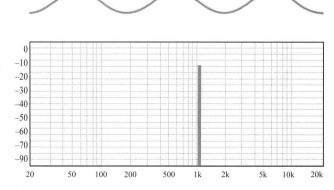

Figure 13.44 Sine wave and its harmonic density

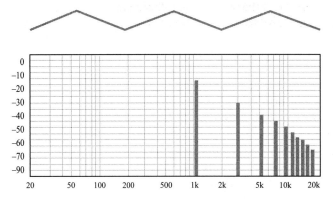

Figure 13.45 Triangle wave and its harmonic density

Triangle wave is slightly more harmonically dense than a sine wave but is still a 'soft' sounding waveform. Triangle waves consist of odd harmonics only, and they roll off faster than other waves. Triangle waves are great for flute or pad sounds (Figure 13.45).

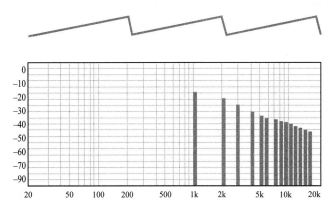

Figure 13.46 Sawtooth wave and its harmonic density

Sawtooth wave is a sharp and bright sound that contains both even and odd harmonics. Sawtooth waves are used as a starting point for a wide variety of timbres such as strings, brass, harder pads and bright basses (Figure 13.46).

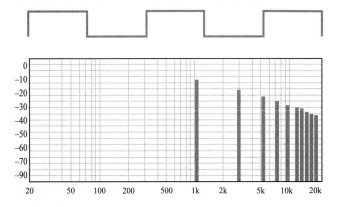

Figure 13.47 Square wave and its harmonic density

Square wave is the brightest and harshest of these simple waveforms and contains odd harmonics. They are most often used for reeds, pads, and bass sounds (Figure 13.47).

As well as these basic types, synthesizers often have functions for creating noise. Noise is useful for creating percussive sounds such as snares or wind-like sounds.

Here is a quick breakdown of the different types of noise found in synthesis:

- *White noise* is the most common noise waveform found on synths. White noise contains all frequencies at equal level around a centre frequency.
- *Pink noise* contains all frequencies, but decreases in level by 3 dB per octave as you ascend the frequency spectrum.
- *Red noise* contains all frequencies, but decreases in level by 6 dB per octave as you ascend the frequency spectrum.
- *Blue noise* is the opposite of pink noise in that it increases in level by 3 dB per octave as it ascends the frequency spectrum.

Filters

The application of filters should be a walk in the park for anyone that knows the operation of EQ. The filters on a synthesizer have one important distinction: they can only act subtractively (with the exception of a resonance parameter). Let's quickly outline the most common and basic filter types:

- *Low-pass filter* or *LPF*: Low frequencies are passed but high frequencies are attenuated (Figure 13.48).

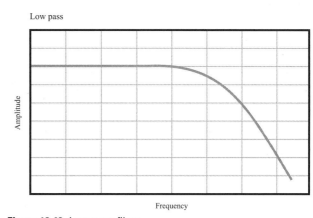

Figure 13.48 Low-pass filter

- *High-pass filter* or *HPF*: High frequencies are passed but low frequencies are attenuated (Figure 13.49).

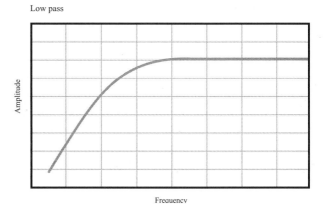

Figure 13.49 High-pass filter

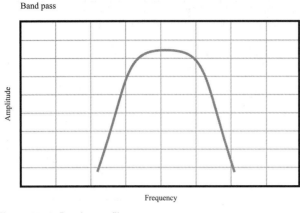

Figure 13.50 Band-pass filter

- *Band-pass filter.* Only frequencies within the frequency band are passed (Figure 13.50).

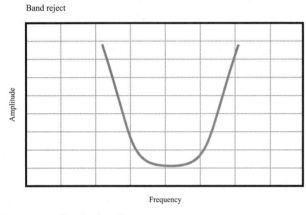

Figure 13.51 Band-reject filter

- *Band-reject filter.* Only frequencies within a frequency band are attenuated (Figure 13.51).

- *All-pass filter.* All frequencies in the spectrum are passed. However, the phase of the output is modified.

On a synthesizer, these filters can be implemented using just a few simple parameters: cut-off, resonance, bandwidth and slope. All filters use the cut-off and resonance parameters, but the bandwidth and slope parameters are used on specific filter types:

- *Cut-off* gets its name from early synthesizers where you might only have a low-pass filter. The cut-off on an LPF is the frequency at which the filter starts to attenuate. In modern synths, the cut-off parameter might have a slightly different function based on the filter type, so I like to think of the cut-off parameter as an 'action frequency'. For instance, on an HPF, the cut-off is the frequency at which lower frequencies are attenuated; on a band-pass filter, the cut-off acts as the centre frequency at which signal is passed.

- *Resonance* parameters emphasize or suppress frequencies around the cut-off parameter. The result of this is a change to the waveform's shape, and therefore its timbre. When you emphasize around the cut-off frequency, you typically get a harsher, more aggressive filter, and when you suppress around the cut-off you get a softer and more smooth filter (Figure 13.52).

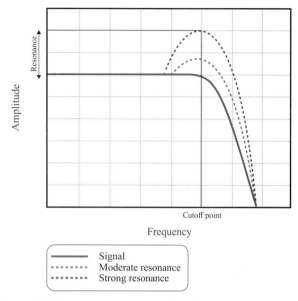

Figure 13.52 Resonance parameter effects

- *Slope* parameters allow the user to control how quickly the filter attenuates from the cut-off frequency. Slope controls can be found on high-pass and low-pass filters, and are typically found in settings of 6, 12 and 24 dB per octave (Figure 13.53).

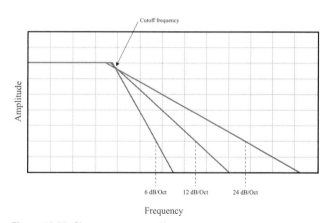

Figure 13.53 Slope parameter effects

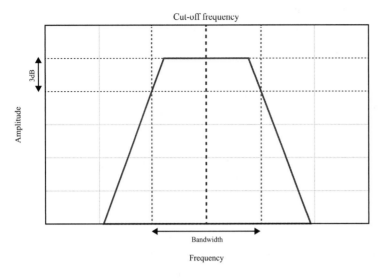

- *Bandwidth* parameters let you control the range of frequencies that are affected around the cut-off parameter. You will find bandwidth parameters on band-pass and band-reject filters (Figure 13.54).

Figure 13.54 Bandwidth parameter effects

Envelope

The parameters involved in changing a sound's envelope should already be clear, as we discussed them in the time domain section of Chapter 6: Arrangement and Orchestration. A synthesizer will have ADSR controls for changing the velocity of the attack, decay, sustain and release phases of the sound's envelope.

Despite the fact that the envelope functions of a synthesizer were designed to be used initially for amplitude control, you can also use envelopes to control other functions too. For instance, if you wanted to slowly raise a note in pitch over the course of a bar, you could achieve this effect by assigning an envelope to the pitch and implementing a slow attack setting. This functionality makes envelopes extremely useful on modern EDM, where artists will frequently automate parameters of a software synthesizer, such as Native Instruments Massive.

Low-Frequency Oscillator (LFO)

One of the most important elements of a subtractive synthesizer's sound is the ability to use additional oscillators not to directly create sound, but to trigger a change in another parameter's values. To avoid FM modulation occurring, this oscillator needs to operate outside of our audible spectrum; this way, it is heard as a cyclical alteration rather than a constant change in tone. Low-frequency oscillators operate below the lowest extreme of our hearing (below 20 Hz), hence their name. Like audio signal oscillators, they can take the shape of sine, sawtooth, triangle or square waves. LFOs can be routed to control many parameters; for instance, routing an LFO to control pitch creates vibrato; when routed to control amplitude, the result is tremolo. The LFOs on modern synths will likely feature several parameters, including waveform shape, routing, tempo sync and modulation rate. A modern use of an LFO is to route it to control a cut-off frequency on a filter. Setting the LFO to a high frequency creates the ripple-like effect you hear frequently in many genres of EDM; used on a bassline, a low-frequency LFO creates a wobble effect, like you hear on many dubstep and drum and bass tracks.

Global Controls
Global controls, or global settings, are parameters that affect the overall output signal of the synth. Some commonly found global parameters include:

- *Master volume* changes the output volume of the synthesizer.
- *Glide* (or *portamento*) is used to set the time it takes for a pitch to slide up or down to the next pitch when a new note is pressed. This is great for emulating the sound of instruments such as brass, which slide from note to note, rather than directly moving from one pitch to another.
- *Pitch-bend* is generally hardwired to a wheel to the side of a synthesizer's keyboard. Moving the wheel up from its central position smoothly raises the pitch, and moving the wheel down lowers the pitch. Pitch-bend wheels will usually allow a player the option to bend up to one octave either side, but traditionally players like to set the extremes of the pitch-bend to three semitones each way. Pitch-bend wheels are extremely expressive and can help keyboardists achieve an effect similar to a guitarist bending his strings during a solo.
- *Voices* defines the number of notes that can be produced simultaneously. The ability of a keyboard to produce notes simultaneously is called *polyphony*. A voice parameter sets an upper limit on the number of notes that can be played at any given time.
- *Unison* allows a second voice to be heard one octave above the frequency of the played note. The advantage of this is that it can make the sound fuller and/or more interesting. Not all synths will have a specific unison function, but most modern synths will have ways to layer or 'stack' voices. A potential problem with a layering function is that it will reduce the amount of polyphony available because each note actually contains more than one voice.

Advanced Parameters
Finally, there are a couple of more advanced parameters that can be used to create interesting sounds:

- *Detune.* When you are working with more than one voice, a detune parameter can help to differentiate between voices. By detuning by coarse amounts, you are creating harmony; fine detuning can result in a creative effect much like flanging.
- *Arpeggiator.* An arpeggiator automatically steps through a sequence of notes, based on a note or chord played, to create an arpeggio. Arpeggiators frequently have controls for speed of arpeggiation, pitch range, and whether the notes ascend or descend in pitch, or sometimes both. Advanced appreggiators can even allow you to step through a pre-programmed set of intervals. Arpeggiators are often combined with other modulation effects to create fast-paced rhythmic patterns that are popular in EDM genres such as trance.

EXERCISES

1. Find ways to get some time with players of as many of these instruments as you can, and take time to really analyse the instruments' sounds.

2. Take time to mic up those instruments in the ways suggested in this chapter, as well as trying to use your own judgement.

3. Experiment with sample libraries and compare them to your experiences with real instruments.

4. Think about occasions in songs you have already recorded that might suit some of this instrumentation, and try adding real instruments and sampled versions of the instruments.

5. Experiment with synthesizers and organs, and try to get them to sound as much like traditional instruments as you can.

6. Use synthesizers and organs to fill the same space as traditional instruments but try to make them unique.

7. Take a loop from a commercial release and try writing/arranging a song in your DAW based on that sample.

13.8 'TAKE HER AWAY' EXTRA INSTRUMENTATION

In the making of 'Take Her Away', we used very little instrumentation that couldn't be played live by the band. This was primarily because the rhythm guitar's use of effects created a dense organ-like sound and there wasn't a great deal of textural space that needed filling. As well as this, the song's use of syncopation and time signatures meant that:

1. it kept the listener's interest more easily than a simple rhythm; and
2. other rhythms tended to clutter the arrangement.

However, during the middle section, where the rhythm is a lot straighter and the music breaks down, we wanted to create something that had a sense of crescendo and achieved a Pink Floyd-esque 'trippiness'. This was done by incrementally increasing the tempo and using voice samples and Foley in a musical way, in a similar vein to parts of 'Dark Side of the Moon'. I can't take credit for this idea; the band had this intention from the start. They had even recorded some samples of themselves doing/saying crazy things (perhaps drunkenly?), which they manipulated and processed with effects to incorporate into the live performance. They did this with the use of the drummer James's sample pad. When it came time to record the track, I was given a stereo file of these samples as they were mixed by the band, as well as the clean multitrack files. I noticed that the more rhythmic elements, such as the heartbeat sample, worked best when I used Elastic Audio to 'pump' in time with the music. I also decided to implement some pitch shifting, modulation effects and auto-panners, as well as a vast amount of automation, to achieve the desired effect.

The only other element of instrumentation that the band couldn't easily play live was the organ section during the second guitar solo, which immediately precedes the final chorus. This was added after we had wrapped up the recording session and I'd done some basic balancing of levels. The reason for its inclusion was to fill out space, as I thought there was a lack of low-mid warmth, or, to put it another way, the track needed an element to support and anchor the upper mids of the distorted lead guitar and rhythm guitar (which was no longer processed with a POG octave effect). I programmed in the organ using the DB-33 organ software instrument plug-in that comes bundled with Pro Tools. I kept this as an ultra-simple padding part (sticking with notes from the triad), and kept it low enough in the mix to be felt missing if it was removed, but not loud enough to be perceived distinctly. Once I showed Wil this new part, he suggested we make it a little more exotic and forward in the mix, so we added some slightly less consonant intervals and raised its level.

Figure 13.55 Pro Tools DB-33 Hammond organ emulation

NOTES

1. How to use Piano Pedals, accessed August 2014, www.dummies.com/how-to/content/how-to-use-piano-pedals.html.
2. Audio frequency charts, *Sound on Sound Magazine*, January 2012.
3. Hugh Robjohns, recording strings, *Sound on Sound*, April 2009.
4. Hugh Robjohns, recording strings, *Sound on Sound*, April 2009.
5. Orchestra, accessed August 2014, http://en.wikipedia.org/wiki/Orchestra.
6. Blair Jackson, orchestral recording, *Mix Magazine*, January 2006.
7. Hammond Organ Company, *Creating Beautiful Colors with the Harmonic Drawbars of the Hammond Organ*, Hammond Organ Co, 1965.
8. How Subtractive Synthesizers Work, accessed August 2014, https://documentation.apple.com/en/logicexpress/instruments/index.html#chapter=A%26section=3%26tasks=true.

BIBLIOGRAPHY

Musical Instrument Range Chart, accessed August 2014, http://solomonsmusic.net/insrange.htm.

David Miles Huber and Robert E. Runstein, *Modern Recording Techniques* (Focal Press, 2013).

14

TRACKING VOCALS

Now that we have covered the accompaniment, let's move on to look at arguably the most important part of any song: the vocals.

Although the technical aspects of vocal recording are important, what really makes or breaks a recording is the producer/engineer's ability to get the best performance out of the singer, both in terms of emotional conviction and technical precision. Before we talk in depth about how to get the singer to deliver the goods, let's get some of the engineering aspects out of the way.

14.1 VOCAL ENGINEERING

You may be glad to know that, compared to other sound sources, there is more of a standard industry practice for recording vocals when it comes to selecting mic, pre-amp and performer/mic placement. But, as ever in music production, these rules can be broken when you have a strong vision.

14.1.1 Microphone Placement in the Room

The biggest trick with mic placement for vocal recording is to minimize the number of early reflections (reflections occurring less than 20 ms after the direct sound) from room boundaries that are captured by the mic. This is because the ear cannot differentiate between direct sound and reflections that are this immediate. Having too many uneven early reflections introduces many negative artefacts throughout the audible spectrums, such as muddiness, harshness, hollowness (comb-filtering), etc. The size, angles and material of the walls all play a part in this.

Despite the fact that you often see small vocal booths being used for vocal work and/or voiceover, large professional studios often use larger rooms to record vocals in. This is not out of a desire to capture room ambience, but rather that the recording won't suffer from the sub-20 ms early reflections I've already warned about. To get early reflections that are greater than 20 ms, you must be at least 3.5 metres away from any boundary (including the ceiling). Reflections from the floor are usually minimized with a rug, as it is incredibly difficult to get 3.5 metres away (unless you are recording the world's tallest man!). The prevalent use of vocal booths is purely to save space and in such small spaces you will have to acoustically treat the whole area to avoid such disastrous comb-filtering. My advice is that such vocal booths are rarely worthwhile.

Most of the time in modern pop and rock records, producers look to capture an intimate vocal sound. In a large room, the RT60 (natural room reverb time) would be larger, which negates the idea of intimacy. Therefore, larger studios will create a specially treated area of their live room called a 'dead area', which is treated much more extensively with acoustic treatment. This will still give enough space to avoid the negative effects of comb-filtering but also 'soak up' sound to allow for a small RT60.

Beware, though, the intimacy required doesn't come from having no room reverb at all (or else we'd be recording in anechoic chambers), but more from a small and controlled RT60 that is as even across the audible spectrum as possible. Spatial effects such as reverbs and delays are added at the mix stage, usually via busses, to create that larger-than-life excitement; they can then be EQ'd and compressed in such a way that they don't compete with the lead vocal.

To find the best spot in the room, try the places far enough away from any wall to avoid early reflections. Bear in mind that the spot directly in the middle of the room (length-, width- or height-wise) is not ideal because of problems with its bass response due to room modes. Remember that you can always get the singer to walk around these areas while performing to find some sweet spots.

14.1.2 Minimizing Room Reflections

Unless you are recording in a specialist 'dead area', you will probably also want to further minimize room reflections. Here are a few ways to enhance your vocal sound:

1. Polar pattern. When recording a vocalist, you are going to want the vocal mic to pick up the sound from one direction only, so switching the polar pattern to cardioid will allow for a cleaner tone and a better direct signal-to-room tone ratio.

Figure 14.1
sE Electronics Reflexion Filter

2. Makeshift 'dead area'. There are professional products on the market such as the sE Electronics Reflexion Filter that help to minimize early reflections from the rear and sides of the microphone and also stop the mic picking up some reverberant noise (Figure 14.1). You can also carefully position gobos in a semi-circle around the rear and side of the mic. A great but cheaper alternative to gobos or reflection filters is to hang duvets around the singer. Doing this in studios that already have well-treated rooms will still improve the tone most of the time (Figure 14.2).

3. When working with a mic with a cardioid polar pattern, the space behind the singer is the place most prone to capturing a lot of early reflections. Here is another space

where a gobos or blankets in a semicircle can help to minimize tonal coloration. Just remember not to get them too close to the singer as to create early reflection problems! This also works well for the singer's sense of freedom and comfort.

Please note that I wouldn't advise plastering your walls with foam absorbers to handle early reflections, primarily because foam isn't a great broadband absorber and doesn't reduce reflections in a tonally neutral way. It often leaves low-mid reflections untouched while significantly reducing high-end reflections, leading to 'boomy' and 'boxy' sounding recordings. This is one of the reasons why many vocal booths are, in my opinion, a poor solution.

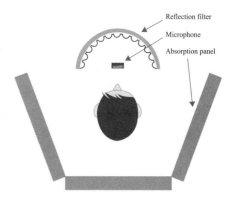

Figure 14.2 Makeshift dead area

14.1.3 Microphone Choice

Before recording the vocalist, it is common to 'shoot out' several microphones that may suit not only the genre of music, but also the tonality of the singer's voice.

A 'shoot-out' from an audio production perspective is a short test of multiple devices (of the same type) to help you decide which to choose for the specific application. The aim with this is to keep every other variable as consistent as possible; for instance, if you are testing vocal mics, then you would keep the whole chain identical apart from the mic. This includes the performance, so wherever possible it would be wise to sing the same part with the same intent (I'd suggest preferably the chorus because it's the focal point) to ensure fair comparison.

A vocal mic shoot-out may feature a variety of dynamic microphones, large diaphragm condensers and tube microphones. While it is true that the tonal characteristics of similar microphones are less significant than manufacturers would have you believe, I have found that my choice of microphone varies massively between sessions. Even sonically (and visually), similar microphones such as the SE4400, AKG414 and AT4050 can provide significant differences in overall quality. When you match the right mic to the right singer, your vocal track should effortlessly fit in the mix without much equalization. Having some knowledge of the gear available before the session is important, as the last thing a singer wants to do is test 10–20 different mics; doing this is likely to tire the vocal cords before you even begin tracking in earnest. My advice is to shoot out two or three of your favourite mics and only resort to trying more if you are still completely unhappy with the tone you have.

A key factor that dictates which mics make the shoot-out is the gender of the vocalist. Women's voices are usually less harmonically complex, are 'sweeter' sounding and have less of a low-mid bass response. You will probably settle on two sets of 'go-to' microphones – one favouring male vocals, the other favouring female vocals.

Figure 14.3 Shure SM7B (picture courtesy of Metropolis)

Dynamic Microphones

Dynamic mics are likely to give vocals an edgy, gritty sound due to their mid-range presence. Although they roll off in the top end a little more than condensers, more expensive dynamic mics still have a considerable high-frequency response. They work well on vocals in genres where you want the music and vocals to mesh together a little more. Dynamic mics are therefore a good choice for rock/punk/alternative and even rap vocalists. In radio and other broadcast mediums, large dynamic mics are also hugely popular. Common choices of dynamic mic are the Shure SM7B and the Electro-Voice RE20 (Figure 14.3).

Figure 14.4 Neumann U87 (picture courtesy of Metropolis)

Condenser Microphones

Condenser mics suit music in which there is a focus on 'sparkly highs'; they produce a more defined, rich and colourful top end. They are also great for EQ'ing 'air' into the recording (13–16 kHz), which gives that exciting, larger-than-life vocal sound. Pop and RnB singers often like the sound of large diaphragm condenser mics on their vocals. These characteristics probably make them the most popular all-round mics for vocal recording. Popular large diaphragm condenser mics are the Neumann U87 and the AKG 414 (Figure 14.4).

Tube Microphones

Tube mics have all the benefits of condenser mics with the addition of a warm bottom end courtesy of the tube. Tube mics suit genres where a warm vocal tone is desired and/or on voices that feel 'airy' and in need of fattening up. They are great for acoustic and country music or when you need a vocal to cut through in the low end of a rock mix. Most popular tube mics are based around three classic designs: the Neumann U47, U67 or AKG C12 (Figure 14.5).

Figure 14.5 Neumann U47 (picture courtesy of Metropolis)

> **Further Resources**   www.audioproductiontips.com/source/resources
>
> Microphone choice examples with male and female voices. (91)

14.1.4 Microphone Settings and Accessories

As well as the microphone itself, there are a number of features and accessories that can help to enhance the results. For instance, many mics will have a high-pass filter (HPF) and a pad button. A shock mount and the use of a pop shield are often required.

High-Pass Filter

An HPF reduces a microphone's low-end response up to a certain frequency (usually 80–100 Hz); this is great for removing unwanted noise and rumble below the range of the human voice. You may even be able to adjust the steepness of the filter slope, as on the SE4400 pictured in Figure 14.6. An HPF may be indicated on the mic by the following symbol: ⌐

Pad

On many microphones, you are able to reduce the mic's sensitivity (usually by 10–20 dB) via a pad switch. This allows the singer to get closer to the mic without the risk of distortion or damage to the mic. Believe it or not, the strongest vocalists are capable of producing sound pressure levels (SPLs) that can break microphones. If you believe the vocalist is strong, you should consider padding the mic, asking the singer to step back or exchanging the mic for one that can withstand high SPLs.

Figure 14.6 High-pass filter on an SE4400 mic

Shock Mounts

Many microphones (particularly condensers) are susceptible to noise caused by vibrations from the floor being transmitted through the mic stand and into the mic. These vibrations can occur due to the vocalist moving around as they sing, stepping 'in and out' of the mic between lines, tapping his or her foot, etc. Shock mounts are a type of mic clip that suspend the mic in an elastic/rubber frame to stop the transference of energy into the mic's casing.

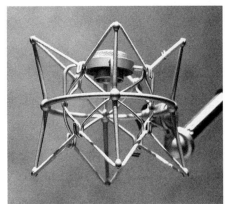

Figure 14.7 Neumann shock mount (picture courtesy of Metropolis)

Pop Shield

Whether you are using a dynamic, condenser or tube mic, there is a high probability that you will need to protect the microphone's capsule from being hit with gusts of air from the singer's mouth while he or she is singing or spit and water, which can oxidize the capsule. Gusts of air are usually most audible on plosive sounds such as 'b' and 'p' and produce a loud 'pop'. You have probably heard of a pop shield being used to

Figure 14.8 Pop shield on a Neumann U87 (picture courtesy of Metropolis)

counteract this. Many engineers even set them up as standard. In practice, though, a pop shield should be used only when needed. For instance, I've recorded many singers where there wasn't a problem with plosives. Even when there is, you can sometimes alleviate this with some experimentation with microphone placement. When this doesn't work, it might be time to turn to the trusty pop shield. Pop shields prevent this blast of air from hitting the capsule and are made from a thin metal or nylon mesh with an adjustable arm that attaches to the mic stand. Pop shields are placed a couple of inches away from the microphone on the side that the singer is singing into. The problem is that a pop shield can colour the signal, and therefore should be used where needed and not just as a staple.

14.1.5 Pre-Amp Selection

The fact that many unchangeable physical attributes will affect a vocalist's tone (e.g. the size and shape of the vocal cords) means that the mic and pre-amp selection is arguably more crucial when recording vocals than it is for other instruments. Between all the technical variables at the disposal of the engineer, I usually find that it is microphone selection and placement that provides the biggest difference to the vocal tone. However, the right mic and pre-amp combination will undoubtedly add extra class and focus, and excite the sound in the right areas. It is a case of thinking about the mic and pre-amp as a complete system. For instance, a Neumann U87 might have the lush highs that you require but not a rounded, larger-than-life bottom end that in some circumstances would be required. One solution would be to try a microphone such as a U47 that has that character, or you could try to pick a mic pre-amp that has a nice warm bottom end.

When deciding on a pre-amp to match with a microphone, I usually consider pre-amps in the two broadest categories: do I want something transparent or something coloured? If I decide to go for a coloured pre-amp, I need to think about in what area I want the colour to be focused. For instance, a Universal Audio LA610 Mk I that I own has the tendency to sound very coloured and gritty in the mid range and treble, while a Neve 1073 is likely to sound coloured in the low to low mids. So in practice, if I want more 'grit' than any of my mics give me (like I often do in a guitar tone), I might reach for the LA610, or if I wanted more body the Neve might be the way to go. This is another occasion where getting to know the tonal attributes of your gear will save a lot of time in session.

So, to summarize, my selection procedure usually starts by getting as close as I can by shooting out a few mics (through an identical mic pre-amp), and then if I notice any weaknesses or ways that I can improve the tone further,

I will consider the attributes of the mic pre-amps available and quickly audition another one that I believe has a high probability of getting me closer to the vision in my head. Since we are working from the mic backwards, I will usually start with a more 'uncoloured' pre-amp for the shoot-out, followed by finding a pre-amp that compensates for anything that is still found to be lacking. However, I will sometimes have a strong enough vision to know a particular type of 'colour' will really suit the style of music and artistic direction, so will reach for that pre-amp first.

14.1.6 Mic Technique

Before we look at techniques for improving the vocalist's performance, we need to consider the singer's mic technique. This is where the technicalities of vocal engineering cross over with the vocalist's performance. During a vocal recording, you might consider the following in relation to the vocalist's position:

- distance from the microphone;
- microphone axis; and
- the singer's movement when singing loudly.

Distance from the Microphone

When considering the distance between the microphone and the singer, you need to consider proximity effect. As discussed in previous chapters, when dynamic and condenser mics have a cardioid polar pattern, the closer the mic is to the source, the greater the bass response. This is a double-edged sword when recording vocals. On the one hand, having a big bottom end is great for 'voice of God' voiceovers or achieving an overly intimate vocal. On the other hand, if you are looking for a natural sound, back off 15–30 centimetres from the mic. This tends to work well for 'conversational' voiceovers, less hyped music or songs that are already busy in the low mid range.

I've read many articles and watched tutorials on the Internet about vocal technique. Generally, I see about a 70/30 split between people advising to back off slightly from the mic compared to utilizing the proximity effect. As with many production techniques, this is a question of personal preference and the demands of the genre that you are working in.

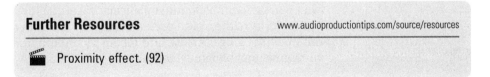

Further Resources www.audioproductiontips.com/source/resources

Proximity effect. (92)

As a starting point, I often get the singer to stand around 10–15 centimetres away from the mic. You can then get the singer to move either closer or farther away in order to create the level of proximity effect or naturalness that is required. Often I like an intimate and slightly hyped sound. Although this is my preference, I know many engineers who start singers further from the mic. Being 20–25 centimetres

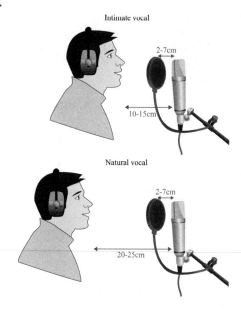

Figure 14.9 Proposed microphone distances

away creates a much less hyped sound, but at this distance in small rooms and those less well treated I tend to hear a little of the room tone being introduced. In large studios with a specifically designed 'dead area', you can get away with being even further away before a vocal starts feeling too ambient (Figure 14.9).

Recording vocals is like recording any other instrument: your choices should reflect your vision. This means that the vocalist's style and the song itself should dictate your approach. If you aren't hearing what you envisaged, experiment until you do. I've also found that the quality and desirability of a mic's proximity effect is largely determined by your mic choice; sometimes the result is perfect but sometimes it is 'boxy'. So if you need a full-bodied, intimate vocal but you're struggling with boxiness, try changing the mic.

Microphone Axis

When plosives are a problem, you should try to move the microphone slightly off-axis before reaching for the pop shield. This can be done by raising or lowering the mic, or by having the singer move half a step either side of the mic. Whichever way you choose, this stops the air from directly hitting the capsule of the microphone, resulting in a more controllable sound. While the plosives will be more controllable in any direction off-axis, there are a few other factors to consider:

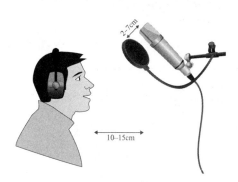

Figure 14.10 Preferred off-axis mic position

Vertical Positioning/Angles

When placing the mic higher or lower than the singer, it is usually recommended that you tilt the angle of the mic to face towards the singer's mouth. As high frequencies reflect off the top of the singer's mouth and through the nose, a mic positioned below the mouth captures these reflections and records a signal richer in top end. A mic placed above the singer is a little less pronounced in the treble but is smooth and not as prone to problematic sibilance or nasal qualities. The majority of the time, I prefer placing the mic above the tip of the singer's nose and facing the mouth at a 45-degree angle (Figure 14.10).

Horizontal Positioning/Angles

Taking a step to the side of the mic or tilting the mic slightly has a very similar effect. It is often easier to ask the singer to take a small step to the side of the mic. However, there is one major issue that makes me favour moving the mic off-axis on the vertical plane: the tonal character of microphones when off-axis can vary greatly, especially for LDCs. From my experience, using a horizontal off-axis position gives a less direct sound with an increase in room tone and a more erratic frequency response compared to on-axis or vertical off-axis placement.

Please note, if you move a mic off-axis, ensure the singer doesn't adjust his or her position to directly face the mic again, cancelling the effect of moving the mic. I've met many singers who feel that they need to sing directly at something and habitually change position to sing into the mic despite a slap on the wrist. In these cases, you can set up a 'dummy mic' that the vocalist can sing into (see Figure 14.11). In fact, this is not only good for directionality, but it also helps the singer to keep an equal distance from the mic for each take.

Figure 14.11 Dummy mic set-up

Singer's Movement When Singing Loudly

If you observe many of the best vocalists, either in a studio or live environment, you will notice that they increase their distance from the mic on loud notes, which has the effect of reducing their dynamic range. A word of warning, though: if a singer does not execute this technique accurately, it can lead to a performance that seems uneven in bass response as the proximity effect varies in strength – or, in extreme cases, it can expand the dynamic range if the vocalist leans too far back on unwarranted notes.

EXERCISES

1. Experiment with recording vocals in more live and dead areas. If you don't have access to a more dead space, try to dampen it by creating a makeshift dead area.

2. Borrow/rent a tube mic, a condenser mic and a dynamic mic suitable for vocals, or alternatively book a couple of hours at a studio (try to get a Neumann U47, U87 and a Shure SM7 or EV RE20 if possible). Record male and female singers, getting used to their sound, and think about the types of applications where they would be useful.

3. Now try them with different types of pre-amp.

4. Experiment with different mic positioning.

5. Listen to the differences between using a pop shield and not.

14.2 GETTING THE BEST VOCAL PERFORMANCE

As well as the technical considerations, it is as important – if not arguably more important – to get the best emotional performance out of the vocalist. This is another area where it's advantageous for producers and engineers to have a combination of musical background, social skills and an ability to motivate. Let's start with the preparation and work from there.

14.2.1 Preparation

For a few days before a recording, the singer's voice should be rested as much as possible. Some light vocal exercises once each day can help to keep the voice in shape, but make sure he or she is not tiring. Dairy products, alcohol, cigarettes or any illicit substances are also best avoided (especially on the day of recording). Honey-based drinks or gargling a tablespoon of honey can also help the vocal cords. When choosing drinks for consumption during recording, make sure they are room temperature as ice-cold drinks tighten the muscles related to the voice.

Singers should know *all* of their lyrics before the day of recording, and although you may want to change a few lyrics upon inspection, it is best that the vocalist can concentrate on his or her performance rather than reading from lyric sheets or having the stress of writing lyrics on the day.

14.2.2 Warm-Up

A thorough warm-up on the day of recording is important; this includes scales, breathing exercises, over-pronunciation and the practice of correct posture. It's beyond the scope of this course to discuss specific exercises, so here are further resources to help advise singers on how to warm up.

> **Further Resources**   www.audioproductiontips.com/source/resources
>
> When it comes to recommending a full 'how to sing series', Brett Manning's courses are always top of the list – www.singingsuccess.com. (93)

14.2.3 Relaxed Atmosphere

Recording vocals is demanding both physically and emotionally, and even seasoned vocalists can feel pressure and suffer from nerves. It is therefore important to make the atmosphere as comfortable as possible. Here are a few tricks I use for achieving this:

- Drop the lighting in the live room so that it feels a little more intimate.
- Limit the number of people in the studio. In fact, the fewer the better, most of the time.
- If nerves are obvious, put as little pressure on the singer as possible. Sometimes singers will benefit from the opposite and a little push in the right direction might be needed. This is one such situation where the social skill of recognizing the needs of a certain personality type can come in handy.

14.2.4 Positive Attitude

There is almost nothing more important for a producer than having a positive attitude while recording, and framing criticism in such a way. This relates not only to vocal recording, but your interaction with the whole band. However, for several reasons, it is absolutely critical to keep in good spirits when recording vocals:

- Vocalists don't have an instrument to 'hide behind', so many criticisms are taken more personally.
- Frustration, worry or lack of confidence is more immediately obvious in a vocalist's performance than on other instruments.
- Vocalists are quite often arrogant, deluded, wishful thinkers or overconfident; knocking them off that pedestal with criticisms can lead not only to them hating you, but also a drop in performance level. On the other end of the spectrum, if a singer is under-confident, it can completely shatter his or her self-esteem and make him or her want to give up.

As well as generally getting excited about the music and sharing ideas enthusiastically, you should also lead into criticisms from a compliment or frame an observation in a positive way. Let's go through a couple of scenarios:

1. The vocalist is flat. Instead of saying, 'That note at the end was a bit flat', you could say, 'Bring that end note a little sharper'. Or even better, 'Man, I loved that take. There was loads of passion, but could you make that last note a little sharper?'
2. The vocalist's performance is lacklustre or boring. Try saying, 'Hey man, that was great, but I still think you have more energy in there. The next one will be golden.' I even jump around and wave my arms when giving this sort of observation. If a musician is a little passive, I even get him or her to do the same. I remember one album session where I literally got a drummer to shout 'I have big balls and attitude' between every take so that he produced the right kind of aggression in his drumming.

14.2.5 Finding the Right Approach

When recording vocals, there is not a 'one size fits all' approach. Some singers like to sing whole takes and then composite (commonly referred to as comping) the best parts together. Others like to meticulously sing each section, line or phrase individually, to give the best performance. If you have worked with the singer before, you should know what produces the best results. When working with a singer for the first time, I suggest the following workflow to help determine the best working method:

1. Tell the singer that you are going to do three whole takes to get the voice warmed up and that they should be treated as a 'proper take'.
2. Record all these takes, giving appropriate feedback and positive observations after each one.
3. After the three takes, invite the singer back into the control room and listen for consistency between takes, consistency throughout each take, and breathing and performance. If any of these elements are significantly stronger at the start than they are during the rest of the song, you may be best recording section by section or line by line. If there are sporadic mistakes, you might

be best doing a series of whole takes and then 'dropping'/'punching' in where there are any remaining issues. As you adjust to this workflow, you can evaluate the performances as they are being recorded. On rare occasions, you might be working with such a special singer that you already have what you need, so don't discount the option of that.

14.2.6 Saving 'Gritty' or 'Screamy' Vocals until Last

This is common sense but worth noting. To get the best performance for every song, softer vocal performances should be done first and songs with screamy/shouty parts should be saved until last, so that the voice doesn't break up during soft tracks. This order produces the greatest longevity from the vocalist.

14.2.7 Allocate Sufficient Time for Vocal Recording

The vocals are often the most important part of the recording, but they are also often the last part to be recorded. When you are starting out as an engineer or producer, you will encounter sessions that run over or have to be rushed due to time constraints. This often means that the vocals have to be rushed and aren't given due care and attention, leading to a less than satisfactory end product. When having to stick to a strict budget, I advise you to schedule a time to start the lead vocals, and keep to it even if other instruments are incomplete. Return to these instruments when the vocals are complete.

14.2.8 Headphone Mix

A key factor for aiding a vocalist's comfort is a good headphone mix. A major concern for vocalists is being able to correctly pitch notes. This usually necessitates a more 'finished mix' in terms of instrument levels. However, timing should still be important so I often have a solid rhythm part (often a rhythm guitar) slightly higher and his or her vocal on top of that. On some occasions, I've done the opposite and buried the vocal so that he or she has to work harder, creating a vocal performance that is often rawer and more aggressive. If neither of these options is getting the results you require, or the performance is often dropping flat or sharp, you may want to ask the singer to take the headphones off one ear so that he or she can hear him or herself in the room. This can sometimes help a singer to improve his or her pitching. However, when doing this, you should always remember to mute the side of the headphones that he or she isn't using to help minimize bleed.

One other point to note when setting up a headphone mix is that the use of spatial effects such as reverb and delay can often be advantageous, but it ultimately comes down to the taste of the singer. I seem to encounter just as many singers who like to sing 'dry' as those who want reverb in their headphone mix (often more reverb than you would use in the final mix). It is easier to critically judge pitch and timing when the vocals are dry, so I prefer to start there and add effects to taste if it produces a better performance. I usually avoid adding long delays with high feedback values, as the echoes tend to affect the accuracy of a singer's timing.

14.3 TECHNICAL ADVICE

As a producer, I often need to give technical advice to singers. To be able to do this, I recommend that you practise singing yourself and, if you can, even have a few vocal coaching lessons. This is so that you can get used to the feelings in your body that helps you fault-find other singers. As vocal coaching is a huge subject, I cannot go into great detail in this course. However, I can give you the real basics, a few rules of thumb, and also advise on some programs to help master this skill. This is one of those areas where personal experience is invaluable.

14.3.1 Posture

First, posture is critically important. If a singer is slouching, he or she will not be able to control his or her diaphragm correctly, which inhibits the breathing necessary for singing. I recommend that you get used to the correct singing posture yourself so that you can easily describe it to others if they are struggling. The best way for me to quickly describe correct posture is to stand with your feet shoulder-width apart and imagine a string pulling you up from the top of your head. This should make you stand up straight, which aligns the ears, shoulders and hips. If this is done properly, you should see the top of the chest near the lungs pop out. You should also concentrate on looking straight forwards while singing, as this provides the best air supply. Now try to relax and feel comfortable while in this position.

14.3.2 Breathing

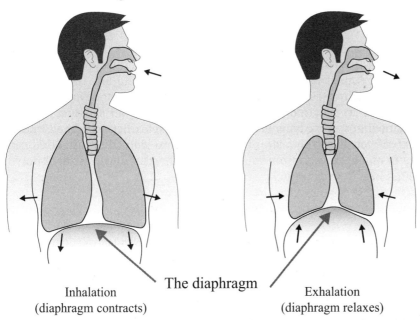

Figure 14.12 The diaphragm[1]

Much like posture, the quickest way to be able to relay correct technique to a singer is to be able to breathe correctly yourself, and then you can show it to him or her while explaining the steps as a series of feelings. The most important technique when breathing is being able to effortlessly take quick and deep breaths, and slowly control your exhale. It might be instinctual to 'gasp' and try to force a lot of air into your lungs quickly; not only is this noisy, but it is also not necessary most of the time. What you should be doing is using the muscles between your ribs (called intercostal muscles) to make your ribcage expand upwards and outwards as the diaphragm contracts, which quickly and quietly expands the lungs. The diaphragm is a sheet of internal skeletal muscle across the bottom of the ribcage. The relationship between the intercostal muscles, the diaphragm and the lungs, and correctly controlling these parts of the body is probably the most important aspect of correct singing technique. On some of the most tricky phrases, you may have to 'snatch' a breath where there isn't sufficient pause to take a full breath. This feels like a gasp but sounds softer, and is therefore preferable for recording.

Inhale

To do this, imagine there is a rubber ring attached to the bottom of your ribcage and when you breathe in through your nose you push the rubber ring outwards. You shouldn't need to force air into your lungs – it should feel comfortable. If this is done correctly, you should get tension in your lower abdominal muscles. You should feel that you have air right in the bottom of your lungs, and your chest and tummy should have expanded outwards. Your head, shoulders and hip tension and posture should remain in the same position – nice and relaxed. This might be a new sensation and cause you to either feel slightly light-headed or even make you yawn – this is normal.

Exhale

When exhaling, you should breathe out of both the nose and the mouth. During the inhalation process the lungs are filled quickly with air; the opposite happens during exhalation, as it is a slow and controlled release of air. This gives you more control on high notes and increases your capacity to sing long and sustained phrases. In order to get used to the sensation of letting air slowly out of the lungs, imagine that you are lying down and blowing a feather up in the air using your mouth. If you blow too fast, it will rise and you will either lose control of the feather or quickly lose air from your lungs and the feather will fall back down.

EXERCISES

1. Practise correct posture and breathing. Get used to the feelings in your body when you do this and try to internalize them so that you can verbalize the process to others.

2. Sing as much as you can. Record yourself and then critique the performances. Where there are problems with pitch, waver or breath control, try to think about whether posture and/or breathing problems may be contributing to this.

However, if you slowly exhale, the feather will stay under your control and in the air for the full duration of your breath. If you do this efficiently, your chest will stay expanded but your lower abdominal muscles will contract.

14.3.3 Sound Creation

As well as the vocalist's posture and breathing, which help to create a comfortable and powerful vocal performance, you also need to have a basic awareness of how the voice box (larynx) and specifically the vocal cords work with the respiratory system and vocal tract to create sound. It is not as important to understand the full mechanism, but you should be mindful of the fact that the positioning of the vocal cords has a dramatic effect not only on pitch, but also on timbre. The properties of a person's vocal cords combined with the uniqueness of the sizes, shapes and positioning of the rest of the system gives each vocalist his or her individuality.

The way that the human voice creates sound is another topic where there is a lot of misinformation, and confusion is widespread. First, we only have one voice. You will often hear vocal coaches saying things such as, 'Use your chest voice' or, 'Use your head voice', which could cause the assumption that we have multiple systems for creating sound when, in fact, we don't. The change in tone and pitch is based on the change in size, shape and tension of the vocal cords and breath control. What is actually happening is that the change in the vocal cords and the pressure created by the air passing through them cause the vocal cords to vibrate in different ways; this phenomenon is called *phonation*. Finally, other areas such as the tongue, teeth, lips and mouth are used to shape the sound; this is called *articulation*. The vibrations caused by the air passing through the vocal cords cause different cavities of the body to resonate, which gives the singer the impression that his or her voice is being cultivated from specific parts of his or her body. So why do vocal coaches still use such terms as 'head voice'? Well, this will be partly because of the fact that these terms have become so engrained into the culture of singing. However, perhaps more importantly, these terms can also help the vocalist to relate to the sensations in his or her body. Being able to recognize and control the vocal cords to create these different sensations is an important skill for all vocalists.

To summarize, the actual sound is *not* generated in the cavities where we 'feel' the resonances; the sound is the result of the respiratory system and the larynx (which contains the vocal cords) along with the tongue, lips and mouth for fine-tuning.

To be able to better understand what the human voice is capable of, we need to categorize vocal range and vocal register and define what each term means.

14.3.4 Vocal Range

Vocal range describes the complete span of notes that a particular vocalist can sing. Different vocalists will have different ranges, dependent on the size and flexibility of the vocal cords. To give the most accurate representation of a vocalist's range, the highest and lowest notes are usually listed by their MIDI note equivalent.

A novice singer is likely to have a range of about an octave and a half, which would potentially increase through proper training and practice. The average trained vocalist can sing approximately two and a half octaves. Vocalists with a truly exceptional range can sing around five octaves; these include Mariah Carey (F2–G7) and Axl Rose (F1–B♭6).[2]

Name	Range
Soprano	C4–A5
Alto	F3–D5
Tenor	C3–G4
Bass	E2–C4

The range of a vocalist can be crudely categorized into several groups, with males usually sitting in lower registers than females. In Chapter 5: Advanced Music Theory, we outlined four vocal ranges used in choral-style four-part writing. These were as shown on the left.

As choral works are predominantly polyphonic (i.e. without a lead voice per se), this simple classification system is sufficient. For other classical voice genres such as operatic works, however, this four-voice system is rather restrictive. Because of this, operatic works use an expanded and slightly different classification system, which also describes vocal tonality. Operatic ranges are often split down further into seven main categories (and more subcategories that further describe the tonality of voice). Of these seven categories, three are for female voices and four are for male voices.

Classifying voices like this isn't as important (or even desirable) in popular music genres as it is in choral or operatic works. In fact, it is sometimes very difficult to classify non-classical singers. However, it can sometimes be a useful tool to help assign 'roles' with the voices you have available, especially within vocal harmony groups. Remember, though, that classical voice classification can be too rigid when applied to popular music, and its notion of restricted ranges can, in turn, restrict creativity. Please note that the following descriptions will omit all details regarding tonality, which we will start to cover in the voice registers section.

Female Ranges

Name	Approximate Range	Description	Example
Soprano	C4–C6	Soprano is the highest vocal range of all the voice types.	Björk, Mariah Carey, Kate Bush
Mezzo-soprano	A3–A5	The mezzo-soprano vocal range lies between the soprano and the contralto and will generally have a heavier, darker tone than soprano.	Beyoncé, Taylor Swift, Whitney Houston
Contralto	F3–F5	The contralto vocal range is the lowest female voice type and falls between the tenor and mezzo-soprano ranges.	Adele, Amy Winehouse, Lady Gaga

Male Ranges

Name	Approximate Range	Description	Example
Counter-tenor	G3–D5	The countertenor is the male vocal range that is equivalent to that of the female contralto or mezzo-soprano. An exceptionally high voice in this range does not really exist for males in popular music. However, some males with lower ranges achieve this sort of effect using falsetto, which we will speak about in the register section.	Artists with notable falsetto usage: Matt Bellamy (Muse), Prince, Jeff Buckley
Tenor	C3–C5	The tenor vocal range is a high-pitched male vocal. However, it is lower than the range exhibited by a typical female.	Freddie Mercury, Bono, Michael Jackson
Baritone	F2–F4	The baritone lies between the tenor and bass but is known as being a deeper-sounding male vocal.	David Bowie, Johnny Cash, Lou Reed
Bass	C2–E4	Bass is the lowest of all vocal ranges. This is again rare in popular music, and is most commonly found in harmonizers rather than lead vocalists.	Barry White, Peter Steele (Type O Negative)

In some of the examples listed above, the vocalist may span other ranges, but I have listed them according to the range I believe the vocal occupies for the most part. For example, I listed Mariah Carey as a soprano, but as her range is so large she would be comfortable in the mezzo-soprano range and could even hit notes in the contralto range too.

Out of all of these ranges, the mezzo-soprano is the most commonly found in the female range and the baritone in the male range. As well as these types, pre-pubescent boys' vocals are often classified as *treble*; examples of this are Michael Jackson and Donnie Osmond in their early years, when they were performing in their respective family groups.

4.3.5 Vocal Register

As well as the range that a vocalist can cover, we also need to consider the tonality of the voice. There is a yin-yang relationship between pitch and tone that depends on how vocal cords are operating. By configuring the vocal cords in different ways, many contrasting tonalities are achievable. The operational methods of the vocal cords are categorized into *vocal registers*. However, there are operational limits to each configuration that means that only certain pitches can be cleanly achieved in each register. The ranges of achievable pitches in each register change from person to person based on their physical attributes.

When experienced vocalists sing, they are able to smoothly move between registers and utilize these registers creatively to add extra X-factor and emotion to the performance and enhance the message of the music. There are several types of vocal registers, and rather confusingly they can have many alternative names. On top of this, there is also some debate, overlap and disagreement about the exact attributes of each. For the purposes of this course, I will refer to them (from lowest to highest in pitch) as: *vocal fry, modal 1 (M1), mixed, modal (M2), falsetto* and *whistle*. Modal and mixed registers are the most heavily utilized by lead vocalists and the falsetto register is also fairly common (particularly with male vocalists).

Modal Register

This is the most widely used vocal register. Most speech and singing resides in this range, and a well-trained singer can produce two octaves or more in this register. As pitch rises in this register, the vocal folds lengthen, their tension increases, and their edges become slimmer.

The modal register can be separated further into two categories, M1 and M2 – more commonly referred to as 'chest voice' and 'head voice', respectively.

M1/Chest Voice
M1 is characterized by its rich, powerful lower-pitched sound and by the 'boominess' from the chest cavity. It is strong, projects well and is probably the default state for most singers. When singing in chest voice, the vocalist should feel the sensation in the upper chest region.

M2/Head Voice
M2 is characterized by its floaty, uplifting and higher-pitched tone without becoming too airy, whispery or breathy. As its name suggests, this is felt in the head and sinuses of the vocalist.

It is important to understand that the terms 'chest voice' and 'head voice' are not intended to be descriptive of any particular technique, but rather of the sensations associated with the two different registers. The techniques used to access each register do not involve the chest or the head themselves; registers are achieved via the manipulation of the vocal cords. Because these two registers produce resonances in distinct areas of the body, giving feedback to the singer, the terms 'chest voice' and 'head voice' are commonly used to describe registers to a singer in terms of experience, rather than the specifics of technique.

Mixed Register/Middle Voice

One of the key skills for a vocalist is being able to smoothly transition between registers. The area between M1 and M2 can often be problematic. In between these registers, vocalists will often try to push their chest voice but the result is a strained or 'shouty' sound, or sometimes they will flip into falsetto, which is not necessarily bad but will often sound like too much of a contrast. Sometimes the answer is to use the mixed register or, as it is otherwise known, middle voice. As its name suggests, the mixed register blends attributes of both M1 and M2. The amount of blend is dependent on the pitch and the emotion required.

Falsetto

The falsetto register lies above the modal voice register. It overlaps the modal register by roughly an octave. Both men and women can achieve falsetto, although it is perhaps a more common effect for male vocalists. Compared to the modal register, falsetto sounds weaker and breathy but when controlled well it can have a haunting effect with fewer overtones. Anatomically, the main difference between the modal and falsetto registers is the type of movement and amount of vocal cord used. The falsetto voice is produced by the vibration of the edges of the vocal cords, rather than the main body as in modal register. The main folds of the vocal cords in falsetto are, in fact, more relaxed. Because of this, the falsetto voice is more limited than the modal voice in both dynamics and tonal variation.

Vocal Fry

Vocal fry or 'glottal rattle' describes the lowest part of the voice. It creates a raspy, rattle-like sound. Vocal fry is achieved by the shortening and fattening of the vocal cords to the point where individual vibrations can be heard, hence the 'rattliness'. You can hear the vocal fry in overly smooth, almost seductive radio voices. When using vocal fry in extremely low registers, it can almost sound like a burping type of sound. Vocal fry might sound destructive to your voice, but is not inherently bad for it; in fact, vocal therapists often utilize vocal fry exercises. However, the effect of vocal fry in extensive and perhaps forcible ways such as in 'scream' genres isn't conclusive yet. In pop music, vocal fry is not used for prolonged periods and is more of a creative effect that can conjure extra emotion and, commonly, feelings of introspection and sadness. Despite the fact that it is not a common register, vocal fry is often utilized in the following situations:

- metal, to create the genre's screamy 'growl' like sound;
- male country singers who create the low raspy vocals at the start of phrases before moving to modal voice; and
- to get extra-low vocal lines in harmonized parts in genres such as gospel.

Whistle

The whistle register is the highest register of the human voice. As its name suggests, it creates a clean, whistle-like sound. Physiologically, it is the least understood of the vocal registers as it is difficult to film the vocal cords while operating in this register. However, it is known that the whistle register uses less of the vocal cord to achieve a shorter vibrating length that favours the production of higher-pitched sounds. Not everyone is able to create the whistle register, and typically female singers are more capable of producing it. You see more females utilizing it in their performances – in fact, it can be a real show-stopper when done correctly and is a calling card of live performances from singers such as Christina Aguilera and Mariah Carey.

Further Resources www.audioproductiontips.com/source/resources

 Examples of each vocal register with male and female voices. (94)

EXERCISES

1. Listen to some of your favourite songs and try to recognize the different registers the vocalists are using. Note them down.

2. Analyse how the different registers can portray different emotive qualities.

3. Practise your own vocal abilities. Try to cultivate each register and pay special attention to the feelings in your body, not only in the vocal cords, but also the resonances in the rest of your body. This way you can verbalize the feelings better to singers when they need help.

14.3.4 Pitching

Correct pitching takes a lot of awareness, concentration and commitment. As the engineer or producer judging the singer's pitching, it is very important that you not only identify pitching problems, but also the probable causes as to why the singer is not pitching correctly.

If you feel that you are not capable of working out whether a note is sharp or flat, then you need to develop that skill quickly as you cannot rely on singers to always correct themselves. Gaining experience with singers will help, as does playing an instrument, but ear-training programs that concentrate on building pitch recognition will help you to recognize tuning issues more quickly.

To be able to quickly give advice to singers on pitch, having either enough confidence in your own voice to pitch correctly or being able to play the melody lines on an electric piano (or other tuned instrument) set up in the control room is very advantageous. As outlined in Chapter 3: Basic Music Theory, I also use the hooks of some nursery rhymes or popular music to help remember the different musical intervals.

If a singer has problems pitching, the root cause could be a lack of training or concentration, or monitoring issues. You should establish whether a vocalist has issues with his or her technique during pre-production; this allows a significant improvement to be made before recording. Usually the following issues are the cause of pitch problems:

- Lack of breath support.
- Not being able to recognize pitch intervals correctly.
- An issue with the pronunciation or mouth/jaw expression.
- Transitional notes between registers (i.e. the way that singers blend between these registers). This is perhaps the biggest problem I run into and is an area that you will often need to pay specific attention to. Telltale signs of this are tension in the vocal cords (stretching their range or pushing).
- Level in the headphone mix being too high or low.

If the singer needs a lot of training, maybe mention hiring a vocal coach in pre-production. Just remember that as singers are often oversensitive souls, you should frame this in a positive way, be polite and tell white lies if necessary.

Saying things such as, 'Wow, you are a good vocalist but if we could just get *this aspect* of your performance sorted, you would be epic!' can help cushion their ego so that they aren't insulted.

If you are left with undesirable results on the day, here are a few tips to help get the best out of the singer:

Singers are usually flat rather than sharp. If they are sharp, it is likely that they are pushing their voice because they're 'trying too hard' or are trying to be heard over a loud headphone mix without a satisfactory amount of vocal. A quick fix would be to ask the singer to relax, raise the vocal in the headphone mix or lower the rest of the music. If this isn't working, the 'one headphone off' trick is good to help a vocalist hear his or her voice.

Singing flat is far more common and is usually caused by not hearing enough of the chordal structure of the music to pitch to. In this case, either turn the vocals down in the headphone mix or the music up. If there is still a problem after this, try turning up the bass guitar and/or rhythm guitar.

Sometimes singers pitch a little flat if they do not have the correct facial expression, especially on sustained or higher notes. If this is the case, ask the singer to pronounce each syllable clearly and to smile. This usually opens his or her mouth wider and he or she has more control on his or her cheek muscles, which can help with pitching.

Quite often the entrance of a line can start off sharp or flat and then he or she settles into the note and corrects him or herself. If this is happening, the singer needs to practise hitting the required note immediately. The singer may find it helpful to sing that note along with an electric piano.

If a singer's voice sounds strained in its normal register at the ends of words, and long notes are fluctuating sharp or flat at the end, this is more than likely down to breath control. To correct this, make sure that the singer is taking big enough breaths in the right places and is not letting the breath out too quickly.

14.3.5 Timing

Usually, pitching is a much greater problem for vocalists than timing. However, if there is not enough drums in their headphone mix, their entry into lines may tend to be late. A lack of rhythm parts in general can also cause timing to waver within phrases. I've been known to play a click-track with a heavy emphasis on the first beat, or on each of the principal beats, to improve the singer's timing.

14.3.6 Personality

I am saving the most important aspect of the vocals until last: the personality of the performance. The personality a vocalist can express is so important that I would forgo some pitching and timing accuracy and also some sound quality/isolation in search of that perfect performance. A great vocal performance is something that is very difficult to teach and also varies wildly due to style, genre, tempo, type of voice and many other factors. While I agree that a vocalist's style is something he or she should develop for him or herself (which will help a band to stand out from the crowd), there are techniques that will help singers deliver the goods when they record. If I were to summarize what I mean in one sentence, I would say: 'Vocalists should aim to express themselves and make their listeners feel what they are feeling, without deliberately copying others.' Of course, everyone has influences and tastes, but does every English singer in a hard rock band have to put on a fake American accent? When a vocalist hasn't yet found his or her identity, it can be very difficult to break bad habits. In fact, a 'fake American accent' or numerous other problems are often a front for an under-confident, fragile individual. In these cases, proper vocal coaching, a rigorous and intense pre-production period and plenty of gig experience will help. However, even with all of this, the vocalist has to be receptive to change.

Sometimes you will have the luxury of working with a band that have a sizeable budget, are open to ideas and want to push themselves. At other times, you will have to work to a strict budget and with people that think they know best.

Whether you have to work to get the best out of the band on the day or have time to help develop the singer during pre-production, the key elements to help a singer get the best performance are still the same:

- Correct posture and breathing should be so ingrained that the vocalist only has to concentrate on the performance. If the singer needs it, then getting a few sessions with a vocal coach pre-recording is massively advantageous.
- The environment should match the mood required. If an artist is singing a ballad, low lighting, and even candles, might help to match the environment to the mood of the song.
- The performer should feel free to perform and the technical considerations should be made based on that. For instance, if a singer is giving an aggressive performance with a lot of grit, it might be best to not only give the singer a dynamic mic but also let him or her hold it like he or she would on stage. Sometimes I've even recorded lead vocals in front of the monitors if that is what gets the vocalist in the right state of mind.
- The vocalist should 'feel' what he or she is singing. I often ask singers to close their eyes and imagine a similar situation that they have been in and sing as if they were in that place now.
- If the singer is feeling the music and the emotion, he or she should be able to sound more like him or herself. If he or she is still singing in a fake or unnatural way, keep trying to develop the singer's emotional reaction to the music.
- Under most circumstances, you should be able to understand every lyric of the song. Sometimes this requires over-pronunciation. A listener will feel a deeper connection to a track if he or she can make out the lyrics, whether he or she is consciously listening to the lyrical content or not. Obviously, in some genres, such as hardcore or 'screamo', this is not as necessary. As the engineer, if you don't understand a line, ask the vocalist to tell you what it is, and, if you feel it necessary, keep recording until you can understand every word. Bear in mind that you should primarily be aiming to get the most emotive performance possible; there are many songs, particularly in alternative genres, that are very moving despite lyrics being hard to distinguish.
- As was discussed in Chapter 5: Advanced Music Theory, I will often ask the singer to remove a couple of syllables from lines of a song to make it flow better.

14.3.7 Backing Vocals

Although the process of recording backing vocals is usually very similar to recording lead vocals, I try to do certain things differently in order to give some distinction between the vocal layers. Here are a few techniques that may help you to make backing vocals distinguishable from the lead vocals, a little subtler, or have them take up less space in the mix. Here is a breakdown of some considerations when tracking backing vocals:

- *Different mic.* Generally, I will try to make the backing vocals take up less space in the mix. If I can do this by using a different mic rather than via EQ, then all the better. For instance, if I record the lead vocal with a tube mic, I might select an LDC for the backing vocal so that it still retains a natural top end but loses a little low-mid warmth.
- *More distance.* Your lead vocal may be intimate and benefit from the proximity effect. However, the backing vocals rarely benefit from such a treatment, and increasing the distance from the mic will give a more natural sound.
- *Off-axis.* Recording backing vocals off-axis can help to soften the vocals and have less high-frequency content. While this is a useful technique, it really depends on the calibre of microphone that you are using, as tonal quality can easily suffer off-axis (particularly on budget or mid-range condensers).
- *Ask the backing singer to hold back a little to soften the performance.* Softening the attack on consonants is particularly advisable.
- *More compression.* To make the backing vocals less forward, you can reduce their impact with a fast attack, fast release compressor. This can quickly sound unnatural, though, and doing this at mixdown is probably the better option as you can add compression to taste during the finishing touches.
- *Polar patterns.* Often, rock productions suit 'gang vocal' shouts and melodies. If it is the sort of song or genre that won't require precise timing and tuning, I will often record these with a group of people standing around one mic in omni mode. If there are only a couple of people, then a figure-of-eight mode is also good.
- *Different pre-amp.* Generally, I don't like backing vocals to stand out too much unless vocal harmonies are a prominent feature of the track. Bearing this in mind, a solid-state, transformerless pre-amp is usually a great choice.

14.4 DOUBLE TRACKING AND VOCAL HARMONIES

As well as the technical differences between the lead vocal and backing vocals, the arrangement of backing vocals and use of double tracking is also worth exploring. Most modern rock and pop tracks can be given extra depth by the layering of extra vocal textures. This can range from a simple double tracking of the lead vocal to a complex arrangement of vocal harmonies, both of which can be performed either by the lead vocalist or others. While it is usually the same vocalist that performs the double tracking, if another singer has a similar or complementary voice, having them sing the harmony can lead to a more interesting end result. Much of Pink Floyd's 'Dark Side of the Moon' was done this way with double tracking performed by guitarist Dave Gilmour and keyboard player Rick Wright. The use of double tracking in the denser areas of a track, such as choruses, can help the illusion of pay-off or help to make the vocal stand on top of the arrangement. As well as double tracking, vocal harmonies can be a prominent feature used to aid the progression of, and add excitement to, a song's arrangement. The writing of vocal harmonies can sometimes be very daunting to an inexperienced producer, so let's first run over some of the key concepts and then we will move on to some tips and tricks.

14.4.1 Constructing Vocal Harmonies

The first thing to consider when writing harmonies is your vision. What is the aim of the track? For instance, if the song is about solitude or loneliness, prominent use of backing vocals might seem incongruent with the message of the song. As well as the overall vision, the genre might dictate how you implement vocal harmonies. A '60s pop-influenced track would likely benefit from a carefully constructed multi-part harmony; a garage rock track might benefit from the occasional use of simple harmony and prominent use of double tracking. If you know the genre you are working in well, then working out whether backing vocals will be advantageous should be fairly straightforward; if not, it is time to get listening critically. Despite knowing that you want to add vocal harmonies, actually implementing suitable harmonies is often difficult. Some talented vocalists will be able to improvise harmony ideas on the spot, in which case the producer's job might be to help arrange those ideas and guide the singer in the right direction. Most of the time, though, the producer might have to write (or co-write with the vocalist) the harmony lines and direct the singer.

To work out vocal melodies quickly, I usually have a keyboard set up in the control room. In case you play another instrument, having that handy can speed up the process. I have also had great success using an autotuner to experiment with harmony ideas. To set this up simply, duplicate your lead vocal channel and insert an autotune plug-in that allows you to move notes around individually, such as Melodyne (there will be more on autotuners shortly). From there, you can drag the notes around in the autotuner to create the harmonies you wish to try out. While the pitch-manipulated versions might seem 'forced' or 'robotic' in isolation, together with the original version it is often good enough to work out whether a particular harmony line would suit the track. This method is also a good way to show the singer what harmony lines to sing, and inexperienced vocalists often lock into it quicker than by giving them the melody line to sing by playing it on your chosen instrument.

For me, the process of writing vocal harmonies happens rather instinctually and emotively through following what my vision is communicating to me. Sometimes, trial and error and playing around on the keyboard can yield great results quite quickly. Just remember to feel your way through it and go with your gut instincts. You will usually find a good solution if you put your mind to it. When carefully crafted and executed, the casual listener might not even perceive the vocal as being harmonized. Used as a prominent musical part, harmonies can add that extra 'sparkle' to a track and are often used to add a lift to chorus sections. The types of harmony part vary between genres. In pop and rock genres, thirds above and sixths below are commonplace; fourths and fifths are used occasionally. Jazz and progressive music sometimes have more unorthodox and complicated intervals. Sometimes in genres such as country, the vocal harmonies are pivotally important to the sound of that style of music. When writing harmony parts, it is important to also understand that the underlying chord in the music affects the suitability of the vocal harmony as much as the lead vocal line. In fact, vocal harmonies can be designed to act as a pad section to the music rather than as a direct harmony of the lead vocal or as a distinct counterpoint harmony.

Let's break down three basic approaches to writing a vocal harmony:

1. *Simple harmony.* Simple harmonies follow the lead vocal line, just above or below it. I define most pop harmonies as simple harmonies because the majority of the time they move in parallel with the lead vocal. This type of movement can easily become predictable and monotonous, so vocal lines will often drop in and out of the track around key phrases (e.g. during every other line of the chorus). At other times, the vocal line(s) will be in parallel the majority of the time but use other types of movement on key 'pivot notes' to allow them to transition between intervals. Most harmonies in popular music will focus on the thirds or sixths (which are inverted thirds). You can construct most of the harmonies that you need in a pop or rock track just by using simple harmonies. When a vocalist is coming up with a harmony on the spot, this is usually the type of harmony that is created.
2. *Padding* or *chordal harmony.* With this approach, you aren't specifically concerned with harmonizing directly to the melody, but instead the chord behind it. In theory, you can choose any chord tones from the current chord to harmonize. When the chord changes, you normally change the harmony note, but sometimes the same note is found within adjacent chords in a sequence and can be carried through. This acts like a bed for the lead vocal to 'sit on'. The age-old harmonized 'ooohs' and 'aaahs' are often an example of chordal harmony. One thing to be aware of when using chordal harmony is that passing notes in the melody may clash with the held harmony note, and adjustments may need to be made accordingly.
3. *Counter-harmony.* This is where any harmony lines have a large amount of independent movement sometimes both melodically and rhythmically. This is trickier to implement and is likely to take more time and forethought. The most common occurrence of this in popular music is a countermelody, which at times intertwines harmonically with the lead melody.

When writing vocal harmonies, remember that songs can utilize any or all of these techniques. You will often hear songs in which different sections use differing techniques; sometimes you will hear different techniques used simultaneously. Harmonies can be easy to overuse or overcomplicate; in a lot of modern arrangements, dense vocal harmonies are used sparingly to give maximum impact where they do appear.

While my process of writing vocal harmonies is more instinctual, a little knowledge of music theory and voice-leading principles will help focus your direction and help you out when you lack inspiration or to get you out of a jam. So let's recap on some of the key voice-leading principles from Chapter 5: Advanced Music Theory, but from a vocal perspective:

1. *Avoid parallel movement of fifths and octaves (and sometimes fourths too).* The problem with these intervals is that they have very little independence from the lead melody. A harmony of a fifth or (especially) an octave can seem to 'unify' to such an extent that the harmony can sound like a single note; contrast this with the harmonies of a third or a sixth, which sound very much like two distinct notes. If, within a harmonization, several consecutive octave harmonies or several consecutive fifth harmonies appear (parallel octaves/fifths), it can cause the sense of distinct voices to be lost; the harmony will seem to 'collapse' into a single voice for the duration of the parallel movement. In many popular music genres, a less distinct harmony is what is required as it provides a type of 'thickening' effect. This is usually done by combining the lead vocal with a version that is either an octave above or below the lead. In some other genres, particularly rock music, a fifth can even be used to create a similar effect while also sounding more menacing. Because the casual listener will often hear the vocal lines as one part, I consider the use of parallel octaves and fifths to be a creative effect that makes the vocal sound thicker or more interesting rather than distinct harmony.

   It is important to stress, in order to avoid any misunderstanding, that the advice against parallel octaves and fifths does not imply that these harmonic intervals themselves are in any way undesirable or to be avoided. In fact, they are very pleasant, stable intervals that occur constantly, especially in popular music. The emphasis is not on the avoidance of octave or fifth harmonies, but rather that they should be approached from and progress to other harmonic intervals. When this occurs, the sense of distinct voices, even as they pass through the octave or fifth relationship, is not lost.

   Because the other harmonic intervals retain a much better sense of distinction between voices, parallel movement of these intervals is not considered problematic.

2. *Keep harmonies close in pitch to the lead vocal.* Just like open and closed voicing of chords, there are two types of harmony spacing. If the vocal lines are spaced within an octave, it is called *close harmony*. If the voices are spread over a span that is greater than an octave, it is called *open harmony*. In popular music, the vast majority of harmony work utilizes close harmony.

3. *Don't cross voices.* This means that once a harmony is established above or below the melody, it should stay there and shouldn't be allowed to cross over it. In practice, this is sometimes hard to avoid, especially if the singer is prone to large leaps.

4. *Utilize other types of contrapuntal motion as well as parallel movement.* When you are looking to construct harmonies, you don't have to use parallel movement. The three other types of motion are *contrary, similar* and *oblique,* and as previously mentioned they can be used as pivots between sections of parallel movement or used more freely to create more complex harmonies. Movement by these means will take a lot more thought and pre-planning in order to avoid clashes or problems with voices becoming too wide or crossing, but the result is often sweeter and more entertaining.

14.4.2 Notable Harmony Examples

One of the best ways to understand the practical applications and the tonality of different harmonies is to listen to and study some well-known songs. If you can figure out each harmony line found in a track and practise singing along with each of the harmonies, then you should find that creating vocal harmonies will become more instinctual rather than theoretical.

Here are three popular artists known for their vocal harmonies, along with a brief overview of an example track:

The Beatles – 'Twist and Shout'. The Beatles are known for their excellent use of vocal harmony. 'Twist and Shout' utilizes a lot of simple harmony with triad relationships in the vocals. In the verses, Paul and George shadow the lead vocal with a 1st and 3rd relationship, as well as some falsetto 'oohs'. The famous 'aah' section happens over an A chord (1:37). A new voice is introduced at the start of each bar and they ascend through the triad, John starting with the root (A), followed by George with the 3rd (C♯), Paul with the 5th (E) and then John with the leading tone (the 7th, a G note) before resolving back to the root an octave higher from where they started.

Simon & Garfunkel – 'Sound of Silence'. The beauty of this song is the way that Garfunkel is taking the lead melody but Paul Simon's harmonies are not only roughly the same volume level, he is also harmonizing underneath him the whole way through in an understated and often monotonous way that blends the two vocals together. It adds a depth and rich harmonic texture to the song. The key to this harmony is that it uses a lot of oblique motion, with Paul Simon's part often droning on the same note through the phrase while Garfunkel's melody is moving more freely before coming together again.

The Beach Boys – 'I Get Around'. Just like with The Beatles, I could have chosen a number of Beach Boys tracks as fine examples of harmony. 'I Get Around' is a great one because of its use of a falsetto countermelody in the chorus. This countermelody intertwines beautifully with a four-part vocal harmony. On the surface, you might think that the four-part harmony uses parallel motion but there is actually some oblique motion in there too due to some of the parts droning in various sections of the chorus.

EXERCISES

1. Analyse the vocal harmonies used in the following songs: 'Go Your Own Way' by Fleetwood Mac, 'Bohemian Rhapsody' by Queen and 'The End of the Road' by Boyz II Men. Get the score or a reliable MIDI file of the songs to check how you did.

2. Add new vocal harmony parts to some of your own productions where they didn't exist previously. Try to experiment using each type of motion while also bearing in mind voice-leading principles.

3. Experiment by thickening lead vocal parts using octaves and/or fifths.

14.5 'TAKE HER AWAY' VOCAL CHOICES

The lead vocals for 'Take Her Away' were recorded with a Violet Globe large diaphragm condenser mic through an Avalon 737 tube pre-amp. This helped to achieve a warm yet modern and punchy sound. I started by asking Wil to stand very close to the mic utilizing the proximity effect to get an intimate sound. However, this wasn't working for the track as well as I'd envisaged. I settled on a position of 10–15 centimetres away, with the mic slightly below the mouth. The backing vocals performed by Wil were also recorded using the Violet Globe at the same distance. However, we used a more sonically neutral non-tube, transformer-less pre-amp that produced less bottom end, which helped de-clutter the overall vocal tone and helped the backing vocals sit behind the lead.

While tracking, we experimented with the use of double tracking and lead vocal harmonies. In the end, we felt that the lyrical content necessitated a certain sense of solitude; therefore, we only used double tracking on certain important lines in the verses and utilized a simple vocal harmony underneath some of the lead vocal in the chorus. By keeping the vocal harmony tucked underneath the melody, it made it more subtle, as well as darker and less uplifting. The song's story is one of a failed relationship with subsequent crazy behaviour performed by the other party. As a way to emphasize that this behaviour was not a unique occurrence (despite it feeling like that to the character at the beginning of the song), we used group vocals that appear in the last chorus. These consisted of two takes using the Violet Globe with the other members of the band crowded around the front of the mic, about 20–25 centimetres inches away. As with the other backing vocals, we used the more neutral pre-amp.

14.6 VOCAL EDITING

When it comes to vocals, the extent of any editing I perform is usually dictated by the precedents set by the genre of the song. For the types of rock/indie and metal I often produce, I usually prefer to leave the performance as natural as I can. On some ultra-modern and clinical rock music, my approach might be different. When producing genres such as pop, country or RnB, a more detailed, rigorous and 'perfect' editing job may be required, even to the point where it becomes noticeable. On occasion, I will even manipulate the editing to the extent that it could be considered a special effect.

With other instruments, any tuning deficiencies can and should be addressed *before* committing the recording. However, with vocals, you are very much at the mercy of the vocalist's skill. So while you should do all you can to minimize pitch problems at source, you will often find yourself having to correct vocal pitch issues post-recording. As well as pitch events, you may also be required to edit:

- the timing of lead vocals;
- the timing relationship between the lead vocals and backing vocals, ensuring they are synchronized; and
- removal of unwanted noise between vocal lines.

There are many autotune plug-ins on the market, but the most famous are *Auto-Tune* by Antares and *Melodyne* by Celemony. Most of these tools offer the ability to adjust timing, amplitude and other factors. However, I still find that it is easier to do any timing and volume editing before hitting the tuning software, mainly because I like to handle one process at a time and am most accustomed to editing in Pro Tools. My standard procedure is to:

1. *Remove any unwanted noise or headphone bleed between lines.* I do this by simply deleting any unwanted audio between lines. I am careful not to remove subtle sounds at the end of words. I am also aware of any rhythmic or important breathing. If you remove breaths before lines, it can often seem unnatural. If one particular breath seems a little too prominent, you may wish to replace it with another. It might be tempting to use strip silence like you might on drum channels, but in my experience the subtleties and gentle nuances of the human voice mean that the ends of words and important breaths can easily get cut off and it takes you longer than working manually. Finally, on a more stripped-down song or exposed section, cutting out noise entirely may seem unnatural, particularly when a lot of compression is being used that brings out the nuances. If there is a particular bleed problem that is intrusive, you may wish to replace it with ambient 'air' noise from another part of the recording to keep the consistency and realism of the performance.
2. *Edit the timing of the lead vocal.* If you have edited the music to be tight before the vocalist has recorded his or her part, you should find that the vocalist sings with more accurate rhythm. However, I always like to look at how close the vocal comes in at the first beat of each bar (particularly the very first lines of each section) and also how any vocal parts align with any 'rhythmic stabs' in the music. It is far more unusual to have a singer that sings habitually out of time than it is to have one that is 'pitchy'. If he or she drifts considerably during vocal lines, you may want to consider whether his or her headphone mix is causing problems – often raising the kick, snare and/or a rhythm guitar can help the vocalist keep time. Doing this can get a more natural result and is generally quicker than spending hours editing the performance.
3. *Synchronize the backing vocals to the lead.* Once the lead vocal has had some timing correction, I align the backing vocals to it. This is one area that really helps make the production seem more 'slick'. If I am going for a '60s or '70s-style production, I might leave it ever so looser. The majority of the time, however, I will use a plug-in called *vocALign* by Synchro Arts to perfectly synchronize the vocals with a few clicks.
4. *Tuning.* Once all this is done, I tune the lead vocals. As mentioned earlier, for a rock or old-school metal track, this might largely be left alone (dependent on the quality of the singer), but a pop, country or RnB track will usually be tuned heavily. Tight pitch cohesion is particularly important where there are backing vocals, as pitch drift here can often have bigger consequences. Remember, what matters is how the vocals *fit together*. Sometimes by tuning one vocal line, it highlights the others being slightly out of tune. I will often tune the vocal by moving in smaller increments rather than snapping the notes more 'perfectly'.

14.6.1 Using Pitch Correction and Autotune

As well as an included video tutorial, I quickly want to outline the basic functions of a pitch correction plug-in – in this case, *Melodyne*. Unlike *Auto-Tune* by Antares, which can work in real time, Melodyne works by first loading and analysing the audio that you wish to correct. The audio can be analysed and processed in different ways depending upon which mode of operation the user chooses. If you are running Melodyne outside of a DAW (called stand-alone mode), it processes the file while importing. If you are using Melodyne from within a DAW, you have to 'transfer' the audio into the plug-in's memory. Once this is done, you can use the following tools to manipulate many different facets of the vocal performance.

Figure 14.13
Melodyne toolbar

Working from left to right, we have:

- *Main 'arrow' tool* – This allows you to do normal mouse procedures such as grabbing, scrolling and zooming.
- *Pitch tool* – As the name suggests, this allows you to change the pitch of the particular note that is sung. However, Melodyne is very sophisticated and allows you to not only edit the natural vibrato of the note by selecting modulation mode, but also the way that the notes glide to one another with the pitch drift mode.
- *Formant tool* – This manipulates the timbre of the vocalist and it is more specifically likened to the differences in the size and shape of the throat. With a quick edit, you can make the throat deeper and more 'soulful' or lighter and thinner like your favourite adolescent pop singer. Remember, though, that the effects of using this tool can be very noticeable and all but the slightest of changes can produce comedic results.
- *Timing tool* – With this tool, you can literally move notes in time, adjust their length and lock them to the grid just like you would with Elastic Audio in Pro Tools.
- *Separation tool* – While Melodyne is pretty accurate at differentiating between every note that the vocalist has sung, sometimes it can make a mistake that can cause the plug-in to incorrectly manipulate specific notes. The note separation tool allows you to either make new separations or join previously separated notes so that Melodyne processes the performance correctly.

As you can see from the main features of Melodyne, it is an expansive and powerful tool to correct issues. This is particularly useful for vocals, but it is also useful for tuning lead guitar lines, bass guitars and other monophonic sources. However, as Melodyne has been updated and expanded, it is now able to tune polyphonic material (multi-note or chordal structures) and adjust rhythmic material such as attack, decay and dynamics of drums.

> **Further Resources** www.audioproductiontips.com/source/resources
>
> Using Melodyne. (95)

14.6.2 Using VocALign

Syncro Art's VocALign is another powerful vocal editing tool. This plug-in allows you to match the timing of vocal performances. This allows you to create seamless double tracks and tightly timed backing vocals. It works by analysing the content of a 'guide track' (in most cases, your lead vocal) and then using that as a template to time-stretch and shift a 'dub' track (usually your double track or backing vocals). This can save a lot of arduous editing and provide more perfect alignment. However, it can sometimes sound too perfect, and on some material I prefer to edit the timing of vocals manually for a more natural and realistic result. I have again provided a short tutorial on operating VocALign.

Figure 14.14 VocALign plug-in window

> **Further Resources** www.audioproductiontips.com/source/resources
>
> Using VocALign. (96)

EXERCISES

1. Use a pitch-correction tool to correct vocal pitch on a track. Experiment with snapping the notes with 100 per cent accuracy and then doing it by ear.

2. Now tune some vocal harmonies and listen to how the tuning affects the vocal lines you have yet to pitch-correct.

3. Manually correct vocal timing and then use VocALign to do the same. What are the similarities and differences? Would there be times when each approach would be more appropriate?

NOTES

1. How to breathe using your diaphragm, accessed September 2014, www.bellissimoliverpool.co.uk/apps/blog/show/42514361-how-to-breathe-using-your-diaphragm.
2. See which of your favorite singers can hit the highest and lowest notes, accessed September 2014, http://time.com/105319/compare-vocal-ranges-of-worlds-greatest-singers/.

BIBLIOGRAPHY

Brett Manning, *Singing Success* (Singing Success, Inc., 2005).
Pamelia S. Phillips, *Singing for Dummies* (John Wiley & Sons, 2010).

EPILOGUE

It has been quite a journey, hasn't it? Well done for getting this far! Some of what follows may tread over ground we've already covered in the Preface and Chapter 1. However, after spending so much time zoomed in on theory, practice, techniques and tricks, zooming out and reasserting the bigger picture should help to keep your perspective on what matters most: your own personal journey.

If you have immersed yourself in the written material, exercises and recommended resources in this book, you will find yourself in a much stronger position for your assault (or a renewed assault!) on the music industry. However, the content found in this course is only a foundation. Keep listening to music, your peers, musicians – and take inspiration from everywhere. Keep learning from books, videos, blogs, forums and your own practice.

This course has primarily covered a single facet of the music production chain: the recording of music. I strongly advise that you also delve into the connected fields involved in music creation, such as mixing and mastering. (My next course, perhaps?) I have, however, included some video files of the 'Take Her Away' mixing process in the online resources of the *Audio Production Tips* website.

Further Resources www.audioproductiontips.com/source/resources

 Take Her Away mixing overview (97)

 Take Her Away mastering interview (98)

I am confident that the material I have presented to you in this book will help you get closer to achieving your goals – but remember, my intention is not to present you with dogma, nor strict rules to obey. Your approach to the practice of creating music should be guided by *what works best for you*. Some of the practices and techniques I've detailed won't suit *your* style, workflow or taste, so the impetus is now on you to practise, experience the process for yourself, and keep questioning. Once you have developed your own taste, bands, artists and labels will start to choose your services *because* of your taste. Take some time to work out which techniques you like and which you don't.

I feel, however, that it is necessary to complement this advice with a caveat: It can be easy to fall into the trap in which you weigh your opinion above all others. Avoid the temptation to impose your will over that of the band or artist, or to utilize the same techniques and principles on every project with the aim that the result 'sounds like you'. The sound of a record should always be guided by the collaborative vision set out by the band or artist and the production team. In my opinion, the best producers find a way to make their vision meet the band's or artist's in a way that is best for everyone. Ultimately, this is a mixture of skilful engineering, psychological prowess and compromise. This can be a fine line, so don't lose sight of the fact that your ideas and taste are probably the reason why you were picked in the first place – and that you shouldn't be afraid to try out new ideas or try to take the band to the next level.

The more obvious musical attributes discussed above are still only part of the jigsaw puzzle. To become a successful producer or engineer, perseverance and determination are arguably just as important, You *will* make mistakes, you *will* have projects where the end result is not what you, the artist or label require. How you dust yourself off and strive to do better next time is critical. As you build experience, satisfactory results will increasingly outnumber unsatisfactory results – just be prepared to experience this exponential curve and ride it out.

There is not enough time in this course to give in-depth career path guidance, but I do see fit to include a few pieces of advice to bring us to a close. First, getting started in the music business is tough, whether you go the freelance or commercial studio route. As a freelancer, it is all about networking, and it can be tough to meet enough artists and bands to get regular work. In my early years, I worked at many local venues providing live sound engineering, which helped gain local artists' trust and friendship, some of which turned into studio work. As for commercial studios, my experience is that the top studios are less bothered about engineering or production prowess, as this skill can be taught. Self-drive, a hard-working nature and common sense are usually valued highest in these entry-level opportunities. You should be able to follow instructions effectively and think on your feet, as well as be reliable and possess good timekeeping.

Even when you are on the career ladder, it is very easy for you to stagnate and remain in the same place. I think this is the number one reason why people quit, as they literally cannot foresee how things are going to improve or in which areas to push forward. This was certainly a big challenge for me, working in small commercial studios in a small city with next to zero industry presence. I could have stayed there and hoped to get lucky – and if you are good enough, and come across good opportunities, this can and does happen. Personally, though, I prefer to have more control over my direction. Upon facing a ceiling to your development, it may be a case of taking one step back to take two steps forward. I moved from a small city to London, where there are more opportunities, before taking a job at the bottom of the pile in a large commercial studio facility that is regularly working with major labels and big names. Creative industries favour the brave, bold and ambitious, and I have had no regrets in taking such steps. If these are not traits you possess naturally, you may want to think about cultivating them.

Good luck! And remember, whatever happens, have faith in yourself.

—Pete

APPENDIX

1 DECIBEL MATHEMATICS

The standard for representing power ratios is the *bel* (B). If we increase an initial power value by a factor of 10, we say that it has increased by 1 bel. If we increase the power by a factor of 100 (= 10 × 10, or 10^2), we say that it has increased by 2 bels. Increasing the power by a factor of 1,000 (= 10 × 10 × 10, or 10^3) would be an increase of 3 bels, and so forth.

As you can see, even with small increases in bels, the power ratios quickly become very large. For example, increasing by 6 bels is the same as increasing by a factor of 1 million! To avoid having to work in fractions of bels, we use the more common decibel (dB) system. 10 decibels equals 1 bel.

1 dB also conveniently corresponds to roughly the smallest volume increase we are able to discern with our ears, making it a very useful unit for us.

The formula below shows you how to calculate an increase in power, in decibels. You must know the starting power (P_0) and the final power (P_1) values first. These can be in watts or any other power unit.

$$L_{dB} = 10 \log_{10} (P_1/P_0)$$

We can rearrange this formula to find out what change in power is achieved by increasing a signal by L decibels:

$$P_1 = 10(L_{dB}/10) P_0$$

In the following example, let's use a starting power (P_0) of 1 watt, and turn it up by 1 decibel to get the new power value (P_1). Here is the formula with those numbers substituted in:

$$P_1 = 10(1/10) \times 1 \approx 1.259$$

So increasing 1.000 W by 1 dB gives us approximately 1.259 W. We can therefore say that a 1 dB increase is to increase power by a ratio of 1.259.

For a 2 dB increase (applied to 1.000 W), this is:

$$P_1 = 10(2/10) \times 1 \approx 1.585$$

So turning something up by 2 dB can be said to multiply the power by around 1.585.

Here is a table showing the power ratios of some common decibel increases and decreases:

dB	Power Ratio
20	100
10	10
6	3.981
3	1.995
2	1.585
1	1.259
0	1
−1	0.794
−2	0.631
−3	0.501
−6	0.251
−10	0.1
−20	0.01

Helpfully, it is easy to remember that an increase of 3 dB roughly doubles the power. A decrease of 3 dB roughly halves the power.

2 FUNDAMENTAL ELECTRONIC PRINCIPLES

To get a good idea of how electricity works on a basic level, I like to imagine a simple circuit of a battery, wire, resistor and a motor.

The particle that travels across a wire when electricity is flowing through a circuit is called an electron, and these are negatively charged. To make a useful analogy of how electricity works, I am going to give electrons a personality. As their negative charge would suggest, let's call them 'grumpy and argumentative'.

If you put two electrons in a room, there would be a personality clash and repulsion would be created between the two of them; therefore, they would want to get as far away from each other as possible. This repulsion can potentially cause movement to occur as long as there is somewhere for the electrons to go. This potential-energy/repulsion is called *voltage*. If you put many of the electrons together in a room, there would be a greater clash of personalities, thus there would be a greater potential energy and – yes, you guessed it – a greater voltage!

Think of the battery as a room with two doors storing all of these electrons. One door is called a positive pole (+)and is locked from the outside, and the other is a negative pole (–).Now think of a magnet. We all learnt at school that there are two poles of a magnet – the north (N) and the south (S) – and that if you have two magnets, opposite poles (i.e. the N pole of one magnet and the S pole of the other) will attract and pull towards one another. Conversely, the like poles (i.e. N and N, or S and S) will repel each other. The two poles in the battery and the electrons act like a magnet; the electrons are repelled from the negative pole and attracted towards the positive pole.

Let's apply this to our angry electrons analogy. As the electrons cannot leave the room through the positive door, they are all sent out of the negative door into a corridor that leads back to the positive door. If the corridor is large, it allows more electrons to escape quicker, and if the corridor is small it impedes the escape of the electrons. The corridor itself is the equivalent of a conductor in an electrical circuit and is simply a medium that allows electrons to travel between two points. The most common example of a conductor is a copper wire. The rate at which the electrons can escape to a certain point in the corridor over a given time is called *current*, and that is measured in amperes or amps (A). The properties of the corridor that contribute to the slowing down of the electrons' escape is called *resistance* or *impedance*, and is measured in ohms (Ω). Everything that is placed between the positive and negative poles of the battery will have a value of resistance. In the case of a copper wire in a circuit, it offers a very small resistance, and is therefore a strong conductor. The electrical component known as a resistor is purposely designed to slow down the electrons; resistors are valued in ohms based on their strength.

So, we have covered voltage, current and resistance. Now let's imagine a revolving door into the corridor. As the electrons move through the revolving door, it moves around. This is similar to how a simple motor works (but it is done with magnets). You can increase the power of the motor in two ways. The first is by increasing voltage (increasing the force at which the electrons hit the door because repulsion is higher). The second is by increasing current (the extra amount of electrons hitting the door). So *power* can be defined as the rate at which electrical energy is fed into or taken from a device or system.

Ohm's law of resistance can help you calculate the voltage, current and resistance in a circuit:

(V = voltage, I = current and R = resistance)

V = I × R

I = V / R

R = V / I

Power can then be calculated by multiplying voltage and current, and is measured in watts (W).

Ohm's Triangle

Cover the variable you want to find and perform the resulting calculation (multiplication/division) indicated.

Figure A.1 Ohm's triangle

INDEX

UK Spelling used in this index

I–IV–V chord progressions 105, 109
12-tone equal temperament tuning (12tET) 54–7, 61, 68, 91
'50s chord progression 105–6

A/B comparisons 286–7
Abbott, Darrell Lance 389
absorption 262–6, 329
accents 38, 40, 42–8, 124
accessories 326–8, 421–3, 483–4
acoustic folk 196
acoustic guitar tracking 421–6
acoustic quality 253–60
acoustic treatments 262–6
active pickups 360–1, 409
ADC *see* analogue-to-digital converters
additive synthesizers 467
advanced music theory 141–71; *see also* music theory
advanced synthesizer parameters 475–6
Aeolian mode 70–2, 74, 169–71
aliasing 226–7, 236
aligning phase 293, 401
all-pass filters 472
alliteration 131
altered chords 88–9
alto 142–3, 494
ambience control 262, 265–6
amplifiers: guitars 355, 366–81, 384–9, 391–8, 403, 411–12; pre-amps 244–8, 342, 367, 484–5; selection techniques 302–3, 366–80, 411–12; solid-state 244–5, 367–8, 372–6, 387, 389, 394; synthesizers 468, 469
amplitude variation 256–7
analogue audio 219–20, 226–9, 231–3, 300–2
analogue meters 222–4
analogue operating levels 221, 223–4
analogue scale headroom 224
analogue synthesizers 466–8
analogue-to-digital converters (ADC) 226–9

anatomy, electric guitars 356
anechoic chambers 266
angle considerations 326–8, 486–7
arpeggiators 475
arrangement 32–4, 173–218, 284; clarity 175, 182, 184, 187; common mistakes 208–12; frequency/time domains 174–90; melody 215–17; muddiness 176, 182–5, 190, 201; punch 175–9, 183, 189–90, 198; rhythm 198–202; structural changes 208–10
articulations 428–30
atmosphere aspects 488
attitude aspects 488–9
augmentation 85–6, 350–2
aurotones 276–7
authentic cadences 94–5
auto-alignment 293
Auto-Tune 507–9
axial modes 257, 486–7

backing vocals 501, 506–7
baffles 380
band-pass/reject filters 472
bandwidth parameters 474
bar (measure) 37–47
baritone 495
basic music theory 35–110
bass, vocals 142–3, 494, 495
bass clef 57–8
bass drums; *see also* kick drums
bass guitars: amplifiers 411–12; arrangement/orchestration 176–7, 182–6, 192, 194–6, 201, 210–11; cabinets 411–12; editing 413; effects 412–13; frets 409; instrumentation 408–10; microphones 413; pedals 412–13; pickups 409; re-amping 393–4, 413; speakers 411–12; strings 409–10; tracking 355, 371, 393–4, 403, 408–14
bass trapping 262, 263–4
Beach Boys 197, 505
Beat Detective 348, 403–4

Beatles 45, 73, 96, 106, 113–15, 121–3, 127, 133, 167, 177, 204, 465, 505
beats, rhythm 37–49
bell sizes 314
'bigs' speakers 273–4
bit depths 219–20, 227–8, 229
Black Label Society 388
bleed 241, 290–1, 296, 303, 336, 339–40, 344, 351–2, 423, 507
blues 112, 193, 385
blues harps (harmonicas) 204, 446, 450, 451–2, 453–5
Blumlein Pair mic'ing 419, 426, 439–40
Borland, Wes 389–90
boundary microphones 240
bows/bowing 428–30
brass tracking 440–5
breathing techniques 491–2, 500
bridge, song structure 112, 113, 114–17, 133
broadband treatment 262
buying equipment 237

cabinets 355, 366–7, 370–80, 386, 389–90, 393, 396–403, 411–12
cadences 93–7, 99, 106–7
calibrating recording levels 230–6
cardioid polar patterns 241–2, 419
Celestion speakers 355, 370, 379, 385–6, 388–9, 411
cello tracking 427–30, 431–40
chest voice 496
chordal harmony 503
chords: definitions 78–9; diatonic substitution 161–2; embellishments 100–3, 164; extensions 156–7; inversions 151–3; leading 98–9; melody 135–8; modulation 78, 165–8; music theory 74–110, 135–8, 147, 149–71; numbering 91–3; passing chords 158–60; pivot modulation 167–8; progressions 76, 90–110,

160, 163, 165–71; simplification 160–1; spacing 147; tones 137–8; tritone substitution 163; voice leading 149–50; voicing 153–5
chorus 112–17, 131–3, 208–9, 383
chromatic scale 54–7, 63–4
church modes 70, 168–71
circle of fifths 103–5, 106–8
circuitry, guitars 365–7, 369, 373, 375, 377, 381, 394
clarity 175, 182, 184, 187
classical string instruments: bows/bowing 428–30; microphones 433–40; sampling 432–3, 466; strings 428–30; tracking 427–40
classification systems 54–7, 246
clean boost pedals 380–1
clef 57–9
click tracks 29
clipping 224–5, 227–32, 235
closed-backed cabinets 379
closed chord voicing 153, 154
clutter avoidance 201
coatings, drumheads 312
coda 113–14, 131, 133
coincidental pair set-up 332, 419–20, 464–5
colour pre-amps 244
comb filtering 256, 258–60, 290–5
combo (combined) amps 366, 369–70
common chord modulation 167–8
common time 38–40, 45, 193, 199
compositional elements 29–30, 32–4, 35–110, 284
compound time signature 41, 43, 46–7
compression: drums 310, 324, 342, 350, 353; guitars 394, 412; recording levels 299–300
computer music; *see also* sampling
concert flutes 447, 448–9
concert pitch 57
condenser microphones 239–40: classical string instruments 435–6; drums 331, 336–7, 338–9; guitars 399–400, 424; pianos 418–20; vocals 482, 485
cone(s), guitar microphones 398–9
connotation 129
consoles 306
consonance 143–4
constructive interference 258–60, 289
constructs, rhythm 40–9

contact points, strings 428
contemporary folk 196
content, lyrics 118–34
contralto 494
contrapuntal movement 504
contrary motion 146–7
control room environments 237–48, 253–60, 262–6, 287–9
converting analogue/digital audio 226–9
core structures 212–13
counter-harmony 503
counterpoints 75
counting rhythm 40–3
crash cymbals 309, 313, 315, 317
critical listening 1, 18–19, 27–31, 253–86, 353–4, 500
crossfading 405–7
crossover frequencies 278–9
cross-voicing 504
crotchets 39–47
current 515
cut-off 472–4
cymbals 309–10, 313–18, 324, 326–8, 331–3, 334–5, 339–46, 350–1, 354

dampening shells 323–5
dBU (decibels unloaded) meters 223
dBV meters 223
dead areas 480
dead notes 365
Decca Tree mic'ing technique 439–40
deceptive cadences 96
decibels 9–18, 223, 513–14
decision-making 282–307
detonation, lyrics 129
detune 475
DI *see* direct-injection boxes
diaphragm microphones 239–40, 331, 336–7, 338–9, 399–400, 435–6, 463–5
diatonic chords 161–2
diatonic circles 106–8
diatonic scales 62, 63–4
diffusion 263, 266
digital audio 11, 219–20, 226–9, 300–2
digital clipping 225
digital meters 224–9
digital operating levels 221
digital synthesizers 467–8
Dimebag Darrell 389
diminished chords 84–5
direct-injection (DI) boxes 247–8
direct modulation 165–7

direction of motion 144–7
dissonance 143–4
distortion pedals 381, 412–13
distribution amps 302
dominant chords 159–60
Dorian mode 70–2, 73, 74, 169–71
dotted notes 39–47, 386
double bass tracking 427–30, 432–40
double-ply drumheads 313
double reeds 450
double tracking 206, 396–7, 501–5
drawbars 460–2
dreadnought acoustic guitars 421, 423
drivers, electric guitars 355, 366, 370–80, 385–90, 393–4, 396, 398
drumheads 312–13, 318–19
drummer performance 344–6
drum rings 324
drums: arrangement/orchestration 177, 182–7, 190, 196, 199, 204, 211; augmentation 350–2; editing 347–52; maintenance 318–29, 340; microphones 331–44, 350, 353–4; placement 329–41; replacement 350–2; selection techniques 309–17; shells 309–11, 317–25, 328, 335–7, 340–3, 352–4; timing aspects 347–50; tracking 309–54; tuning 318–29, 340
drumsticks 315–16
drywall live rooms 248, 250
duple metre 42–4, 46
dynamic microphones 238, 335–6, 338–40, 399–400, 463–5, 482, 485
dynamics, drums 345–6

ear fatigue 15–17
early reflections 260, 262, 264–5, 479
EarMaster ear training program 69
ears 8–17, 283
easy listening music 191–3, 196
editing 347–52, 403–7, 413, 506–9
EDM *see* electronic dance music
effects: guitars 355, 380–90, 394–5, 412–13; Hammond organ 462
effect tubes 367
Elastic Audio 348, 349–50, 403–5, 413
electric guitars: amplifiers 355, 366–81, 384–9, 391–8, 403; cabinets 355, 366–7, 370–80, 386, 389–90, 393, 396, 398–403; circuitry 365–7, 369, 373, 375, 377, 381, 394; double tracking 396–7;

INDEX

drivers 355, 366, 370–80, 385–90, 393–4, 396, 398; editing 403–7; effects 355, 380–90, 394–5; enclosures 355, 366–7, 370–80, 386, 389–90, 393, 396, 398–403; gain 395–6, 407; insider information 390–8; instrumentation 356–60; maintenance 363–6; microphones 398–402, 405; multing 397–8; pedals 355, 380–90, 394–5; pickups 360–2; power matching 374–5, 407; room acoustics 400–1; speakers 355, 366, 370–80, 385–90, 393–4, 396, 398–9; strings 362–3; tracking 355–407; tuning 390
electronic dance music (EDM) 111–12, 198–200, 465, 474–5
electronic principles/electrons 514–15
emotion 63, 73, 79
emulation 232–3, 245, 301, 391–3, 403, 459, 462, 475, 476–7
enclosures, guitars 355, 366–7, 370–80, 386, 389–90, 393, 396, 398–403
engineering vocals 479–87
ensembles 432–3, 436–40
envelopes: sound 178–9, 474; time curves 270
environment properties 237–48, 253–81, 287–9, 423–4
EQ: bass guitars 412; drums 336–7, 343, 350–1, 353; production philosophies 1, 3–4, 6–7, 19; recording 299–302; room acoustics 266, 271–2
equal gain/power 407
equal loudness contours 12–13
ES-335 electric guitars 356–7, 360
exhaling techniques 491–2
exponential fade 406
extended chords 156–7

faders 230–2, 285, 303–4, 405–7
falsetto register 497
familiarity processes 28–9, 284
female vocal ranges 494–5
Fender: amps 355, 369–71, 379–80, 384, 385–9, 455, 457; guitars 355–8, 361, 363–4, 367–71, 379–80, 384–9, 402–3, 408–9
fictional lyrical modes 120–1
figure of eight polar patterns 242
figure of speech 130

filters 468, 469, 471–4, 483–4
fipple flutes 447–8, 449
flanger effects 382–3
Fletcher-Munson curves 12–13
flutes 446–9, 461
flutter echo 260
focus, music theory 175, 177, 179–80
folk 196
forms, song structure 114–16
foundation organ tone family 461–2
four-voiced progressions 142–3
French horns 442–3
frequency 35–6, 49–57, 61, 68, 174–90; arrangement/orchestration 174–90; classical string instruments 427; modulation 467; response graphs 267, 268
fretbuzz 364–5
frets 364–5, 409
full band drum recordings 344
full track analysis 150
fundamental frequency 12, 36, 50–5, 469
Fuzz Measure 266
fuzz pedals 381, 413

gaffer tape 324
gain, electric guitars 395–6, 407
gain staging/structure 221–2, 229–31, 284, 303–4
gauge, strings 362–3, 410, 418, 421–2
genre precedents 177–8, 191–200, 284
Ghoulish Records 21
Gibson electric guitars 355–7, 358–60, 361–3, 367, 388–9, 402
Gilmour, David 386–7
global synthesis control 468, 475
glockenspiel 456, 457
Glyn Johns drum mic'ing 334–5
grand auditorium acoustic guitars 421–3
grand pianos 416, 417–21
granular synthesis 467
graphical software 266–71
Greenwood, Jonny 387–8
gritty vocals 490
groove, drums 345–6
grot-boxes 276–7, 280
guitars: acoustic guitars 421–6; arrangement/orchestration 176–8, 182–6, 192, 194–6, 201, 210–11; bass guitars 176–8, 182–6, 192, 194–6, 201, 210–11, 355, 371, 393–4, 403, 408–14; editing 403–7, 413; voice leading 149–50; see also electric guitars

Haas effect 14
Hammond B3 organ 459–65, 476–7
hand and foot coordination 346
hand percussion 455–6
hardware, drums 316–17, 326–8
harmonic minor 64, 66, 97–8
harmonic series 49–55, 59–68, 71, 75–6, 79, 85–6, 97–8, 373–4
harmonicas 204, 446, 450, 451–2, 453–5
harmonizers, guitars 384–5
harmony 35, 49–55, 59–90, 97–8, 117–18, 134–9: arrangement/orchestration 215–17; melody 117–18, 134–9, 215–17; reharmonization 141, 155–64; vocal harmonies 501–5
headphones 277–8, 280, 302–7, 490, 507
head plus cab amps 366
headroom 221, 224, 229, 231–2
head voice 496
hearing 8–17, 283
Hendrix, Jimi 385–6
'Hey Jude' 45, 69, 73, 113–14, 127, 133
high frequency utilization 187–9
high-pass filters (HPF) 471–2, 483
hi-hat cymbals 309, 313–14, 317, 327, 334–5, 339, 341, 343–6, 350
hip-hop 197–8
hitting skills, drums 345
Hi-Z input boxes 247
Hofner V3 402
hooks 131–2, 137, 211
horn section tracking 440–5
HPF see high-pass filters
humbucker pickups 360–2
hybrid recording systems 231–3, 245, 300–2
hyper-cardioid polar patterns 241–2

'I Get Around', The Beach Boys 505
impedance 246, 247, 375–6, 515
inhaling techniques 491–2
insider information 390–8
instrumentation: bass guitars 408–10; electric guitars 356–60; selection techniques 190–8, 206, 209; song structure 114

instrument registers 191
integrated software 306
interference 258–60, 289–99
internalizing rhythm 48–9
interrupted cadences 96
intervals 53–76, 79, 91, 97, 100–7, 143–4
in-the-box (ITB) mixes 230, 233, 295, 302
intonation 356, 361, 365–6, 437
intro lengths 208
intro section 113
inversion, chords 151–3
Ionian mode 70–2, 74, 168–71
irregular time signatures 48
ITB *see* in-the-box mixes

Jaguar HH 402–3
jazz 112, 191–3, 409
Jazz Bass guitars 409
Johns, Glyn 334–5
journey analogy of songs 31, 119, 136, 202–5

Katz, Bob 234–6
keeping listener's interest 6, 27, 174, 202, 205–8
kick drums:
  arrangement/orchestration 177, 182–7, 190, 196, 199, 204, 211; tracking 309–13, 317–24, 328–9, 331–9, 342–3, 346, 353
K-Metering 235–6
K-System 234–6

latency, headphones 302–7
lead melody 117–18, 134–9, 216–17
lead vocals 141–50, 151–5, 211
ledger lines 58
Les Paul electric guitars 355–7, 358–60, 364
Leslie, Donald, speakers 462–5
LFO *see* low-frequency oscillators
Limp Bizkit 389–90
line input boxes 247
linear faders 406
listening skills 1, 18–19, 27–31, 253–86, 353–4, 500
literary devices 129–31
live rooms 248–51
Locrian mode 70–2, 73–4, 169–71
logarithmic faders 406
loudness perception 11–13
low frequency muddiness 182–4
low-frequency oscillators (LFO) 474–5

low frequency utilization 187–9
low-pass filters (LPF) 471–2
LPF *see* low-pass filters
Luna Kiss; *see also* '*Take Her Away*', Luna Kiss
Lydian mode 70–4, 169–71
lyrics 117–34, 136–7, 211, 216–17: content 118–34; hooks 131–2, 137, 211; literary devices 129–31; melody 117–19, 136–7, 216–17; modes 120–3; rhyming schemes 126–9; rhythm 124–6

'magic-bullet' solutions 1–2
magnets, speakers 371
'mains'/'bigs' speakers 273–4
maintenance aspects 318–29, 340, 363–6
major chords 76–86, 89, 91–2, 161–3
major scales 63–4, 65
major triads 76–86, 89, 91–2
male vocal ranges 494–5
mallet percussion 456–9
managing projects 23–6
manual edits 347, 403–5
march time 43–4
marimbas 456, 457
Marr, Jonny 387
Marshall amplifiers 355, 368–70, 374, 380, 385–6, 388, 403
masking 210–11
master cadences 93–4
mastering 7
matching amps & speaker sizes 371–8
materials: cabinets 379–80; cymbals 314, 354; drums 311–12, 354; electric guitar strings 363
Mayer, John 386
melodic minor 64, 66–7, 97–8
melody 35, 49, 53, 59–68, 72–6, 90–1, 97–8: arrangement/orchestration 215–17; harmony 117–18, 134–9, 215–17; lyrics 117–19, 136–7, 216–17; vocal harmonies 502
Melodyne 404, 413, 502, 507–9
Mesa/Boogie amplifiers 369–70, 389–90
metaphors, lyrics 130
metering recording levels 221–2
metre 38, 40, 42–4, 48, 213–15
metronomes 28–9, 36, 41–9
microphones 237–48, 287–9: brass 444–5; classical string instruments 433–40; control rooms 237–48, 287–9; cymbals

341; drums 331–44, 350, 353–4; flutes 448–9; guitars 398–402, 405, 413, 424–6; Leslie speakers 463–5; percussion 455–6, 458–9; phase differences 290–5; pianos 418–21; placement 287–9; polarity 237, 241–2, 289, 295–9; pre-amps 244–8; reeds 452–5; types 238–40; vocals 479–87
middle eight 112, 113, 114–17, 133
middle vocal register 496
mid-field monitors 279
mid frequency importance 184–7
MIDI 48, 56–9
Mini-Moog synth 467
minims 39, 201
minor chords 77–86, 92–3, 161–3
minor scales 63, 64–7
minor triads 77–86, 92–3
mix 4
mixed vocal register 496
mixing 7, 302–7, 490
Mixolydian mode 70–2, 73, 74, 169–71
modal frequencies 256–7
modal vocal register 496
modelling: guitar amps 391–4; microphone pre-amps 245
modes 62–3, 69–74, 120–3, 168–71
modulation 78, 141, 165–8, 382–3, 467–9
monitoring levels 17–18, 233–6
monitors 253, 260–2, 272–81; *see also* speakers
Moog synth 467
Moongel 324
'more me' amps 303
motion directions 144–7
mounting speakers 380
mouth organs 204, 446, 450, 451–2, 453–5
muddiness 176, 182–5, 190, 201
multi mic placement 425–6
multi-tracked drum recordings 344
multing, guitars 397–8, 413
musical stress 124
musical value expression 57–9
music theory: advanced elements 141–71; arrangement 32–4, 173–218; basic elements 35–110; cadences 93–7, 99, 106–7; chords 74–110, 135–8, 147, 149–71; compositional elements 32–4, 35–110; frequency 35–6, 49–57, 61, 68, 174–90; harmony 35, 49–55, 59–90, 97–8, 117–18, 134–9;

INDEX

intervals 53–76, 79, 91, 97, 100–7; lead vocals 141–50, 151–5, 211; lyrics 117–33, 136–7, 211, 216–17; melody 35, 49, 53, 59–68, 72–6, 90–1, 97–8, 117–19, 134–9, 215–17; modes 62–3, 69–74, 120–3, 168–71; modulation 78, 141, 165–8; need for 31–4; orchestration 32–4, 173–218; overtones 50–4; pitch 49–62, 65, 69, 147–8; reharmonization 141, 155–64; rhythm 35–49, 124–6, 198–202, 213–15; scales 59–79, 84–6, 91–3, 97–107; song structure 111–17, 208–12; time domains 174–90; triads 76–86, 89–93, 100–3; voice leading 141–55, 211; Western notation system 35, 53–9

narrative lyrical modes 120, 121–3
natural minor 64–7
nearfield speakers 274–6, 280–1
noise generation 468, 469–71
non-chord tones 137–8
note length 37–41, 125, 457
Nyquist frequency 226–7

oblique modes 257
oblique motion 147
octaver pedals 384–5, 412
Ohm's law of resistance 515
omnidirectional polar patterns 241
on-/off-axis 242–4, 400
open-backed cabinets 379
open cadences 95–6
open chord voicing 153–5
optimizing lyrical flow 124–6
orchestral seating plans 436–8
orchestration 32–4, 173–218: clarity 175, 182, 184, 187; frequency domain 174–90; muddiness 176, 182–5, 190, 201; punch 175–9, 183, 189–90, 198; time domain 174–90
organ tracking 459–65, 476–7
organization techniques 284
ORTF overhead set-up 333
Osbourne, Ozzy 388
oscillators 468, 469–71, 474–5
Oswinski, Bobby 356
outro 113–14, 131, 133
overdrive pedals 381
overhead microphones 331–5, 342, 354
overtones 50–4

pad switches 483
padding vocal harmonies 503
panning 206
Pantera 389
parallel keys 165, 166–7
parallel movement 145, 504
parallel speaker wiring 377–8
parent scales 70
partials; see also overtones
passage form 115
passing chords 158–60
passive pickups 360–1, 409
peak programme meters (PPM) 223
pedals: drums 316–17, 328; guitars 355, 380–90, 394–5, 412–13; pianos 418–19
pentatonic scales 62, 138
percussion tracking 455–9
perfect circles 106
perfect intervals 65, 106
perfect ratios 54, 68
performance aspects 344–6, 487–90, 500, 507
personality considerations 499–500
personalized mixers 303
personifying lyrics 131
phase differences 258–60, 289–95, 307, 401
phase distortion synthesis 467
phase-rotation 293, 401
phase shifters 383, 401
phrasal cadences 93
phrase lengths 209
phrase modulation 165–7
Phrygian mode 70–4, 169–71
piano tracking 415–21
pickups 360–2, 409
piece forms 115–16
Pink Floyd 386–7
pitch: harmonic series 49–53; music theory 49–62, 65, 69, 147–8; shifters 384–5; vocals 495, 498–9, 508–9; voice leading 147–8
pivot modulation 167–8
placement: classical string instrument mic's 433–40; control rooms 253, 260–2; drums 329–41; guitar mic's 398–402, 424–6; microphones 287–9; organ mic's 463–5; piano mic's 418–21; vocal mic's 479–81, 485–7
plagal cadences 95
plectrums 53, 366, 408, 410, 412, 423
polar patterns 237, 241–2, 480, 485
polarity 237, 241–2, 289, 295–9, 307
pop chord progressions 106

pop shields 483–4
popular music 111–17, 148–50, 197
porting nearfield speakers 275–6
post faders 303–4
posture aspects 491, 500
power, definition 515
power amp tubes 368
power chords 89–90, 157–8
power matching 374–5, 407
power ratios 223
PPM see peak programme meters
pre-amps 244–8, 342, 367, 484–5
pre-chorus 114, 116
Precision Bass guitars 408
preparation techniques 284, 488
pre/post faders 303–4
pre-production 27–31
pressure zone microphones (PZM) 240, 337
processes, pre-production 28–31
production philosophies 1–8, 205–8
production team's mission 6–8
programming synthesizers 468–75
progression problems 211–12
project management 23–6
properties, environment 253–60
prosody 207–8
protecting hearing 16–17
proximity effects 242, 243
pulses 36–49
punch 175–9, 183, 189–90, 198
Pythagoras 53, 54–5
PZM see pressure zone microphones

quadruple metre 42–3, 45, 47
quartet seating plans 437–8
quavers 39–42, 46–7

Raconteurs, The 388–9
Radiohead 387–8
range, vocals 493–5
rattle 325–7, 329, 338, 497
re-amping 391–4, 413
Recorderman mic'ing set-up 334
recorders 447–8, 449
recording: decision-making 283–307; levels 219–36; see also tracking
rectifier tubes 368–9, 370
reed instruments 446, 449–55, 461
reflection-free zones (RFZ) 262, 264–5
reflections 290, 294, 329–30, 479–81
refrain 112–13, 131
reggae 193–5
registers 191, 495–7

reharmonization 141, 155–64
related modulation keys 165
relative major/minor scales 65, 162–3, 166
repetition avoidance 208–9, 210
replacing drums 350–2
resistance 515
resolution 219–20
resonance 371, 458–9, 473
resonator tubes 458–9
reverb time (RT60) 266, 269, 480
RFZ *see* reflection-free zones
rhyming schemes 126–9
rhythm 35–49, 124–6, 198–202, 213–15
ribbon microphones 238–9, 331, 399–400, 434–5
ride cymbals 309, 313, 315, 317, 332, 341
rock 196–7, 385
rocksteady 193–5
room acoustics 253–60, 266, 271–2, 400–1
room analysis software 266–71
Room EQ Wizard 266
room microphones 340–1, 343, 433–40, 479–81
room modes 255–8
room treatments 262–6
root position 151–5
root/tonic of scales 62
RT60 *see* reverb time

sampling: classical string instruments 432–3, 466; pianos 416, 466; recording levels 219–20, 226–9, 236; tracking techniques 465–6
sawtooth waves 470
saxophones 450–1, 452–3
scale degrees 60, 63–5, 68–78, 91–2, 97
scales: decibels 10–11; music theory 59–79, 84–6, 91–3, 97–107
screamy vocals 490
S-curve fade 406
seating plans 436–8
secondary chords 159–62
sectional cadences 93–4
sectional lyrics 131–3
selection techniques: amplifiers 302–3, 366–80, 411–12; drums 309–17; guitars 366–80, 398–402, 411–13, 424–6; headphones 302–3; instruments 190–8, 206, 209; microphones 246–7, 287–9,
398–402, 413; piano mics 418–21; pre-amps 246–7; vocal mics 481–2
semibreves 39, 202
semiquavers 39, 41, 202–3, 211
Senior, Mike 275, 307
sensitivity, speakers 371
sentence stress 124
series speaker wiring 376–8
session files 30–1
settings, vocal microphones 483–4
seventh chords 81–4, 135
SG electric guitars 356–7, 359–60, 402
Shannon-Nyquist theorem 226–7, 228, 236
shells, drums 309–11, 317–25, 328, 335–7, 340–3, 352–4
shock mounts 483
'shoot-out' 481
sibilance 188–9
signal to noise ratio 221
similar motion directions 144–5
Simon & Garfunkel 73, 115, 122, 196, 204, 505
Simon and Patrick SP6 acoustic guitar 421
simple time signature 41, 42–5
simplifying chords 160–1
sine waves 226–7, 230–2, 258–9, 289, 296–9, 469
singe-coil pickups 360–1
single classical string players 432–3
single mic placement 425
single-ply drumheads 313
single reeds 449–51
sixth chords 79–81, 135
size aspects: amps & speakers 371–8; cabinets 379; cymbals 314; drums 311
ska 193–5
skins 312–13
slap echo 260
slope parameters 473
small tracking rooms, drum placement 329–31
Smiths, The 387
snare drums 309–13, 317, 320–6, 328, 332–5, 338–9, 343, 345–6, 349–51
social skills 5
socks, dampening shells 324–5
software 266–71, 306
solid-state amplifiers 244–5, 367–8, 372–6, 387, 389, 394
solo sections 114

song analysis 31
song structure 111–17, 208–12
soprano 142–3, 494–5
sound creation, vocals 492–3
sound envelopes 178–9, 474
sound intensity 11
sound localization 13–14
sound pressure levels (SPL) 11–13, 295–9, 335–9, 444
'Sound of Silence' 505
soundboards 418
space considerations 175, 177, 180–6, 191, 200–4, 211
spaced pair microphone set-up 332, 419–20, 425–6, 458–9, 464
speakers: bass guitars 411–12; control rooms 253, 260–2, 272–81; electric guitars 355, 366, 370–80, 385–90, 393–4, 396, 398–9; Leslie 462–5; mounting 380; placement 253, 260–2; transducers 337; wiring 376–8
special effects 207
SPL *see* sound pressure levels
spot-mic'ing 434–6
square waves 470
stable intervals 143–4
stacking thirds 102–3
standing waves 256–7
stands 326–7
stave system 57–9
stereo-imaging deficiencies 260
Steven Slate Trigger 2, drums 352
stone live rooms 248, 249
Stratocaster electric guitars 356–8, 385–7, 403
strings: classical instruments 428–30; guitars 362–3, 409–10, 418, 421–2; Hammond organ 462; pianos 418
structural arrangement changes 208–10
sub-bass frequencies 188
subgenre precedents 191–200
substituting chords 161–3
subtractive synthesizers 467
subwoofers 278–9
super-cardioid polar patterns 241–2
suspended chords 86–8, 157–8
sympathetic notes 60–2
synthesizers 466–76

'*Take Her Away*', Luna Kiss 20–2: arrangement/orchestration 212–17; drums 310, 321–2,

329–30, 341–6, 353; electric guitars 402–3; extra instrumentation 476–7; microphones 341–4; vocals 506
tangential modes 257
technical vocal advice 491–501
Telecaster electric guitars 355–8, 387–8, 402
tempo: arrangement/orchestration 183, 193, 195, 199–201, 210, 213–15; rhythm 37, 44–5, 48–9
tenor 142–3, 494, 495
thickness, cymbals 314
tightening drums 347–50
timbre 50–1
time delay 290–1
time domains 40–9, 174–90
time signatures 38–49, 193, 199
timing aspects 347–50, 490, 499, 506–7
tom-tom drums 309–13, 317, 320–5, 328–9, 334, 337, 339–40, 343–6, 349
tone 29, 378, 460–2, 495
tonic, of scales 62
Tozzoli, Rich 356
tracking 7: acoustic guitar 421–6; backing vocals 501; bass guitars 355, 371, 393–4, 403, 408–14; brass 440–5; classical string instruments 427–40; decision-making 283–307; drums 309–54; electric guitars 355–407; levels 219–36; organs 459–65, 476–7; percussion 455–9; pianos 415–21; sampling 465–6; synthesizers 466–76; vocals 479–510
track optimization 32–4, 173–218
track sheets 25–6
transducers 337

transformers 245
transient response 237
transparency 150, 175, 182, 184, 190
transparent pre-amps 244
transposition 69, 441
treble clef 57–9
treble frequencies 187–8
tremolo 384, 430
triads 76–86, 89–93, 100–3
trial and error techniques 135, 136–7
triangle waves 470
trigger devices 351–2
triple metre 42–3, 44–5, 46
triplets 41–2
tritone substitution, chords 163
trombones 442–5
troubleshooting 271
trumpets 442–3, 444–5
tubas 442–3
tube amplifiers 367–70, 372–6, 381, 387, 391, 403
tube microphones 240, 244–5, 482
tuning 54–7, 61, 68, 91, 318–29, 340, 390, 507
'Twist and Shout', Beatles 96, 505

unisons 148
unstable intervals 143–4
upper mid frequency range 186–7

valve amplifiers 367–70
verse 112, 114–17, 132–3, 208–9
vertical pianos 417–21
vibraphone 456, 457
vibrato 384, 430
viola tracking 427–30, 431, 432–40
violin tracking 427–31, 432–40
virtual synthesizers 467–8
vision aspects 3, 250–1
vocal fry 497

vocal harmonies 501–5
vocal registers 495–7
VocALign 509
vocals: double tracking 501–5; editing 506–9; engineering 479–87; lead vocals 141–50, 151–5, 211; microphones 479–87; performance aspects 487–90, 500, 507; pitch 495, 498–9, 508–9; pre-amps 484–5; range 493–5; sibilance 188–9; technical advice 491–501; tracking 479–510
voice leading 141–55, 211
voltage polarity 289, 295–9, 514–15
volume matching 286
VU (volume unit) meters 222–3

wah-wah pedal 382
waltz time 44–5
warm-up 488
Washburn J28SCE/BK guitar 422
waterfall plots 267, 269, 276
wattage matching 374–5
weight, cymbals 314
Western notation system 35, 53–9
whistle 497
White, Jack 388–9
White Stripes The 388–9
wiring speakers 376–8
wooden live rooms 248, 250
woodwind tracking 440–2, 445–55
word stress 124
work schedules 24–6
Wylde, Zakk 388

XY pair microphone set-up 332, 419–20, 464–5
xylophones 456, 457

Yamaha NS10 monitors 274–6